W9-CSP-050

ACS SYMPOSIUM SERIES 266

Materials for Microlithography

Radiation-Sensitive Polymers

L. F. Thompson, EDITOR
AT&T Bell Laboratories

C. G. Willson, EDITOR
International Business Machines Corporation

J. M. J. Fréchet, EDITOR
University of Ottawa

Based on a symposium cosponsored by the
Division of Polymeric Materials Science and Engineering
and the Division of Polymer Chemistry
at the 187th Meeting
of the American Chemical Society,
St. Louis, Missouri,
April 8–13, 1984

American Chemical Society, Washington, D.C. 1984

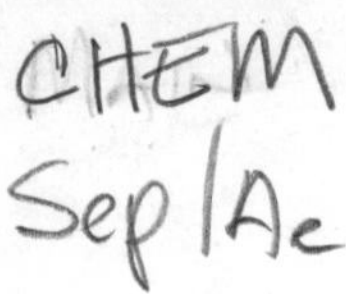

Library of Congress Cataloging in Publication Data

Materials for microlithography.
(ACS symposium series, ISSN 0097-6156; 266)

Includes bibliographies and indexes.

1. Polymers and polymerization—Congresses. 2. Photoresists—Congresses. 3. Lithography—Congresses. 4. Microelectronics—Materials—Congresses.

I. Thompson, L. F., 1944- . II. Willson, C. G. (C. Grant), 1939- . III. Fréchet, J. M. J. IV. American Chemical Society. Division of Polymeric Materials: Science and Engineering. V. American Chemical Society. Division of Polymer Chemistry. VI. American Chemical Society. Meeting (187th: 1984: St. Louis, Mo.) VII. Series. VIII. Title: Radiation-sensitive polymers.

TK7871.15.P6M37 1984 621.381'73 84-21744
ISBN 0-8412-0871-9

PRINTED IN THE UNITED STATES OF AMERICA

ACS Symposium Series

FOREWORD

The ACS SYMPOSIUM SERIES was founded in 1974 to provide a medium for publishing symposia quickly in book form. The format of the Series parallels that of the continuing ADVANCES IN CHEMISTRY SERIES except that in order to save time the papers are not typeset but are reproduced as they are submitted by the authors in camera-ready form. Papers are reviewed under the supervision of the Editors with the assistance of the Series Advisory Board and are selected to maintain the integrity of the symposia; however, verbatim reproductions of previously published papers are not accepted. Both reviews and reports of research are acceptable since symposia may embrace both types of presentation.

CONTENTS

PREFACE

THE CONTINUING MINIATURIZATION OF ELECTRONIC DEVICES shows little sign of abating. Currently, devices containing over 1,000,000 transistors are manufactured with minimum circuit features of 1.0–1.25 μm. The circuit elements are fabricated by a series of processes known collectively as lithography. During the past several years the term microlithography has been used increasingly to denote the trend toward decreasing feature sizes. The technology used today continues to be dominated by traditional photolithographic techniques, but most scientists and engineers in the semiconductor industry are convinced that devices with dimensions less than 1 μm will be required in the near future, and that new technologies will be required to fabricate them. These new technologies will likely be a combination of entirely new exposure techniques such as electron-beam, X-ray, or deep-UV lithography coupled with significant modifications of conventional UV processes. These modifications will most likely use multilayer schemes.

Polymeric materials, especially the radiation-sensitive polymers, lie at the heart of all of these new technologies. For the first time, it will be necessary for the development engineer as well as the scientist to possess a thorough understanding of the chemistry of radiation-sensitive polymers. This book provides the foundation for such an understanding. Although the book is based on research papers presented at a symposium, it has been constructed with considerable introductory material and considerable editing with the result that it is more of a unified text than a compilation of research papers.

Three extensive introductory chapters by Everhart, Broers, and Bowden provide a solid foundation in the physics and chemistry of the lithographic process together with an overview of current resist systems. These 3 chapters, coupled with 20 chapters from outstanding radiation polymer chemists throughout the world, provide a firm basis for understanding the important fundamental concepts in radiation chemistry as applied to design, development, and application of resist materials.

The authors and editors are indebted to the AT&T Bell Laboratories Text Processing Group, especially Susan Pope and Carmela Patuto. Special thanks are due to AT&T Bell Laboratories Art Department and Roy T.

Anderson who coordinated all of the art work. Special acknowledgment also goes to Bonnie Cahill who handled most of the administrative aspects in preparing this volume.

L. F. THOMPSON
AT&T Bell Laboratories
Murray Hill, NJ 07974

C. G. WILLSON
IBM Research
San Jose, CA 95193

J. M. J. FRECHET
University of Ottawa
Ottawa, Ontario
K1N-9B4, Canada

August 1984

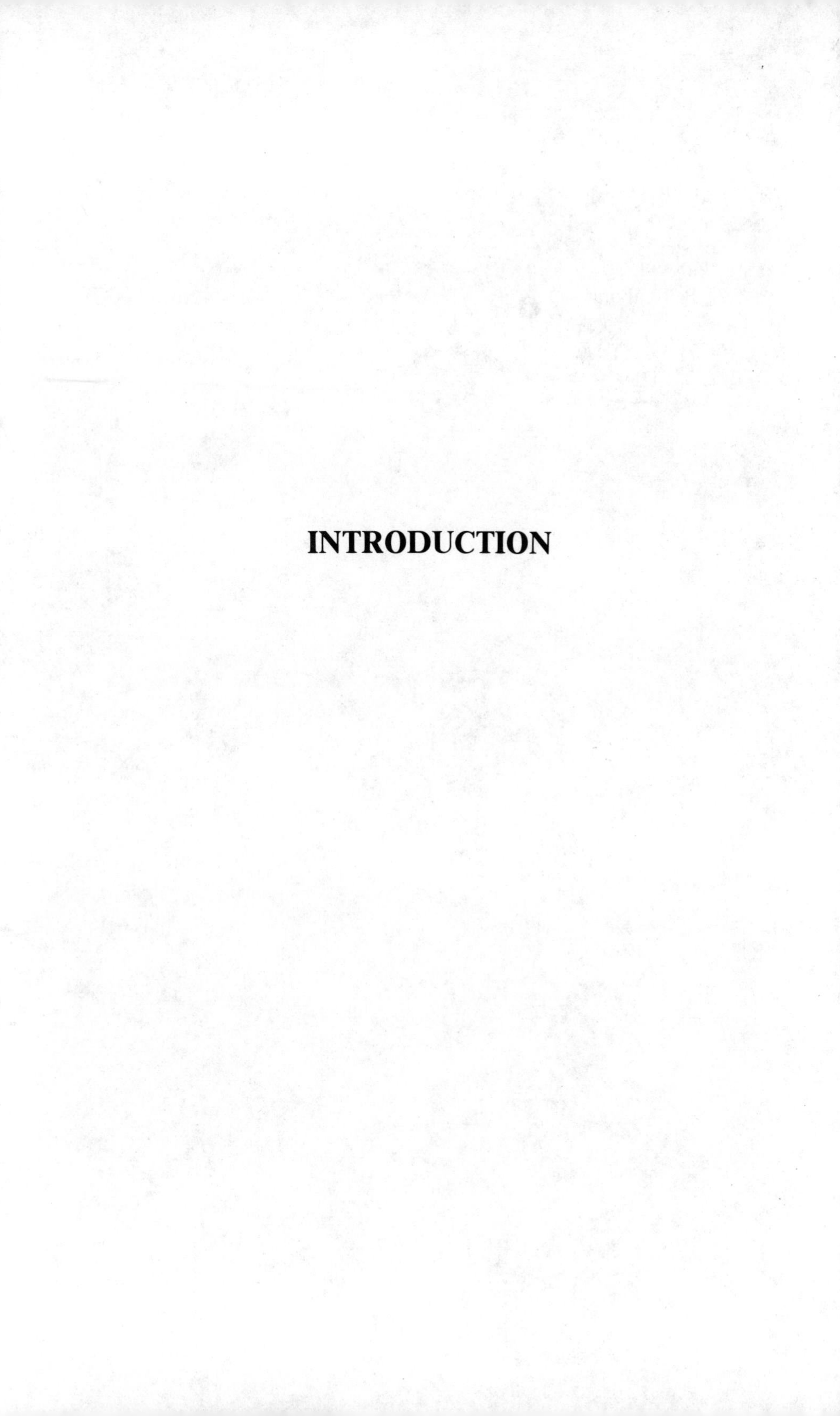

INTRODUCTION

Introduction

Within the next few years, production quantities of devices will be commercially available which contain structures less than 1.0 microns. To accomplish this, new microlithography technologies will be required and these new technologies will require new resist (radiation sensitive polymers) systems and/or processes. The purpose of this book is to provide the scientist and engineer with the fundamental knowledge base that underlies resist design and present some of the newer approaches being investigated. This may be better appreciated through an understanding of the physics and chemistry of the lithographic process along with a historical perspective on materials for fine line lithography which forms the subject of the first three chapters of this book.

In Chapter 1, Tom Everhart (Cornell) reviews his work on the fundamental limits of lithography based on a statistical theory. This theory provides the basic limitation of sensitivity and defects.

In Chapter 2, Alec Broers (IBM) discusses fundamental limitations that are imposed by engineering considerations of hardware design. These limitations affect the economical feasibility of a particular lithographic strategy, an understanding of which is necessary in order to choose between the many complex and competing alternatives.

In Chapter 3, Murrae Bowden (Bellcore) presents a comprehensive perspective on resist materials currently used in microlithography research and development.

The reader is strongly urged to read these three chapters first as they provide the context within which the following chapters which deal with specific new research in radiation chemistry and application to resist design may be better understood.

1

Fundamental Limits of Lithography

T. E. EVERHART

College of Engineering
Cornell University, Ithaca, NY 14853

Fine lithographic patterns are produced when radiation (photons, electrons, ions, or neutral particles) interacts with a material, and produces physical or chemical changes (or both) in that material selectively, as required to replicate the desired pattern. The type and energy of radiation is important to characterize the process, the amount of radiation (exposure) generally must be controlled, and the resulting removal of material by physical or chemical means must also be controlled. Photons generally interact with materials somewhat differently than electrons, which are again different than ions, as momentum of the incoming radiation increases from photon to electron to ion. The incoming radiation excites electrons in the material, and these generally produce the chemical changes "developed" by dry or wet chemical processes in many forms of lithography. The volume of material exposed by the incoming radiation depends on how it penetrates the material being exposed, how it is scattered by the material it is penetrating, and how far secondary electrons that are excited by the incident radiation travel in the material while they are energetic enough to produce exposure.

Patterns can be produced in a mask, and that mask used to define the pattern to be produced on the material to be exposed. Alternatively, particle beams can be deflected in patterns over the substrate to be exposed, and the patterns are produced by turning the beam off and on, changing its shape and size, or both. Regardless of how the pattern is produced, and regardless of the type of radiation used, there are some fundamental aspects of exposure that hold for all lithographic systems, and these are the topic of this chapter.

If photons are the exposing radiation, the energy loss of the photons as they penetrate the material will determine the excitation of secondary electrons, and subsequent exposure of the resist material. The energy loss per unit length will depend upon the energy of the photons, and must be known. Often the useful exposure produced by incoming photons is determined experimentally, and it is assumed here that such dose information to produce correct exposure is known. If electrons are the exposing radiation, the energy loss is again related to the incoming energy of the electrons, and the transverse scatter also affects not only the transverse resolution of the fine lines of the pattern, but also energy dosage per unit depth delivered to the material to be exposed. Similar statements can be made for incoming ions. Ions generally do not scatter

Current address: University of Illinois, Urbana, IL 61801

0097-6156/84/0266-0005$06.00/0

transversely in nearly the same degree that lighter electrons do, and they lose energy in much shorter distances, hence for a given incident energy, ions penetrate much less into the material, but deliver much more dose per incident particle.

We shall develop a fundamental relationship between the dose Q delivered to the resist, the number of particles N striking the pixel (picture element or elementary area to be exposed in the resist), the flux of radiation delivering the exposure dose, the time required to expose a pixel, and the dimension of the pixel. To be specific, we shall consider the case of electrons exposing a resist, and shall strive to provide intuitive insight, rather than an absolutely precise mathematical formulation.

In order to be certain we have produced a significant difference between an exposed pixel and a pixel that is not exposed, a certain number of incident photons, electrons, or ions must each produce an effect in the volume of resist underlying that pixel. This means that if the resist requires a dose of Q coulombs/cm^2 for correct exposure, it also requires a minimum number of electrons N_m to strike and lose their energy in each pixel so that each pixel receives the minimum necessary exposure on statistical grounds. Therefore, if l_p is the pixel linear dimension and e is the electronic charge,

$$Q/e \geqslant \left[N_m/l_p{}^2\right] \tag{1}$$

Equation 1 states a very important point, namely, the exposure dose must always exceed a minimum value, and the smaller the pixel dimension, the larger that minimum value must be. This is a well-known phenomena in photography, for example, since very high resolution film is much less sensitive than lower resolution, or "fast" film. Put another way, Equation 1 explains why large grain size goes with "fast" film, and high definition goes with "slow" film. Since the dose must increase as the pixel dimension decreases for the probability that each pixel will be correctly exposed to remain constant, high resolution resists are expected to be "slower" (i.e., require larger doses) than lower resolution resists. Put yet another way, based only on pixel signal-to-noise ratio considerations only, the minimum total number of electrons needed to reliably expose a pattern of a given complexity is independent of the size of the pixel. More sensitive resists are useful for larger pixels, and less sensitive must be used for smaller pixels. This argument assumes that an electron's energy is lost within a pixel, i.e., that the transverse scattering is considerably smaller than a pixel, and that the beam size is at least as small as a pixel.

Next, consider the time required to expose a pattern with a focused electron beam. The electron beam with current density $J\,(A/\text{cm}^2)$ must strike a pixel for time τ(sec) to produce exposure Q (coulombs/cm^2) = $J\ \tau$. The beam current density is

$$J = J_c(\text{eV/kT})\ \alpha^2 \tag{2}$$

after Langmuir, where J_c, T, and V are cathode current density, temperature, and beam accelerating voltage, e and k are the electronic charge (1.6×10^{-19} coulombs) and Boltzmann's constant (1.38×10^{-23} J/°K), and α is the beam convergence angle.

By increasing α, the current density exposing the pattern increases, which is desirable. However, if α is increased too far, the beam spot diameter increases because of the spherical aberration of the focusing system. An optimum value of α occurs when the diameter of the disk of confusion due to spherical aberration, $d_s = 0.5\ C_s\alpha^3$ (C_s is the spherical aberration coefficient), is set equal to the gaussian spot diameter, $d_s = d_g = \ell_p/\sqrt{2}$. Using the normal approximation of adding spot diameters in quadrature, the total spot size then is $d = \left(d_s^2 + d_g^2\right)^{1/2} = \ell_p$, the pixel dimension. The optimum convergence angle is then

$$\alpha_{opt} \simeq \left[\frac{\sqrt{2}\,\ell_p}{C_s}\right]^{1/3}, \qquad (3)$$

and the exposure in time τ is

$$Q = J\tau = J_c\,\frac{\mathrm{eV}}{\mathrm{kT}}\left[\frac{\sqrt{2}\,\ell_p}{C_s}\right]^{2/3}\tau = \frac{\beta\pi 2^{1/3}}{C_s{}^{2/3}}\,\ell_p{}^{2/3}\tau\,, \qquad (4)$$

where β is the electron optical brightness ($J_c\mathrm{eV}/\pi\mathrm{kT}$). Equation 4 gives the change density deposited in a spot of diameter ℓ_p in time τ. For resist exposure, this charge density must equal the resist sensitivity under the exposure conditions used.

To ensure that each pixel is correctly exposed, a minimum number of electrons must strike each pixel. Since electron emission is a random process, the actual number of electrons striking each pixel, n, will vary in a random manner about a mean value, $\bar{n}$. Adapting the signal-to-noise analysis found in Schwartz (1959) to the case of binary exposure of a resist, one can show straightforwardly that the probability of error for large values of the mean number of electrons/pixel $\bar{n}$ is $e^{-\bar{n}/8}/[(\pi/2)\bar{n}]^{1/2}$. This leads to the following table of probability of error of exposure:

$\bar{n}$	50	100	150	200
Probability of error	2.2×10^{-4}	3×10^{-7}	4.7×10^{-10}	7.8×10^{-13}

To be conservative, we choose $\bar{n} = 200$, which should mean that, on average, no pixels in a field of 10^{10} pixels are incorrectly exposed due to randomness, as long as each electron striking a pixel causes at least one exposure event in the resist. For a pixel of dimension ℓ_p, the minimum number of electrons striking it (= 200 here) to provide adequate probability of exposure is N_m, and the charge density is then $Q = N_m e/\ell_p{}^2$. Substituting into Equation 4 gives

$$N_m e = \frac{\beta\pi 2^{1/3}}{C_s{}^{2/3}}\,\tau\ell_p{}^{8/3} \qquad (5)$$

To determine how noise limits pixel dimension, arrange Equation 5 so that normalized exposure time depends on pixel dimension; note that $2^{1/3}\pi \simeq 4$:

$$\left[\frac{4\beta}{N_m e C_s{}^{2/3}}\right] \tau = \ell_p{}^{-8/3} . \tag{5a}$$

A corresponding equation for real resist exposure is

$$\left[\frac{4\beta}{N_m e C_s{}^{2/3}}\right] \tau_R = \frac{Q}{N_m e} \ell_p{}^{-2/3} . \tag{4a}$$

Here the same normalization was chosen for τ to facilitate plotting Equation 4a and 5a on the same figure of τ vs. ℓ_p.

The above paragraphs give the fundamental considerations of electron beam formation and focusing that cause the time, τ, required to expose a pixel to N_m electrons to increase as the pixel linear dimension ℓ_p decreases. As shown by the left-hand curve in Figure 1, $\tau \alpha \ell_p{}^{-8/3}$. To correctly expose a real resist of sensitivity Q coulombs/cm^2, a fixed number of electrons per unit area must strike the resist, and the time required to expose such a resist is $\tau_R \alpha \ell_p{}^{-2/3}$. A family of curves corresponding to such real resist exposure is also shown in Figure 1. For a given probability that each pixel will be correctly exposed, these curves for a real resist cannot extend to the left past the limiting curve. As we proceed to the right of the limiting curve along a curve for constant sensitivity, Q, the number of electrons striking each picture element increases, improving the pixel signal-to-noise ratio. Because the normalization factor on the ordinate of Figure 1 includes N_m, the vertical positioning of the τ_R curves depends on the value of N_m actually chosen. For binary exposure, the probability that a pixel struck by 200 electrons is not correctly exposed is less than 10^{-12}; if struck by 100 electrons, a pixel has a probability of incorrect exposure of 3×10^{-7}, enough to cause many errors in a pattern of 10^{10} pixels. Hence we have set $N_m = 200$ in the τ_R curves of Figure 1.

These curves predict that for $Q = 10^{-8}$ coulombs/cm^2, pixels smaller than $\ell_p = 1.0$ μm should be possible, and for $Q = 10^{-6}$ coulombs/cm^2, pixels below $\ell_p = 0.1$ μm should be attainable, based on a signal-to-noise ratio considerations alone. Resists such as polymethyl methacrylate processed for high resolution by the correct choice of developer have demonstrated linewidths less than 0.1 μm. The fundamental point emphasized here is that slow resists are necessary to get higher resolution, a result familiar to all photographers. Note that if all electron energy is not dissipated within the pixel (due to lateral scattering, for example), the exposure time per pixel increases and the solid curve in Figure 1 moves toward the dashed curve. Inclusion of quantitative information on scattering and aberrations in addition to spherical aberration will cause the actual limiting curve to move toward the right at small pixel dimensions, as shown in Figure 1.

Equations 4a and 5a both show that exposure time can be shortened by increasing the beam brightness, or decreasing the spherical aberration C_s. By plotting these equations in normalized form in Figure 1, absolute pixel

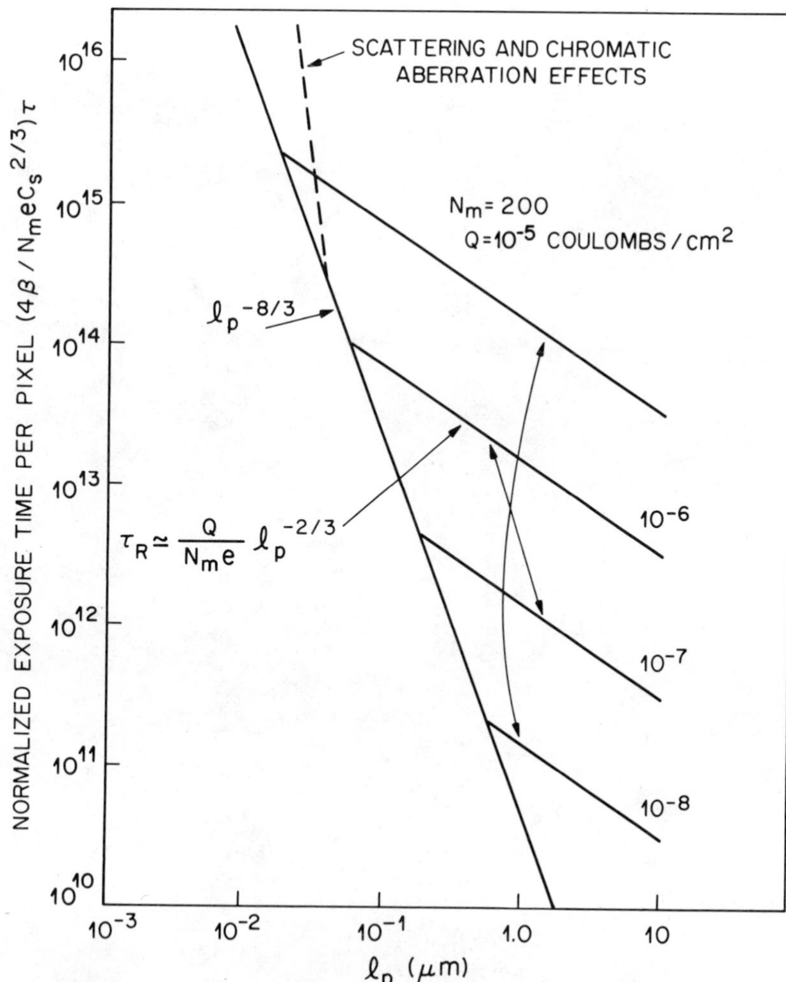

Figure 1. Normalized exposure time/pixel vs pixel dimension.

resolution for resists of a given sensitivity are obtained. Such information can also be useful for ion beams if the brightness is known, and the assumptions made here are satisfied.

Literature Cited

1. The basic arguments presented here may be found in "Basic Limitations in Microcircuit Fabrication Technology"; Sutherland, I. E.; Mead, C. A.; Everhart, T. E., Eds.; R-1956-ARPA, November 1976, a report prepared for DARPA by Rand, Santa Monica, California 90406.

2. Schwartz, M. "Information Transmission, Modulation, and Noise"; McGraw-Hill Book Company, Inc.: New York, 1959; pp. 382-384.

RECEIVED October 3, 1984

Practical and Fundamental Aspects of Lithography

A. N. BROERS

IBM Corporate Headquarters
Armonk, New York

Smaller microelectronic components are generally faster, lower power, and cheaper. Lithography is the most universal microcircuit fabrication step, and plays a major role in gating progress toward further miniaturization. Resolution and overlay are important in choosing a lithography process, but in production, cost, complexity, speed, yield and reliability are equally significant. Optical lithography predominates in manufacturing because it is the lowest cost and simplest of the methods. If these manufacturing constraints were removed, however, other factors would limit the rate of miniaturization, such as device design, material deposition, doping, and etching. Devices would be smaller, but not as small as the alternative, higher resolution, methods would allow. This chapter discusses both the manufacturing, and the fundamental limits of microlithography.

Linewidth and Overlay

Lithography tools and processes for manufacturing integrated circuits must produce statistically good control of dimensions. Minimum linewidth is set by the ability to control linewidth and not simply by the ability to reproduce a given linewidth. The device designer chooses the minimum linewidth that gives him an acceptable ratio (~10:1) of linewidth to the experimentally measured standard deviation ($\sigma(1)$) of linewidth. Devices are typically made tolerant to $\pm 3\sigma(1)$ variations in linewidth. To obtain the data base necessary to make these decisions, patterns containing a range of linewidths are exposed on many wafers, and hundreds of measurements made on all the surfaces encountered in the process. Linewidth has to be maintained over worst-case topography.

Errors in the position of one pattern with respect to another are called overlay errors, and must also be treated statistically. Any systematic offset error (δ) due to faulty adjustment of the alignment system is determined and subtracted to leave the random errors. The error distribution is again fitted to a Gaussian and the standard deviation ($\sigma(a)$) calculated. Devices are designed to be tolerant to alignment errors of up to $\pm(\delta + 3\sigma(a))$. Two kinds of error contribute to overlay errors; those contributed by the exposure tool, such as alignment errors and image distortions, and those in the mask. Again, it is important to make measurements on all the surfaces encountered in the processes under consideration.

Current address: Electrical Division, Engineering Department, Cambridge University, Trumpington Street, Cambridge, CB2 1PZ, England

0097–6156/84/0266–0011$08.25/0

UV Shadow Printing

Contact or proximity printing is no longer widely used because of defects caused by mask/wafer contact. It is still used in non-critical situations when such defects can be tolerated, or when the resolution required is low enough to allow an adequate gap to be left between mask and wafer.

Resolution in a contact image is set by diffraction between the mask and the bottom of the resist. Thick resists, or gaps between mask and resist, degrade resolution. In practice, the minimum usable linewidth $W(m)$ can be approximated from

$$W = \sqrt{1.13\lambda S} \tag{1}$$

where λ is the wavelength of the radiation used to expose the resist (it is assumed that the wavelength is the same in the resist as in the gap), and S is the distance between the mask and the bottom of the resist. This is the condition where the intensity at the center of an isolated line matches the intensity in a large area. This criterion is derived from the degradation in resolution due to Fresnel diffraction (*1*). More rigorous computations of exposure profiles for proximity printing have been made by Lin (*2*).

Linewidth versus gap for deep UV radiation and soft x-rays is plotted in Figure 1. Good agreement has been established in practice, at least in the region of 0.5 μm to 2 μm. For example, 0.5 μm linewidth has been produced in 1 μm of PMMA with 200nm-260nm radiation, and with a method for maintaining intimate contact between mask and wafer (*2*). It is obviously

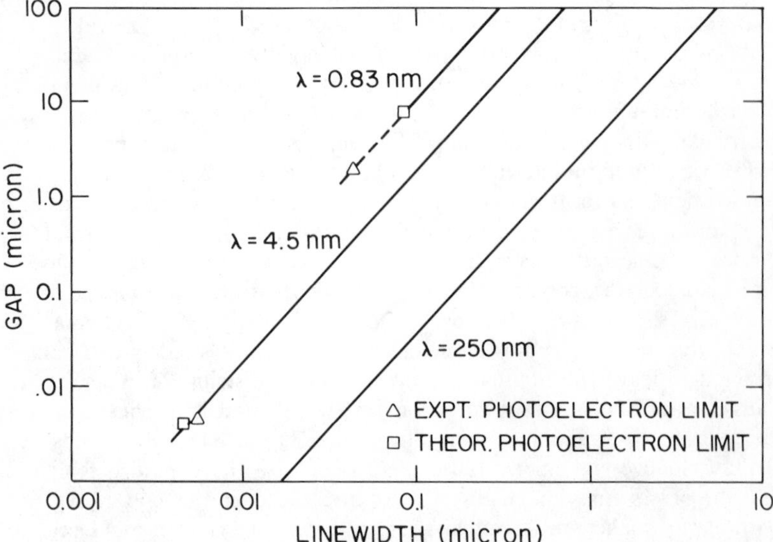

Figure 1. Linewidth versus gap for Deep UV and X-ray proximity printing. Theoretical points correspond to the Gruen range for the maximum energy photoelectrons. Experimental points were measured by Feder and Spiller (27).

difficult to perfect contact over large areas because of contaminating dust particles, and a few square centimeter is probably a practical limit even for experimental devices. Ultimately, linewidth is limited only by the thickness of the imaging layer. For example, it should be possible to produce 250 nm dimensions in 100 nm thick resist provided intimate contact is maintained between mask and resist. Multilayer resist schemes allow this resolution to be transferred into thicker resist layers.

Optical Projection

Two types of optical projection camera are used for integrated circuit fabrication; 1:1 full-wafer scanning, and step-and-repeat (S/R) reduction. At present, the scanning cameras use reflecting lenses and the S/R cameras refracting lenses, but there is no particular reason for this and ultimately it may be best to combine mirror lenses with S/R. The Ultratech S/R camera (*3*) already uses a modified Dyson mirror lens (0.35 N.A.). Mirror lenses can be operated at shorter wavelength than present refractive lenses, and provide higher resolution for the same N.A.. Refractive lenses, however, presently have higher numerical aperture (N.A.).

Throughput

Throughput is highest with the full-field cameras, which can expose more than 100 125 mm wafers per hour. Step-and-repeat cameras with lenses that cover 1 cm^2 expose about 25 wafers per hour, and those with lenses that cover >2 cm^2, about 50 wafers per hour. The larger field lenses generally operate at 5X reduction rather than 10X. In either case, the masks are easier to make than the 1X masks needed for the full-field cameras. Throughput of step-and-repeat cameras can be seriously degraded if reticles must be changed to include test-sites and/or a variety of chip images.

Resolution

Resolution is set by the numerical aperture of the lens and by the wavelength of the exposing radiation. Contrast at a given resolution is typically assumed to be given by the modulation transfer function (M.T.F.), where

$$\text{M.T.F.} = \frac{2}{\pi} (\phi - \cos\phi \sin\phi)$$

$$\phi = \cos^{-1} \frac{\lambda}{4L.(\text{N.A.})}$$

and L (nm) is the linewidth. Strictly, this is not correct as the M.T.F. gives contrast in a sinusoidal image of a sinusoidal object, whereas an integrated circuit mask is a square wave transmission object. However, the approximation gives useful rules of thumb and an M.T.F. of 60% is considered sufficient for typical applications, and 80% for cases where image size control of a tenth of the minimum linewidth is required.

Higher contrast can be obtained for relatively large linewidths by using partially coherent illumination; that is by arranging that the image of the source only partially fills the pupil of the projection lens. 30% to 50% filling of the pupil has proven optimum and, for example, can increase contrast from 60% for the incoherent case, to more than 80% (*4*). Partial coherence increases depth of field but increases exposure time. Too high a degree of partial coherence gives rise to undesirable interference effects between lines. Contrast for the Micralign (0.16 N.A.) optical system and a 0.35 N.A. refractive lens are shown in Figure 2.

Depth of field depends on substrate reflectivity, the degree of partial coherence and the minimum feature size (*5*). In practice, however, the classical depth of field for the incoherent case $\pm(\lambda \div 2(\text{N.A.})^2)$ gives a reasonable approximation. Two layer resist processes in which the image is formed in a thin, flat, resist layer on top of a much thicker planarizing layer, alleviate the need for a large depth of field and make it easier to form high resolution, high aspect ratio, resist patterns (*6,7*). Satisfactory results can be obtained at contrast levels as low as 40%.

Scanning Mirror Systems

Scanning mirror systems cover a whole wafer with a single exposure scan and have high throughput, but their lenses have relatively small numerical aperture. For example, the mirror lenses in the Micralign series (100, 200, 300 and 500) of cameras built by Perkin Elmer (*8,9,10*) have a numerical aperture of 0.16. Exposure can be made at short wavelengths, however, which compensates for the numerical aperture. Three wavelength regimes have been used,

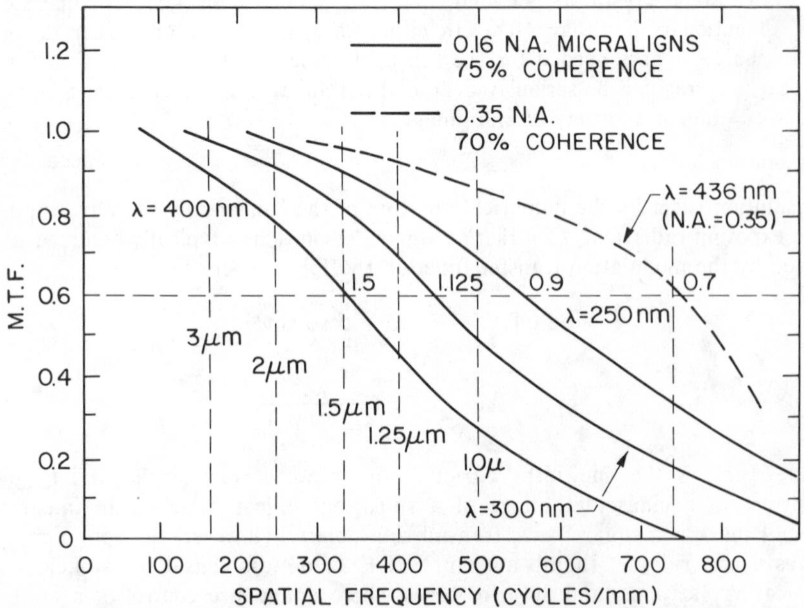

Figure 2. Modulation Transfer Functions (MTF's) for the Perkin Elmer Micralign cameras operating at 250 nm, 300 nm and 400 nm, and for a step-and-repeat camera lens with a numerical aperture of 0.35.

conventional ultra-violet (UV $\lambda = 400$ nm), mid-UV ($\lambda = 300$ nm) and deep UV ($\lambda = 250$ nm).

Major advances with the newest of the Perkin Elmer Cameras, the 500, are the ability to expose larger wafers (up to 150 mm, compared to 100 mm for the earlier cameras), and to adjust magnification. The latter makes it possible to compensate for the inevitable microscopic changes in wafer size that arise during device processing. Image size in the direction of the slit is changed by small axial motions of the 'stronger' pair of refractive elements. Correction in the opposite direction is made by micro-scanning the mask during exposure in order to slightly increases or decrease the scan length for the mask compared to that for the wafer.

If distortions of the sample are truly isotropic, the only errors that remain after magnification is corrected are residual distortions in the optics, and mask errors. The easiest way to reduce mask errors is to go to 5X or 10X step and repeat systems, but eventually, it should be possible with electron beams to reduce mask errors to insignificant levels. Distortion in the optics of the model 500 is already below 0.1 μm and $\pm 3\sigma(a)$ overlays of <0.5 μm have been obtained.

As already mentioned, resolution with mirror lenses can be improved by operating at shorter wavelengths, the only difficulty is to find resist/lamp combinations that give short enough exposure times. Several resists have already been found that are satisfactory for 300 nm exposure and there are a number of others with potential for operation at 250 nm (*11*). The major difficulty with deep UV resists is not sensitivity but the need to find resins that are transparent enough to allow uniform exposure to the bottom of the resist.

Present full-wafer mirror scanning systems are limited to a numerical aperture of about 0.16 by the size of the optical components, and by the need to cover a 125 mm wafer in a single scan and yet avoid having the mask or wafer vignette the optical system. If the full wafer capability is sacrificed, and a system designed to only expose a portion of the wafer in each scan, then it should be possible to increase the numerical aperture to 0.35 or higher. Eventually it may be possible to build a system with an numerical aperture of 0.4 which, at a wavelength of 250 nm, would produce 60% contrast for linewidths of 0.5 μm. Depth of field would be very small (± 0.8 μm for the incoherent case), but capacitive sensing could be used to monitor the wafer surface to $< \pm 0.1$ μm, and piezo-electric manipulators to correct focus.

Step and Repeat Camera

Step-and-repeat (S/R) cameras use refractive lenses with numerical apertures of 0.2 to 0.4. The best lenses expose a field of about 2 cm × 2 cm at a numerical aperture of 0.3. Higher numerical aperture is available for smaller field-size. The field-size and numerical aperture for many of the lenses used in microcircuit cameras are shown in Figure 3. Most lenses are designed to operate at a single wavelength that corresponds to a strong line in the mercury spectrum (365 nm, 405 nm, or 436 nm), but lenses have also been corrected for two wavelengths (405 nm and 436 nm) in order to reduce the effects of

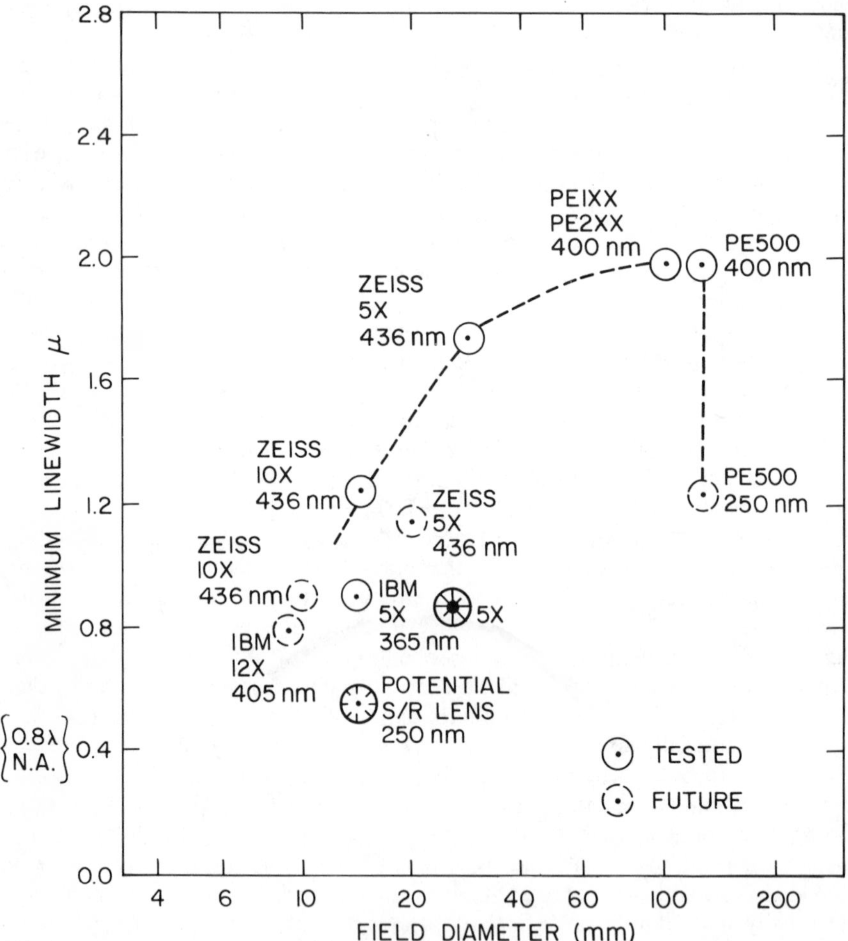

Figure 3. Field size and minimum linewidth for microcircuit lenses. IBM lenses were designed by Wilczynski and Tibbetts (63).

standing wave interference in the resist. With two layer resists, double wavelength exposure is not really necessary.

As dimensions get smaller it is increasingly important to improve overlay and this places stringent requirements on the distortion of the lenses. The distortion of the best lenses remains at about 0.1 μm and this error contributes significantly to final overlay when different cameras are used for different layers.

Refractive lenses must always be used in the step-and-repeat mode because the field size they cover is very much smaller than a silicon wafer. Sample position is either tracked by a laser interferometer, after an initial reticle to wafer alignment, or the reticle and sample are aligned with respect to each other at every chip site (*12-17*). Alignment at every chip avoids errors due to drifts between the wafer and the interferometer reference point, but has proven very difficult. Almost all automatic alignment schemes suffer from the difficulty that for certain resist thickness and alignment mark type, light is reflected equally from the mark and from the background surface, and the mark 'disappears'. To avoid this difficulty, it is necessary to carefully control resist thickness (<10 nm control is needed in some cases) or to remove the resist from the marks. Both solutions complicate the process. The alignment marks can be examined with dark-field or bright-field illumination, and it is best to have both available.

In principle, resolution of optical lithography can be similar to that of optical microscopy. A lens with a numerical aperture of 0.95 would produce linewidths <0.4 μm over an area of about 0.2 mm × 0.2 mm. The small field would not be an inherent limitation because the image could be stepped or scanned, but the depth of field would only be about ±0.2 μm restricting resist thickness to <0.2 μm. Improvement could be obtained by operating with mid-UV or deep-UV light, but refractive lenses are difficult to build for these wavelengths because of the lack of a range of glasses that are adequately transparent and yet have relatively high refractive indices. If such lenses can be made, then excimer lasers would provide more than enough illuminating power at these frequencies.

Alignment in Optical Lithography

In the absence of the problems discussed above, optical methods for positioning an image with respect to a sample can be extremely precise (<0.1 μm). Accuracy beyond the Rayleigh criterion. (0.6λ/N.A. (cm)) is possible by threshold detection because in general, S/N ratios in the detected signal are favorable. Methods that employ diffracting components on mask and wafer have also proven effective in the laboratory (*18*), although in some instances errors arise when the mask to wafer spacing varies (*19*). Overlay accuracy over the whole sample will of course depend on temperature control and on the ability to control mask and wafer distortion. No fundamental limits can be identified in this instance.

X-Ray Lithography

The short wavelength of x-rays makes them attractive for high resolution lithography even though resolution approaching diffraction limits has yet to be obtained with x-ray imaging systems. This is because patterns can be shadow-printed with x-rays with very high resolution. Soft x-rays of wavelength between 0.4 nm and 5 nm (*20,21*) are used so that the adequate energy is absorbed in the resist. The wavelength is still so short that diffraction effects are negligible down to linewidths of 0.2 μm, even for a mask to wafer spacing of 50 μm. The mask cannot be made on a quartz plate as is used for optical lithography, because the plate would absorb too much energy, so it is formed on a thin (<10 μm) membrane. The membrane is generally made from a combination of two materials with one chosen for dimensional stability (silicon, boron nitride, etc.), and the other for ruggedness (e.g. polyimide). The absorber pattern is generally written with electron beam.

X-ray lithography offers one outstanding advantage over the other methods used for fabricating sub-micron dimensions. The x-rays are neither scattered in the resist nor are they diffracted, so very high aspect ratio resist patterns are produced, and there is no proximity effect as there is with electrons.

On the other hand, the mask is fragile and may not be dimensionally stable, and the large divergence of conventional electron bombardment x-ray sources gives rise to image distortions if mask and wafer are not perfectly flat (see Figure 4). In order to keep pattern distortion below 0.1 μm for a 125 mm diameter wafer and a source to wafer distance of 50 cm, the mask and wafer must be held flat to better than 1 μm. Wafer buckling that arises during many standard integrated circuit processes typically exceeds 10 μm, making it unlikely that full wafer exposure can be accomplished for wafers larger than about 7.5 cm.

The run-out error could be avoided if resist sensitivity was improved to the point that exposure could be made at a point further removed from the source. This requires a sensitivity approaching 1 mJ/cm^2, however, and while there are some resists with sensitivity at this level (*22*), they do not yet produce high aspect ratios and therefore lack the major advantage of x-ray lithography over optical lithography. High aspect ratio can be obtained by using a tri-layer resist process, but this option is also open to optical lithography.

The run-out error could also be avoided by going to a S/R approach and limiting mask size to a few centimeters. This would also alleviate problems due to mask and/or wafer distortions but exposure times with presently available resists and conventional sources are too long for this to be economical. Exposure times of a few seconds would be needed for a source to wafer distance of about 50 cm which infers similar sensitivity to that required for the full-wafer case.

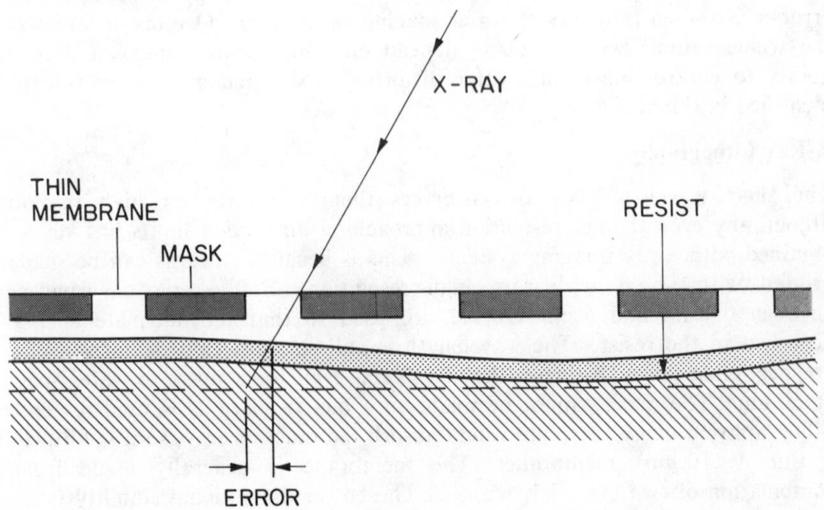

Figure 4. Run-out error encountered with conventional point source x-ray lithography.

Electron Storage Ring

One source that does produce an x-ray beam with all the properties needed for S/R exposure is the electron storage-ring (*23*). The x-ray beams emitted from a storage ring are so intense, and have such small divergence, that the problems of run-out error and long exposure time are removed. The problems are complexity and cost, but these can potentially be overcome, as discussed later. A storage ring consists of a circular stainless-steel vacuum tube in which electrons circulate at speeds close to the speed of light. The electrons are injected into the ring from an accelerator, and their energy increased, and subsequently maintained, by accelerating them trough a microwave cavity each time they pass around the ring. The electrons are held in orbit by a series of magnets. Each time they are deflected by one of these magnets, they emit radiation with a spectrum that extends from visible light to hard x-rays. By choosing the appropriate electron energy (~1 GeV), copious soft x-rays suitable for x-ray lithography are emitted.

The radiation from a storage ring is emitted in a broad sweep with a very narrow vertical spread. The beam is typically 1 cm high and 30 cm wide at a distance of about 10 m from the ring. An oscillating, grazing incidence, mirror is used to broaden it to the desired height.

An alternative method for generating the x-rays is to "wiggle" the electrons while they pass down the straights in the ring between the bending magnets. Wiggling is accomplished with multi-pole magnets that produce a sequence of alternating magnetic fields. The bending radius can be smaller and the x-ray energy at each deflection adds up to yield increased total output for a given electron energy. Lower energy reduces the cost of the ring.

Enough radiation is emitted at each of the bending magnets, or wigglers, to expose 40-50 125 mm diameter wafers per hour (*24*), a rate similar to that of present optical S/R cameras (this requires a stored current approaching 1 ampere, but at present maximum currents are only about a third of an ampere). Therefore, a fully utilized storage ring can produce the equivalent of 6 to 12 optical cameras. A S/R camera costs $500K to $1M, so fully loaded, a storage ring with a cost of $5-10M becomes competitive with optics, and, of course provides better resolution and higher aspect ratio resist patterns. Several laboratories are working on designs for 'compact' storage rings that are potentially suitable for production environments (*25,26*).

X-ray lithography with a storage ring offers high throughput, and a S/R approach that is relatively tolerant to distortions of mask or wafer, but alignment accuracy will be no better than with optics, because the alignment methods being developed today are optical. This is not necessarily a problem, because, as we have already seen, high accuracy can be obtained with optics. Two basic methods have been tried for x-ray lithography; the first uses simultaneous imaging of mask and wafer with an optical system operating either with two different wavelengths or with two different optical path lengths, and the second examines the superposition of diffracted beams from gratings or zone plates placed on the mask and wafer. Both methods are theoretically capable of achieving accuracies of <0.1 μm, and this has been demonstrated in the laboratory, but is yet to be demonstrated in a device process.

In addition to alignment errors, overlay accuracy depends on mask errors and mask and wafer distortions. Errors and distortions are likely to be relatively serious because the mask is fragile and has the same dimensions as the wafer. With the conventional sources, this problem is somewhat alleviated because isotropic changes in wafer or mask size can be corrected by changing the mask-wafer gap, but adequate overlay for sub-micron devices has yet to be demonstrated.

Resolution of X-Ray Lithography

Two factors set the resolution for x-ray proximity printings: 1) diffraction between mask and wafer, and 2) the range of the photoelectrons formed when the x-ray photons are absorbed in the resist. The same diffraction criterion used above for optical proximity printing can be applied for x-rays to give the relationships shown in Figure 1. The range of photoelectrons in resist has been measured experimentally by Spiller and Feder (*27*). The Gruen depth dose relationship has also been used to estimate the photoelectron range although Gruen only confirmed this relationship over the energy range of 5 keV to 54 keV. It has been suggested (*28*) that the minimum linewidth, as set by the photoelectron range, will be equal to the Gruen range (R_G)

$$R_G(m) = \frac{2.57\times10^{-11}E\,(\mathrm{eV})^{1.75}}{(\mathrm{g/cm^3})} \simeq 10^{-23}\lambda^{1.75}(m) \qquad (3)$$

where E(ev) is the maximum photoelectron energy associated with $\lambda(m)$, the wavelength of the exposing x-rays, and (g/cm^3) is the density of the resist (1.2 g/cm^3 for PMMA). The Gruen range and Feder and Spiller's experimental results are shown in Figure 1. As can be seen, the highest resolution predicted is about 5 nm using carbon characteristic radiation (4.5 nm). In support of this prediction, 4.5 nm x-rays have been used to shadow-print biological samples into PMMA with a spatial resolution of about 10 nm (*29*). In this case, only a relief profile was obtained in the resist rather than a fully developed image. The resist was metallized and the sample examined in surface SEM. Lines of about 20 nm have also been replicated with x-ray lithography using a mask that was fabricated by alternately shadowing tungsten and carbon onto the side of a surface step (*30*), see Figure 5.

High aspect ratio (ratio of resist thickness to minimum linewidth) in the resist is a significant advantage with x-ray lithography when compared to optical and electron beam methods. The advantage is very large at dimensions greater than ~0.1 μm, but disappears compared to electron beam lithography for dimensions below 0.1 μm (see Figure 6), where degradation due to diffraction with x-rays exceeds that due to scattering of high energy electrons (~100 keV). Aspect ratios are nonetheless satisfactory with x-rays down to linewidths of a few tens of nanometers.

For integrated circuit applications, the significant factor to be observed from Figure 6 is that a gap in excess of 100 μm can be tolerated for 1 μm resolution. With conventional sources, penumbral blurring due to the finite source size is more likely to limit resolution. X-ray lithography is perhaps the most promising method for large production of sub-0.75 μm devices because it

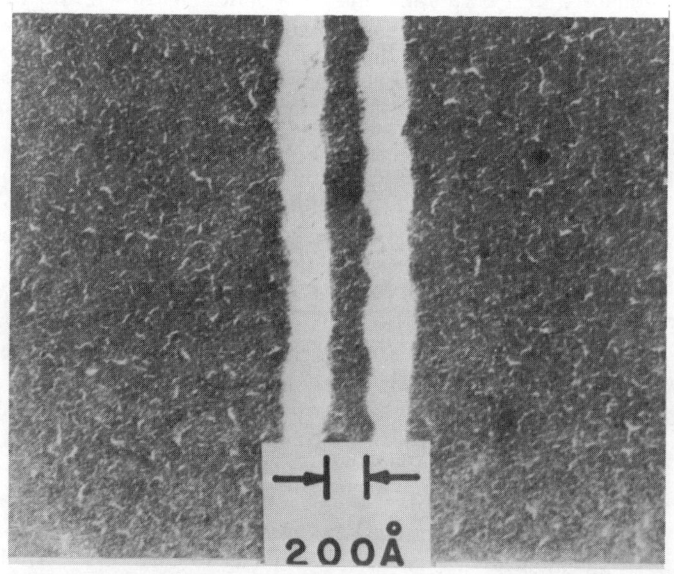

Figure 5. 20 nm lines produced by Flanders (30) in PMMA using x-ray lithography.

Figure 6. High aspect ratio 1 μm resist pattern obtained with x-ray lithography (Spiller et al. (18)).

should produce wider process margins than optics and lower cost than scanning electron beams. Many new materials and methods are required, however, and the margin of improvement will have to be considerable before the change will be worthwhile.

Scanning Electron Beams

Scanning electron beam lithography is important in manufacturing because it is the fastest method for generating patterns. It is becoming the predominant method for making masks, and is already established for customizing gate-arrays. In research and development it has the additional advantage that it produces better resolution and overlay than any other method. Many early micron and sub-micron integrated circuit devices were made first with electron beams, and structures as small as a few tens of nanometers have been made for electrical characterization (*31,32*). Throughput remains too low, and system cost too high for general direct wafer writing.

Electronic versus Mechanical Scanning

Sub-micron electron beams cannot be scanned over a whole wafer or mask because of deflection aberrations and deflection system noise. Mechanical movement is combined with electronic scanning to cover a wafer. The position of the beam with respect to the sample can be maintained in several ways. In the conventional 'vectorscan' type of system, a beam to sample reference is made before each complete section (chip) of the pattern is exposed. To sense beam position, scattered electrons are collected as the beam scans across four marks located at the corner of each field. High energy electrons are detected rather than low energy secondaries, so that the alignment marks can be detected through thick resist layers. Position, magnification, rotation, and orthogonality are checked. After registration, the chip is written with a two-dimensional electronic scan. A laser interferometer is also included in many cases to provide a back-up, or alternative, to the every chip alignment.

In the 'EBES' electron beam mask-maker, the beam is electronically scanned in one direction only, and the sample continuously moved in the other direction (*33*). Chips are written strip by strip, the same strip on every chip being written before proceeding to the next strip. The position of the beam is checked initially with a direct beam to sample measurement, but after this a laser interferometer keeps track of the sample. Errors in position are corrected by feeding signals to the electron beam deflection coils.

Shaped Beam Versus Round Beam

Two alternative approaches maximize electron beam current and minimize writing time. The first uses a rectangular beam, the size of which can be changed to fill the pattern features (*34-36*). Current in the round spot is maximized by using a high brightness cathode. Full pattern flexibility is retained with a round beam (i.e. the ability to write curved and angled lines) but the current cannot be as high as it is with the shaped beam, and the number of beam addresses is very much greater.

With shaped beams the source for the electron optical column is an image of a square aperture that is formed in the plane of a second square aperture. Non-square apertures have also been used to produce angled shapes. The final beam takes on a broad selection of shapes as the position of the image is varied with respect to the second aperture.

Provided the electron gun produces adequate current at a given brightness, the brightness in a shaped beam is the same as that in a round beam, in fact the average brightness is higher. This means that for the same beam aperture and beam brightness, the beam current in the shaped beam is higher by at least the ratio of the spot areas.

The larger the shaped spot, the larger the beam current, and the smaller the number of beam addresses. This approach cannot be taken too far, however, because as the current increases, electron electron interactions become excessive and edge definition degrades. For example, in EL3, which has a beam current density of 50 A/cm^2 and a maximum spot size of 4 $\mu m \times 4$ μm (16 μm^2), spot edge definition degrades beyond 0.25 μm for spots larger than about 6 μm^2 (*34*). Larger spots could be used if the current density was reduced, but this would increase the exposure time for the majority of shapes, which are generally smaller than 6 μm^2. The optimum spot size is that which maximizes the time average beam current. In order to minimize exposure time, relatively high current is used, and the problem of poor edge definition on large shapes is overcome by outlining these shapes with small narrow rectangles.

For the time average current in a round beam to equal that in a shaped beam, the spot current density must exceed that in the shaped beam by the average number of resolution elements projected in the shaped beam, assuming the same deflection/focusing system and pattern generation concept (raster/vector etc.) are used. This number is typically 30-100. The current in the shaped beam is limited by electron-electron interactions, however, and for example, this limit is reached in EL3 at a gun brightness of about 3×10^5 A/cm^2.steradian (25 kV). So a brightness of $1\text{-}3 \times 10^7$ A/cm^2.steradian would be adequate to produce the same net current in a round beam.

Such brightness is available from thermal-field emitters, and workers are using them in prototype high throughput systems (*39*), but the ir reliability at the high total currents needed to produce a final beam current of several microamps is not proven for production environments.

Lanthanum hexaboride cathodes fall short of the required brightness for a round beam to compete with a shaped beam only producing $\sim 1 \times 10^6$ A/cm^2.ster. (25 kV) with adequate reliability for unattended production applications (*40*). In summary, highest throughput in production is obtained with shaped beams.

Round-beam Vectorscan Systems

In vectorscan systems, the beam is only directed to those areas of the chip that require exposure. With round beams, the beam diameter must be less than or equal to one quarter of the minimum linewidth, and to obtain adequate pattern fidelity, the beam must be incremented to at least sixteen positions to write a single minimum image square. Chip writing time can be estimated from the beam incrementing rate, and the average fraction of the chip area that needs to be exposed. For example, with a 1 μm image, a 5 mm × 5 mm chip contains a total of 25×10^6 minimum images, or $16\times25\times10^6$ beam positions. If, on average, 25% of the chip must be exposed, then 10^8 beam positions must be written per chip. At 10 MHz, which is the maximum rate used so far to successfully to make integrated circuits, this will take 10 seconds. Time taken for table stepping, registration, wafer loading, and data transfer must be added to the writing time to come up with the overhead times are below 1 second per chip, and a 125 mm wafer containing 500 chips can be exposed in about 2 hours.

From an electronic control system point of view, it should be possible in the near future to reach incrementing rates of 50-100 MHz, and provided adequate current density can be delivered to a practical resist, a throughput of about 5,125 mm wafers per hour will be obtained.

EBES Systems

EBES systems also use a small round beam with a diameter about one quarter of the minimum linewidth, but unlike the vectorscan systems, they write the pattern in a raster manner. This means that the beam has to be deflected to every element on the wafer whether or not exposure is required (advanced EBES systems skip kerf areas and areas where no exposure is required but still write the rest of the areas in a raster manner). If the grid on which the pattern is written is finer than the beam diameter, as is typically the case for masks where the ratio of minimum image to design grid can be as high as 20:1, then the number of beam addresses becomes greater than would be expected from the beam diameter. This reduces throughput. In most vectorscan system, each shape can be placed with a precision that is smaller than the beam diameter without impacting throughput, only the precision of the pattern control electronics is affected.

Throughput in an EBES system scales inversely with the square of the image design grid. For example, it takes about 3 hours to write a 1 μm minimum linewidth mask for a 125 mm diameter wafer using a design grid of 0.1 μm and an incrementing rate of 80 MHz (the maximum rate available today). This is more than an order of magnitude faster than the opto-mechanical mask-makers that preceded electron beam systems, but is an order of magnitude slower than variable-shaped beam systems.

EBES-like systems are used extensively for mask making. They can make masks with dimensions as small as 0.5 μm, and 3σ overlay accuracy approaching 0.1 μm. They generally use a small round beam, although shaped beam versions have also been described (*40*).

EL3 — Variable-Shaped Beam Vectorscan System

The key specifications of EL3 are shown in Table I. EL3 combines a high current variable shaped beam column with a dual deflection system (*42*). Maximum beam size is 4 μm × 4 μm and shapes smaller than this are available in 0.1 μm increments. Spot edge definition is better than 0.2 μm. A highly accurate but relatively slow magnetic deflection coil deflects the beam in a raster sequence to the center of an array of 75 μm subfields. The variable

Table I. EL-3 Key Specifications

Minimum image	1 μ	2 μ
Field size	5 mm	10 mm
Beam edge resolution	0.25	0.5
Overlay (Mean + 3σ)	0.4 μm	0.7 μm
Wafer/mask capabilities	(57 → 165 mm)	
Current density	(50 A/cm^2)	
Throughput W/h at 10 $\mu C/cm^2$, 3 in. wafers	10-20	20-45

shaped beam is vector-addressed inside the sub-fields with electrostatic deflection. The sub-fields overlap each other so that there are no discontinuities, and they are so small that super-precision deflection is not needed. The array of sub-fields can be as large as 10 mm × 10 mm. Errors in both magnetic and electrostatic deflection are 'learn-corrected' by scanning a reference grid, storing errors in the control computer, and feeding back corrections during writing.

Throughput of EL3 (see Table I) is limited almost equally by exposure time (current density/resist sensitivity), mechanical stepping and control system speed. Further improvement will require advances in all areas.

Future Direct-Write Electron Beam Systems

The most obvious way to improve throughput in scanning electron beam systems is to combine a variable shaped beam column, with a continuously moving table. The shaped beam ensures maximum beam current, and the continuously moving table potentially eliminates many overhead times. Registration can be accomplished without stopping the table, either by means of a laser interferometer, or through direct beam to sample reference.

Resolution for Electron Beam Exposure

Ultimately, resolution in electron beam lithography is set by the range over which the primary electrons interact with the resist. That is by the distance over which the low energy secondary electrons are created (the resist is exposed mainly by secondaries) and by the range of the secondaries in the resist. For thin resists, and thin substrates (thin compared with the primary electron penetration) this resolution limit has been measured to be about 12 nm (*43*).

For thick substrates, backscattered electrons from the substrate decrease contrast and the minimum dimension increases to about 20 nm. For thick resists, and samples thick compared to the primary electron penetration range, electron scattering in the resist (forward scattering) and backscattering of electrons from the substrate, become more important than the electron interaction range. In these cases, exposure dose is sometimes altered according to the local pattern density to compensate for variations in the backscattered dose (proximity effect) (*44*). Compensation for this effect is valuable even at dimensions of 1 μm - 2 μm, particularly for thick (>500 nm) resist layers.

As already discussed, there is an important case where resolution is determined by fundamental limitations of the electron optical system and not by electron scattering. This occurs with the high current shaped electron beams used in high throughput direct-write tools. The Coulomb interaction between electrons in these columns displaces the electrons from their intended trajectories and blurs the edges of the spot. As discussed above in connection to throughput, this effect, which is related to the Boersch effect (*45*) forces a compromise between throughput and resolution.

Resist Contrast

The fundamental resolution limit of electron beam resist exposure has been measured for PMMA under conditions where electron scattering effects are negligible (*43*). Thin layers (<0.1 μm) on thin substrates (<0.1 μm) were exposed with a 1 nm diameter electron beam using a scanning transmission electron microscope (STEM). The samples were subsequently examined in the same instrument (see Figure 7). The experimental data indicate that the exposure distribution can be approximated by a Gaussian distribution with a sigma (σ_f) of 12.5 nm. It is assumed that this spread arises from straggling of secondary electrons into the resist. Using this distribution, and the data on resist exposure due to backscattered electrons (*46*), the contrast for electron beam exposure can be calculated over a broad range of linewidths, assuming that the backscattered distribution is also Gaussian (sigma σ_b) (see Figure 8).

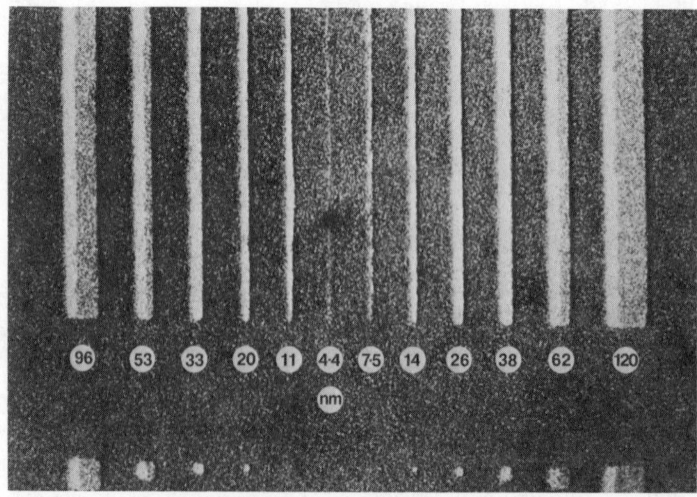

Figure 7. Scanning transmission scanning electron micrograph of a test pattern developed in a 30 nm thick layer of PMMA. Pattern was written with an electron beam with a diameter below 1 nm. The narrowest lines are 10 nm wide. Minimum linewidth and center-to-center spacing is limited by straggling of secondary electrons into the resist.

The contrast (K) is calculated for a pattern of equal lines and spaces from

$$K = \frac{Q_L - Q_S}{Q_L + Q_S} \tag{4}$$

where Q_L is the dose at the center of a line, and Q_S is the dose at the center of a space.

$$\begin{aligned} Q_L = {} & erf\left\{S/4\sigma_f\right\} + \sum_{n=1}^{n=m}\left[erf\left\{(4n+1)S/4\sigma_f\right\} - erf\left\{(4n-1)S/4\sigma_f\right\}\right] \\ & + \eta\left\{erf S/4\sigma_b + \sum_{n=1}^{n=m}\left[erf\,(4n+1)S/4\sigma_b - erf\,(4n-1)S/4\sigma_b\right]\right\} \end{aligned} \tag{5}$$

$$\begin{aligned} Q_S = {} & \sum_{n=1}^{n=m}\left[erf\,(4n-1)S/4\sigma_f - erf\,(4n-3)S/4\sigma_f\right] \\ & + \eta\left\{\sum_{n=1}^{n=m}\left[erf\,(4n-1)S/4\sigma_b - erf\,(4n-3)S/4\sigma_b\right.\right\} \end{aligned} \tag{6}$$

S is the center to center spacing between the lines and m is the number of lines taken into account on either side of the line under consideration. Here η is the ratio of the exposure due to incident electrons to the exposure due to backscattered electrons, and $(1-\eta)$ is the exposure dose in the center of a large shape.

The definition of contrast given in (4) is similar to that of the MTF for an optical system except in the optical case, the pattern distribution is assumed to be sinusoidal rather than square wave as assumed for the electron beam case.

Contrast (K_G) for an isolated space is assumed to be given by

$$K_G = \frac{Q_P - Q_G}{Q_P + Q_G} \tag{7}$$

where Q_P is the exposure dose received at a point half the gap dimension away from the gap, and Q_G is the exposure dose received at the center of the gap.

$$\begin{aligned} Q_P = {} & 1 + \eta - 0.5\left\{erf\,\frac{3L}{2\sigma_f} - erf\,\frac{L}{2\sigma_f}\right\} \\ & - \frac{\eta}{2}\left\{erf\,\frac{3L}{2\sigma_f} - erf\,\frac{L}{2\sigma_f}\right\} \end{aligned} \tag{8}$$

$$Q_G = 1 + \eta - erf\,\frac{L}{2\sigma_f} - erf\,\frac{L}{2\sigma_f} \tag{9}$$

where $L\,(m)$ is the gap width.

It is important to note in Figure 8 that with a thick substrate the

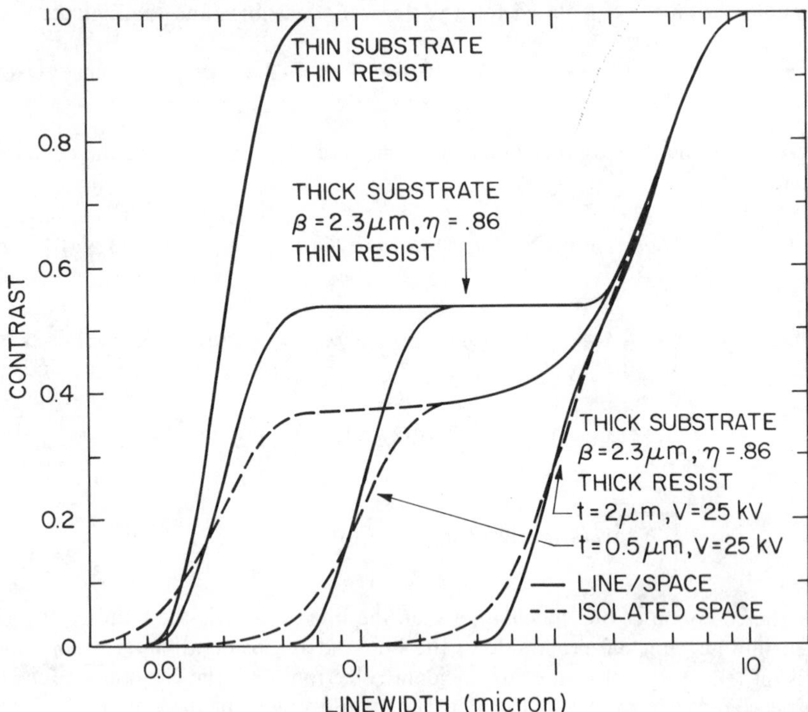

Figure 8. Contrast for line/space patterns and for isolated spaces written with 25 kV electrons (43). Thin substrate improves contrast rather than resolution.

contrast is the same for 0.05 μm lines as it is for 1 μm lines provided the resist is thin (<0.1 μm). This is because the backscattered distribution is much wider than either the secondary distribution or the forward scattered distribution. With multilayer resist processes, where the image is formed in a thin top layer that is backed by a thicker underlayer, the exposure due to backscattered electrons is reduced even further. The underlying resist is of lower atomic weight than silicon and consequently backscatters fewer electrons. The contrast for the multilayer siloxane process described by Parasczak et al. (*47*), is shown in Figure 9. It can be seen that the contrast is improved for all linewidths.

For relatively thick resist (<0.25 μm) and low accelerating potentials (<50 kV), lateral scattering of high energy electrons in the resist, rather than straggling of secondaries, determines resolution. This forward scattering can again be approximated by a Gaussian and here we obtain the standard deviation of the distribution (σ_f) from reference (*48*), Figure 1. The data in this figure are determined from Monte Carlo simulations, but are in good agreement with experimental measurements reported in reference (*49*). We have used the expression

$$\sigma_f = \left\{ \frac{9.64 \times Z\,(\mu)}{V\ (\mathrm{kV})} \right\}^{1.75} \qquad (10)$$

to approximate lateral scattering when calculating contrast (Figure 8) and aspect ratio (Figure 6). Equation 10 closely fits the Monte Carlo data up to the point that the number of collisions suffered by an electron passing through the resist exceeds 100.

In general, higher accelerating potential offers higher resolution and higher contrast. Higher energy electrons are scattered less in the resist and are backscattered from the substrate over a larger area. The overall effect is that contrast is increased for all linewidths except those close to the width of the backscattered distribution, and contrast at these relatively large linewidths is so high that this is not significant. Figure 10 shows the improvement to be gained by going from 25 kV to 50 kV. The effective resist sensitivity is reduced at higher voltages because less of the electron energy is dissipated in the resist, but the electron brightness is increased in proportion to the voltage, and the increase in beam current partially compensates for the loss. The overall improvement at higher electron energy should make it easier both to correct for proximity effects and to write high-resolution patterns in thick resist layers.

Normalized Aperture Exposure

The resist contrast function described above is useful in estimating ultimate resolution but does not provide a good indication of the width of the 'process window' for a given lithography process. An array of lines may be clearly

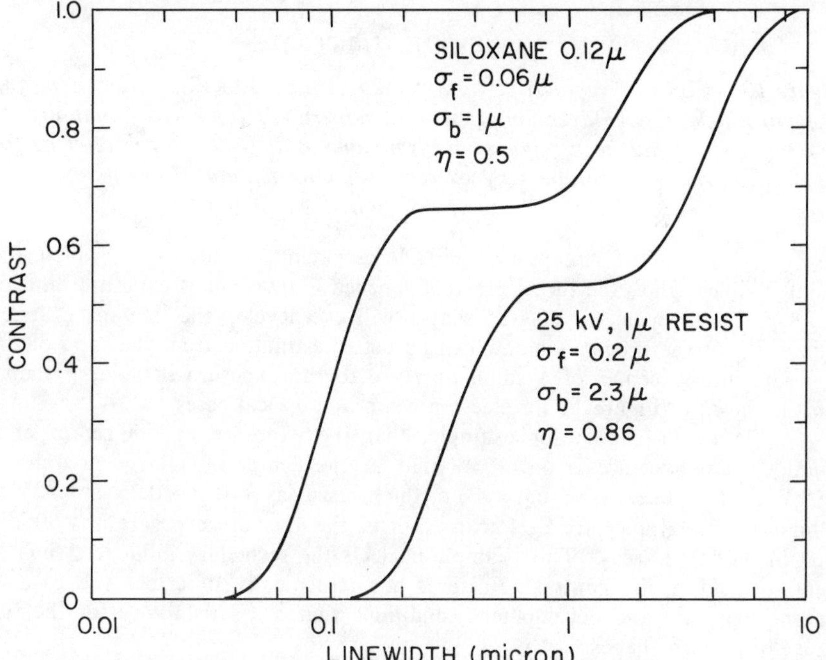

Figure 9. Contrast for line/space patterns for the double layer siloxane resist process (47) is higher than for the single layer case for all linewidths.

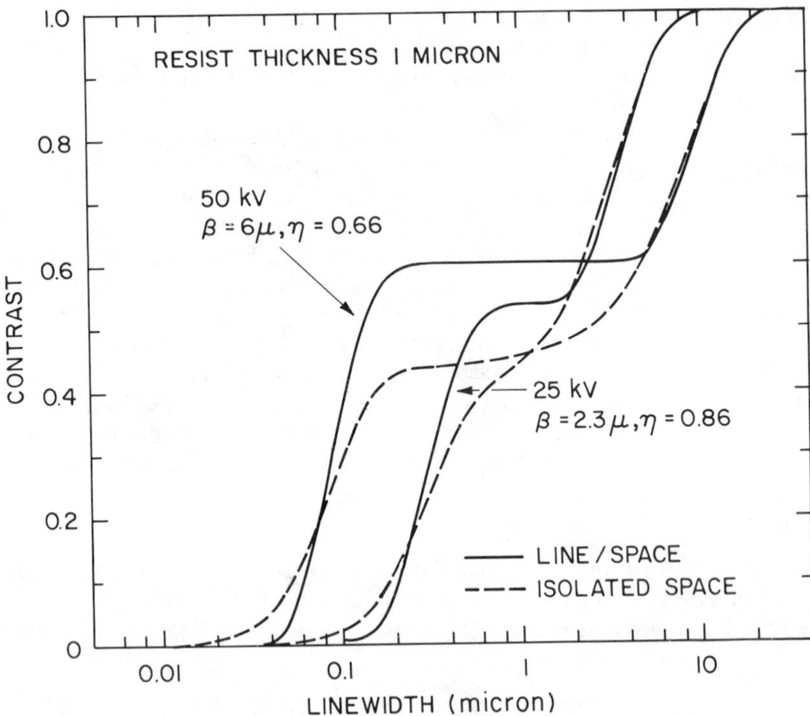

Figure 10. Contrast versus linewidth for 25 kV and 50 kV electrons exposing a 1 μm thick resist layer on a silicon substrate. Improved immunity to proximity effect has been reported by Neill and Bull (64). Backscattering for 50 kV electrons was obtained by extrapolation from data given in references (64) and (65).

resolved, but only for exposure and development conditions at which the isolated lines and apertures are not opened. Increasing exposure and/or development to open the isolated shapes will overdevelop the line/space array. The width of the process window can be better estimated from the ratio of the exposure at the center of a small aperture to the exposure in a large shape. This is shown in Figure 11 for electron beam and optical cases.

Figure 11 shows, for example, that the exposure at the center of a 1 μm × 1 μm aperture is only 55% that at the center of a large area for a 25 kV electron beam exposing a 1 μm thick resist layer ($\sigma_f = 0.25$ μm). With a high numerical aperture S&R camera lens, the aperture exposure is >80% of that in the large areas. This example neglects the secondary influences such as depth-of-field, resist contrast etcetera, but realistically describes the ease of finding exposure and development conditions that are satisfactory for the full range of pattern shapes.

Figure 12 shows the more favorable case of electron beam exposure of the multilayer siloxane resist. Here, the exposure due to backscattered electrons is reduced from 86% to 50% of the primary exposure, and the imaging layer is very thin. Exposure in the 1 μm × 1 μm aperture exceeds 70% and is

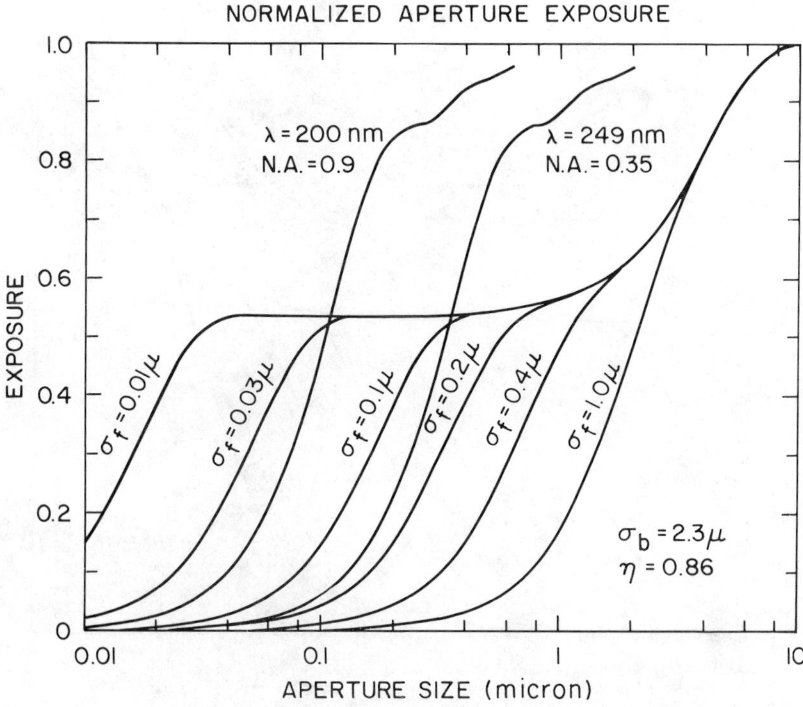

Figure 11. Exposure received at the center of an aperture, normalized to the exposure at the center of a large shape, for electron beam and optical cases.

greater than the exposure for the optical case below 0.4. It is clear that it will be much easier to find acceptable exposure and development conditions with the siloxane process. The penalty is that more processing steps are required making the process more expensive.

Alignment Accuracy with Electron Beam

Ultimately, registration accuracy on solid samples can be better than ± 0.1 μm, and there is no fundamental reason why registration accuracy equal to the resolution of a good surface SEM, 5 nm, cannot be reached provided suitable alignment marks can be used and the resist overcoat is thin enough.

With thin film substrates, it should be possible to locate the beam to an accuracy corresponding to the resolution of a transmission electron microscope (<0.5 nm).

Fabrication of Structure with Dimensions Below 10 nm

In principle, electron beams should be able to produce structures with dimensions as small as the beam diameter in a STEM (0.5 nm), but before this can be done techniques must be found which are sensitive to the high energy primary electrons, rather than the low energy secondaries electrons. This is

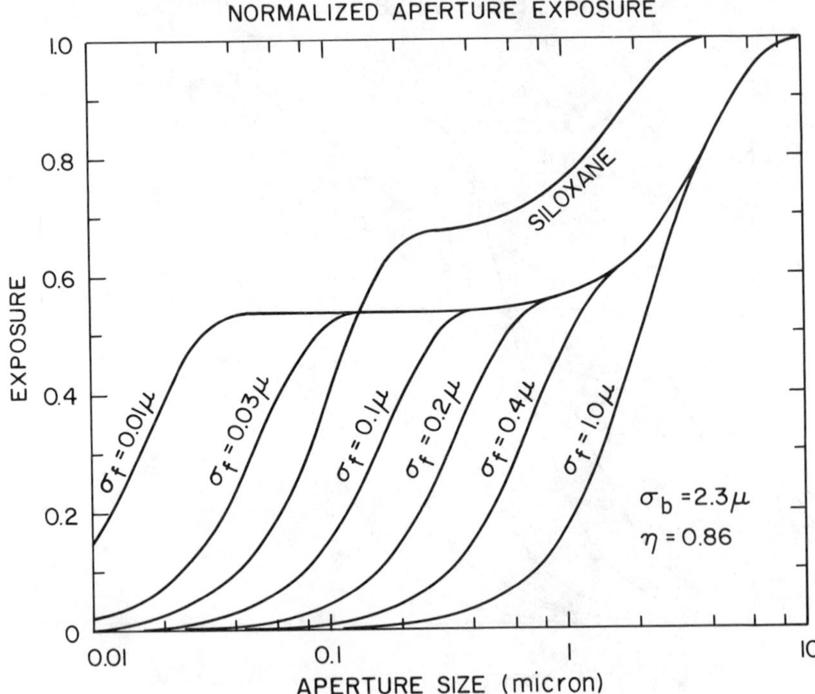

Figure 12. Aperture exposure for multi-layer siloxane electron beam resist process.

necessary to avoid the blurring of the image which occurs because the secondaries are delocalized from the primary beam.

Several methods have been discovered which offer this possibility, two of which are briefly described here. The first is the direct sublimation of ionic crystals such as NaCl, MgF_2, and LiF (*49*). Figure 13 shows holes drilled through a NaCl crystal by a 1 nm diameter 50 kV electron beam. The crystal was estimated to be about 0.25 μm thick. The convergence half-angle of the beam was 10^{-2} rad., and the beam, therefore, formed a conical shaped hole with the base of the cone being about 5 nm in diameter, assuming the beam is focused on one face of the crystal. This explains the discrepancy between the apparent size of the hole and the beam diameter, and suggests that the resolution of the process is much better than 5 nm. Isaacson et al. (*50*) confirmed this by writing structures in thinner NaCl films as small as 1.5 nm.

The second method is the exposure of multilayer Langmuir-Blodgett films (*51*). Lines, as seen in transmission electron microscopy (Figure 14), can be written in these films with much lower dosage, $\sim 10^{-4}$ C/cm^2, than for the ionic crystals where almost 0.1 C/cm^2 is required. Minimum linewidth was about 10 nm, but it was difficult to confirm the linewidth without further exposing the stearate. The lines were confirmed to be slots by low angle shadowing of the samples with AuPd, but we were not successful in using the

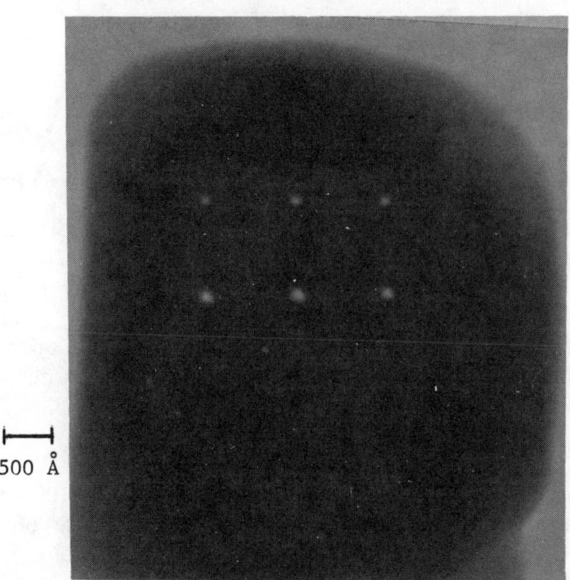

Figure 13. 5 nm diameter holes drilled through a 0.25 μm thick NaCl crystal with a 1 nm diameter electron beam (49). Physical process for this very high resolution fabrication technique is not understood, but structures as small as 1.5 nm have been made (50).

Figure 14. Sub-10 nm lines written in a stack of Langmuir-Blodgett layers with an electron beam with a diameter of less than 1 nm. Micrograph was taken with the same scanning transmission electron microscope that was used to write the lines (51).

patterned stearate layers to protect underlying layers in standard transfer processes.

Vapor resist has been used to produce 8 nm structures (*52*) and operating devices with dimensions of 25 nm (Figure 15) (*53*), but suffers from similar resolution constraints as conventional resists.

Ion Beam Lithography

Focused ion beams can be used to expose resist, to write directly diffusion patterns into semiconductor substrates, and to repair masks. These techniques can potentially simplify semiconductor device production and perhaps reduce cost. Many of the technological challenges with ion beams are similar to those encountered with electron beams, but the development of ion sources and focusing/deflection systems are at a much earlier stage of development so application to manufacturing is several years away.

Ion Beam Resist Exposure

High energy ions can be used to expose resist materials (>100 keV ions are needed to penetrate typical resist layers) with the promise of less image blurring due to lateral scattering (proximity effect) than electrons.

Experiments indicate that ions are almost two orders of magnitude more efficient than 20 keV electrons (*54*). This is because the energy of the ions is more completely absorbed in the resist layer. The high sensitivity can be a problem, however, for very high resolution because of inherent noise fluctuations created by the small number of ions needed to expose an image element (*55*). At an exposure dose of 10^{-8} C/cm^2, only 6 ions will expose a 0.1 μm $\times$ 0.1 μm element resulting in an exposure uncertainty of 40%. This is not satisfactory.

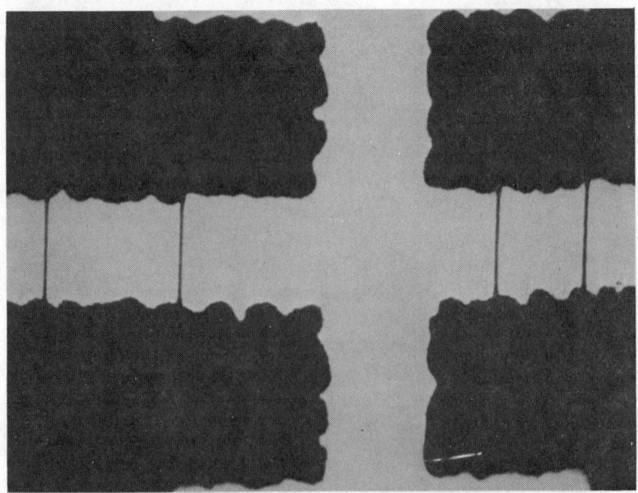

Figure 15. Transmission scanning electron micrograph of two nanobridge SQUID's (Superconducting Quantum Interference Devices). A SQUID consists of a superconducting ring containing two 'weak-links'. In this instance, the weak links are niobium wires 25 nm wide fabricated by electron beam.

Typically a hundred times this exposure is needed to reduce the uncertainty to an acceptable level of <5%. In other words the minimum usable sensitivity for ion exposure of any resist when 0.1 μm edge definition is required is 10^{-6} C/cm^2. This in effect removes the sensitivity advantage of ions over electrons because resists with a sensitivity of 10^{-6} C/cm^2 are available for electrons.

Even without the problem of statistical fluctuations the performance of ion beam systems is marginal for economical exposure of resist because of constraints similar to those of electron beams. In the electron beam case, as discussed above, a combination of shaped-beam optics, large-angle highly corrected lens/deflection systems, and high brightness electron guns only make it economical for special applications such as mask making and gate-array personalization. The brightness of ion sources approaches that of electron sources, although with poorer reliability, but the lens/deflection systems are not yet able to operate at comparable aperture. Electron beam systems have been corrected to the point that beam half-angles of 10^{-2} radian can be maintained for field sizes up to 10 mm × 10 mm. Half-angles for ion optics are only a few milliradians.

Two types of ion source produce high enough brightness ($\geqslant 10^6$ A/cm^2.ster., 20 keV) for them to be considered for semiconductor fabrication applications; the field ion source (*56*) and the liquid metal source (*57,58*). The field ion source produces relatively small energy spread ($\leqslant$ 3 eV) and when combined with a short focal length ($\leqslant$ 1 cm) electrostatic focusing system should be able to produce beam sizes as small as 10 nm with adequate current (10^{-11} amp) for laboratory microfabrication experiments. As with field emission electron sources, the field ion source only produces a limited total current and the maximum beam current is limited to about 10^{-10} amp.

The liquid metal source produces higher currents (~1 nA into ~0.25 μm) making it the leading contender for fabrication applications, but the energy spread is higher (10-12 eV), and the minimum beam diameter is limited to about 100 nm. 40 nm wide slots have been machined in thin gold films with a 100 nm diameter ion beam (*59*). The higher beam current and brightness have been produced with liquid metals, mainly gallium. When other species are needed, for example for implantation, or for greater penetration into resists, eutectic alloys are used but the brightness is reduced (*60*) Boron, arsenic and silicon ion beams have been formed in this way.

As already mentioned, the electrostatic focusing and deflection systems needed for ions are in a much earlier stage of development than those for electrons. Significant correction of aberrations can be accomplished in this way. Multi-pole lenses (quadrupoles and octopoles) are used for focusing high energy ion beams, and some combination of these may offer a solution to this difficulty, but this is yet to be accomplished. The easiest way to cover a large field will be to move the sample mechanically and join together smaller fields. Laser interferometry can be used to track the sample position and errors in position can be fed back to the ion deflection in the same way used in electron beam mask makers. This approach, in principle yields unlimited field size.

Ion Beam Proximity Printing

Ions can be used for proximity printing just as UV light, electrons and x-rays, however the mask is more difficult than for any of the other alternatives. A bright ion source is not needed, only one that produces adequate current. If necessary, the beam can be scanned across the mask to complete the exposure. Masks have been made from silicon membranes which have adequate transparency to protons, and it is possible in principle to use the same stencil masks developed for electron beam proximity printing. With silicon membranes the ions are scattered and the resolution depends on the mask to wafer separation. Image blurring is estimated to be 0.1 μm for a mask to resist separation of 15-25 μm (*61*). This separation is already marginal from the point of view of mask to wafer damage. Problems might be encountered because of sputtering with the metal stencil masks.

These mask problems would seem to limit the chances of this particular method succeeding in competition with UV, x-ray, or electron alternatives.

Resolution and Ion Beam Processes

Ultimately, resolution in a sputtering process is limited by ion penetration into the substrate and/or by the range over which momentum can be transferred effectively enough to remove atoms from the sample surface. The diameter from which atoms can be sputtered has been reported to be about 10 nm for incident ion energies up to 12 keV (*62*). This means that it should be possible to fabricate structures of about 10 nm which also happens to be the size of the smallest ion beam that can be produced with present ion sources.

For resist exposure, the resolution limit will be set by the range over which the ions interact with the resist. As with electron beam exposure, ions create secondary electrons up to several nanometers away from the beam, and these electrons can travel further before their energy is absorbed. Ultimate resolution will probably be about 10-20 nm, as it is with electrons. At present this limit is beyond the capabilities of the ion optical systems.

Literature Cited

1. Tischer, P. "Electronics in Microelectronics"; Kaiser, W. A. and Proebster, W. E., Eds.; North Holland, 1980.
2. Lin, B. J. "Microcircuit Engineering 81"; Swiss Federal Instit. Technol.: Lausanne, Switzerland, 1981, 47.
3. Hershel, R. *SPIE Conference*, Santa Clara, California, 1982.
4. Offner, A. *Photogr. Sci. Eng.* 1979, *23*, 374.
5. King, M. C. "Principles of Optical Lithography"; Einsbruch, Ed.; Academic Press: 1981; Chapt. in VLSI Electronics Microstructure Sci., Vol. 1&2.
6. Lin, B. J. *SPIE Proc.*, 1979, *174*.
7. Tai, K. L. *J. Vac. Sci. Technol.* 1979, *216*, 1977.
8. Markle, D. A. *Solid State Tech.* 1984, *217*, 50.
9. Greed, J. J.; Markle, D. A. *SPIE Conf.*, Santa Clara, California, 1982.

10. Zernike, F.; Sewell, H.; Barosi, N.; Gansfried, M. "Microcircuit Engineering 83"; Ahmed, H.; Cleaver, J.; Jones, G., Eds.; Academic Press: Cambridge, England, 1983; p. 225.
11. Willson, C. G.; Ito, H. "Microcircuit Engineering 1982"; Sitecmo Dieppe, Paris; 1982, p. 261; and "Photopolymer Principles"; Ellenville, 1982; p. 331.
12. Wittekoek, S. "Microcircuit Engineering 80"; Kramer, R. P., Ed.; Delft Univ. Press: Holland, 1980; p. 155.
13. Lauria, J. "Microcircuit Engineering 80"; Kramer, R. P., Ed.; Delft Univ. Press: Holland, 1980; p. 171.
14. Dubreoueq, G. "Microcircuit Engineering 80"; Kramer, R. P., Ed.; Delft Univ. Press: Holland, 1980; p. 181.
15. Mayer, H. E. "Microcircuit Engineering 80"; Kramer, R. P., Ed.; Delft Univ. Press: Holland, 1980; p. 191.
16. Dey, J. "Microcircuit Engineering 80"; Kramer, R. P., Ed.; Delft Univ. Press: Holland, 1980; p. 211.
17. Wilczynski, J. S. *J. Vac. Sci. Technol.* 1979, *216*, 1929.
18. Flanders, D. C. *Applied Physics Letters* 1977, *231*, 426.
19. Kern D.; Nelson, D. *9th Int. Conf.* On "Electron & Ion Beam Sci. & Technol."; Bakish, R., Ed.; Electrochemical Society: Princeton, NJ, 1980; 491.
20. Spears, D. L.; Smith, H. I. *Electron Lett.* 1972, *8*, 102.
21. Spiller, E.; Feder, R. *Sci. Am.*, Nov., 1978.
22. Taylor, G. N.; Wolf, T. M. "Microcircuit Engineering 81"; Swiss Federal Instit. Technol.: Lausanne, Switzerland, 1981; 381.
23. Grobman, W. D. "IEDM Digest"; IEEE: NY, 1980; 415; and Spiller, E. *J. Appl. Phys.* 1976, *47*, 5450.
24. Grobman, W. D. *J. Vac. Sci. Technol., B* 1983, *1*, 1300.
25. Gronid, E. "Microcircuit Engineering 81"; Swiss Federal Instit Technol.: Lausanne, Switzerland, 1981; 122.
26. LeDuff, J. "Microcircuit Engineering 81"; Swiss Federal Instit. Technol.: Lausanne, Switzerland, 1981; 130.
27. Spiller, E.; Feder, R. "X-Ray Lithography"; In Queisser H. J., "X-Ray Optics"; Springerverlag, Berlin, 1977; 24.
28. Gruen, A. E. *Z. Naturforsch* 1957, *12a*, 89-95.
29. Feder, R. *Science* 1977, *197*, 259.
30. Flanders, D. C.; Smith, H. I. *J. Vac. Sci. Technol.* 1978, *15*, 1001.
31. Laibowitz, R. *Appl. Phys. Lett.* 1979, *35*, 891.
32. Voss, R. *Appl. Phys. Lett.* 1980, *37*, 656.
33. Herriott, D. R. *IEEE Trans. Electron Devices* 1975, *ED-22*, 385.
34. Pfeiffer, H. C. *J. Vac. Sci. Technol.* 1978, *15*,887.
35. Goto, E.; Soma, T.; Idesawa, M. *J. Vac. Sci. Technol.* 1978, *15*, 883.
36. Thomson, M. G. R.; Collier, R. J.; Herriott, D. R. *J. Vac. Sci. Technol.* 1978, *15*, 891.
37. Chang, T. H. P. *Proc. 7th Int. Conf.* On "Electron & Ion Beam Sci. & Technol."; Bakish, R., Ed.; Electrochem Soc. Inc.: Princeton, NJ, 1974; p. 97.

38. Varnell, G. *J. Vac. Sci. Technol.* 1979, *16*, 1787.
39. Eidson, J. C. *J. Vac. Sci. Technol.* 1981, *19*, 932.
40. Broers, A. N. "SEM 1974"; ITT Research Inst.: Chicago, Ill., 1975, p. 9.
41. Sano, S. "Microcircuit Engineering 80"; Kramer, R. P., Ed.; Delft Univ. Press: 1981; 99.
42. Moore, R. D. *J. Vac. Sci. Technol.* 1981, *19*, 950.
43. Broers, A. N. *J. Electrochem. Soc.* 1981, *128*, 166.
44. Chang, T. H. P. *J. Vac. Sci. Technol.* 1975, *12*, 127.
45. Boersch, H. *Z. Phys.* 1954, *139*, 115.
46. Grobman, W. D.; Speth, A. *Proc. 8th Internat. Conf.* On "Electron & Ion Beam Tech."; Baksih, R., Ed.; Electrochemical Soc.: Princeton, NJ, 1978; 276.
47. Parasczak, J. *J. Vac. Sci. Technol., B* 1983, *1*, 1372.
48. Greeneich, J. S. *J. Vac. Sci. Technol.* 1979, *16*, 1749.
49. Broers, A. N. "Electron Microscopy 1978"; Sturgess, J. M., Ed.; Microscop. Soc. of Canada: Toronto, 1978; Vol. III, 343.
50. Isaacson, M.; Muray, A. *J. Vac. Sci. Technol.* 1981, *19*, 1117.
51. Broers, A. N.; Pomerantz, M. *Thin Solid Films*, *99*, 323.
52. Broers, A. N. *Appl. Phys. Lett.* 1976, *29*, 596.
53. Laibowitz, R. and Broers, A. N. *Treatise on Materials Science and Technology* 1982, *24*, 285.
54. Hall, T. M. *J. Vac. Sci. Technol.* 1980, *37*, 656.
55. Wagner, A. *Proc. SPIE Conf.*, Santa Barbara, California, 1979, 167.
56. Hanson, G. R.; Siegel, B. M. *J. Vac. Sci. Technol.* 1979, *16*, 1875.
57. Clampitt, R. *J. Vac. Sci. Technol.* 1975, *12*, 1208.
58. Krohn, V. E.; Ringo, G. R. *Appl. Phys. Lett.* 1975, *27*, 479.
59. Seliger, R. L. *J. Vac. Sci. Technol.* 1979, *16*, 1610.
60. Gamo, K. "Microcircuit Engineering 80"; Kramer, R. P., Ed.; Delft Univ. Press: Holland, 1980; 283.
61. Rensch, D. B. *J. Vac. Sci. Technol.* 1979, *16*, 1897.
62. McHugh, J. A. "Secondary Ion Spectrometry"; Wolsky, S.; Czanderna, A., Eds.; Elsevier Pub. Co.: Amsterdam, 1975; Chapt. in "Methods of Surface Analysis".
63. Tibbets, R.; Wilczynski, J. S. 2fProc. Inter. Lens Design Conf. "SPIE"; Fischer, R. E., Ed.; Bellingham: Washington, D. C., 1980; *23*, 321.
64. Neill, T. R.; Bull, C. J. "Microcircuit Engineering 80"; Kramer, R. P., Ed.; Delft Univ. Press: 1981; 45.
65. Parikh, M.; Kyser, D. F. *J. Appl. Phys.* 1979, *50*, 1104.

RECEIVED September 28, 1984

3

A Perspective on Resist Materials for Fine-Line Lithography

M. J. BOWDEN

Bell Communications Research, Inc.
Murray Hill, NJ 07974

Microcircuit fabrication requires the selective diffusion of tiny amounts of impurities into regions of a semiconductor substrate such as silicon, to produce the desired electrical characteristics of the circuit. These regions (together with the metal conductor paths that link the active circuit elements) are defined by lithographic processes in which the desired pattern is first generated in a resist layer (usually a polymeric film ~0.5-1.0 μm thick which is spin-coated onto the substrate) and then transferred via processes such as etching into the underlying substrate. In silicon integrated circuit manufacture, the latter is usually SiO_2 which is present as a thin layer on top of the silicon and which functions as the actual mask for the diffusion process. The definition of the pattern in the resist layer is achieved by exposing the resist to some suitable form of patterned radiation such as ultraviolet light, electrons, x-rays or ions. The resist contains radiation-sensitive groups which chemically respond to the incident radiation forming a latent image of the circuit pattern which can subsequently be "developed" by solvent or plasma treatment to produce a three-dimensional relief image, i.e., the radiation-induced chemical reactions such as cross-linking and/or chain degradation enable exposed regions to be differentiated from unexposed regions either by solubility differences or by differences in plasma etch rate. The areas of resist remaining after development (the exposed or non-exposed areas as the case may be depending on whether irradiation causes the exposed areas to become more soluble or less soluble than the unexposed areas) must now protect the underlying substrate during the variety of additive and/or substractive processes encountered in semiconductor processing. If, for example, the underlying substrate were SiO_2, immersion of the structure into an etchant such as buffered hydrofluoric acid would result in selective etching of the SiO_2 in those areas that were bared during the development step.

The basic steps of the lithographic process are shown schematically in Figure 1. The example shown corresponds to photolithography in which the photosensitive resist or photoresist is applied as a thin film to the substrate (SiO_2 on Si) and subsequently exposed in an image-wise fashion through a mask. The mask contains clear and opaque features that define the circuit pattern. The areas in the photoresist that are exposed to light are made either

0097-6156/84/0266-0039$19.05/0

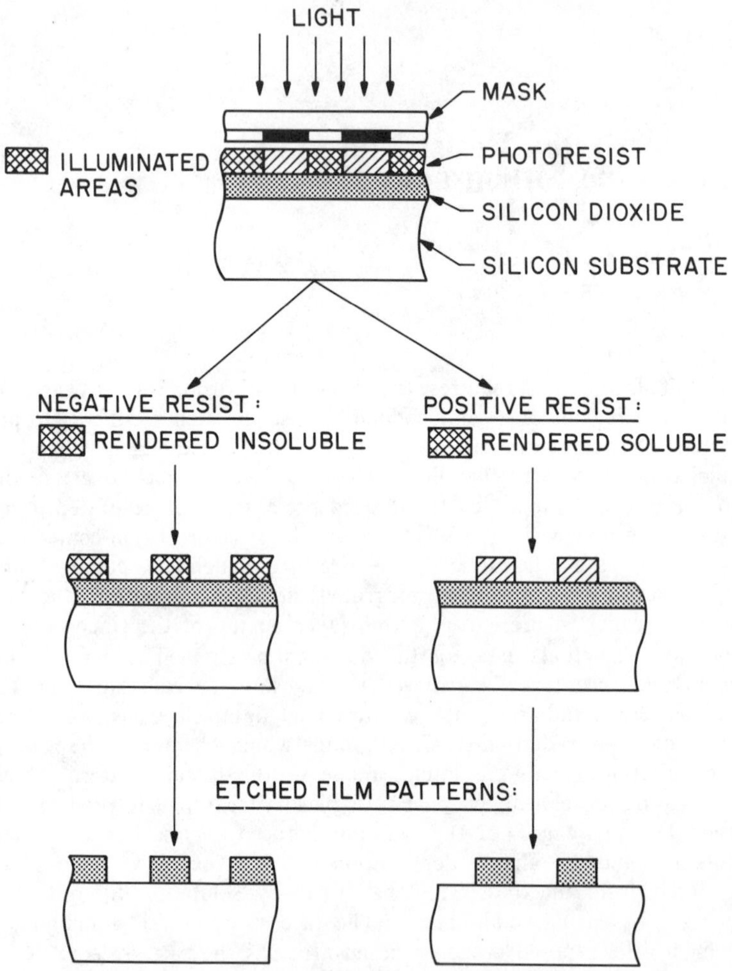

Figure 1. Schematic diagram of the photolithographic process.

soluble or insoluble in a specific solvent termed a developer. In the first case, the *irradiated* regions are dissolved leaving a positive image of the mask in the resist. Therefore, the photoresist is termed a positive resist. In the second case, the *unirradiated* regions are dissolved leaving a negative image; hence, the resist is termed a negative resist. Following development, the exposed SiO_2 areas are removed by etching, thereby replicating the mask pattern in the oxide layer. The underlying silicon is exposed by this etching procedure, and impurities or dopants such as boron or arsenic can now be diffused into these exposed regions. Complex circuits are built into the silicon by a succession of such procedures. The polysilicon and/or metal conductor interconnections together with the terminals that make the circuit accessible to the outside world are also fabricated by lithographic techniques.

The role of the resist in this process is seen to be twofold. First, it must respond to the exposing radiation to form a latent image of the circuit pattern described in the mask which can subsequently be developed to produce a three dimensional relief image. Second, the areas of resist remaining after development (the exposed or non-exposed areas as the case may be) must protect the underlying substrate during subsequent processing. Indeed, the generic name *resist* no doubt evolved as a description of the ability of these materials to "resist" etchants.

Circuit design trends are toward increasing the scale of integration (device complexity), thereby reducing fabrication costs per circuit element and increasing performance. This continuing push toward higher packing densities requires the feature size of the circuit elements to be correspondingly reduced, a trend which has continued linearly with time since the early 70's and which has resulted in doubling the number of components per chip every year. This trend was first pointed out by Moore and is illustrated in Figure 2 for the minimum feature size in MOS random access memory devices. If this trend continues, it will ultimately take us beyond the resolution capability of photolithography which is primarily limited by diffraction considerations. Realization of this limitation has resulted in the active and continuing development of other

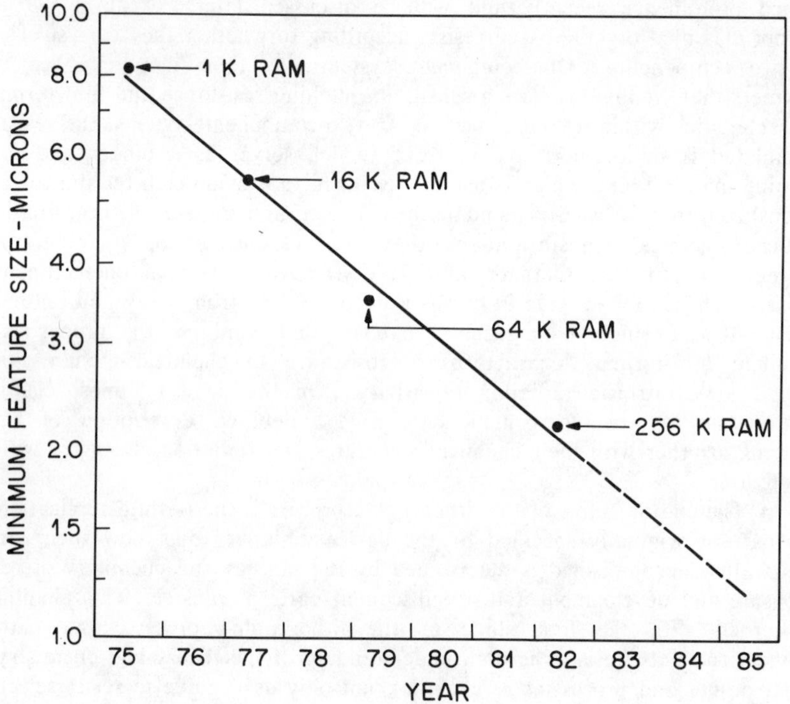

Figure 2. Minimum feature size on MOS random access memory devices as a function of year of commercial availability. (Reproduced with premission from Ref. 1.)

potentially higher resolution technologies, *viz.*, deep-UV photolithography, electron-beam lithography, x-ray lithography and ion-beam lithography. Many of these techniques are still in the development stage although electron-beam lithography is already in commercial practice where it is rapidly becoming the dominant technology for fabricating chromium master masks for photolithography. It is also used by companies such as IBM and Texas Instruments for direct wafer fabrication.

Irrespective of what particular lithographic technology is being employed, whether it be conventional photolithography or the other high resolution techniques mentioned above, a resist system is required which enables the exposed regions to be differentiated from the unexposed regions. As shown earlier, there are two types of resist depending on the tone of the image following exposure and development. Positive resists produce positive-tone images in which the exposed areas are removed (either by solvent as mentioned earlier or by dry processes). Conversely, negative resists give rise to negative-tone images in which the non-irradiated areas are removed. Both types of resist are used in circuit manufacture.

Nearly all resists are organic polymers which are designed to respond to the exposing radiation (resists based on inorganic, chalcogenide glasses have gained prominence recently and will be discussed later). Willson (*1*) has further classified organic-based resists according to whether they consist of one or two components. One-component systems, as their name implies, are polymers that combine radiation sensitivity, etching resistance and film forming characteristics within a single species. In two-component systems, the resist is formulated from an inert matrix resin (which serves as a binder and film-forming material, and which is usually designed to provide etch resistance) and a sensitizer that is usually monomeric in nature and undergoes the radiation-induced chemical transformations that are responsible for imaging. For example, poly(methyl methacrylate) (PMMA) is a classical one-component resist which has found wide utility as a deep UV, electron, x-ray and ion-beam resist. It is a single homogeneous material that combines the properties of excellent film-forming characteristics, resistance to chemical etchants, and (albeit low) intrinsic radiation sensitivity. On the contrary, most classical photoresists are two-component systems. A detailed description of these systems together with their chemical structures and radiation chemistry will be given later.

The end objective of the lithographic process is the faithful replication of the pattern originally specified by the device designer. Just how successfully this will be accomplished is determined by the physics and chemistry of resist exposure and development and of subsequent pattern transfer. This chapter is concerned with the first stage of the lithographic process, i.e., pattern delineation in the resist where an understanding of the theory and chemistry of resist design and processing is essential, not only as a guide to resist selection and process optimization, but also as an aid in the design and development of new resist systems.

Resist Requirements

Resist requirements for microfabrication are similar, regardless of the exposure technology. While the general requirements have been discussed in many articles (*2,3*) it is instructive here to review the more important requirements, particularly in view of the special limitations imposed by the need to fabricate fine line VLSI devices.

Solubility. Since films are normally deposited on a substrate by spin-coating from solution, solubility in organic solvents is a necessary requirement. Solubility in aqueous media may be undersirable if wet etching is contemplated since many of the etching solutions are aqueous. Although other coating techniques such as plasma deposition have been used, such resist systems generally employ dissolution techniques to differentiate between exposed and non-exposed regions. One goal of resist research is the design of material systems capable of all-dry processing in which the resist would be applied to the substrate by CVD or plasma deposition techniques, exposed, and then developed by plasma treatment. Such systems would remove solubility requirements in resist design. There have been reports of all-dry processed systems, (*4*) but none have been reduced to practice yet. Recent advances in process development of inorganic resists come close to achieving this goal. They are deposited by vacuum deposition and may, in addition to regular wet development, be plasma-developed following exposure using fluorocarbon etching gases, e.g., CF_4. The thin radiation-sensitive layer is usually deposited in a wet process.

Adhesion. A variety of substrates are encountered in semiconductor processing. These include metal films such as aluminum, chromium, and gold, insulators such as silicon dioxide and silicon nitride, and semiconductor materials. The resist must possess adequate adhesion to withstand all processing steps. Poor adhesion leads to marked undercutting, loss of resolution and, in the worst case, complete destruction of the pattern. Pattern transfer to the underlying substrate has been conventionally accomplished using liquid etching techniques which require tenacious adhesion between resist and substrate in order to minimize undercutting and maintain edge acuity and feature size control. In some cases, adhesion between the resist and substrate may be enhanced with adhesion promoters such as hexamethyldisilazane. As geometries are reduced, liquid etching is being replaced by dry-etching techniques which place less stringent adhesion requirements on the resist. It should be noted, however, that there still has to be sufficient adhesion to withstand development, and this becomes increasingly difficult as feature sizes decrease.

Etchant Resistance. Etchant resistance refers to the ability of the resist to withstand the etching environment during the pattern transfer process. The most common method of pattern transfer is wet chemical etching which places emphasis on the adhesion and chemical stability of the resist. Etchant solutions may be either acidic or basic, depending on the type of substrate to be etched. For example, buffered hydrofluoric acid is used to etch SiO_2. However, lateral penetration of the chemical etchant is significant for thick substrate films and results in unacceptable line width control for feature sizes <2 μm. For thin substrate films, as found for example in chromium photomasks where the thickness of the chromium is on the order of 50-100 nm, liquid etching is capable of delineating sub-micron features. The need to pattern fine features in thick substrates has led to the development of anisotropic etching methods such as reactive-ion etching, plasma etching, ion milling, and sputter etching. While not requiring a premium in terms of adhesion, these techniques place other very stringent requirements on the resist. Most dry-etching techniques rely on plasma-induced gas reactions in an environment of high radiation flux and temperatures in excess of 80 °C. The physical and chemical properties of the resist therefore play an important part in determining dry-etching resistance and may present a fundamental limitation. For example, dry-etchant resistance is generally observed in polymers that possess thermal and radiation stability such as obtain in polymers that exhibit high glass transition temperatures and contain radiation stable groups, e.g., aromatic structures. High sensitivity in positive resists, on the other hand, usually requires polymers that are unstable to radiation. Thus the requirement of high sensitivity coupled with adequate resistance to dry-etching environments presents an apparent dichotomy to the resist designer.

Sensitivity. Sensitivity is conventionally defined as the input incident energy (measured in terms of energy or the number of photons or particles (fluence) per unit area) required to attain a certain degree of chemical response in the resist that results, after development, in the desired relief image. This represents an operational, *lithographic* definition of sensitivity.

Typically a plot is made of normalized film thickness remaining as a function of log(dose) as shown in Figs. 3a and 3b. Note that sensitivity *increases* as the dose required to produce the lithographic image *decreases*. The sensitivity of a positive resist is the dose (D_c in Figure 3a) required to effect complete solubility of the exposed region under conditions where the unexposed region remains completely insoluble. In the case of a negative resist, sensitivity is defined as the dose at which a lithographically useful image is formed. Although gel (insoluble cross-linked resist) begins to form at the interface gel dose (D_g^i in Figure 3b), no lithographically useful image has been formed and additional dose is required to build the relief image to the desired thickness (usually to $D_g^{0.5}$ which is the dose at which 50% of the original film thickness has been retained after development).

The practice of referring to sensitivity in terms of incident energy per unit area is somewhat ambiguous since this value depends on a variety of exposure parameters. For example, in the case of electrons, the sensitivity expressed as Coulombs/cm^2 (C/cm^2) depends on such factors as accelerating voltage and resist thickness. Similarly in photolithography, a resist responds only to radiation in the spectral regions which encompass its absorption band. The sensitivity will therefore vary depending on whether the full output energy of the lamp is measured or whether a given output region is isolated by the use of filters. It will also differ for different lamps and optical systems depending on their spectral output characteristics. Obviously, it would be more practical

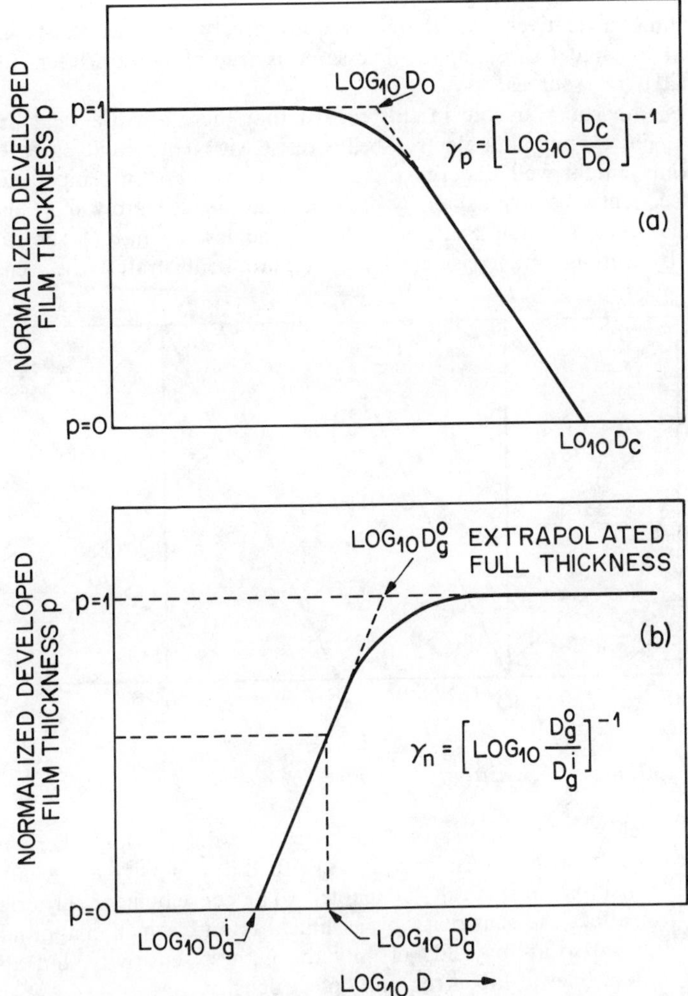

Figure 3. Typical lithographic response or contrast plots for (a) positive resists and (b) negative resists in terms of the developed thickness normalized to initial resist thickness (p) as a function of log (dose).

to speak of absorbed energy density. However, this value is generally not easily obtained, and so in this text we will follow the accepted convention and use the former terminology.

Clearly, the sensitivity of a resist should be commensurate with machine design parameters to allow optimized throughput. For example, an electron beam exposure system writing at a modulation rate of 20 MHz (dwell time of 50 nsec), a beam current of 5×10^{-8} amps at 10 kV, and an address structure (spot size) of 0.25 μm^2 would require a resist with a sensitivity of 10^{-6} C/cm^2 (1 μC/cm^2) or better in order to write the maximum number of wafers per hour of which it is capable. The same argument also applies to other exposure tools.

The power output of photolithographic projection printers in the deep UV range ($\lambda < 300$ nm) is relatively low thereby requiring very sensitive resists in order to achieve economical throughput. The same is true of x-ray printers utilizing conventional x-ray sources.

It is important to note in this regard that there are practical limits to sensitivity. For example, an electron-beam resist with a sensitivity greater than $\sim 10^{-8}$ C/cm^2 might well undergo thermal reactions at room temperature and would consequently be unsatisfactory because of an unacceptably short shelf-life of both the resist solution and spun films. The lower limit of sensitivity is governed by throughput considerations. Figure 4 illustrates the sensitivity

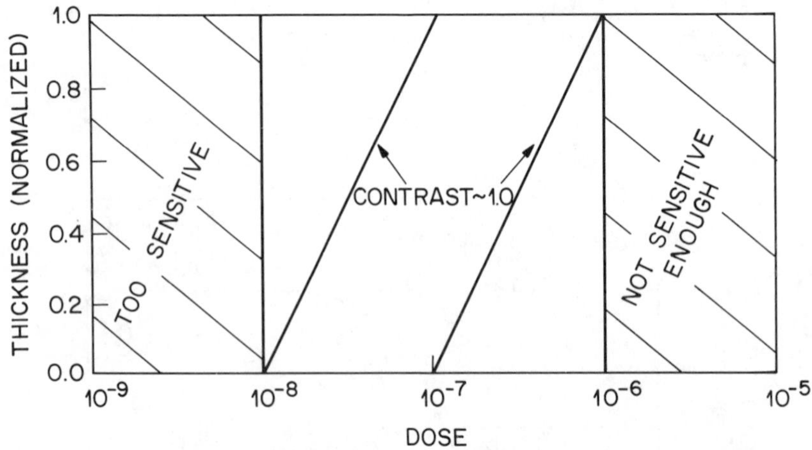

Figure 4. Sensitivity "window" for electron beam resists.

"window" for electron resists; similar limits can be defined for other types of resist. Clearly, processing must provide us with the sensitivity and contrast of which the resist is intrinsically capable. The development criteria and conditions can affect the sensitivity by as much as an order of magnitude and must be optimized so as to yield not only the highest sensitivity, but also the desired resolution, linewidth control, and defect density.

Contrast (Resolution). The pattern resolution attainable with a given resist for a particular set of processing conditions is determined in large part by the resist contrast (γ). Contrast in the case of a negative resist is related to the rate of cross-linked network (gel) formation and, for a positive resist, to both the rate of degradation and the rate of change of solubility with molecular weight with the latter being markedly solvent dependent. The numerical value of γ is obtained from the slope of the linear portion of the response curves shown in Figures 3a & 3b. The contrast of a positive resist is expressed mathematically as

$$\gamma_p = 1/(\log D_c - \log D_o) = \log \left[\frac{D_c}{D_o} \right]^{-1} \quad (1)$$

where D_o is the dose at which the developer begins to dissolve the irradiated film and is determined by extrapolating the linear portion of the film thickness remaining vs. dose plot to a value of 1.0 normalized film thickness; D_c is the complete development dose.

The contrast of a negative resist is expressed as

$$\gamma_n = 1/(\log D_g^o = \log D_g^i) = \log \left[\frac{D_g^o}{D_g^i}\right]^{-1} \quad (2)$$

where D_g^o is the dose required to produce 100% initial film thickness and is obtained by extrapolating the linear portion of the thickness vs. dose plot to a value of 1.0 normalized film thickness; D_g^i is the interface gel dose.

High contrast is important since it minimizes the deleterious effects due to scattering of radiation in the resist film. All exposure techniques result in some energy being deposited outside the primary image area. Diffraction of light in photolithography, scattering of primary and secondary electrons in electron-beam lithography and of photoelectrons following x-ray absorption or scattering of ions in ion-beam lithography all contribute to energy deposition in areas remote from the desired exposure area. These effects present fundamental physical limitations to resolution, although the situation may be improved by high contrast resists which do not respond significantly to low levels of scattered radiation. In projection printing, for example, at feature sizes on the order of 1 μm, diffraction of light at the edge of an opaque feature results in the projected pattern or image in the photoresist exhibiting gradual (rather than sharp) transitions from light to dark, causing the edges of the projected features to appear blurred (see Figure 5). The quality of the projected image is expressed in terms of the modulation transfer function (MTF) whose value is dependent on the optics (numerical aperture of the exposure system, wavelength, etc.) as well as the spatial frequency of the mask pattern to be imaged. A value of MTF equal to 1 corresponds, to a first approximation, to perfect image transfer. For organic resists with contrast ~2, a satisfactory resist image can be obtained with a projection printer that has an MTF of approximately 0.6, i.e., although the physics associated with image projection degrades the quality of the transferred image, we can still compensate for this via chemistry and processing, provided image degradation is not too great. With lower contrast resists or at higher spatial frequencies (MTF < 0.6) image degradation is too severe and acceptable patterns cannot be obtained. An alternative approach towards improving image quality involves improving the illumination contrast. Griffing and West (*7*) described the use of photobleachable materials in conjunction with conventional photoresists. The bleachable layer is applied on top of the photoresist and serves to enhance the contrast of the illumination reaching the photoresist surface. In general, though, the greater the contrast of the resist, the greater will be the extent of image degradation that can be tolerated.

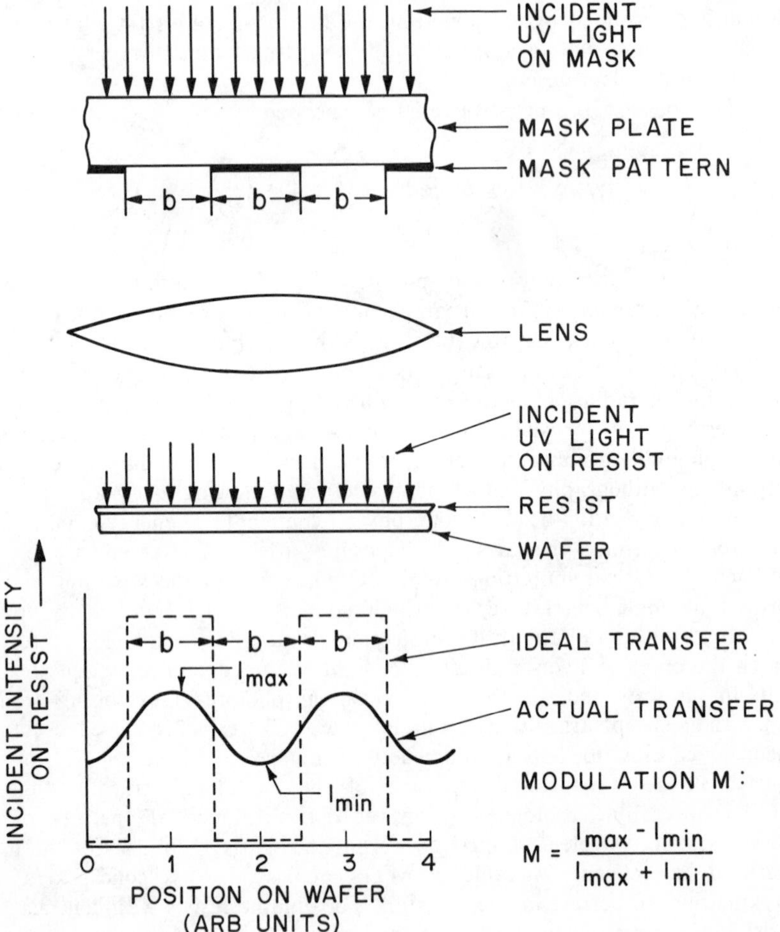

Figure 5. Schematic representation of image transfer efficiency for a 1:1 projection printer.

Fundamental Considerations

We have seen that two important lithographic parameters of a resist are sensitivity and contrast. This leads to a consideration of the design features that must be incorporated into the resist in order to optimize these parameters and, in turn, requires a fundamental understanding of the interaction of radiation with matter and how the polymer molecular parameters affect lithographic response. These aspects have been extensively covered in the literature, (*5,6*) and only the conclusions relating to lithographic performance will be summarized.

From the fundamental standpoint, the primary concern in resist sensitometry is the measurement of resist response to radiation which involves measurements designed to determine the intrinsic radiation-sensitivity of the resist. For photochemical reactions, the response is expressed quantitatively in terms of photo-efficiency or quantum yield (Φ) defined as

$$\Phi = \frac{\text{Number of photo-induced events}}{\text{Number of photons absorbed}} \quad (3)$$

and, for high energy radiation, in terms of the G-value defined as the number of radiation-induced events per 100 eV of absorbed energy. Thus positive resist systems designed to operate by chain scission should possess a high Φ or a high G(scission) (the higher the better) as the case may be, with negligible cross-linking. Optimum sensitivity and contrast for a given resist usually require high molecular weight and narrow molecular weight distribution. The choice of developer is also critical for optimum performance. Preferably the glass transition temperature (T_g) should be greater than the dissolution temperature in order to minimize swelling effects.

Likewise, negative resists should possess high values of Φ or G(cross-linking) with negligible G(scission). Sensitivity is directly proportional to molecular weight (the higher the molecular weight, the greater the sensitivity, i.e., the lower the exposure dose required) and molecular weight distribution. Contrast, on the other hand, appears to decrease with increasing molecular weight, possibly due to increasing contribution from intra-molecular cross-linking. Contrast does improve, however, as the molecular weight distribution is narrowed. In many polymers, both scission and cross-linking occur simultaneously. In such cases the predominating effect is determined by the relative magnitudes of G(scission) and G(cross-linking). For example, if $G(\text{scission}) < 4\,G(\text{cross-linking})$ the sample will always cross-link.

The presence of aromatic groups in the polymer chain would be expected to lead to increased radiation stability and hence to improved resistance to dry etching environments. Unfortunately the lithographic sensitivity of resists designed to respond to high-energy radiation is adversely affected because of the radiation protective effect of such groups, suggesting that a compromise between the two requirements must be made.

Photoresists

Conventional Photolithography. Photolithography using light in the 350-450 nm wavelength region is the lithographic technology currently used to fabricate almost all commercial solid state devices and will remain dominant, at least for this decade. Several exposure techniques are in use, *viz.*, contact and proximity printing, projection printing on a 1:1 scale, and reduction step-and-repeat projection printing. Parallel projection (1:1) using 350-410 nm light is currently the dominant technology for VLSI device fabrication and, at present, is routinely capable of reproducing features in the region of 2 μm. Resolution on the order of 1.0 μm appears feasible with reduction step and repeat systems operating at 350-410 nm wavelength.

Positive Photoresists. All positive photoresists used for conventional photolithography are two-component systems and operate on a mechanism that involves destruction of a dissolution inhibitor. These resists are formulated from three ingredients: solvent, an acidic polymer that dissolves or disperses in aqueous base, and a photosensitive dissolution inhibitor (20-50% by weight) that forms a solid solution with the acidic polymer and prevents, or inhibits, the dissolution of the latter in an aqueous base.

The positive photoresist materials evolved from work performed at the Kalle Corporation in Germany which led to the development of the first positive acting photoresist based on the use of a novolac matrix resin (condensation polymer of formaldehyde with phenol and/or substituted phenols, e.g., cresols) with a quinonediazide dissolution inhibitor acting as the photoactive compound or sensitizer. These materials were designed originally to produce photoplates for the printing industry and were subsequently adopted for semiconductor manufacture. A schematic representation of the exposure and development processes is shown in Figure 6.

A selection of commercially available positive photoresists is listed in Table I. There are a wide variety offered with many companies claiming superior properties with respect to dimensional stability under high temperature conditions, striation free coatings, adhesion, sensitivity to variations in developing temperature, etchant resistance, development uniformity, etc. Resists can be obtained in custom dilutions at almost any viscosity, thereby permitting selection of specific viscosities for the coating thickness desired. Great strides have been made by the manufacturers in improving the quality control of the product with respect to exposure and process reproducibility and defect density.

While a variety of subtle functional and chemical differences separate these photoresists, many of their primary imaging and etching characteristics are similar. This stems from the fact that all current positive photoresists are based upon the original formulation and photochemistry of the Kalle material. Novolac resins possess desirable dissolution characteristics in aqueous bases in that they show no swelling. They are soluble in a wide variety of common organic solvents, and can be coated from solution to form uniform films of high quality. The resins used in resist formulations are typically of relatively low molecular weight ($M_n < 2,000$) with broad dispersivities and melting points of 90-120°C.

The photoactive compounds (PAC) or sensitizers are based, in all cases, on the o-quinonediazide group in which the substituent, R, in Figure 6 is generally an arylsulfonate. (In some cases, the sensitizer is not a second component per se, but is actually chemically attached to the novolac). The chemical structure of this R-substituent influences the solubility characteristics of the sensitizer molecule and also influences the light absorption characteristics of the chromophore. The sensitizer molecule acts as a dissolution inhibitor for aqueous-base development of the novolac resin. Upon absorption of light, the quinonediazide undergoes a photochemical rearrangement (known as the Wolff rearrangement) followed by hydrolysis to a base-soluble indenecarboxylic acid that no longer inhibits dissolution of the novolac resin matrix in aqueous base, thereby facilitating development of a positive image (*8*). The sequence of photochemical transformations is outlined in Figure 7.

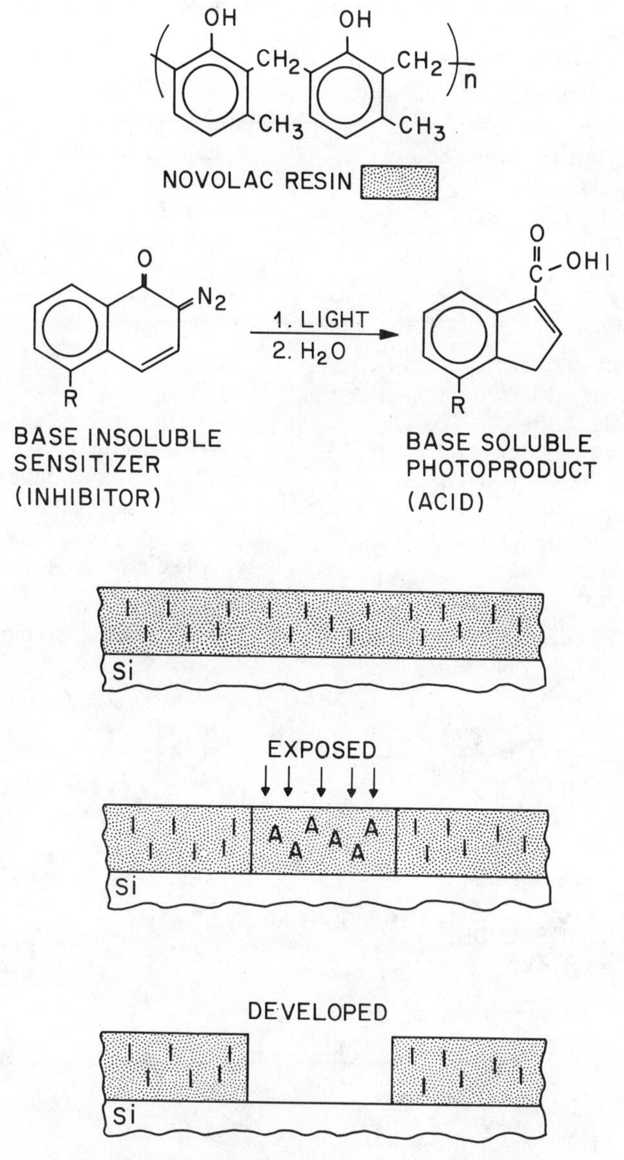

Figure 6. Schematic representation of the mechanism of positive resist action in quinonediazide – novolac resists. (Reproduced with permission from Ref. 1.)

As indicated earlier, the structure of the quinonediazide determines the absorption characteristics of the resist which should be tailored to the desired spectral range. The 5-arylsulfonates are most commonly used, and are characterized by an absorbance maximum at approximately 400 nm and a second slightly stronger maximum at approximately 340 nm. These absorption

Table I. Commercially Available Positive Photoresists

RESIST	SUPPLIER
Micropositive Resist, 800 Series (809, 820)	Eastman Kodak
Baker 1-PR	J. T. Baker
*AZ-1300 Series	American Hoechst
*AZ-4,000 Series	American Hoechst
*AZ-1400 Series	Shipley
*AZ-2,400 Series	Shipley
*Microposit Series	Shipley
Waycoat HPR-200 Series (204,206)	Phillip A. Hunt
Ultramac PR 73,74	MacDermid
Dynalith OFPR-800	Dynachem (Tokyo Ohka)
*Selectilux P Series	E. M. Chemicals
Isofine Positive Resist	Microimage Technology
Acculith	Allied Chemical

* A wide variety of resists are available in these series covering different solids contents, viscosities, etc.

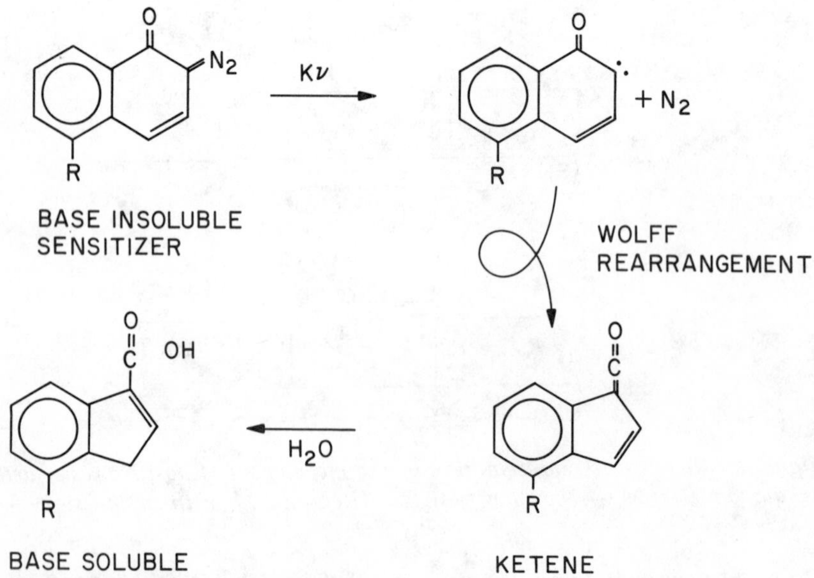

Figure 7. Sequence of photochemical transformations of the quinonediazide sensitizer (dissolution inhibitor).

characteristics may differ somewhat for different positive resists depending on the nature of the particular sulfonate ester. In certain instances, e.g. AZ 2400 resist, the absorption characteristics are suggestive of a 4-sulfonate analog. An understanding of the effect of substituents on the absorption characteristics is important if one is to design a resist for a particular exposure tool. Such studies coupled with semi-empirical calculations have enabled Miller et al. (*9,10*) to specifically design o-quinonediazide sensitizers for mid-UV application.

Several interesting process variations have been reported using positive photoresists. Most notable is the ability to reverse the tone of the image so that the resist functions as a negative resist (*11*). The so-called "monazoline process" involves introducing a small amount of a basic additive such as monazoline (1-hydroxyethyl-2-alkylimidazoline) into a diazoquinone-novolac resin photoresist. The "doped" resist is exposed through a mask in the usual way resulting in the formation of the indenecarboxylic acid. Instead of developing at this stage as in conventional positive tone processing, the resist is first post-exposure baked during which the base catalyzes a thermal decarboxylation of the quinonediazide photoproduct, i.e., the indenecarboxylic acid, producing a substituted indene. The latter is insoluble in aqueous base and thus functions as a dissolution inhibitor. The resist is now flood-exposed with UV light causing the previously unexposed areas (where quinonediazide still remains) to become soluble via transformation of the dissolution inhibitor to the indenecarboxylic acid. During dissolution in aqueous base, the areas which were originally subjected to image-wise exposure dissolve at a much slower rate because of the presence of the substituted indene in these areas. The process is shown schematically in Figure 8.

As a final comment, we note that an increasing share of the total worldwide photoresist market is falling to positive photoresists where important

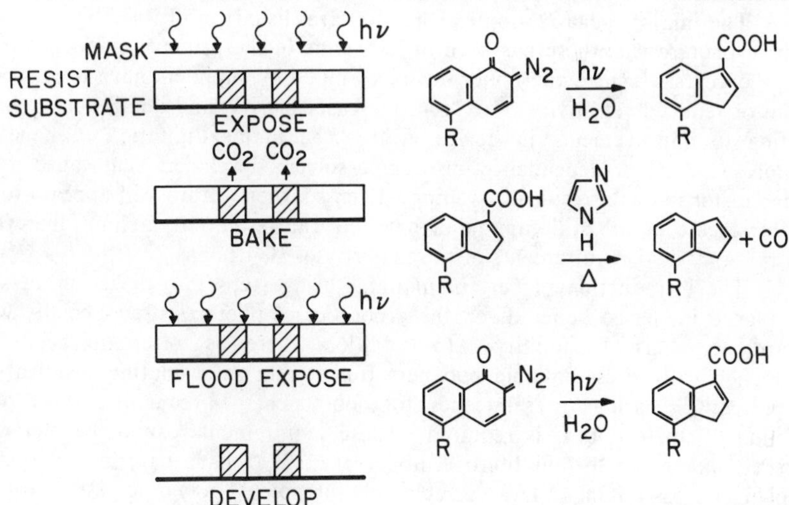

Figure 8. Schematic representation of monazoline process for image reversal of positive photoresists. (Reproduced with premission from Ref. 1.)

advantages in resolution and plasma etching resistance are realized. Significant advances have been made in modeling the exposure and development processes, thereby reducing much of the "art" of processing to mathematical modeling data consistent with empirical data (*12,13*). Such schemes have greatly facilitated the successful production use of positive photoresists.

Negative Photoresists. Traditionally, negative photoresists have been the workhorse of the microelectronics industry, and it has only been in recent years where resolution below 2 μm is required that positive photoresists have come into prominence. All negative photoresists function by cross-linking a chemically reactive polymer via a photosensitive agent that initiates the chemical cross-linking reaction. The most commonly used matrix resin is cyclized cis- 1,4-polyisoprene. This material is prepared from 1,4-polyisoprene by treatment with a variety of reagents under conditions that are proprietary to the manufacturers. The cyclized rubber matrix materials are soluble in non-polar organic solvents such as toluene, xylene, or halogenated aliphatic hydrocarbons. They spin uniform films that show excellent adhesion to a wide range of substrates, and as such, are ideal for wet liquid etching in relatively harsh environments.

The cyclized rubber matrix resins are generally sensitized with bisazide photosensitive cross-linking agents that initiate the formation of an insoluble 3-dimensional cross-linked network. The photochemical transformations associated with the generation of the network are summarized in Figure 9. Absorption of light leads to the elimination of nitrogen with formation of an extremely reactive nitrene intermediate that can undergo a variety of reactions including insertion into double bonds or into carbon-hydrogen bonds to form amines. These reactions result in cross-linking of the matrix resin. The sensitizers may be structurally modified in order that their absorption characteristics match exposure tool spectral output.

The major negative optical resists are listed in Table II. As with positive photoresists, there has been little change in the primary photochemistry over the years. Instead the major advances made by suppliers have been in the realm of reduced sensitivity to oxygen, increased purity, and quality control, all leading to improvements in device yield. The major limitation of negative photoresists is their dependence on organic solvent developers that cause resist image distortion as a result of swelling. Even so, their future still appears to be bright since they are still capable of handling the resolution, etching, linewidth control, and yield requirements of the majority of IC designs.

The introduction of certain manufacturing steps into device processing has placed increased demands on the properties of photoresists, especially with regard to thermal stability. In addition there is often a need for photochemically cross-linkable polymers functioning as insulating, passivating, or protective coatings in semiconductor components. Frequently, short term stability of up to 450°C is required. These requirements cannot be met with current photoresist formulations. Rubner et al. (*14*) have described a system involving cross-linking of a soluble photoreactive polyamic ester. After development, the imaged areas are thermally converted to a polyimide by cleavage and volatilization of the cross-links. The resulting image has the high temperature stability characteristic of polyimides.

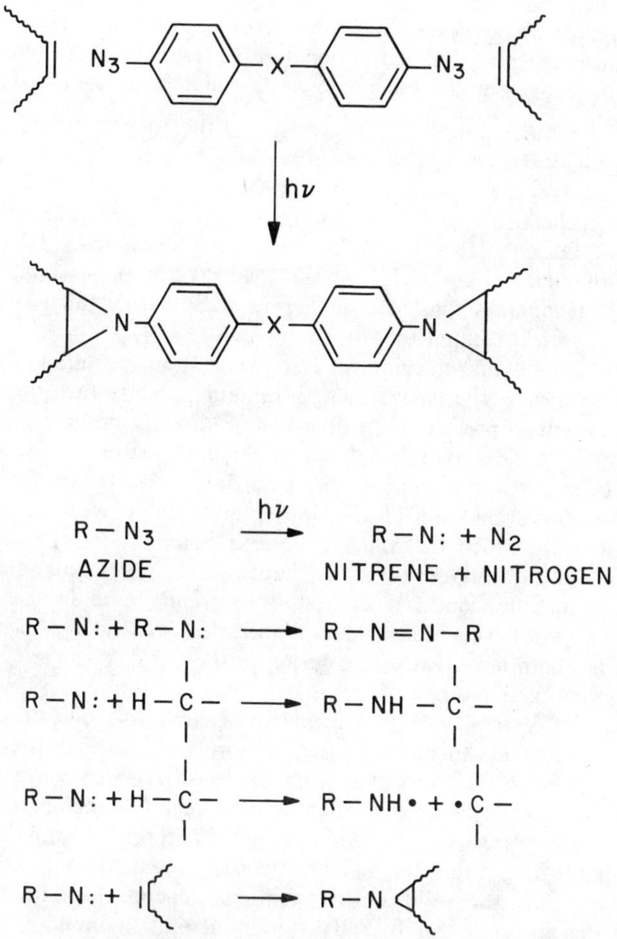

N_3—X—N_3

$h\nu$

N—X—N

$$R-N_3 \xrightarrow{h\nu} R-N: + N_2$$

AZIDE → NITRENE + NITROGEN

$$R-N: + R-N: \longrightarrow R-N=N-R$$

$$R-N: + H-C- \longrightarrow R-NH-C-$$

$$R-N: + H-C- \longrightarrow R-NH\bullet + \bullet C-$$

$$R-N: + \text{C=C} \longrightarrow R-N\langle$$

Figure 9. Photochemical transformations of a bisazide sensitizer in negative photoresists based on cyclized polyisoprene.

Table II. Commercially Available Negative Photoresists

RESIST	SUPPLIER
Microresist 700 Series	Eastman Kodak
Waycoat HNR & HR Series	Phillip A. Hunt
Selectilux N	E. M. Chemicals
OMR Series	Dynachem (Tokyo Ohka)
ONNR Series	Tokyo Ohka
Isopoly Negative Resists	Microimage Technology

Although other resist systems have been described in the literature, it does not appear that these newer resist formulations will displace the older more established resists for conventional wavelengths. Much more interest is being shown in dry-developable photoresists which will be discussed later.

Deep-UV Photolithography. There is little doubt that conventional photolithography employing wavelengths > 365 nm can be considered to be a mature technology. Resist needs are met by a variety of competent suppliers, and the photochemistry by which these resist systems operate is well known and understood. On the other hand, the continuing drive towards higher circuit density with concomitant reduction in feature size, is creating a need for lithographic techniques that have higher resolution than can be achieved with conventional optical technology.

One method of achieving an improvement in resolution is through the use of higher energy, shorter wavelength radiation. Since the diffraction-limited resolution of optical projection printing tools is directly proportional to exposure wavelength, it is clear that a reduction in the exposure wavelength from the near-UV to the 313 nm mid-UV emission line of a Hg arc lamp, or to the 254 nm line (or even lower if the appropriate source were available) in the deep-UV would produce a significant increase in resolution. This approach also has the added advantage that the hardware requirements are merely an extension or modification of those systems currently used for near-UV optical imaging; e.g., the Perkin Elmer Macralign series 300 and the recently released 500 machines both have reasonable output in the mid-UV range.

Conventional positive photoresists only have limited utility in the mid-UV and deep-UV range. While it appears feasible to modify the structure of the o-quinonediazides to optimize their absorption characteristics in the mid-UV range, (*8,9*) or even to use appropriate sensitizers for conventional resists, (*15*) no such simple solution is available in the deep-UV region.

There are a variety of problems with respect to both the resist and exposure hardware. From the standpoint of the resist, there are two important factors that limit the utility of conventional chemical positive photoresists. Although the absorption of the photoactive dissolution inhibitor (sensitizer) is relatively intense at 250 nm, it does not bleach upon exposure; i.e., the photoproducts also absorb at 250 nm. Secondly, the novolac matrix resin strongly absorbs at 250 nm, and the combined absorption of the PAC and the resin renders films more than 1 μm thick essentially opaque at 250 nm so that little or no radiation reaches the photoactive compound at the resist-substrate interface. Attempts to photodecompose the PAC at the resist/substrate interface by increasing exposure time serve only to overexpose the upper levels of the resist which leads to degraded profiles on development. Such effects are readily demonstrated in modeling studies such as SAMPLE, (*11,12*) a computer program which simulates lithographic performance based on inputs of modulation transfer function, wavelength, dose and resist performance parameters (*16*).

Hardware limitations in the deep-UV region revolve around lens design (and fabrication) and source brightness. The lens problem is a materials-related issue in that the number of optical-grade materials with acceptable transmission characteristics below 300 nm is limited. Source brightness is another problem, particularly for Hg arc lamps which are the most common exposure source. The amount of power at 254 nm (at which there is an allowed intense emission line) is actually very low due to efficient self absorption. This situation is further exacerbated when we realize that sensitivity is fundamentally related to the *number* of photons per cm^2 impinging on the sample and one will necessarily need to have more J/cm^2, i.e., energy per unit area, to have the same photon flux as the wavelength is decreased. Thus deep-UV lithography will require more light energy (Joules) in a region of decreasing lamp output to cause the same photochemical effect that would be observed with existing materials at conventional wavelengths.

A major consequence of these considerations is that new exposure sources and/or very sensitive resist materials must be developed in order to realize the resolution enhancement offered by deep-UV lithography without the penalty of extremely long exposure times. Considerable advances have been made on both fronts.

There is considerable interest at present in excimer lasers (*17-19*) as deep-UV sources, particularly lasers based on KrCl and KrF which have outputs at 222 and 249 nm respectively. The principal advantage of excimer lasers is that their steady state power output in the deep-UV region can be in excess of 10 watts, as opposed to a few tens of millwatts of output power available from a 1-kW Xe-Hg lamp in the desired spectral region. High-quality contact-printed images have been produced using excimer laser radiation in 1 μm thick diazo-type photoresists such as AZ-2400 where advantage is taken of the minimum in the absorption curve around 250 nm. In spite of the extremely high intensity, reciprocity failure does not appear to be significant, and such sources therefore promise very high throughput with even relatively insensitive resists. Implementation of the technology will require extensive redesign of the projection optics.

Research in the resist arena has concentrated on developing highly sensitive resists based in part on the assumption that the high pressure mercury arc lamp currently being used will continue to be the best exposure source for the near future. An ideal deep-UV resist should possess the following properties: (1) realistic sensitivity and processing characteristics to achieve reasonable exposure and development times, (2) sharp cut-off in the longer wavelength region to eliminate inefficient spectral filtration of the illumination, (3) optimum optical density for high aspect ratio in the developed resist, and (4) chemical and physical characteristics compatible with microfabrication. The bulk of the research has been directed towards positive resists, principally because of the superior resolution of these materials compared to that of negative resists whose resolution is limited in general by swelling during development which leads to pattern distortion below ~1.5 μm. This problem has been circumvented to some extent by some unique approaches to resist design as we shall see later.

Positive Resists for Deep UV. Resists based on poly(methyl methacrylate) (PMMA) have been widely used (*20,21*). The spectral absorption of a variety of methacrylates is similar, showing absorption below 260 nm with a peak at about 215-220 nm. This absorption is due to the n-π^* transition of the carbonyl group, with a maximum absorption coefficient at 215 nm of 0.27-0.47 μm^{-1}. This low value of the absorption coefficient coupled with the low energy output of conventional sources in this wavelength range leads to intolerably long exposures. Irradiation of PMMA in this wavelength region leads to chain scission as follows:

$$\left[-CH_2-\underset{\underset{\underset{CH_3}{|}}{\underset{O}{|}}}{\underset{C=O}{|}}{\overset{CH_3}{\overset{|}{C}}}-CH_2-\underset{\underset{\underset{CH_3}{|}}{\underset{O}{|}}}{\underset{C=O}{|}}{\overset{CH_3}{\overset{|}{C}}}-\right]_n \xrightarrow{h\nu} \left[-CH_2-\overset{CH_3}{\overset{|}{\dot{C}}}-CH_2-\underset{\underset{\underset{CH_3}{|}}{\underset{O}{|}}}{\underset{C=O}{|}}{\overset{CH_3}{\overset{|}{C}}}-\right]_n + CO + CO_2 + \dot{C}H_3 + \dot{O}CH_3$$

PMMA

$$\xrightarrow{\beta-\text{SCISSION}} -CH_2-\overset{CH_3}{\overset{|}{C}}=CH_2 + \cdot\underset{\underset{\underset{CH_3}{|}}{\underset{O}{|}}}{\underset{C=O}{|}}{\overset{CH_3}{\overset{|}{C}}}- \longrightarrow \text{FURTHER POLYMER DEGRADATION}$$

Sensitization is not possible since there is no low-lying excited state of the ester chromophore.

Since the match between the absorption spectrum of a resist and the output spectrum of the exposure tool is of paramount importance, most of the research has been aimed at developing materials that absorb in the 230-280 nm region which represents a compromise between a desire to work at the shortest possible wavelength and yet have reasonable light output and exposure time. For example, the same type of cleavage reaction that occurs in PMMA can occur in poly(vinyl alkyl ketones), e.g., poly(methyl isopropenyl ketone). However because of their weak absorption (ϵ for carbonyl chromophore at 280 nm (λ_{max}) is only ~25), sensitizers are required. An alternative approach has been to incorporate photosensitive chromophoric groups into PMMA with a view to enhance absorption in the spectral region of interest. The amount of chromophore incorporated is governed by absorbance considerations and a desire to keep a low content so as to maintain the excellent properties of PMMA. In general it is desirable to have as uniform an exposure through the resist film as possible. This suggests an absorbance limit of ~0.3 (50% transmission) although a higher initial absorption could be acceptable if it were bleached substantially during the exposure. Table III summarizes the lithographic properties of several modified methacrylate deep-UV resist systems which were recently reviewed by Chandross et al. (*22,23*). Notable among these are copolymers of methyl methacrylate and indenone which require an exposure of only 60 mJ/cm^2 at 248 nm when 3% indenone is present, and terpolymers of methyl methacrylate, 3-oximo-2-butanone methacrylate and methacrylonitrile. When sensitized with t-butyl benzoic acid, these terpolymers require an exposure of less than 30 mJ/cm^2. These materials provide a substantial increase in sensitivity over PMMA and are capable of submicron resolution. Their detailed lithographic behavior has not yet been thoroughly explored.

A similar approach has been used to modify poly(olefin sulfones) as shown in Table IV. The inclusion of aromatic groups into the chain serves a twofold purpose. First, they provide a convenient chromophore for energy absorption. Second, the aromatic moiety greatly enhances the plasma etching resistance of these materials compared with that of the pure poly(olefin sulfones), making them suitable plasma etching masks. They are also superior in this regard to methacrylate polymers. The quantum efficiency is low, however, although low sensitivity might not necessary be detrimental if high power sources become available.

An interesting variation on this theme is the lithographic response of copolymers of glycidyl methacrylate and methyl methacrylate (*33*). These resists are negative acting under electron irradiation as a result of ring-opening polymerization of the epoxide groups on the side chain. However, the oxirane ring does not absorb in the deep-UV region, and hence the lithographic behaviour under UV irradiation is determined by the photochemistry of the methacrylate group, i.e., the resist is positive acting. This resist is reported to have a sensitivity of 250-350 mJ/cm^2 depending on the GMA content. This is considerably higher than the sensitivity of PMMA, an observation which is somewhat surprising since the spectral absorption characteristics are identical and the same scission mechanisms are operable in both resists. The resist does have the advantage of being thermally cross-linkable at temperatures above 170°C after pattern delineation, thereby improving the adhesion and thermal stability of the resist image.

Several attempts have been made to redesign the traditional two-component near-UV positive resist systems to make them compatible with the deep-UV. Recall that the major problems associated with deep-UV exposure of conventional resists are related to non-bleaching of the o-quinonediazide sensitizer on exposure because of photoproduct absorbance, and strong absorption of the novolac resin. Willson and coworkers[34] attempted to solve this problem using dissolution inhibitors based on 5-diazo Meldrums acid, which undergoes photochemical decomposition as follows:

N_2 O O O O $\xrightarrow{h\nu}$ [O C=O O O + N_2] $\xrightarrow{h\nu}$ 3CO + O

Table III. Properties of Selected Deep UV Positive Photoresists

RESIST	STRUCTURE OF REPEAT UNIT	EFFECTIVE SPECTRAL SENSITIVITY RANGE (nm)	RELATIVE SENSITIVITY	REF. #
POLY (METHYL METHACRYLATE) (PMMA)	$-CH_2-C(CH_3)(COOCH_3)-$	200 - 240 (max 220)	1‡	20,21
POLY (METHYL ISOPROPENYL KETONE) (PMIPK) (ODUR-1010 (TOKYO OHKA)	$-CH_2-C(CH_3)(COCH_3)-$	230 - 320 (max 290)	5	24
PMIPK - S* (ODUR-1013, 1014 (TOKYO OHKA))		230 - 270 (max 250)	15	24
POLY (METHYL METHACRYLATE - CO - 3 - OXIMINO - 2 - BUTANONE METHACRYLATE) P (MMA - OM) (84:16)	$-CH_2-C(CH_3)(COOCH_3)-CH_2-C(CH_3)(COOR)-$ $R = -N=C(CH_3)(COCH_3)$	240 - 270 (max 220)	30	25, 26

POLY (METHYL METHACRYLATE - CO - 3 - OXIMINO - 2 - BUTANONE METHACRYLATE - CO - METHACRYLONITRILE) P (MMA - OM - MAN) (69:16:15)	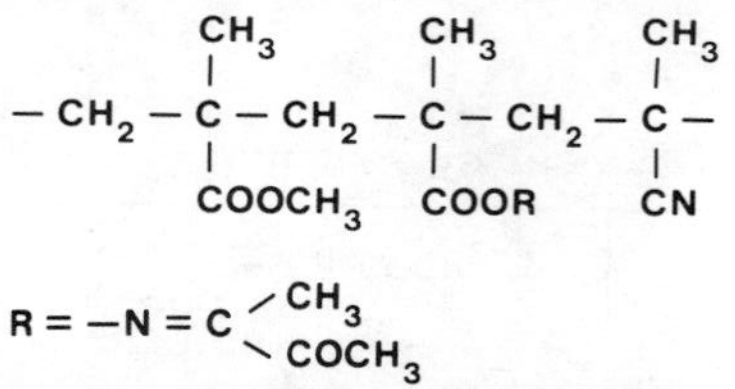 $-CH_2-C(CH_3)(COOCH_3)-CH_2-C(CH_3)(COOR)-CH_2-C(CH_3)(CN)-$ $R = -N=C(CH_3)(COCH_3)$	240 - 270 (max 220)	85	26, 27
P (MMA - OM - MAN) - S*		250 - 280 (max 250)	170	26
POLY (METHYL METHACRYLATE - CO - INDENONE) P (MMA - I)	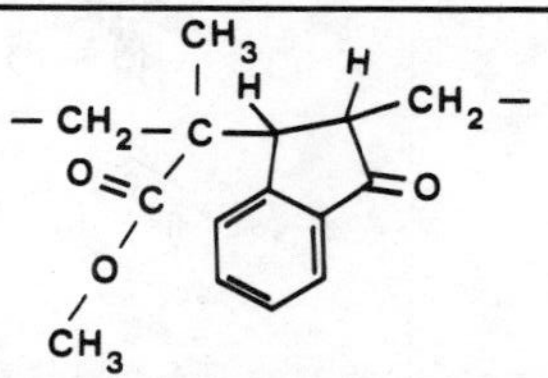 $-CH_2-C(CH_3)(C(=O)OCH_3)-$ [indanone unit, H, H] $-CH_2-$	230 - 300 (max 250 and 290)	35	28
POLY(p - METHOXYPHENYL ISOPROPENYL KETONE - CO - METHYL METHACRYLATE) (15:85)	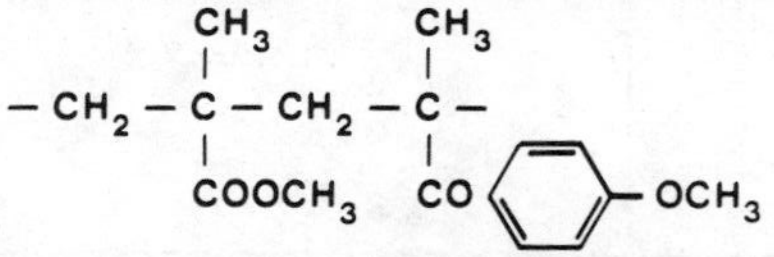 $-CH_2-C(CH_3)(COOCH_3)-CH_2-C(CH_3)(CO-C_6H_4-OCH_3)-$	220 - 360 (max 290 and 340)	166	29

‡ ABSOLUTE SENSITIVITY OF PMMA OVER THE WAVELENGTH REGION 200 - 250 nm USING MIBK AS THE STANDARD DEVELOPER IS 700 mJ/cm^2 FOR RESIST EXPOSED WITH A DEUTERIUM LAMP.

* SENSITIZED WITH p - t - BUTYL BENZOIC ACID.

Table IV. Properties of Polysulfone Deep UV Resists

RESIST	STRUCTURE OF REPEAT UNIT	EFFECTIVE SPECTRAL SENSITIVITY RANGE (nm)	SENSITIVITY* (mJ/cm^2)	REF. #
POLY (BUTENE - 1 SULFONE) (PBS)	$-CH_2-CH-SO_2-$ with CH_2CH_3 on CH	180 - 200	5 (at 185 nm) (MONOCHROMATIC)	30
POLY (STYRENE SULFONE) (STYRENE(2): SO_2(1)) (PSS)	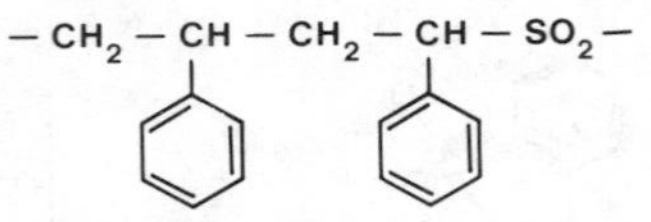 $-CH_2-CH-CH_2-CH-SO_2-$	240 - 280 (max 265)	1,000 (at 265 nm) (MONOCHROMATIC)	31
POLY (STYRENE - CO - ACENAPHTHALENE SULFONE) (VINYL ARENE(2): SO_2 (1) 25% ACENAPHTHALENE)	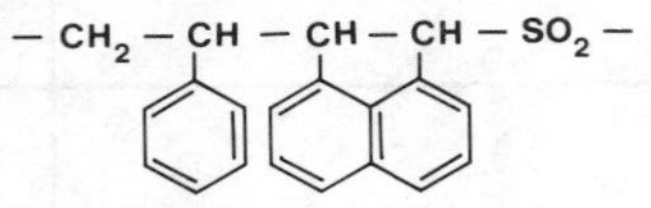 $-CH_2-CH-CH-CH-SO_2-$	250 - 330 (max 290) 320	500 (220 - 400)	31
POLY (5 - HEXENE - 2 - ONE SULFONE)	$-CH_2-CH-SO_2-$ with $CH_2CH_2COCH_3$ on CH	230 - 320 (max 280)	—	32

* MONOCHROMATIC SENSITIVITY OF PMMA AT 220 nm USING MIBK AS THE STANDARD DEVELOPER IS 420 mJ/cm^2

The Meldrum's acid derivative provides a bleachable chromophore at 250 nm with a sharp cutoff in the longer wavelength region, and it is converted to volatile species upon irradiation. Diazo Meldrum's acid is an effective dissolution inhibitor for novolac resins. The system shows reasonable sensitivity and appears capable of high resolution. However, the novolac resin itself has a strong absorption below 310 nm with only a shallow minimum at ~250 nm (O.D. = 0.4 for a 1 μm film). The absorption rises sharply at both shorter and longer wavelengths. This results in poor profiles because of strong attenuation of the light reaching the lower portion of the film. Difficulties were also experienced with sensitizer volatility and solubility.

An alternative system was recently described by Chandross et al. (*35,36*) based on optically transparent aqueous-alkali-soluble methyl methacrylate-methacrylic acid copolymer as the inert resin. The dissolution inhibitor was an o-nitrobenzyl carboxylate which undergoes photochemical decomposition to o-nitrosobenzaldehyde plus a carboxylic acid as follows:

$$\text{o-}NO_2C_6H_4CH(R)-O-C(=O)-R' \xrightarrow{h\nu} \left[\text{cyclic intermediate: N(OH)-O-C(R)(O-C(=O)-R')}\right] \longrightarrow \left[\text{o-}NO\text{-}C_6H_4C(R)(OH)-O-C(=O)-R'\right] \longrightarrow \text{o-}NO\text{-}C_6H_4C(=O)-R + R'CO_2H$$

This system has been reported to show good photosensitivity (~100 mJ/cm^2 in the 230-300 nm range) with extremely high contrast. Values of $\gamma > 5$ were observed, compared to ~2 for conventional systems. Chandross et al. believe the high contrast stems from the unusual dissolution characteristics of the resist in that the basic "developer" appears only to "prime" the irradiated regions for subsequent development in the aqueous rinse. The resist formulation is essentially aliphatic in nature and would be expected to be less stable to dry etching environments than the aromatic-based novolac resin materials.

Negative Resists for Deep UV. There has been considerable effort recently devoted to the design of negative resists for deep-UV application. Iwayanagi and co-workers (*37*) have reported on the properties of resists composed of cyclized polyisoprene and several bisazides whose absorption maxima lie within the deep-UV region. Since the sensitizers do not absorb in the visible region they are referred to as "white resists" and are claimed to be 60-450 times more sensitive than PMMA. One of these resist is available commercially as

ODUR-110-WR, but no detailed lithographic evaluation has been published. Since the matrix resin is cyclized rubber, one would expect the same swelling limitation on resolution evident in conventional negative photoresists.

A novel approach to avoiding the swelling problem has been developed by Iwayanagi et al. (*38*). The resist, known as MRS, consists of a poly(vinylphenol) matrix resin sensitized with a bis-arylazide, *viz*., 3,3-diazidodiphenyl sulfone. Under irradiation, the photolytically generated nitrenes insert into carbon-hydrogen bonds, forming a cross-linked matrix. The insertion reaction presumably takes place on the backbone C-H bonds since the energetics do not favor insertion at the aromatic C-H bonds. The unexposed portion of the resist is subsequently developed in aqueous base in an etching type dissolution mechanism that is devoid of swelling problems. The resist is strongly absorbing in the deep UV (O.D. ~2 for a 1 μm film) so that most of the photochemical reaction occurs in the surface layer, thereby eliminating standing wave effects arising from reflection at the resist/substrate interface. Since light is purportedly absorbed strongly at the surface, it is not clear why there is not significant undercutting with development. Possibly, there is some exposure at 313 nm where the resist is much more transparent (it was not reported in the published data whether these longer wavelengths were filtered out or not). Nevertheless, some undercutting does occur during development which causes a reasonably rapid change in image profile and linewidth with development time, resulting in a fairly narrow processing window. Consequently, reproducibility and satisfactory line width control requires extremely careful control of development and process conditions. The resist has been patterned in 1:1 deep UV projection printing (*39*) in which steep profile images of 1 μm lines in 1 μm thick resist were resolved. This resist is commercially available from Hitachi Chemical under the name of RD-2000N. The strong absorption in the surface layers of the resist is also reported to minimize problems with substrate reflectivity and topography.

Iwayanagi (*40*) recently extended this system into the mid-UV range by changing the sensitizer to an aromatic monoazide compound (4-azidochalcone). Insolubilization of this mid-UV resist does not result in a cross-linked matrix as occurs with the bisazide sensitized MRS. The primary reaction appears to be insertion of the reactive nitrene into a C-H bond on the ring, forming a secondary amine.

$$\underset{\text{NITRENE}}{\text{RN:}} \; + \; \underset{\text{POLYMER}}{\text{H}-\overset{|}{\underset{|}{\text{C}}}-} \longrightarrow \underset{\text{SECONDARY AMINE}}{\text{R}-\text{NH}-\overset{|}{\underset{|}{\text{C}}}-}$$

Apparently, the solubility of novolac resin in aqueous base is sufficiently altered by the pendant amine to facilitate image development.

A modification of these systems involving replacement of poly (4-vinylphenol) with poly(methyl isopropenyl ketone) (PMIPK) was reported by Nakane and co-workers (*41*). Although PMIPK is a positive-acting single-component resist in the deep UV, it functions as a negative resist when mixed with an aromatic bisazide such as 2,6 di(4-azidobenzylidene)-4-

methylcyclohexanone in a two-component formulation. The resist design capitalizes on the low swelling associated with the matrix resin during development, and a resolution of 1 μm has been reported although the sensitivity is approximately one-quarter that of cyclized rubber systems. The resist is available from Tokyo Ohka as ONNR-20.

Considerable interest is currently being shown in halogenated aromatic polymers, both as electron resists and as deep-UV resists. Chloromethylated polystyrene (CMS), prepared by chloromethylation of monodisperse polystyrene, was initially developed as a high-contrast, high-resolution negative electron resist and has since been shown to be sensitive in the deep UV where it is reported to be 40 times more sensitive than PMMA (*42*). The absorption spectrum of polystyrene shows a weak absorption peak at ~255 nm and strong absorption below 230 nm. The absorption is enhanced somewhat and shifted to longer wavelength by the introduction of the chloromethyl group. Similar shifts have been observed in poly(4-chlorostyrene) (*43*). The absorption coefficient in the deep-UV region increases as the chloromethylation ratio increases, and finally a distinct peak at 230 nm appears at virtually complete chloromethylation. The absorption is such that much of the light is absorbed in the upper part of a thick film in much the same way as is reported for MRS resist.

Resolution in CMS, like sensitivity, depends on the molecular weight, molecular weight distribution, developer solvent, and light absorption characteristics. With low degrees of chloromethylation (<0.14), resolution of 1 μm lines and spaces has been demonstrated. As the chloromethylation ratio is increased, the resolution is degraded primarily because of overexposure of the top part of the resist layer.

Other halogenated resist systems include partially chlorinated, narrow dispersity poly(vinyl-toluene). The latter differs from the CMS resist above in that chlorine is substituted on the main chain as well as on the pendant methyl group (*44*).

Electron Resists

Electron Beam Lithography. Computer-controlled electron beams have been used for many years to generate high-resolution patterns in polymeric resists. Electron-beam lithography has been the most actively pursued advanced technology over the past ten years, and most resist research has been directed towards developing resists for this application. There are two specific areas of use for electron beam lithography: (1) primary pattern generation (mask making), and (2) direct writing, in which the circuit configuration is transferred directly onto the wafer. In general, the ideal resist, positive or negative, should be a glassy polymer with little or no tendency to swell during dissolution. It should be capable of exhibiting submicron resolution, and its sensitivity to electron irradiation at 10-30 kV should be on the order of 10^{-6} c/cm^2. In addition, it should be sufficiently stable to withstand a variety of etching processes such as plasma and sputter etching, ion milling, etc., as well as conventional wet etching. These properties are of course in addition to the usual requirements discussed earlier.

In spite of the large amount of work over the past ten years, the ideal resist is yet to be developed. Most materials are deficient in one or more requirements so that resists have tended to be designed for specific applications such as mask making and usually for a specific set of exposure conditions dictated by machine design. For example, the electron beam mask maker EBES, designed at Bell Laboratories, requires resists with sensitivity better than 10^{-6} C/cm^2 at 10 kV for maximum machine throughput.

Positive Electron Resists. Much research has been directed towards the development of positive resists, primarily because of their inherently higher resolution capability. Many resist systems have been designed on the known principle that vinyl polymers containing quaternary carbon atoms in the backbone show a marked tendency to degrade when subjected to high energy radiation. This approach is exemplified by PMMA which was one of the first positive resists investigated and is still used as a experimental resist in many investigations where advantage is taken of its excellent submicron resolution. The main deficiencies in this resist are its low sensitivity (50 μ C/cm^2 at 15 kV) and its tendency to flow and degrade at elevated temperatures. Its resistance to various dry-etching environments is also marginal.

Several approaches have been used to enhance the sensitivity of PMMA, (*45*) either by introducing chemical and/or steric configurations which tend to weaken the main chain stability of the polymer (Table V), or by substitution on the quarternary carbon with polar electronegative substituents (Table VI) which also has the effect of weakening the main chain. The α-substituents are believed to enhance the capture of secondary electrons followed by resonance dissociation (*57*).

Many of the sensitivity claims, although impressive, remain to be substantiated and accordingly should be treated with caution. In nearly all cases, the development criteria such as development time, degree of thinning of the unexposed film, etc. are not available, nor are precise processing details given. In many instances, the literature is contradictory, causing a good deal of confusion. For example, Tada's claim of 4.5 μ C/cm^2 for the sensitivity of poly(trifluoroethyl methacrylate) (*48*) is considerably higher than the value of 20-30 μ C/cm^2 reported by Pittman et al. (*52*) (Table V). Such differences may be due to different development conditions. Another example of this kind of discrepancy is the report by Tada (*48*) that electron beam sensitivity is *decreased* by cyano substitution at the α-position, yet the high sensitivity of the FMR family of resists such as poly (ethyl α-cyanoacrylate) purportedly derives from α-cyano substitution.

One feature of positive resists which makes comparison difficult is that sensitivity is determined not only by changes in molecular weight produced by irradiation but also by changes in solubility characteristics of the polymer. Indeed, fundamental studies on the radiation degradation of a variety of substituted methacrylates by Pittman and co-workers (*52,57,58*) have revealed only small effects on G(scission) for a variety of alkyl and haloalkyl substituted methacrylates. Substitution at the α-position has a somewhat greater effect; G(scission) for example for poly(methyl α-chloroacrylate) (PMCA) is reported to be 5.8 compared to 1.3 for PMMA, although the advantage of a high

Table V. Lithographic Properties of Ester Derivatives of Methacrylate — Based Electron Resists

RESIST	STRUCTURE OF REPEAT UNIT	MW (Tg)	SENSITIVITY *AT 20 kV ($\mu C/cm^2$)	REF. #
POLY (METHYL METHACRYLATE) (PMMA)	$-CH_2-C(CH_3)(COOCH_3)-$	5×10^5 (104°C)	80	46
POLY (t - BUTYL METHACRYLATE)	$-CH_2-C(CH_3)(COOC_4H_9)-$	— (19°C)	0.5	47 48
POLY (HEXAFLUOROBUTYL METHACRYLATE) (FBM - 110 (DAIKIN KOGYO))	$-CH_2-C(CH_3)(COOR)-$ $R = -CH_2CF_2CHFCF_3$	5×10^5 (50°C)	0.4	49

Continued on next page.

Table V (continued)

RESIST	STRUCTURE OF REPEAT UNIT	MW (Tg)	SENSITIVITY *AT 20 kV ($\mu C/cm^2$)	REF. #
POLY (DIMETHYL TETRAFLUOROPROPYL METHACRYLATE) (FPM (DAIKIN KOGYO))	$-CH_2-C(CH_3)(COOR)-$, $R = -CH(CHF_2)_2$	$10^7 - 10^6$ (93°C)	3 – 12	50
POLY (TRICHLOROETHYL METHACRYLATE) (EBR – 1 (TORAY INDUSTRIES))	$-CH_2-C(CH_3)(COOCH_2CCl_3)-$	5.7×10^5 (138°C)	1.25	51
POLY (TRIFLUOROETHYL METHACRYLATE)	$-CH_2-C(CH_3)(COOCH_2CF_3)-$	9.2×10^5	4.5 (20 – 30)	48 52

* NEGLIGIBLE THINNING DURING DEVELOPMENT.

Table VI. Lithographic Properties of α-Substituted Methacrylate Electron Resists

RESIST	STRUCTURE OF REPEAT UNIT	MW (Tg)	SENSITIVITY* AT 20 kV ($\mu C/cm^2$)	REF. #
POLY (METHYL α-CHLOROACRYLATE) (PMCA)	$-CH_2-C(Cl)(COOCH_3)-$	1.6×10^6 (130°C)	46	53
POLY (ETHYL α-CYANOACRYLATE)	$-CH_2-C(CN)(COOC_2H_5)-$	2×10^5	1.5	54

Continued on next page.

Table VI (continued)

RESIST	STRUCTURE OF REPEAT UNIT	MW (Tg)	SENSITIVITY *AT 20 kV ($\mu C/cm^2$)	REF. #
POLY (ETHYL α - CARBOXAMIDOACRYLATE)	$-CH_2-C(CONH_2)(COOC_2H_5)-$	2.5×10^5	1.4**	54
POLY (TRIFLUOROETHYL α- CHLOROACRYLATE) (EBR - 9 (TORAY INDUSTRIES))	$-CH_2-C(Cl)(COOCH_2CF_3)-$	1.2×10^6 (133°C)	0.8 - 6.4	55
POLY (METHACRYLONITRILE) (PMAN)	$-CH_2-C(CH_3)(CN)-$	1.3×10^6 (120°C)	30	56

* NEGLIGIBLE THINNING DURING DEVELOPMENT

** POSSIBLY FORCE DEVELOPED

G(scission) for PMCA is offset to some degree by the occurrence of a concurrent cross-linking reaction (*57*). Although sensitivity does scale with *G*(scission), the increases in sensitivity that have been reported are often much greater than what can be accounted for on the basis of *G*(scission) considerations alone. The same is true for the sensitivities claimed for a variety of methacrylate copolymers (Table VII), e.g., *G*(scission) for the terpolymer of methyl methacrylate, methacrylic acid and methacrylic anhydride is 4.5 compared to 1.5 for PMMA. This suggests a three-fold increase in sensitivity, yet the sensitivity of the terpolymer is improved by a factor of 5-10 over that of PMMA. It is possible that the enhanced sensitivity in many of these cases results more from the solubility characteristics of the polymers rather than molecular weight changes. The dissolution rate (S) is related to molecular weight (M) through the expression

$$S = K M^{\alpha} \quad (4)$$

where K and α are solvent-dependent parameters that are specific to the particular polymer under consideration. Thus by choosing developer solvents for which the value of α is high, we can enhance the solubility rate relative to that of the unirradiated polymer and so enhance sensitivity for a given exposure dose. This solvent and time dependence of development further complicate comparisons of sensitivity and contrast of different systems. Physical factors such as micropore formation resulting from evolution of gaseous radiolysis products during irradiation also complicate the issue (*69*). The presence of these micropores can enhance the dissolution rate and consequently the solubility rate ratio. This mechanism presumably contributes to the overall enhanced sensitivity of the terpolymer resist mentioned above where *G*(total gas) is greater by a factor of 4 than that in pure PMMA.

Attempts to improve the thermal stability of PMMA have centered around an approach first described by Roberts, (*70,71*) of introducing cross-links into the matrix, typically via the formation of acid anhydride bonds. By copolymerizing methacryloyl chloride and methacrylic acid with methyl methacrylate, Roberts was able to form intermolecular cross-links *in situ* on the substrate by heating the resist after spin-coating. The anhydride linkage is broken on subsequent irradiation, thus restoring solubility to the irradiated region. Similar resists have been evaluated by Kitakohji et al. (*47*). Anhydride structures can be formed directly from methacrylic acid by heating above 200°C. This approach has been exploited to generate cross-linked matrices from copolymers of methacrylic acid and t-butyl methacrylate (*68*) by prebaking spun films at elevated temperature (250°C for 1 hr). Under these conditions the t-butyl ester side group also decomposes with the formation of inter- and intra-molecular anhydride linkages. Similar observations have been made on copolymers of butyl methacrylate and methyl methacrylate (*66*). In spite of the large amount of work in cross-linked positive resists, both practical and theoretical, (*72*) they have not found wide utility with serial exposure systems primarily because of low sensitivity ($\sim$10 μ C/cm^2) although they may find application in parallel exposure systems.

As seen in Tables V, VI and VII, a number of these methacrylate-based resist systems are available commercially. Some, like IBM's "terpolymer" resist,

Table VII. Lithographic Properties of Methacrylate Copolymer Resists

RESIST	STRUCTURE OF REPEAT UNIT	MW	SENSITIVITY* AT 20 kV ($\mu C/cm^2$)	REF. #
POLY (METHYL METHACRYLATE - CO - METHACRYLIC ACID) P (MMA - MAA) (4:1)	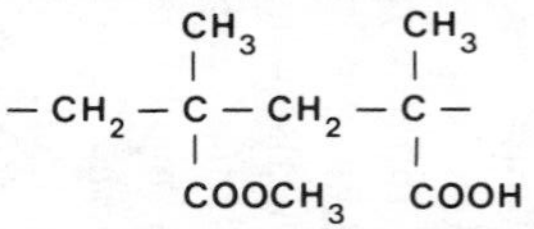 $-CH_2-C(CH_3)(COOCH_3)-CH_2-C(CH_3)(COOH)-$	—	20	59
POLY (METHYL METHACRYLATE - CO - METHACRYLIC ACID - CO - METHACRYLIC ANHYDRIDE) P (MMA - MAA - MAH) (70:15:15)	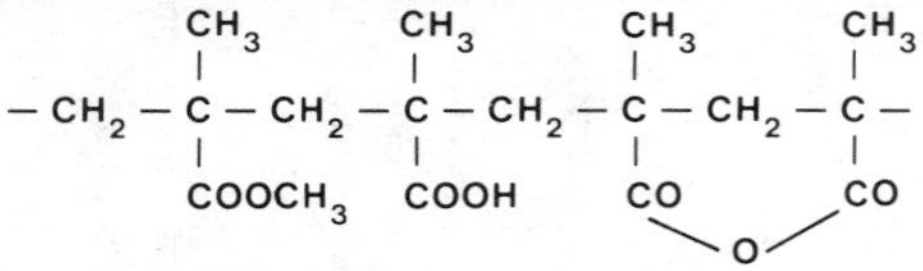 $-CH_2-C(CH_3)(COOCH_3)-CH_2-C(CH_3)(COOH)-CH_2-C(CH_3)(CO)-CH_2-C(CH_3)(CO)-$ (the two CO groups joined through O)	$1-3\times10^5$	$6.5^{\ddagger}$	60
POLY (METHYL METHACRYLATE - CO - ACRYLONITRILE) P (MMA - AN) (89:11) (OEBR - 1013 (TOKYO OHKA)	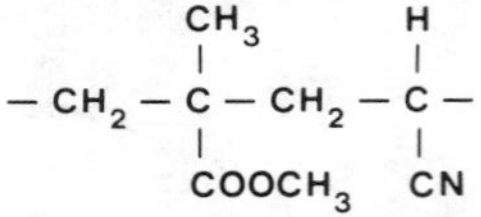 $-CH_2-C(CH_3)(COOCH_3)-CH_2-C(H)(CN)-$	6×10^5	6**	61

POLY (METHYL METHACRYLATE - CO - METHYL α-CHLOROACRYLATE) P (MMA - MCA) (1.6:1)	$-CH_2-C(CH_3)(COOCH_3)-CH_2-C(Cl)(COOCH_3)-$	1.2×10^5	7.5**	62
POLY (METHYL METHACRYLATE - CO - METHACRYLONITRILE) P (MMA - MAN) (1:1)	$-CH_2-C(CH_3)(COOCH_3)-CH_2-C(CH_3)(CN)-$	10^6	8	63
POLY (METHYL α - CHLOROACRYLATE - CO - METHACRYLONITRILE) P (MCA - MAN) (1:1)	$-CH_2-C(Cl)(COOCH_3)-CH_2-C(CH_3)(CN)-$	10^6	24**	53
POLY (ETHYL α - CYANO ACRYLATE - CO - ETHYL α - CARBOXAMIDO ACRYLATE) (9:1) (FMR - E101 (FUJI CHEMICAL))	$-CH_2-C(CN)(COOC_2H_5)-CH_2-C(CONH_2)(COOC_2H_5)-$	2×10^5	1.5	54
POLY (METHYL METHACRYLATE - CO - ISOBUTYLENE) P (MMA - IB) (75:25)	$-CH_2-C(CH_3)(COOCH_3)-CH_2-C(CH_3)(CH_3)-$	10^5	6**	64

Continued on next page.

Table VII (continued)

RESIST	STRUCTURE OF REPEAT UNIT	MW	SENSITIVITY* AT 20 kV ($\mu C/cm^2$)	REF. #
POLY(HEXAFLUORO BUTYL METHACRYLATE - CO - GLYCIDYL METHACRYLATE) P(FBM - GMA) (99:1) FBM - 120 (TOKYO OHKA)	$-CH_2-C(CH_3)(COOR)-CH_2-C(CH_3)(COOR')-$ $R = -CH_2CF_2CHFCF_3$ $R' = -CH_2-CH-CH_2$ (epoxide, bridged by O)	—	0.4	65
POLY(METHYL METHACRYLATE - CO — t — BUTYL METHACRYLATE) P(MMA - t - BMA) (CP-3)	$-CH_2-C(CH_3)(COOCH_3)-CH_2-C(CH_3)(COOC_4H_9)-$	—	1.6	66
POLY(METHACRYLONITRILE) - CO - TRICHLOROETHYL METHACRYLATE) P(MAN - TCEMA) (2:1)	$-CH_2-C(CH_3)(CN)-CH_2-C(CH_3)(COOCH_2CCl_3)-$	4×10^5	7** (40% THINNING OF UNEXPOSED AREA)	67
POLY(METHACRYLIC ACID - CO - t - BUTYL METHACRYLATE) P(MAA - t BMA) (2:1)	$-CH_2-C(CH_3)(COOH)-CH_2-C(CH_3)(COOC_4H_9)-$	6×10^5	15 (<10% THICKNESS LOSS)	68

* NEGLIGIBLE THINNING DURING DEVELOPMENT

** ESTIMATED FROM PUBLISHED VALUE AT 15 kV

‡ ESTIMATED FROM PUBLISHED VALUE AT 25 kV

are used solely "in-house". Others, such as the halogenated alkyl derivatives, appear to find limited use, mainly in Japan.

The other major class of positive resists is based on poly(olefin sulfones) which are alternating copolymers of sulfur dioxide and the respective olefin having the general structure.

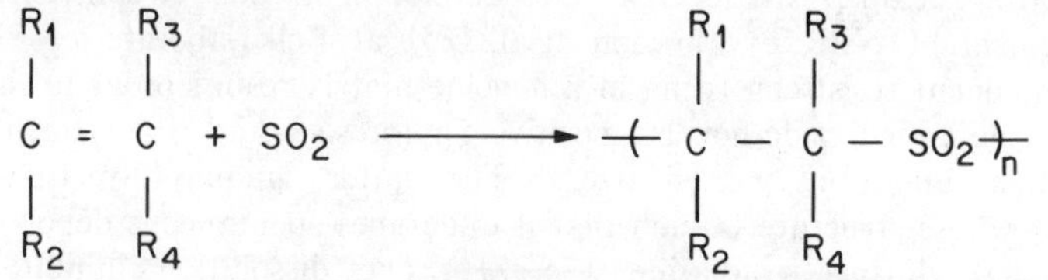

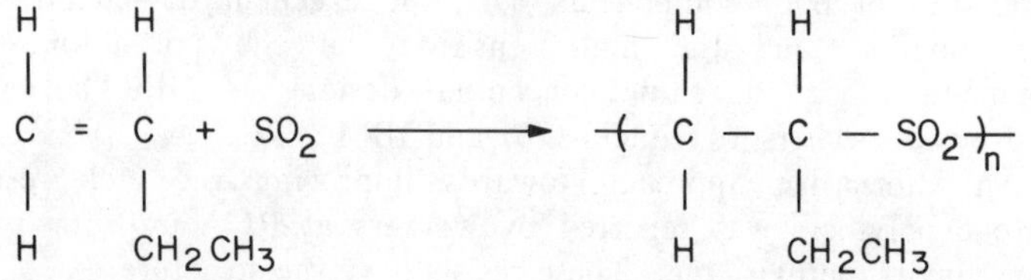

$$R_1R_2C{=}CR_3R_4 + SO_2 \longrightarrow \left(C(R_1)(R_2) - C(R_3)(R_4) - SO_2 \right)_n$$

$$H_2C{=}CH(CH_2CH_3) + SO_2 \longrightarrow \left(CH_2 - CH(CH_2CH_3) - SO_2 \right)_n$$

BUTENE — POLY (BUTENE – 1 SULFONE) PBS

SENSITIVITY – 1.6 $\mu C/cm^2$ AT 20 kV
RESOLUTION – < 0.5 μm.
SUPPLY – COMMERCIAL (MEAD CHEMICAL CO)

They derive their sensitivity from the selective cleavage of the weak carbon-sulfur bond whose bond energy is on the order of 60 kcal/mole, compared with 80 kcal/mol for a carbon-carbon bond. The lithographic properties of these resists have been described by workers from Bell Laboratories, RCA, and IBM and were recently reviewed by Bowden and Thompson (*3*). Of the many materials investigated, poly(butene-1 sulfone) (PBS), developed at Bell Laboratories, was considered to exhibit the best properties and was subsequently made commercially available as a resist for mask making where it exhibits a sensitivity of 1.6 μ C/cm^2 at 20 kV (*73,74*). It is the only readily available commercial (Mead Chemical Co., Rolla, Mo. and Chisso Corp. in Japan) positive resist at present whose lithographic properties are commensurate with machines requiring sensitivity better than 2 μ C/cm^2 at 20 kV. It is therefore the positive resist of choice for the current generation of machines designed on the EBES concept. Processing technology is available under license through the American Telephone and Telegraph Co.

The major limitation of poly(olefin sulfones) is their poor resistance to dry etching. Schemes have been devised to overcome this deficiency, but these have been far from satisfactory and the design of a highly sensitive positive resist capable of withstanding dry etching continues to pose a major challenge to the resist designer.

The problem of etch resistance in the application of poly(olefin sulfones) to electron beam resist design and development was circumvented in the development of NPR by Bowden et al. (*75*) at Bell Laboratories. NPR is a two-component resist consisting of a novolac matrix resin similar to that used in common quinonediazide-novolac positive photoresists and a poly(olefin sulfone) dissolution inhibitor or sensitizer. The latter is poly(2-methyl-1-pentene sulfone) whose structure is such that it undergoes spontaneous depolymerization (unzipping) during irradiation (*76,77*). The dissolution inhibitor is thus vaporized, i.e., removed, during the exposure process with a concomitant increase in solubility of the remaining film. The sensitivity of this resist is 3-5 $\mu C/cm^2$ at 20 kV. The resist succeeds in expressing the favorable characteristics of both components, *viz.*, the excellent dry-etch resistance of novolac polymers and the high sensitivity of the polysulfone sensitizer. Developments along the same conceptual designs as NPR have also been reported by researchers at Hitachi (*78*) and IBM (*79*).

An alternative approach towards improving the etch resistance of polysulfone polymers was reported by workers at RCA (*80*) for polysulfones based on vinyltrimethylsilane. These resists have the structure

$$\left[\begin{array}{c} -CH_2-\underset{\displaystyle SiR_3}{\underset{|}{CH}}-SO_2- \end{array} \right]_n$$

where oxygen plasma resistance is presumably derived from the conversion of Si to SiO_2. It is claimed that the material will withstand etching of 2.5 μm thick polyimide films, although very few processing details have been reported.

Another approach to obtaining plasma etching resistance lies in modifying the conditions. It was mentioned before that PBS shows poor plasma resistance. Nevertheless, Bowden et al. (*74*) were able to etch silicon in a CF_4-O_2 plasma to form fiducial marks 3.8 μm deep using PBS as a mask by maintaining the temperature of the substrate below 15°C. Lowering the substrate temperature reduces the rate of the depolymerization reaction, thereby reducing the rate of film loss. Yamazaki et al. (*81*) also reported recently on a reversal gas plasma etching technique for chromium films using PBS. The technique depends on the presence of a layer of tungsten oxide on the surface of an antireflective chromium film. The latter consists of a thin layer of chromium oxide on a chromium film on a glass substrate. The WO_3 layer acts as a masking layer in the gas plasma environment (mixture of CCl_4, N_2 and O_2). Thus, the areas exposed following development of PBS are masked during plasma treatment, whereas the residual resist is rapidly removed in the unexposed areas. Ordinarily, plasma etching would halt at the interface, i.e., at the WO_3 layer. However, apparently the WO_3 layer is removed in this region by interaction with the decomposition products of the resist and thus etching continues through the chromium film resulting in a negative tone mask pattern.

Negative Electron Resists. Compared to the extensive research on positive resists, much less effort has been expended in developing satisfactory negative resists. This is due primarily to the fact that the inherent sensitivity of the reactive moities incorporated into the polymer chain to promote cross-linking is well within the required exposure range of most electron beam exposure machines. These groups include epoxy, vinyl (allyl), and episulfide moities, all of which cross-link by chain mechanisms. For example, G(cross-linking) for representative epoxy polymers is about 10. Epoxy-containing resists have been extensively studied with those based on glycidyl methacrylate and its copolymers reaching commercial prominence.

The major thrust of research has been directed towards improving dry-etchant resistance and resolution of these resists. This has generally been accomplished by incorporating aromatic groups into the polymer chain. Although the addition of such groups adversely affects sensitivity, a satisfactory compromise can usually be reached between sensitivity and dry etching resistance. The lithographic properties of several systems that are based on these design principles and that have gained commercial significance are listed in Table VIII.

We note from Table VIII a strong interest in halogenated resists, particularly those substituted with chlorine. The addition of chlorine to the aromatic structure of polystyrene has a marked effect on cross-linking efficiency. Monodisperse polystyrene, for example, has a sensitivity on the order of 50 μ C/cm^2, yet with as little as 20% chloromethyl groups substituted on the ring, the sensitivity is improved to 2 μ C/cm^2 for comparable molecular weight and distribution.

These materials show other favorable properties compared to chain-propagating resists. Their high glass transition temperature (T_g of α-M–CMS for example is 170°C (*91*) and localized cross-linking reaction mechanism eliminate the post-polymerization effects which plague line width control in resists such as poly(glycidyl methacrylate) (PGMA). Their high aromatic content ensures good stability in plasma etching environments. It is interesting to note in this regard that the addition of as little as 20% 3-chlorostyrene into PGMA results in a twofold improvement in oxygen plasma etching resistance (*95*).

The primary factor limiting resolution in negative resists is swelling during development. This is almost an inescapable fact of life in the wet development of negative resists since the solvent which is used to dissolve, i.e., develop, the unexposed areas of the film also penetrates and swells the cross-linked regions. We will see later that where the polarity change between exposed and unexposed regions can be made sufficiently great, no swelling occurs, but this represents a special limiting case. The swelling phenomenon leads to an increase in the volume, i.e., dimensions, of the image leading to distortion. In particular, closely spaced lines can swell to the extent that they touch, forming "bridges" between the lines which are retained even after

Table VIII. Lithographic Properties of Selected Negative Electron Resists

RESIST	STRUCTURE OF REPEAT UNIT	MW	SENSITIVITY* AT 20 kV ($\mu C/cm^2$)	REF. #
POLY (GLYCIDYL METHACRYLATE) (PGMA) (OEBR - 100 (TOKYO OHKA))	$-CH_2-C(CH_3)(COOCH_2CH-CH_2 \text{ [epoxide, O]})-$	1.25×10^5	0.5	82
POLY (GLYCIDYL METHACRYLATE - CO - ETHYL ACRYLATE) P (GMA - EA) (COP (MEAD CHEMICAL))	$-CH_2-C(CH_3)(COOCH_2CH-CH_2 \text{ [epoxide, O]})-CH_2-C(H)(COOC_2H_5)-$	1.8×10^5	0.6	83
POLY (GLYCIDYL METHACRYLATE - CO - CHLOROSTYRENE) P (GMA - CℓS) (GMC - I AND II (MEAD CHEMICAL))	$-CH_2-C(CH_3)(COOCH_2-CH-CH_2 \text{ [epoxide, O]})-CH_2-CH(C_6H_4Cl)-$	2×10^5 (4×10^5)	4 (2)	84

CHLOROMETHYLATED POLYSTYRENE (40% CHLOROMETHYLATED) (CMS - EX (TOYO SODA))	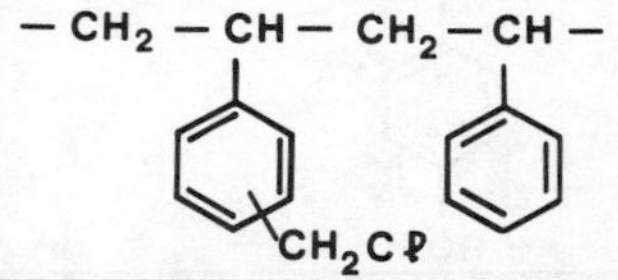 $-CH_2-CH(C_6H_4CH_2Cl)-CH_2-CH(C_6H_5)-$	6.8×10^3 – 5.6×10^5	39 - 0.4	85
POLY(2 - HYDROXY - 3 [METHYL FUMARATE] PROPYL METHACRYLATE - CO - 3 CHLORO - 2 - HYDROXY PROPYL METHACRYLATE) (SEL - N (SOMAR CORP.))	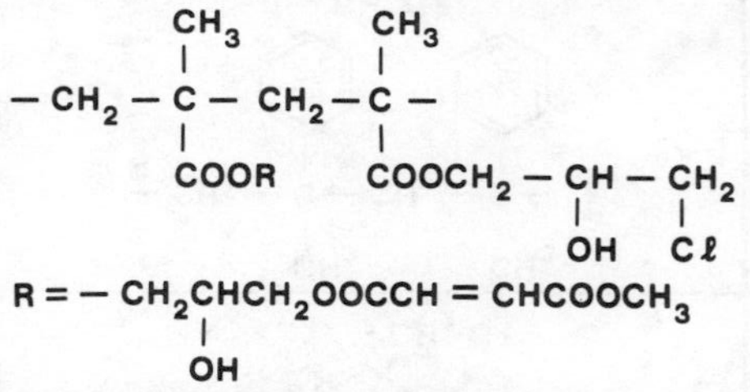 $-CH_2-C(CH_3)(COOR)-CH_2-C(CH_3)(COOCH_2-CH(OH)-CH_2Cl)-$ $R = -CH_2CH(OH)CH_2OOCCH=CHCOOCH_3$	—	0.8	65
POLY IODOSTYRENE) (IPS) (~70% IODINATED) (RE - 4000N (HITACHI CHEMICAL))	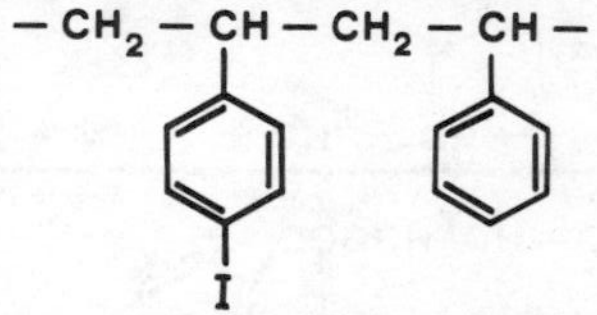 $-CH_2-CH(C_6H_4I)-CH_2-CH(C_6H_5)-$	3.8×10^5	2	86
POLY (ALLYL METHACRYLATE - CO - 2 - HYDROXYETHYL METHACRYALTE) (3:1) (EK-88 (KODAK))	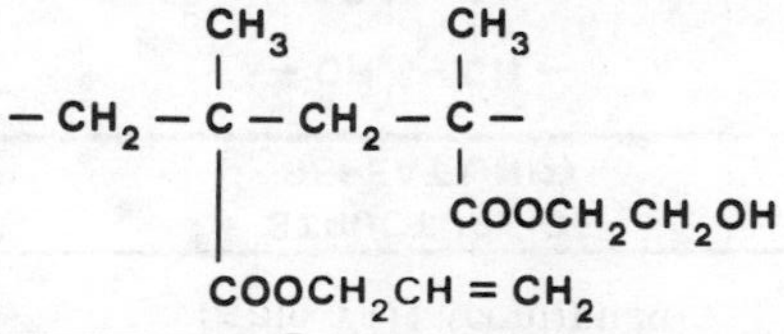 $-CH_2-C(CH_3)(COOCH_2CH=CH_2)-CH_2-C(CH_3)(COOCH_2CH_2OH)-$	3.5 - 7 $\times 10^4$	0.4	87

Continued on next page.

Table VIII (continued)

RESIST	STRUCTURE OF REPEAT UNIT	MW	SENSITIVITY* AT 20 kV ($\mu C/cm^2$)	REF. #
POLY(DIALLYL ORTHO PHTHALATE) (PDOP)	$-CH_2-CH-$; $COO-CH_2$; $COOCH_2CH=CH_2$	$1.1-11.1 \times 10^4$	56 – 0.9	88
POLY(CHLOROMETHYL STYRENE) (PCMS)	$-CH_2-CH-$; $-CH_2Cl$	2×10^4— 4.5×10^5	7 – 0.35	89
CHLOROMETHYLATED POLY(α-METHL STYRENE) (~95% CHLORO-METHYLATED) (αM - CMS)	CH_3 ; CH_3 ; $-CH_2-C-CH_2-C-$; CH_2Cl	8×10^3— 1.9×10^5	42 – 3	90, 91
POLY(4-CHLOROSTYRENE) (PCS)	$-CH_2-CH-$; Cl	3×10^5— 7×10^5	2.5 – 1.5	43

CHLORINATED POLY(VINYL-TOLUENE) (CPMS)	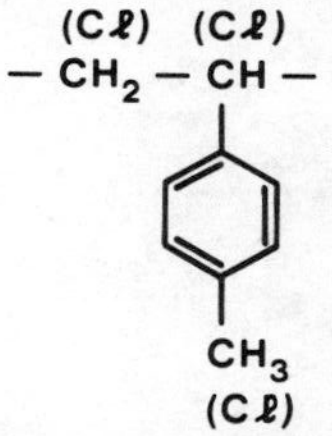 (Cℓ) (Cℓ) $-CH_2-CH-$ (para-substituted benzene ring) CH_3 (Cℓ)	1.2×10^5– 5.8×10^5	0.4 - 1.6	92
EB - 46 (GAF)	—	—	0.4	93
POLY(VINYL METHL SILOXANE) (PVMS)	CH_3 $-Si-O-$ $CH=CH_2$	2.9×10^5	1.5	94

* $Dg^{0.5}$

removal from the developer (Figure 10a). Another phenomenon is the snake-like distortion of long narrow lines shown in Figures 10b and 10c resulting from adhesion and stress relief considerations. Note in Figure 10b how snaking is

a

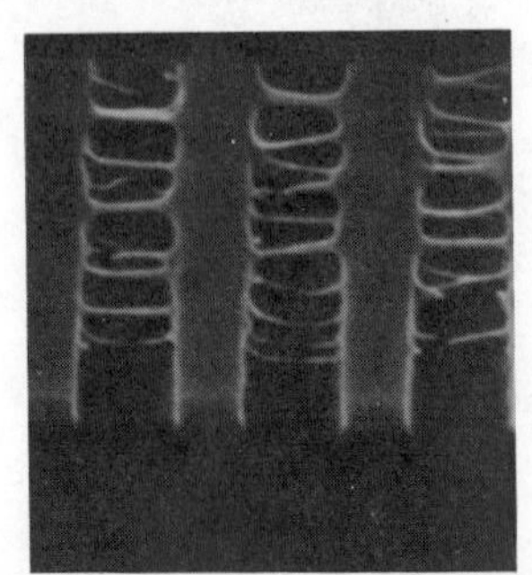

b

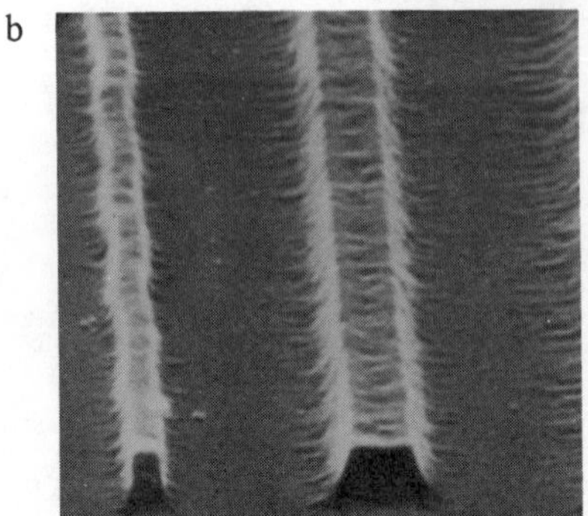

c

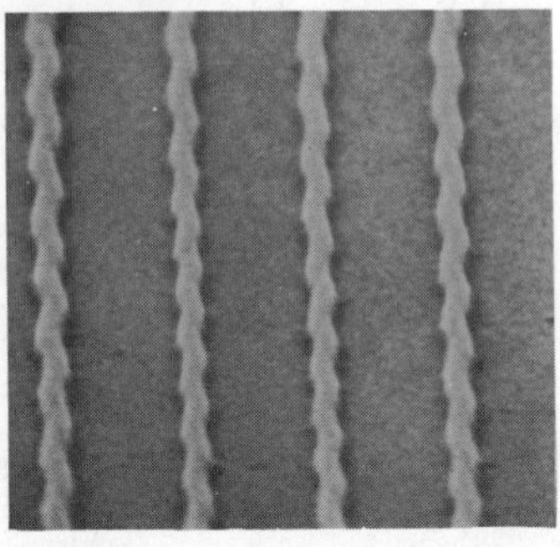

Figure 10. (a) "Bridging" between adjacent features of a negative resist. (b) 0.5 and 1.0 μm lines in negative resist showing distortion in submicron line. (c) "Snake-like" distortions of narrow lines in negative resist.

evident in the submicron line but not in the 1.0 μm line. Incorporation of aromatic groups into the polymer chain tends to minimize the problem by raising the T_g of the polymer. Even so, swelling still limits resolution in negative resists to about 0.75-1.0 μm although, again, there have been some interesting developments on this front.

Sugawara et al. (*91,92*) have extended their work on chloromethylated polystyrene to chloromethylated poly(α-methylstyrene). Poly(α-methyl styrene) has a T_g of 170°C compared to 100°C for polystyrene and, although it undergoes chain scission under electron irradiation (G(scission) ~0.3), the chloromethylated derivative cross-links. Although less sensitive than chloromethylated polystyrene, (CMS), the new resist has a contrast of 2.5 to 4.1 depending on its molecular weight. Using 0.2 μm thick films with M_w of 19,000 and a dose of 3.3 μ C/cm^2, Sugawara et al. were able to resolve a 0.4 μm line and space pattern. Resolution does degrade as the thickness and molecular weight of the polymer increase although sensitivity improves with increasing molecular weight.

The recent results of Shaw, Hatzakis and their co-workers (*94*) on poly(vinylmethylsiloxane) (PVMS) and related siloxane polymers are interesting in that these polymers show sensitivities as high as 1-2 μ C/cm^2 with contrast in the range of 1-2. Moreover, their high silicon content allows them to function in bilevel applications (see Section 9) as plasma etching masks. Using thin resist films (<200 nm) over a 2 μm-thick planarizing layer, 0.5 μm lines in the planarizing film have been resolved.

For the majority of negative electron resists, resolution is less than that attainable with positive resists. The latter generally show higher contrast, and when coupled with the fractional dissolution mode of development during which swelling can be kept to a minimum by judicious choice of developer solvents, show resolution better than 0.5 μm. One approach to improving the resolution of negative resists has been to effect an image reversal of positive photoresists based on quinonediazide chemistry (*96*). Recall that in the normal operation of these resists, UV exposure in the presence of water leads to the destruction of the photoactive compound (PAC) and the formation of 3-indenecarboxylic acid. This compound enhances the solubility of the resin in aqueous basic developer solutions, resulting in positive resist behavior. If the resist is water free during UV exposure such as would occur with exposure in a vacuum environment, the destruction of the PAC produces an ester with the novolac resin and a somewhat cross-linked base resin. By exposing the resist in the vacuum environment of an electron beam writing machine and subsequently flood exposing the resist with UV in a wet atmosphere, a negative resist process is possible. Resolution of 0.5 μm has been claimed although the sensitivity is too low for useful application.

X-ray Resists

X-ray lithography is an extension of near-contact photolithography for replication of 1 μm and smaller geometries. Typical wavelengths that have been used for resist exposure include 4.2 Å ($Pd_{L\alpha}$), 5.4 Å ($Mo_{L\alpha}$), and 8.1 Å ($Al_{K\alpha}$). Although, significant advances both in resist chemistry and on the technological aspects associated with exposure hardware, viz., x-ray source and mask, have been made, further developments in these areas are still required if x-ray lithography is to become a commercial reality. This is particularly true of mask fabrication which is an extremely complex process. X-ray masks are structurally complicated and difficult to fabricate with acceptable yields, particularly at overall mask dimensions that would make x-ray lithography competitive with optical lithography. Probably the greatest stumbling block to implementation of x-ray lithography is the continuing improvement in optical photolithographic art. It now appears, for example, that photolithography will be capable of submicron resolution — possibly as low as 0.5-0.7 μm. The "window" for x-ray lithography is thus being pushed lower and lower with attendant constraints on hardware design and fabrication.

Several factors need to be kept in mind concerning x-ray resist design. Firstly, source design has mainly utilized electron bombardment of the target material to generate x-rays. Power output is limited by heat transfer considerations, and when account is also taken of the separation between source and exposure plane (wafer) as necessitated by geometrical considerations, it turns out that the energy flux at the wafer plane is low, necessitating the use of highly sensitive resists (<10 mJ/cm^2) for economical wafer throughput.

The conventional approach has been to take resists which were developed for electron lithography and apply them to x-ray lithography. To a first order, there is a strong correlation between the sensitivity of resist systems (positive or negative) to electron beam radiation and their corresponding sensitivity to x-ray radiation. Figure 11 shows a plot of the 20 kV electron beam sensitivity in

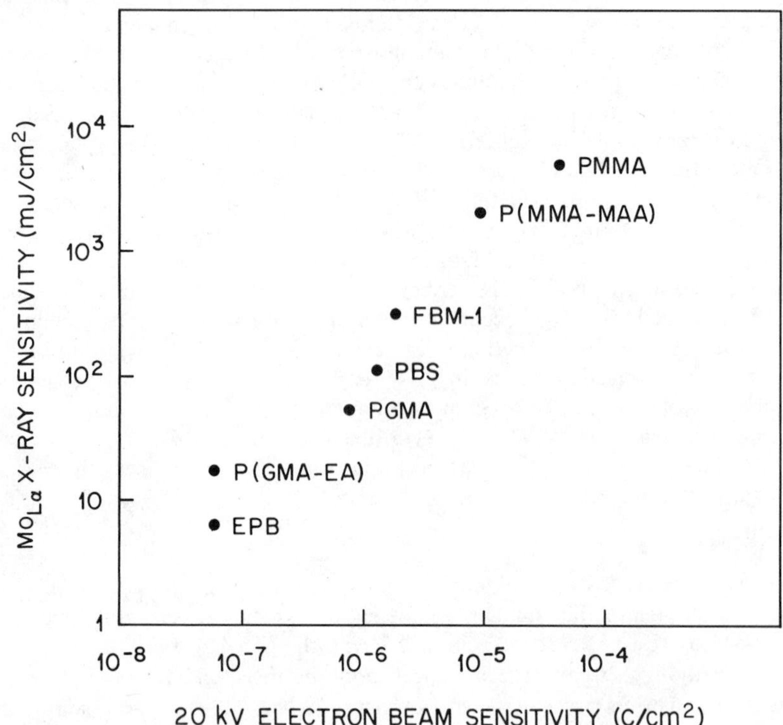

Figure 11. A plot of sensitivity to $Mo_{L\alpha}$ (5.4 Å) x-ray radiation and 20 kV electrons for several resists (see Tables 3-6 for explanation of symbols; EPB is epoxidized polybutadiene). (Reproduced with premission from Ref. 1.)

C/cm^2 plotted against the Mo soft x-ray sensitivity in mJ/cm^2 of incident x-ray flux. The unit slope of the plot suggests that the basic radiation chemistry is the same in both cases.

As a general rule, the sensitivity of conventional electron beam resists is not sufficient for economic throughput in an x-ray lithographic system. This is particularly true of positive electron resists such as PMMA, the most widely used x-ray resist for experimental purposes, whose sensitivity of >500 mJ/cm^2 is some 100 times too slow for practical application. Even PBS only shows a sensitivity of 94 mJ/cm^2 to $Pd_{L\alpha}$ x-rays. Consequently, the major research effort has concentrated on negative resists because of their higher inherent sensitivity.

One of the problems with x-ray exposure is the low absorption by organic materials which necessitates the use of highly sensitive resists. The best approach to improving sensitivity is to enhance the absorption of x-rays in the resist either by inclusion of heavy metal atoms or alternatively by matching the characteristic x-ray emission wavelength of the exposure source to the absorption edge of an element contained in the molecular structure of the resist. Taylor, (*98*) for example, incorporated chlorine atoms into alkyl acrylates to take advantage of the coincidence of the chlorine K-absorption edge with $Pd_{L\alpha}$ x-rays. Similarly, fluorine is suitable for exposure with $Al_{K\alpha}$ radiation. The sensitivities of some typical resists are listed in Table IX.

The highest x-ray resist sensitivity that has been reported is for a plasma-developable formulation based on poly(2,3 dichloropropyl acrylate) and a silicon containing acrylate — *bis*-acryloxybutyltetramethyldisiloxane, the processing details of which were recently discussed by Taylor et al. (*101*). The sensitivity is reported to be 1.5 mJ/cm^2, allowing ~20 sec exposures with a 4 kW Pd source. With optimized post-exposure treatment (fixing), O_2 reactive ion etching, and trilevel SiO_2 etching conditions (Section 10), Taylor et al. have been able to delineate 0.5 μm lines and spaces in 300 nm final resist thickness.

The sensitivity demands on the resist could be substantially alleviated if the power output of the source could be greatly enhanced. Whereas this presents a formidable problem for electron bombardment sources, it is one very real benefit of synchrotron radiation. Synchrotron radiation is emitted by high energy relativistic electrons which have been accelerated in a synchrotron or storage ring by a magnetic field. Sufficient power is available from such sources to expose even the relatively insensitive PMMA with economic throughput. As might be imagined though, storage rings, long the domain of the high-energy physicist, are relatively esoteric (and expensive) when it comes to an exposure tool for lithographic application although IBM has been seriously exploring such a possibility.

One feature which is still lacking in x-ray resists is a highly sensitive positive resist. It seems unlikely that a breakthrough will be achieved using traditional chemical approaches although recent work with plasma-developable systems is encouraging (see later section).

Ion Beam Resists

Ion-beam lithography is a technology which although still very much in its infancy, has many potential advantages, notably absence of proximity effects, that have spurred a great deal of interest in this field (*102*). As a result of the higher linear energy transfer of ions as they penetrate the resist, a lower incident exposure dose is needed to expose the resist than is required for electrons. As with electron lithography, there is a linear correspondence between ion-beam sensitivity and sensitivity to electrons (Figure 12), again suggesting that the radiation chemistry is similar. This is in line with accepted notions of ionizing-radiation chemistry where it is believed that chemical events stem from interaction with secondary electrons generated by interaction of the matrix with the primary photon or particle, be it an x-ray photon, high-energy electron, or ion. The primary difference lies in the distribution of energy within

Table IX. Lithographic Properties of Selected X-Ray Resists

RESIST	STRUCTURE OF REPEAT UNIT	TONE	MAJOR ABS. ELEMENTS	TARGET	SENSITIVITY mJ/cm^2	REF. #
POLY(METHYL METHACRYLATE) (PMMA)	$-CH_2-C(CH_3)(COOCH_3)-$	P	O	Aℓ	500 - 1,000	59
POLY(METHYL METHACRYLATE - CO - METHACRYLIC ACID) P(MMA-CO-MAA)	$-CH_2-C(CH_3)(COOCH_3)-CH_2-C(CH_3)(COOH)-$	P	O	Aℓ	150	59
POLY(BUTENE-1 SULFONE) (PBS)	$-CH_2-C(H)(CH_2CH_3)-SO_2-$	P	S	Pd	94	98

POLY(HEXAFLUOROBUTYL METHACRYLATE) (FBM)	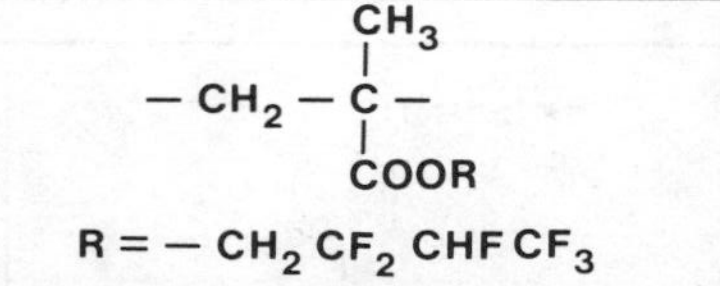 $-CH_2-C(CH_3)(COOR)-$ $R = -CH_2CF_2CHFCF_3$	P	O	Mo	52	49
POLY(GLYCIDYL METHACRYLATE - CO - ETHYL ACRYLATE) (COP)	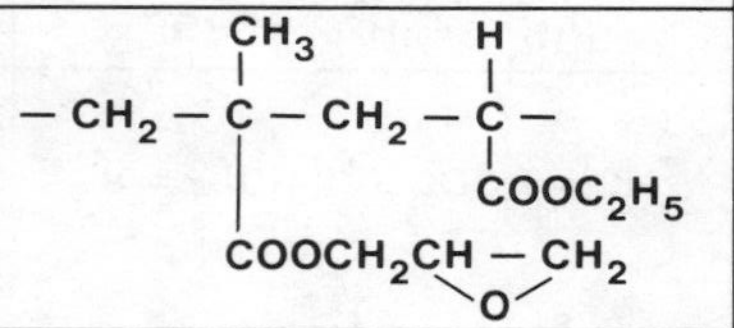 $-CH_2-C(CH_3)(COOCH_2CH-CH_2)-CH_2-C(H)(COOC_2H_5)-$ (epoxide: $CH-CH_2$ bridged by O)	N	O	Pd	175	98
POLY(2, 3 DICHLOROPROPYL ACRYLATE) (DCPA)	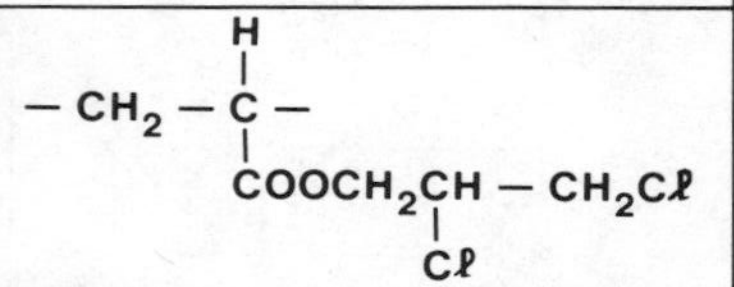 $-CH_2-C(H)(COOCH_2CH(C\ell)-CH_2C\ell)-$	N	Cℓ	Pd	7	98
POLY(2, 3 DICHLOROPROPYL ACRYLATE + N VINYL CARBAZOLE	DCPA 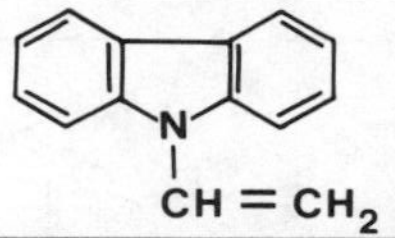 + N-vinylcarbazole (N–$CH=CH_2$)	N	Cℓ	Pd	4.5	98

Continued on next page.

Table IX (continued)

POLY(ALLYL METHACRYLATE - CO - 2 HYDROXYETHYL ACRYLATE) (EK-88 (KODAK))	$-CH_2-C(CH_3)(COOCH_2CH=CH_2)-CH_2-C(CH_3)(COOCH_2CH_2OH)-$	N	O	W	9	88
CHLOROMETHYLATED POLYSTYRENE (PARTIALLY CHLOROMETHYLATED) MW $1.8 \times 10^5 - 10^6$	$-CH_2-CH(C_6H_4CH_2Cl)-CH_2-CH(C_6H_5)-$	N	Cl	M_o	8-29	42
POLY (2. 3 DICHLORO-PROPYL ACRYLATE) + BISACRYLOXYBUTYL-TETRAMETHYL DISILOXANE	DCPA + $R-Si(CH_3)_2-O-Si(CH_3)_2-R$ $R = -(CH_2)_4\ COOCH = CH_2$	N	Cl	Pd	1.5	99
POLY(CHLOROMETHYL STYRENE) (MW - 3×10^5)	$-CH_2-CH(C_6H_4CH_2Cl)-$	N	Cl	Pd	25	100

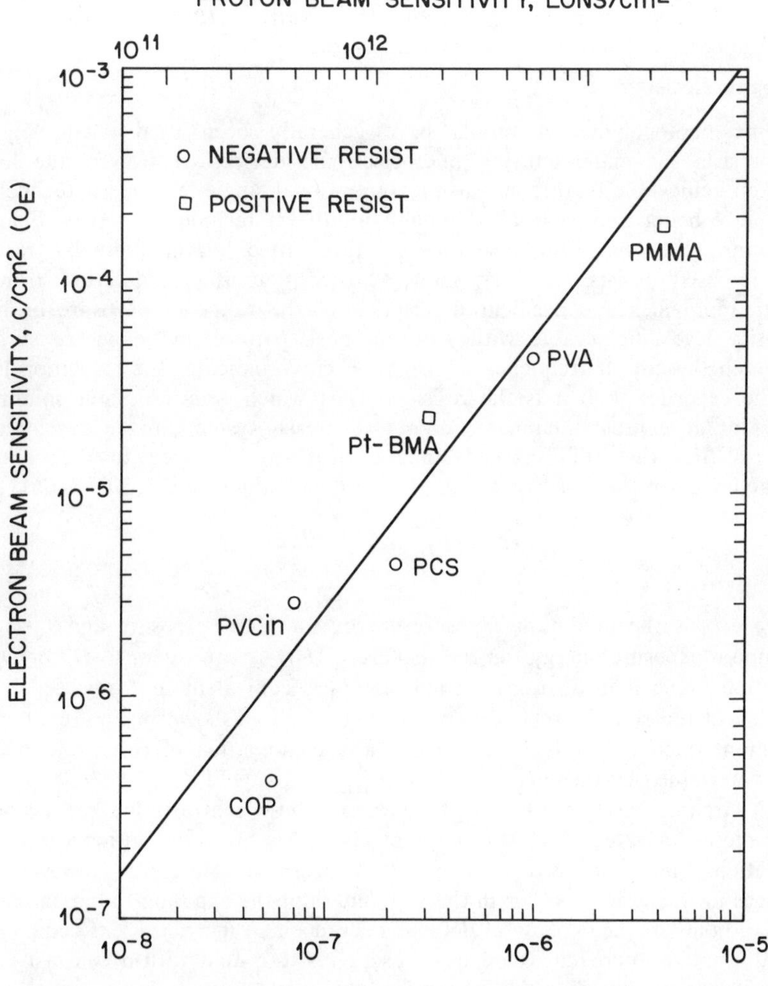

Figure 12. A plot of sensitivity to 100 keV protons and 20 kV electrons for several resists. (Reproduced with premission from Ref. 1.)

the absorbing matrix as expressed in the linear energy transfer (LET) value which is a measure of the absorption cross-section. In general, resists show the same sensitivity irrespective of the type of radiation when sensitivity is expressed in terms of energy *absorbed* per unit volume, i.e., assuming dose rate effects stemming from the different LET values are not operative. Indeed at such dose rates, reciprocity effects may well be observed in ion-beam lithography (*103*).

Statistical considerations necessitate a minimum dose for uniform exposure and, when coupled with energy transfer considerations, dictate the use of relatively insensitive resists resulting in considerable flexibility in resist choice. Resists such as polystyrene and its derivatives (*104*) or inorganic resists (*105*) appear well-suited to ion-beam exposure.

Inorganic Resists

Organic photopolymer materials have generally been used exclusively as photoresists for manufacturing microelectronic devices. However, the lower limit on achievable feature size using conventional single-level resist technology is rapidly being approached. Although multilevel techniques may well lower this limit to below 1 μm, resolution is still limited by the finite contrast of organic-based resists. It was shown earlier that the modulation transfer function provides a convenient description of the range of exposure or light intensity levels achievable with a given mask pattern and exposure system. Although the cut-off frequency of the MTF curve indicates the resolution limit for the exposure tool, it is the resist contrast which determines the minimum MTF for acceptable imaging. For a given resist system, image development requires that the MTF exceed some critical value related to the contrast characteristic of the resist material. This critical value is defined by (*106*)

$$(\mathrm{CMTF})_{\mathrm{resist}} = \frac{I_{100} - I_o}{I_{100} + I_o} \tag{5}$$

where I_{100} is the minimum exposure energy for 100% exposure and I_0 is the maximum exposure energy for zero exposure (I_{100} is equivalent to D_g^0 or D_c in Equation 1 and 2 in Section 2.5; likewise I_0 is equivalent to D_g^i or D_0). The contrast of the resist is related to the $(\mathrm{CMTF})_{\mathrm{resist}}$ as shown in Figure 13. The minimum sized printable feature using any combination of resist and lens is then determined by the condition $(\mathrm{MTF})_{\mathrm{optical}} \geqslant (\mathrm{CMTF})_{\mathrm{resist}}$.

Organic-based polymeric photoresists, whose contrasts are on the order of 2, require a value of $(\mathrm{CMTF})_{\mathrm{resist}}$ of about 0.6. In order to achieve higher resolution, one would need to increase NA, decrease λ or both. However such changes in the system NA or in the wavelength of the exposing radiation can be achieved only at the expense of defocus tolerance and increased lens complexity. An alternative approach would be to use a resist system with a contrast value substantially greater than that of conventional photoresists.

The practical feasibility of such a resist system was demonstrated in 1976 with the observation by Yoshikawa and his co-workers (*107-9*) that silver-doped amorphous chalcogenide glasses behave as resists under UV and electron irradiation. They demonstrated that these so-called inorganic resist systems possess extremely high contrast (γ = 6.8 for electron irradiation) and consequently very high resolution. The first inorganic resist systems were based on Ge-Se chalcogenide glasses and made use of the well-known phenomenon of light-induced silver migration known as photodoping. The technique, developed by Yoshikawa et al., involves deposition of films of $GeSe_x$ by vacuum evaporation or rf sputtering. A thin layer of Ag approximately 100 nm thick is then deposited onto the film surface by dipping the Ge-Se film into an aqueous solution of $AgNO_3$. Amorphous chalcogenide films dissolve easily in appropriate basic aqueous solutions. However, when the resist system is

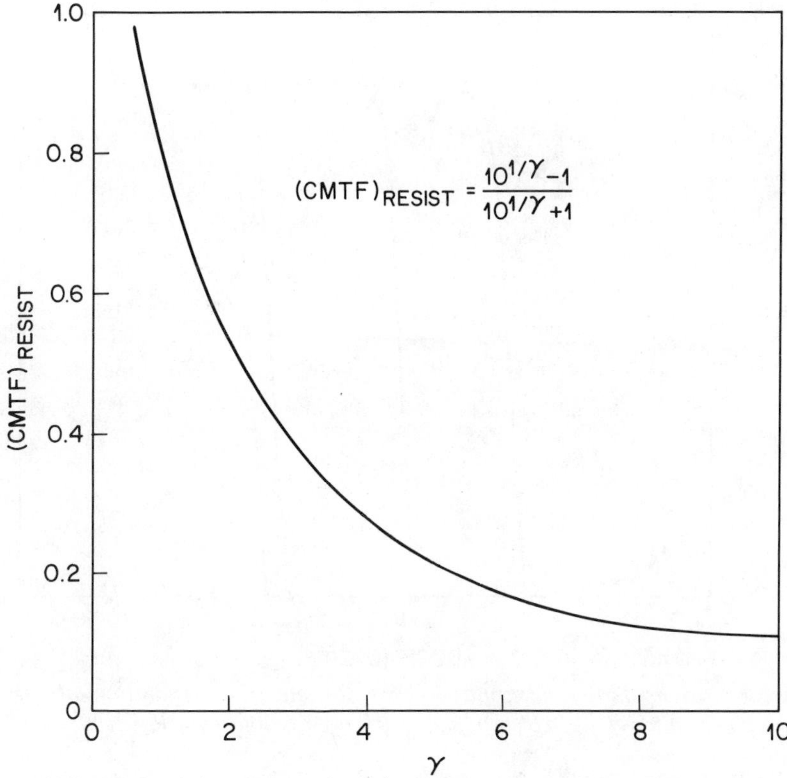

Figure 13. $(CMTF)_{resist}$ as a function of resist contrast. (Reproduced with permission from Ref. 106.)

exposed to UV light or electron irradiation, silver diffuses into the matrix rendering the material almost insoluble in alkaline solutions. This process thus gives rise to negative resist action. Figure 14 shows the sensitivity and contrast curves for this resist under electron irradiation. We should add here that sensitivity to electrons is too low for practical application although this is not necessarily true of ultraviolet and ion-beam exposure. The bulk of the research on these systems has been directed towards application as a photoresist.

Similar work on the Ge-Se system was reported by Tai et al. (*110-12*) using $GeSe_2$ as the chalcogenide glass and Ag_2Se as the photodoping layer. The latter was put down as a 10 nm thick film on vacuum-evaporated Ge_xSe_{1-x} films ~2-300 nm thick by dipping the latter into a basic 0.1M aqueous solution of $KAg(CN)_2$ (pH ~ 10). The proposed reaction for the formation of the Ag_2Se layer is:

$$4Ag(CN^-)_2 + GeSe_2 + 8OH^- \rightarrow 2Ag_2Se + GeO_3^{4-} + 8CN^- + 4H_2O$$

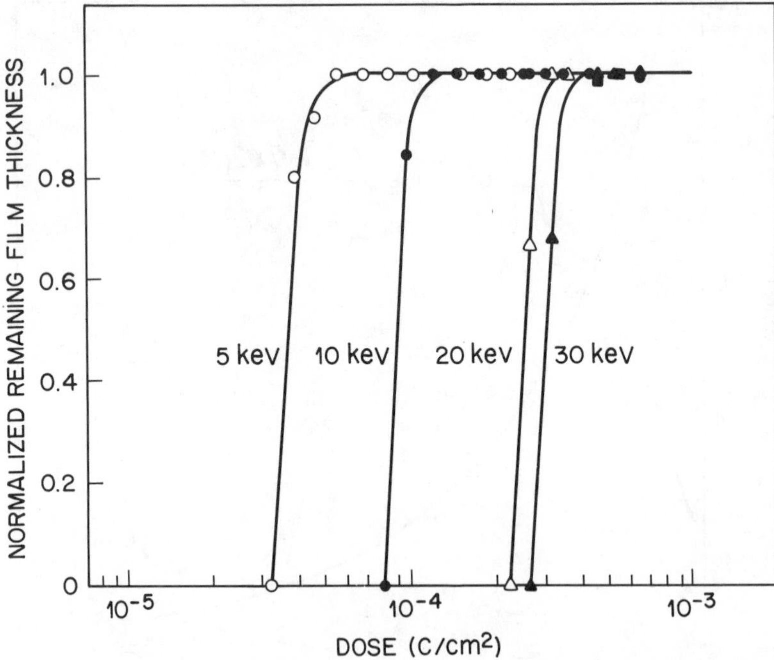

Figure 14. Sensitivity and contrast plot for electron irradiation of Ge-Se inorganic resist. (Reproduced with premission from Ref. 109.)

While the majority of work has been carried out on Ge-Se chalcogenide glasses, other glasses have been used, notably As_2S_3 (*113-15*). All schemes involve a source of silver on the surface of the chalcogenide glass either as pure evaporated Ag or as a silver compound such as Ag_2Se or AgCl. The compounds are applied either by evaporation or by chemical reaction in solution. During exposure by UV, e-beam, or x-ray radiation, Ag diffuses into the underlying chalcogenide glass forming Ag-doped regions that are essentially insoluble in alkaline solutions. A two-stage development procedure is used to define the pattern. First, Ag is removed from the unexposed areas by immersion in either an acid or a fixer solution. Next, the pattern is transferred into the glass by immersion in an aqueous solution containing bases such as $(CH_3)_2NH$, NaOH, or KOH. Dry development in CF_4 plasma is also possible.

These resist systems embody several desirable features. Firstly, there is no swelling during the development process which is more akin to surface etching. The resist is not attacked by the various acids commonly used in microelectronics processing, and its resistance to oxygen plasmas renders it suitable for bilevel applications. We shall see later that such multilevel schemes are important to the successful implementation of submicron lithography on substrates exhibiting topography and standing-wave-related problems. The sensitized resist exhibits a broad-band spectral response to all regions of the UV spectrum, is opaque to light of all wavelengths up to 450 nm, and is transparent to visible light. The broad spectral response is convenient since it provides a unique opportunity to have one resist system suitable for all present and future optical exposure systems. The opacity of the chalcogenide glass below 450 nm eliminates standing wave problems on reflective substrates, while its transparency above 450 nm allows precise alignment to previously defined patterns.

The mechanism of resist action of these inorganic resists is complex. Exposure to radiation of suitable wavelength causes silver to diffuse into the chalcogenide film resulting in a depletion of Ag in the exposed areas of the Ag_2Se layer. The silver diffuses into the upper layer of the Ge-Se film to a depth of ~10-20 nm for a typical exposure with UV-radiation. Yet, in spite of such a shallow doping profile, the etching process is highly anisotropic with the thin latent image acting as an etching mask for the undoped lower layer. Tai et al. (*117*) have shown that the photodoping rate (R) is given by

$$R = \frac{I_o e^{-\alpha T} T}{\tau} \tag{6}$$

where α is the absorption coefficient of Ag_2Se_3, T is the thickness of the Ag_2Se layer, I_o is the intensity of the incident radiation, and τ is a constant related to the sensitivity of photodoping. When $\alpha T < 1$ (very thin Ag_2Se layer) this expression reduces to

$$R = \frac{I_o T}{\tau} \tag{7}$$

A consequence of Equations 6 and 7 is that the effective light intensity increases as photodoping takes place because there is less Ag to absorb the light. This is referred to as photobleaching and serves to amplify the image contrast.

A more important effect occurs as a result of lateral diffusion of Ag within the Ag_2Se layer from the unexposed areas into the exposed areas resulting in enhanced photodoping at the edges. This non-linear resist response at the image edges is referred to as the *edge sharpening effect*. Not only does the lateral diffusion of Ag near the image edge enhance the effect of exposure on the exposed side of the edge but it also suppresses the effect of exposure on the unexposed side (see Figure 15) resulting from diffraction. The combination of lateral Ag diffusion, photobleaching, a thin imaging layer, and high contrast makes the Ag_2Se/Ge-Se system a very high-resolution imaging medium.

Tai et al. believe that the anisotropic nature of the liquid etching process stems from the structure of the amorphous Ge-Se film which has a phase-separated columnar structure containing two types of columns, 20-30 nm in diameter which are Ge-rich or Se-rich, respectively. This separation may relieve the internal stresses in the covalent atomic network and may also reflect the kinetics of film growth. Figure 16 shows the model proposed by Tai and co-workers to explain this phenomenon. It is a schematic cross section illustrating the phase-separated columnar microstructure of the Ge-Se film. An edge of the image is shown in the center of the figure. The exposed region is to the left where the top surface is protected by a thin photodoped (Ag) Ge-Se layer. Pattern transfer to the chalcogenide is accomplished in a two-component

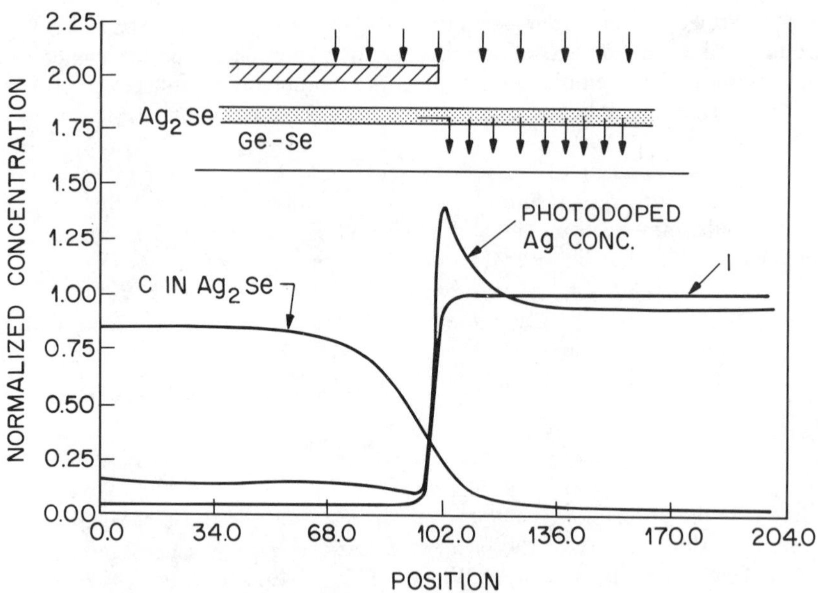

Figure 15. Mechanism of edge sharpening effect. The diagram shows the profile of light intensity (I), photodoped silver concentration in GeSe film, and the concentration of silver remaining in the Ag_2Se layer. (Reproduced with permission from Ref. 117.)

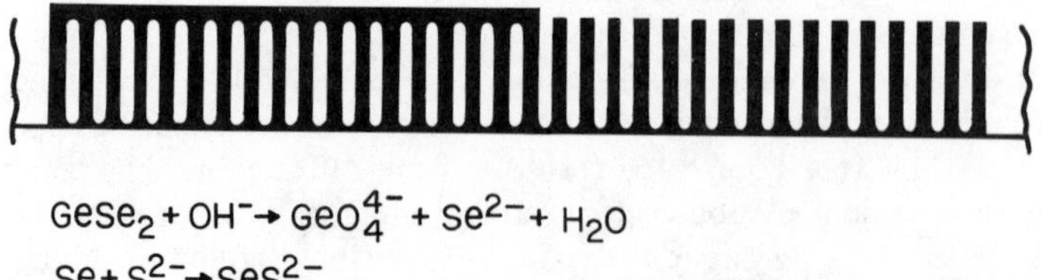

$GeSe_2 + OH^- \rightarrow GeO_4^{4-} + Se^{2-} + H_2O$
$Se + S^{2-} \rightarrow SeS^{2-}$

THE PHASE SEPARATED COLUMNAR STRUCTURE CAUSES ANISOTROPIC ETCHING, MAKING WET DEVELOPING, WITHOUT UNDERCUTTING POSSIBLE.

Figure 16. Schematic diagram illustrating phase-separated columnar-structure-induced anisotropic wet chemical etching. (Reproduced with permission from Ref. 117.)

etchant consisting of a mildly concentrated solution of NaOH and diluted Na_2S solution. The latter species is required for glasses containing more than the stoichiometric amount of Se. The general development reactions are:

$$GeSe_2 + 8\,NaOH \rightarrow Na_4\,GeO_4 + 2\,Na_2Se + 4\,H_2O$$

$$Se + Na_2S \rightarrow Na_2\,[SeS]$$

This latter reaction is slow due to the low concentration of Na_2S. When immersed in the etchant, the Ge-rich columns will be rapidly etched in the unexposed regions since they are not protected by the photodoped layer. As etching proceeds, the Se rich columns are exposed and can be etched laterally. However etching of the region underneath the photodoped layer proceeds quite differently. The relatively insoluble Se-rich columns must be etched one-by-one starting at the side wall of the pattern edge. Therefore, the lateral etching (undercutting) should be very slow compared with the vertical etching rate from the top down and should give rise to anisotropic etching. Tai et al. presented corroborative evidence of this hypothesis using a PMMA resist pattern as an etching mask on top of a Ge-Se film. After etching by the two component developer, a nearly vertical-walled 0.5 μm Ge-Se film pattern was obtained, indicating that anisotropic-etching is inherent to the as-deposited Ge-Se films rather than related to photodoping.

Tai (*117*) has summarized the advantages of these Ag_2Se/Ge-Se resist systems as follows:

1. Edge sharpening effect to compensate for diffraction effects.
2. Photobleaching effect to enhance image contrast
3. Anisotropic wet chemical etching.
4. High absorbance of UV radiation thereby minimizing reflection from the substrate.
5. Resistance to oxygen plasmas making bilevel schemes practical.
6. Transparent to visible wavelengths facilitating optical alignment.
7. Broad band spectral response.

The quality of the images generated in (Ag_2Se) Ge-Se resists is truly impressive. Some representative examples of high-resolution patterns are shown in Figures 17 and 18. But, in spite of the demonstrated success and potential attraction of inorganic resist systems, they have not been adopted by the industry for several reasons. Firstly, conventional photoresists (with which the industry is totally familiar) possess more than sufficient resolution to meet today's device needs and no doubt will continue to do so for several years. Inorganic resists require new deposition techniques with attendant capital equipment costs etc. Also information concerning potential defects (type and concentration) in these vacuum-deposited films is scarse.

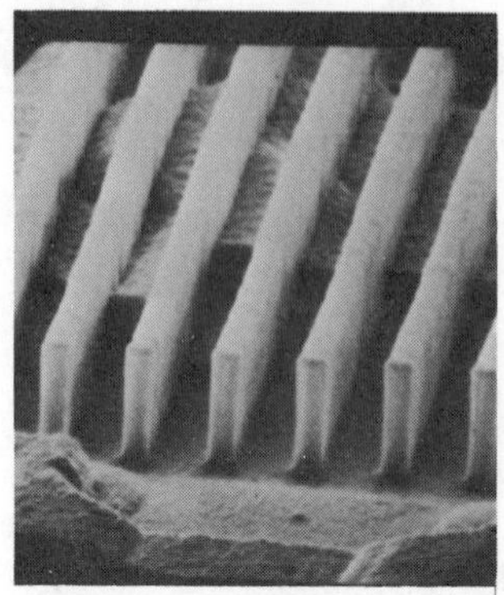

Figure 17. SEM photo miocrographs of GeSe/HPR206 bilevel patterns generated by contact printing over the polysilicon level of a 16K MOS RAM wafer. (Reproduced with permission from Ref. 111).

Figure 18. SEM photomicrographs of $Ge_{0.1}/Se_{0.9}$/HPR 206 bilevel patterns generated by projection printing over the polysilicon level of a 16K MOS RPM wafer. (Reproduced with permission from Ref. 111.)

Multilevel Schemes

Irrespective of exposure technology, a common problem with resist pattern generation is the simultaneous achievement of good linewidth control, high resolution, and good step coverage. The latter requirement stems from the fact that topography generated on the wafer can be as high as 0.8 μm and must be adequately covered by the resist during subsequent pattern definition. Often, such requirements appear to be mutually exclusive. Good step coverage requires thick resist; high resolution, however, is more easily obtained in thin resist. This dichotomy applies to all resists, both positive and negative.

With any resist, the ideal conditions for obtaining high resolution and good linewidth control are a flat surface and a thin resist (<400 nm). The flat surface means that the resist will have very little variation in thickness and, as a result, there will be little variation in resist linewidth. However linewidth variations stemming from resist thickness variations do occur when lines traverse a step as a result of exposure and development effects. Indeed, such effects are the leading cause of poor linewidth control. The realization that improved resolution can be obtained with thin resists and that thick resists are essential to device fabrication has led to the introduction of multi-layer resist (MLR) systems.

As shown in Figure 19, the surface of the wafer is first planarized by spinning down a thick organic layer which conforms to the topography of the wafer and provides a planar surface upon which, after first prebaking the bottom layer, a thin imaging layer can be coated uniformly. In some applications, a thin intermediate masking layer of some suitable material, e.g., SiO_2, is deposited on the thick first resist layer prior to deposition of the thin imaging layer. We should point out that the thick film does not produce a perfectly planar surface as depicted, but in reality serves to smooth out the discontinuities.

High resolution patterns are generated in the thin top layer followed by pattern transfer into the bottom layer. This is accomplished by using the delineated imaging layer as a blanket exposure mask or as an etching mask to pattern the planarizing layer. Since this mask is always in intimate contact with the planarizing layer, it has been called a portable conformable mask or PCM (*118*). When a third layer is used between the imaging and planarizing layer, the process is referred to as a tri-level scheme in which the imaging and masking functions of the top layer are done with two different materials.

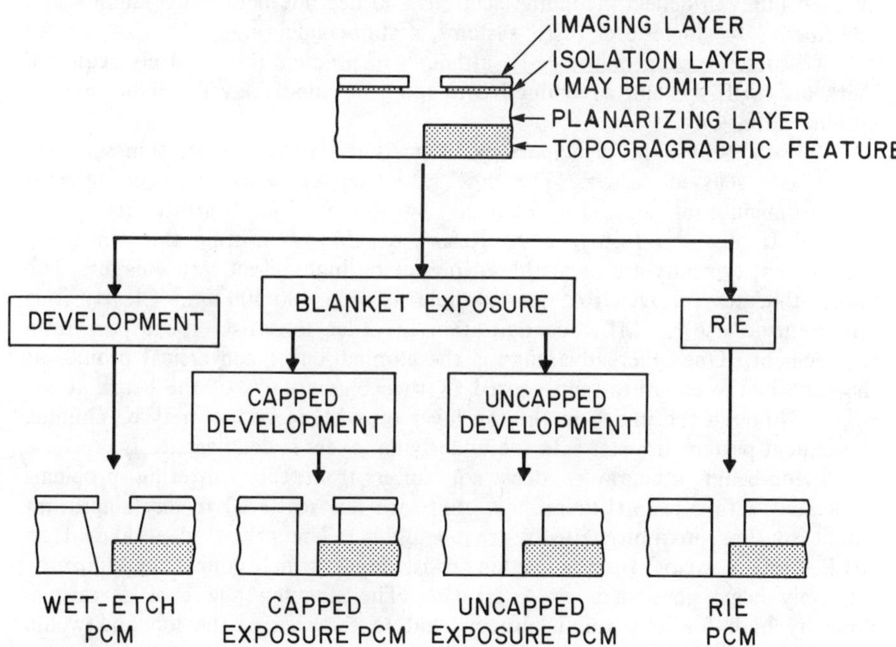

Figure 19. Schematic representations of multi-layer resist processes. (Reproduced with permission from Ref. 118.)

There are a variety of advantages to multilevel processing (*118*) apart from the obvious one discussed earlier, viz., planarization of topography. Many of these advantages are specific to the particular exposure technology, but it would appear that all exposure technologies will benefit be they optical, e-beam, x-ray, or ion-beam. In general, the use of thin resist relieves somewhat the stringent contrast/resolution requirements on the resist since resolution and aspect ratio requirements are separated in a MLR system. For optical lithography, interference effects caused by reflection from the substrate and non-uniformity of reflectivity can be eliminated by using a bottom layer that either is inherently sufficiently absorbing to prevent light reaching the substrate or, alternatively, that contains an absorbing dye (*119*). Depth of focus limitations are also minimized.

For electron-beam lithography, the effects of forward scattering are reduced because the resist is thin, as are the effects of backscattering. The backscattering coefficient is proportional to the atomic number (Z) of the substrate and will be less significant for the thick organic layer which is typically an organic photoresist consisting of C(Z=6), H(Z=1) and O(Z=8) than for Si(Z=14); Al(Z=13) or gold(Z=79). Consequently there will be a reduction of proximity effects with attendant improvement in resolution and linewidth control. One possible problem associated with multilayer resists in electron lithography is the "charging effect" which refers to the accumulation of charge resulting from electrons incident on a thick insulating resist layer. The charged film can deflect incoming electrons and degrade both image quality and placement. In single-level resist systems, a thin-conducting layer of Al or Au may be coated on top of the resist although in practice this is rarely required. With an MLR system, a conductive imaging or middle layer can be used to eliminate charging.

With x-rays, the principal advantage is the ability to use thin sensitive resists as discussed earlier. The dose rate from conventional electron-beam bombardment x-ray sources is relatively low so that highly sensitive resists are required (typically 1-10 mJ/cm^2). Resists capable of meeting this sensitivity requirement typically are incapable of producing high aspect ratio images. The usable thickness of DCOPA, for example, is only 200-300 nm. Clearly it is important to use an MLR configuration in order to satisfy the aspect ratio requirement. One other advantage is the elimination of nonvertical profiles at the edge of the exposure field caused by the oblique entry of the beam at the edge. Although the image in the top layer of a MLR system may be oblique, subsequent pattern transfer into the underlying layers is done vertically.

Ion-beam lithography does not suffer from the scattering problems associated with e-beam lithography. Ions are not scattered to any significant extent so that proximity effects are negligible. The prime advantage of an MLR system stems from the thin resist layer which allows exposure by relatively heavy ions such as Ar or Ga. The stopping power of a resist is typically high for ions and is proportional to the size of the ion and would therefore preclude exposure by the heavy ions mentioned above for thicknesses on the order of a micron. The bottom layer also serves as a barrier against ion bombardment damage to the substrate.

We have indicated that imaging thick single-layer resists presents a formidable problem for fabrication of small feature sizes and that multilayer techniques present a solution to this difficulty. A simple remedial approach is to modify the thick single layer resist in some way so as to simulate multilayer effects and thereby produce better imaging characteristics. Hatzakis et al. (*120*) treated thick layers of AZ1350J by immersion in an aromatic solvent such as chlorobenzene or toluene for a fixed time before or after exposure, but before development. The solvent penetrates ~300 nm into the resist layer reducing the solubility in the penetrated region, possibly by removing residual solvent and low molecular weight resin from this region (*121*). Since the treated layer develops more slowly than the untreated layer, the developed resist profiles have the appearance of two distinct layers. The process appears to be suitable only for lift-off processing and, in general, these systems are inferior to multilayer resist systems with respect to resolution, aspect ratio, and linewidth control.

In principle, bilevel structures are inherently simpler although each exposure technology imposes its own constraints. A common concern for all technologies is the solubility and compatibility of the two resist components. The developer and spinning solvents for the individual layers must be mutually exclusive. Intermixing of the two levels will occur during spinning if the bottom layer is slightly soluble in the spinning solvent used to apply the top layer. In a system where the pattern is transferred via flood exposure and development, an interfacial region may attenuate light reaching the lower layer and interfere with the exposure. Similarly, if the pattern is transferred by wet etching techniques, even a thin interface may render development difficult. Factors affecting the choice of developer include whether the upper masking layer should be retained or removed or whether or not an undercut profile is desirable or acceptable. The glass transition temperatures of the two polymers are important. If a baking step is required after deposition of the top resist, the temperature should preferentially be kept below the T_g's of the two resists in order to minimize interfacial mixing.

Perhaps the simplest bilevel scheme is the wet etchant PCM described by Hatzakis (*122-23*) and shown schematically in Figure 20. It involves forming a composite layer of two resists with vastly different polarities spun from mutually exclusive solvents. For this purpose Hatzakis used PMMA spun from chlorobenzene as the thick bottom layer and a copolymer of MMA and methacrylic acid spun from ethyl Cellosolve acetate as the imaging layer. The electron-beam-exposed pattern is first developed in the top layer using a developer (mixture of ethoxyethanol and 2-propanol) that stops at the interface with the bottom layer. Subsequently a developer that dissolves only the bottom layer (chlorobenzene or toluene) is used and development is continued until sufficient undercutting (as required for lift-off processing) in the composite structure has been obtained. Since the developer for the bottom layer does not dissolve the top resist layer, no further development of the top image takes place. The metal linewidth after evaporation and lift-off is determined by the opening in the top thin resist layer; therefore high resolution and pattern fidelity

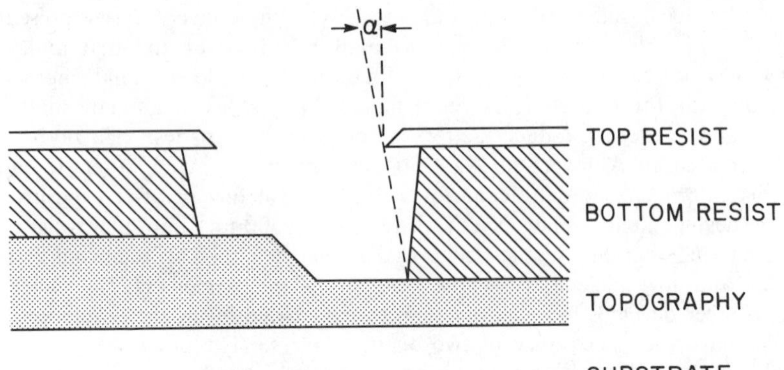

Figure 20. Schematic of the bilevel scheme proposed by Hatzakis. Also indicated is the maximum evaporation angle (α) for lift-off metallization. (Reproduced with permission from Ref. 123.)

are assured by the process. Development of the bottom layer has to be carried out to the point where the angle α in Figure 20 is larger than the evaporation angle.

In the example just cited, the primary lithographic tool was used to expose both resist layers so that only one exposure and two development steps are required. When the sensitivity of the bottom layer is much lower than that of the top, development of the bottom layer can be nearly isotropic and will limit the resolution. This problem can be eliminated by using a secondary exposure source to re-expose the bottom layer. This leads us to the next level of complexity of multilayer systems, *viz.*, that in which the delineated imaging layer is used as a mask for exposure and subsequent development of the thick planarizing layer. The system most commonly associated with bilevel resist processing is that developed by Lin (*124-25*) in which PMMA is used as the planarizing layer and conventional positive photoresists such as HPR-204 or AZ1350J are used as the top imaging and masking layers. While HPR and AZ resists are sensitive in the 400 nm region, they are opaque to 210-250 nm radiation and hence serve as a PCM for flood exposure of PMMA in the deep-UV region.

The resolution limit of this system is dictated by the resolution capability of the thin novolac-quinonediazide resist which can be patterned by UV light using conventional photolithographic techniques ($\lambda > 365$ nm) or by electrons. Since PMMA is capable of very high resolution, excellent pattern transfer with very high aspect ratios and straight wall profiles can be achieved using the developed imaging layer as a contact mask. The bottom planarizing layer can be developed with the AZ cap retained or removed, resulting in a capped or uncapped image, respectively (see Figure 19). Profiles of 0.8 μm lines and spaces in this bi-level resist on a silicon substrate are depicted in Figure 21. With HPR-204 as the imaging layer, difficulties can arise due to intermixing of the two layers since PMMA is slightly soluble in the HPR solvent. Even when the solvent for the resist being spun on has little effect on the underlying layer,

Figure 21. SEM photo showing 0.85 μm lines spaced 2.4 μm apart generated by deep-UV exposure through a 0.2 μm thick AZ1350J conformable mask. (Reproduced by permission from Ref. 125.)

interfacial layers can still form during prebaking. The main cause of intermixing in this case is thermal diffusion and the problem can be minimized by control of processing conditions such as keeping the prebake temperature for the top layer below the T_g's of both resist systems. Other methods for removing this interface include longer exposure and development times, use of a stronger developer, and/or pretreatment with an oxygen plasma prior to PMMA exposure. The key to the adoption of this type of bilevel process by Hewlett Packard for fabrication of their NMOS III 32-bit CPU chip as well as all other chips that go into their model 9000 computer lies in their choice of Kodak 809 positive photoresist for the imaging layer (*118b*). The solvent system for this resist causes little or no interfacial mixing.

Difficulties associated with interfacial mixing can also be eliminated using an inorganic resist as the imaging layer. Since these resists are deposited by evaporation, no interfacial mixing can occur. The Ge-Se inorganic resist system is opaque below 300 nm and can act as a mask for PMMA flood exposures. Recall that these resists are capable of extremely high resolution due to edge sharpening and photobleaching effects, enabling submicron lines and spaces to be replicated into the PMMA. Moreover since the Ge-Se resist system is highly resistant to oxygen plasmas, pattern transfer can be conveniently accomplished by direct reactive ion-beam etching, thereby reducing the number of steps involved in pattern transfer. When using reactive-ion etching for pattern transfer, the thick planarizing layer can be any conveniently processable photopassive material. Tai et al. (*111*) have demonstrated half micron resolution in isolated lines, spaces, and gratings in 2.5 μm hard-baked HPR photoresist using their Ge-Se inorganic resist as the top layer.

The inorganic resist system is unique with respect to optical sensitivity coupled with O_2 plasma resistance. Most organic polymers do not possess sufficient plasma etching resistance to withstand oxygen plasma etching of 2.5 μm of hard-baked photoresist as might be required in a typical bilevel application. This deficiency has spurred interest in metal-containing resists. Hatzakis and co-workers (*94*) have obtained impressive results by combining polysiloxanes with a conventional planarizing layer to form a two-layer reactive ion etch PCM with high sensitivity to e-beam and deep UV. Electron resists based on poly(vinylmethylsiloxane), poly(dimethylsiloxane) and, for ultraviolet sensitivity, poly(dimethyldiphenylvinylsiloxanes), have been shown to have very high resolution capabilities (~0.5 μm) and high contrast. The sensitivity of poly(vinylmethylsiloxane) to electron beam exposure is 2 $\mu C/cm^2$ with $\gamma = 2$. Unlike PVMS, the phenyl derivatives of polysiloxane have an absorption at 250-280 nm and are sensitive to the UV at 253.7 nm. They show sensitivity on the order of 200 mJ/cm^2 with contrast as high as 4. The ratio of the etch rate of AZ1350J to polysiloxane in oxygen plasma is about 50:1 enabling images to be transferred into thick planarizing layers of the former using the polysiloxane as a etch mask. High resolution in these systems is attributed to their high contrast coupled with the reported lack of swelling in the developer solvent. Only thin imaging layers need be used because of the excellent plasma etching resistance of these materials, a fact which also maximizes resolution.

This work on organometallic resists was extended by MacDonald et al. (*126*) to include poly(trimethylstannylstyrene) and poly(trimethylsilylstyrene) together with copolymers of these materials with p-chlorostyrene. The sensitivity of the stannyl derivative to electrons is reported to be 0.5 $\mu C/cm^2$, and the copolymer with p-chlorostyrene could be patterned in the deep UV at a dose of 10 mJ/cm^2 at 254 nm. The materials show excellent oxygen RIE resistance due to the formation of involatile oxides of the metals on the surface of the pattern.

When the etching resistance of the imaging layer is insufficient to allow pattern transfer to the thick planarizing layer, it is necessary to use an intermediate layer to act as the mask for pattern transfer. Such systems are referred to as trilevel systems whose versatility and process compatibility were first demonstrated by Moran and Maydan (*127-28*). These workers used a thin (0.1 μm) layer of plasma-deposited SiO_2 in between the planarizing and imaging layers. E-beam, x-ray, refractive and reflective optical projection techniques were used to delineate the top imaging layer in a variety of resists. In this process, the pattern is first transferred to the intermediate layer which is then used as the mask for subsequent RIE pattern transfer to the thick planarizing layer. Etching of the intermediate layer can be accomplished by wet or dry techniques. Since the SiO_2 layer is very thin, resists with only moderate etch resistance can function satisfactorily; e.g., even PBS has been used as a plasma etching mask for this purpose.

A variety of intermediate layers have been used. In addition to SiO_2, silicon, Si_3N_4, Ge and Ti have been used. In spite of the attractiveness of trilevel processing, *viz.*, that it allows the imaging layer to be chosen purely on the value of its lithographic response rather than pattern transfer capabilities, there are a number of disadvantages relative to the simpler bilevel approach. Extra processing steps are required, the intermediate layer must be defect free, and it must be etched in a medium that does not attack any of the resist layers. Progress is being made on these fronts. For example, SiO_2 can now be conveniently deposited in a spin-on process (*129*) which eliminates the scattering problems associated with plasma deposition. Developments such as this may well make tri-level resist processes more attractive for production applications.

Dry-Developable Resists

Dry-developable resists offer many advantages related to resolution and defect density as well as possible ecological advantages. Since the development process is totally dry, distortion due to resist swelling is obviated suggesting that high resolution should be attainable, even with negative resists. Further, the possibility of all-dry processing is attractive from the point of view of defects since all pattern transfer steps could be done in a vacuum environment, thereby reducing exposure to contamination normally found in processing lines.

The simplest dry-developing scheme is that provided by the self-developing resists. The first example of this was reported by Bowden et al. (*130*) who observed that certain poly(olefin sulfones) underwent scission with concomitant depolymerization (unzipping) from the fractured chain ends, resulting in spontaneous relief image formation. Ito et al. (*131-33*) later demonstrated a similar phenomenon in polyphthalaldehydes sensitized with certain cationic photoinitiators. In the latter case, irradiation generates an acidic species from the sensitizer which catalyzes cleavage of the polyacetal main chain which subsequently depolymerizes to monomeric species. Similar systems have been reported by Hiroaka (*134*) using polymethacrylonitrile, and Hattori and co-workers (*5,135*) on plasma-deposited copolymers of methyl methacrylate and tetramethyltin. These latter two systems require elevated temperatures either during exposure or during post-exposure baking to develop the image. Hattori et al. have coined the term "vacuum lithography" to describe their system since all processing from deposition to pattern transfer is performed by dry processes in a vacuum.

An alternative approach to dry development is to carry out the development procedure itself by a dry technique such as plasma etching. This requires that exposure convert the matrix into a form that enables exposed and unexposed areas to be differentiated on the basis of plasma resistance. The general principle underlying design of plasma-developable resists has been described by Taylor and co-workers (*98,136*). Their scheme, which is shown schematically in Figure 22, involves a resist system comprised of a host polymer (*P*) and a lower concentration of monomer guest (*M*) which is polymerized (locked) by the absorbed radiation. A degree of grafting to the matrix polymer no doubt also occurs. Treatment by heat (fixing) under vacuum removes the unreacted monomer, producing a negative tone relief image whose thickness in dependent on materials and dose. In the third step, plasma treatment removes material from the imaged and non-imaged areas. However because of the different chemical nature of the two regions, the rates of removal are substantially different. In the case of Plasma A which is typically an oxygen

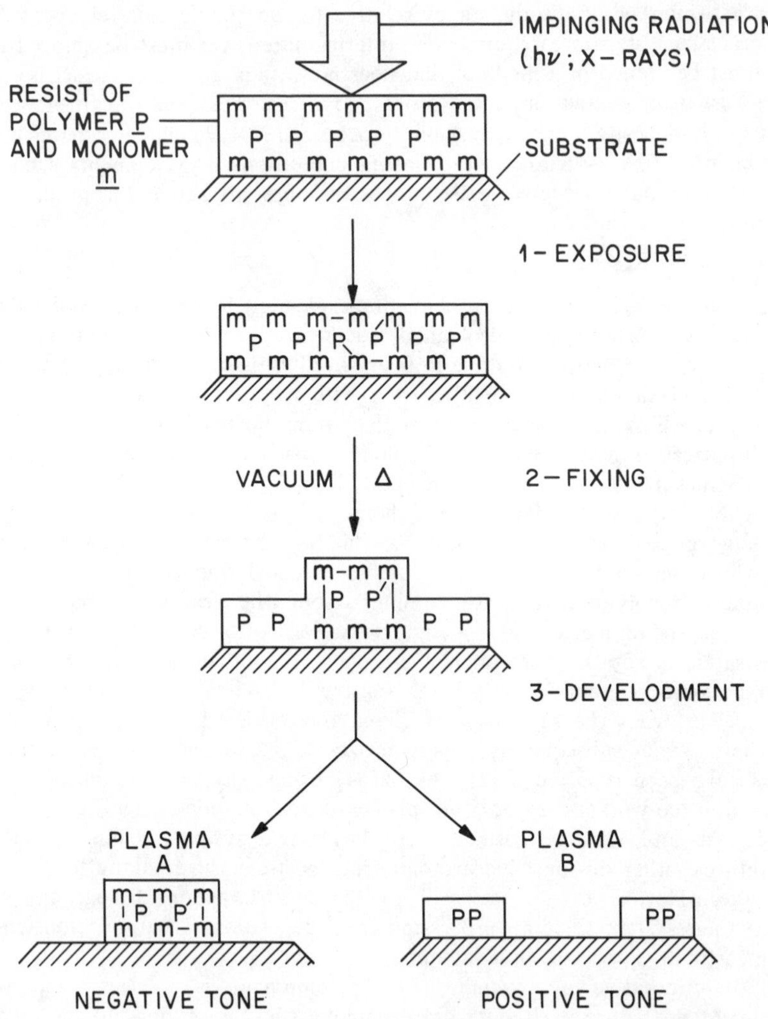

Figure 22. Schematic plasma-development process showing exposure, fixing, and both negative and positive tone plasma development. (Reproduced by permission from Ref. 98.)

plasma, the imaged areas are removed at a lower rate than the non-imaged regions, thus enhancing the initial relief image obtained after fixing and resulting finally in a negative tone pattern. Starting with a 1 μm thick film of poly(2,3-dichloro-1-propyl acrylate) (host) containing 20% N-vinylcarbazole (guest), Taylor et al. have obtained resolution of 0.3 μm in 300 nm residual film thickness using x-ray lithography. Resolution on the order of 1 μm was obtained for UV exposure of the same resist sensitized with quinone sensitizers. Under different conditions (plasma B) the imaged area can be removed at a significantly faster rate than the non-imaged area, thus inverting the tone obtained at the fixing stage and providing a positive-tone pattern upon complete development.

Similar approaches to design of dry-developable photoresists have been described by workers at Motorola (*137*). One disadvantage of their systems appears to be low resist contrast, resulting in resolution less than that of conventional positive resists. One possible way of improving the contrast may involve changing the chemistry from amplified locking by polymerization to discreet locking in which only one or two locking reactions result from a single absorption event. Such an approach would presumably have a detrimental effect on sensitivity.

The major research thrust is aimed at increasing the residual or final film thickness after development while maintaining sensitivity at 10 mJ/cm^2. Taylor and co-workers (*99*) extended their work to include organometallic monomers, particularly those based on silicon. Such monomers, when locked into suitable hosts by incident radiation, are converted to CO_2, H_2O, and metal oxide by oxygen plasma etching. Since the oxides are nonvolatile, a protective shield or coating of metallic oxide is left which prevents etching of the organic material beneath it. Taylor et al. have investigated a variety of silicon containing acrylic monomers. A formulation consisting of 90 parts poly(2,3-dichloro-1-propyl acrylate) and 10 parts bis-acryloxybutyltetramethyldisiloxane has been reported to be extremely sensitive to x-ray exposure, requiring 1.5 mJ/cm^2 for $Pd_{L\alpha}$ x-rays and has resolution better than 0.5 μm with final resist thickness of 300 nm (starting from 1 μm).

A different approach has been used by Tsuda, Nakane and their collaborators (*138,139*) who showed that mixtures of poly(methylisopropenyl ketone) (PMIPK) containing a bisazide sensitizer such as 4,4′-diazodiphenyl sulfide or 4-methyl-2,6-di(4-azidobenzylidene) cyclohexanone function as negative dry-developable resists. It is claimed that the bisazide photodecomposition products, *viz.*, amines and azo compounds, are effective quenchers of the energy of excitation. Thus during subsequent plasma etching, the excited state of PMIPK is effectively quenched in the previously irradiated areas thereby retarding the plasma etching rate relative to the unexposed areas. A factor of 5 difference in rate is claimed. It does not appear that the quenching species are formed directly by irradiation, as would also occur during plasma treatment, but are formed during prior post-exposure annealing at 140°C during which the matrix polymer is also cross-linked. A residual thickness of 0.8 μm compared to 1.0 μm initial thickness has been claimed.

The most recent addition to the list of dry-developable systems involves vapor phase grafting of styrene onto silicone resins in a multilevel approach (*140*) The silicone resin is spun down on a thick planarizing layer, irradiated with electrons, and then exposed to styrene vapor. The reactive sites in the irradiated regions of the silicone film initiate graft copolymerization forming a grafted pattern on the surface which is transferred into the silicone resin by RIE in a CF_4 plasma. The silicone then acts as a mask for O_2-RIE into the thick planarizing layer. In order to obtain sufficient grafting, relatively high doses must be used.

The challenges in this area are many. There is no doubt that differential etching rates need to be improved in order to enhance the thickness of the residual image, although this requirement may not be as stringent if multi-layer processing is contemplated. Very few processing details relating to development uniformity, yield, defect densities, etc. have been released. Very little work has appeared on positive-tone patterning although preliminary work by Taylor (*148*) on some new mechanistic approaches to positive-tone generation suggests very high differential etching rates might be possible (see next section). Most of the approaches described involve removal of a volatile component which would present difficulties in the high vacuum environment of an electron beam exposure system. Either the second component needs to be involatile or alternatively, following the suggestion of Lai, (*141*) one could use a barrier polymer in a trilayer resist system to prevent sublimination and vaporization of the monomer. The barrier polymer must of course be removed following exposure prior to baking (fixing) and plasma development. Lai used poly(vinyl alcohol) as a typical barrier polymer to prevent loss of N-vinylcarbazole from a poly(trichloroethyl methacrylate) matrix and obtained 0.3 μm lines in 0.25 μm of resist.

New Mechanisms

So far we have concentrated on what we might call traditional approaches to resist design. The energies involved in electrons, x-rays and ion beams are sufficient to non-selectively break every bond. Thus resist research has traditionally concentrated on incorporating weak links into polymer chains to promote scission or radiation sensitive cross-linkable groups either into the main chain or on the side chain to promote cross-linking. In recent years resist research has moved into a phase which has seen the emergence of several new mechanisms which may be applicable to several exposure technologies.

The recent development of several new classes of practical photoinitiators for cationic polymerization has now made it possible to utilize this chemistry in a number of ways to produce highly sensitive photoresists (*142-144*). The facile synthesis of onium salts (I-III)

$$\underset{\text{I}}{Ar_2\, I^+\, X^-} \qquad \underset{\text{II}}{Ar_3\, S^+\, X^-} \qquad \underset{\text{III}}{Ar\text{–}\overset{\overset{\displaystyle O}{||}}{C}\text{–}CH_2\text{–}S^+R_2X^-}$$

together with their ease of structural modification to manipulate their spectral absorption characteristics makes it possible to use this chemistry in a number of ways to produce highly sensitive photoresists. Photolysis of these materials produces strong Bronsted acids which can initiate cationic polymerization of a variety of polymers or oligomers containing oxirane or thiirane groups or vinyl ethers to produce negative images. Alternatively, they may be used to catalyze the degradation of sensitive polymers containing acid labile groups. This latter approach was used by Ito et al. (*133*) to catalyze the transformation of poly(p-t-butoxycarbonyloxystrene) (PBOCST) which is, in effect, t-butoxycarbonyl protected poly(vinylphenol). In the presence of sensitizers such as diphenyliodonium hexafluoroarsenate, irradiation of PBOCST produces an acid species which catalyzes the decomposition of the carbonate to liberate CO_2 and isobutylene while, at the same time, freeing the phenolic hydroxyl group to produce poly(p-hydroxystyrene) in the exposed areas of the resist film.

$$\text{SENSITIZER (S)} \xrightarrow{h\nu} \text{ACID (A)}$$

$$-CH_2-\underset{\underset{\text{C}_6\text{H}_4-\text{OCO-}t\text{-Bu}\,(\text{C=O})}{|}}{\overset{H}{\overset{|}{C}}}- \xrightarrow{A} -CH_2-\underset{\underset{\text{C}_6\text{H}_4-\text{OH}}{|}}{\overset{H}{\overset{|}{C}}}-$$

Post-exposure baking of the film is required for a few seconds at 100°C. This process is referred to as *catalytic deprotection.* The deprotection of the poly(vinylphenol) results in a large change in polarity and one would therefore expect similar changes in solubility characteristics. Ito et al. were able to generate either positive or negative tone images, depending on whether polar solvents such as aqueous basic developers or nonpolar solvents such as a mixture of dichloromethane and hexane were used. Development of the exposed resist in nonpolar solvents selectively removes unreacted resist, giving rise to a negative image, whereas development in aqueous base selectively removes the poly(vinylphenol) generated in the acidolysis reaction to give a positive image. Figure 23 shows patterns in both tones obtained by changing the solvent polarity. It is not yet known what effect, if any, the sensitizers may have an device performance or defect density.

Ito has also extended this type of photochemistry to the electron-beam-induced catalytic acidolysis of acid-labile main chain acetal linkages in polyphthaldehyde. These polymers, like the poly(2-methylpentene-1-sulfone) (PMPS) sensitizer in NPR resist described earlier have ceiling temperatures on the order of -40°C. As normally used, the polyaldehydes are end-capped by acylation or alkylation and are thus quite stable. The main chain bonds are very sensitive to acid-catalyzed cleavage which in turn allows the whole chain to revert to monomer in an unzipping sequence similar to that occuring in irradiated PMPS. Irradiation of polyphthaldehyde containing 10% of a suitable sensitizer such as triphenylsulfonium hexafluoroarsenate with either deep UV (2-5 mJ/cm^2 at 265 nm) or electrons (1 $\mu C/cm^2$ at 20 kV) has resulted in clean positive relief patterns in 1 μm thick films with resolution below 1 μm. There is considerable interest in self- or vapor-developing resist systems and the implications of these results are far reaching.

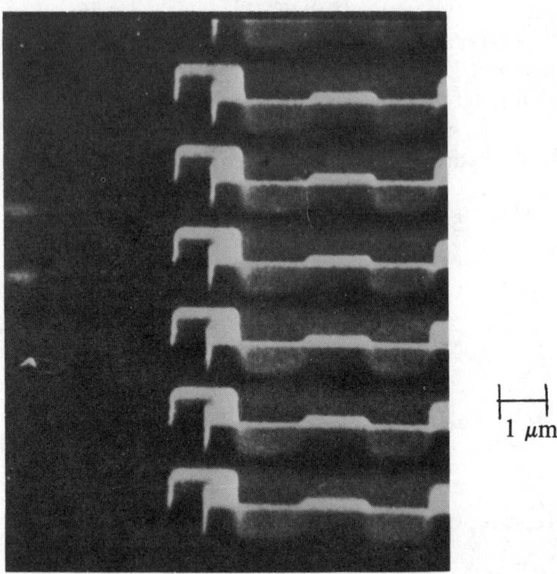

Figure 23. SEM photomicrographs showing positive and negative tone patterns in PBOCST.

It was indicated earlier that swelling limits resolution in solvent-developed negative resists. It was also intimated that swelling effects could be minimized if there were a sufficient polarity change between the exposed and non-exposed areas of the type mentioned in the previous discussion of the PBOCST system. A similar principle was utilized by Hofer et al., (*145-146*) at IBM, based on ion pair formation. The resist consists of a polystyrene polymer to which tetrathiofulvalene (TTF) units have been attached. When spun down with an acceptor such as CBr_4, a complex is formed which, on irradiation, undergoes an electron transfer reaction to form an ion pair:

$$TTF + CBr_4 \xrightarrow{hv} TTF^+Br^-$$

The irradiated areas are rendered insoluble to non-polar solvents due to the formation of these polar ionic species. Swelling does not take place during development to any marked degree, possibly due to the slow rate of diffusion of the low polarity developer solvent into the ionic matrix. The absence of swelling permits very high resolution patterns to be formed as evidenced in Figure 24 which shows a resist pattern from an electron beam exposure of 10 $\mu C/cm^2$ at 20 kV. The sensitivity to $Al_{K\alpha}$ x-rays is 44 mJ/cm^2 for 50% residual thickness and the resist contrast (γ) is also high at 2.5. The sensitivity to electrons is 5.6 $\mu C/cm^2$ for comparable residual thickness and contrast is reported to be 3.3. The resist also shows excellent CF_4 plasma etch resistance. It should be noted however that CBr_4 is both volatile and quite toxic.

It was pointed out earlier that one of the main problems with ion-beam exposure of resists is the penetration depth of the ions. Penetration through 0.5 μm of a polymer, essential for conventional wet chemical development, requires either low mass ions or ions of medium mass having many hundred

Figure 24. SEM of poly(TTF-styrene) resist patterns produced by exposure (10 $\mu C/cm^2$) to 20 kV electrons. (Reproduced by permission from Ref. 145.)

keV energy. Kuwano et al. (*147*) and Venkatesan and co-workers (*148*) have attempted to circumvent this problem with low-energy ion beams by depositing the ions only in the surface regions and then to use this implanted region as a mask for subsequent reactive-ion etching. It was assumed that the deposited species in the ion implant region would form a stable non-volatile compound on the resist surface during etching. Conceptually, the approach is similar to the dry-development scheme of Taylor in that an involatile etch mask forms at the surface during plasma etching and subsequently protects the underlying regions. Using conducting resists, e.g., poly(vinylcarbazole)-trinitrofluorenone complex, to prevent charging, Venkatesan et al. were able to fabricate 250 nm wide lines using low energy In^+ ions, although the incident dose required to provide acceptable plasma etching resistance was quite high (10^{16} ions/cm^2) even though the implanted metal layer mask was only 2-4 monolayers thick.

Taylor et al. (*149*) have recently reported on an extension of this concept which they have called *gas-phase functionalized plasma-developed resists*. The process is envisioned as outlined in Figure 25. Irradiation of the resist (with either electrons, photons, x-rays, or ions) containing reactive groups (A) results in the formation of production groups (P) which are formed at the expense of A. Next, the exposed film is treated with a reactive gas (MR) containing reactive groups (R) and metallic atoms, (M) capable of forming nonvolatile compounds (MY) which are resistant to removal under reactive ion etching conditions using gas Y. Assuming the metal containing gas can diffuse through the matrix and react with either A or P, selective functionalization can occur, allowing either positive-or negative-tone images to be produced on subsequent plasma development. Preliminary results with electron beam exposure using

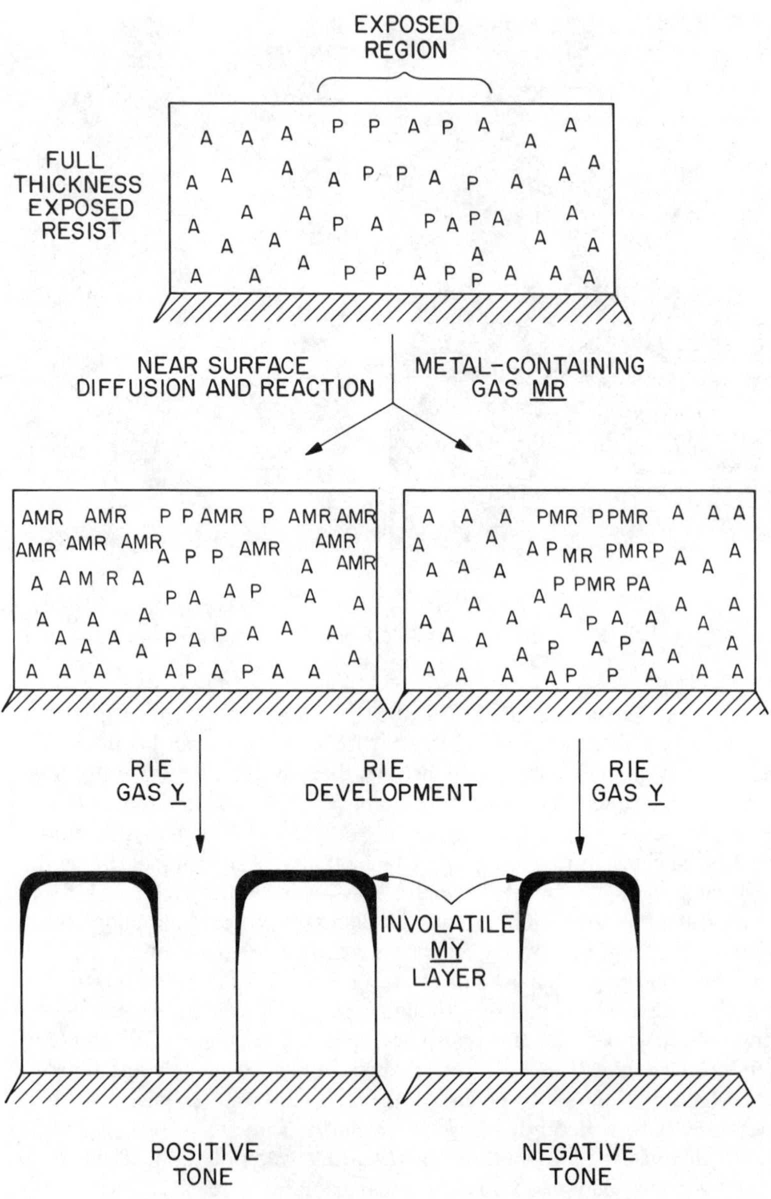

Figure 25. Schematic representation of the gas phase functionalization concept.

poly(2,3-dichloropropyl acrylate) irradiated and treated with B_2H_6 yielded negative tone patterns on subsequent reactive ion etching in O_2, although high doses were required.

These examples serve to illustrate the wide scope and diversity of some of the newer approaches to resist design.

Conclusions

The resist arena today is a highly specialized sphere of activity in which the ingenuity of the synthetic chemist is coupled with the pragmatism and skill of the process engineer to produce an ever increasing array of useful resist systems.

We appear to be at the forefront of a new and exciting era of resist research. The multilevel schemes now allow high aspect ratio structures with submicron resolution to be easily fabricated. Inorganic resist systems show extremely high contrast and may allow the practical application of photolithography well below 1 μm. Plasma-developable resists now enable the advantage of an all dry-processing technology to be realized. The traditional approaches to resist design continue to be refined to meet the increasingly stringent requirements of the semi-conductor industry. Clearly such endeavors can only be pushed so far and totally new mechanisms will no doubt continue to appear which may well lead to useful application.

Literature Cited

1. Willson, C. G. "Introduction to Microlithography"; Thompson, L. F.; Willson, C. G.; Bowden, M. J., Eds.; ADVANCES IN CHEMISTRY SERIES No. 219, American Chemical Society: Washington, D.C., 1983; p. 87.
2. Kerwin, R. E.; Thompson, L. F. *Ann. Rev. Mat. Sci.* 1976, *6*, 267.
3. Bowden, M. J.; Thompson, L. F. *Solid State Technol.* 1979, *22*, 72.
4. Hattori, S.; Morita, S.; Yamada, M.; Tamano, J.; Ieda, M. *Proc. Reg. Tech. Conf.* on "Photopolymers: Principles, Processes and Materials"; Mid-Hudson Section, SPE: Ellenville, New York, Nov. 8-10, 1982, p. 311.
5. Bowden, M. J. *CRC Crit. Rev. in Solid-State Sci.* 1979, *8*, 223.
6. Charlesby A. In "Atomic Radiation Polymers"; Pergamon Press: Oxford, 1960.
7. Griffing, B. F.; West, P. R. *Proc. Reg. Tech. Conf.* on "Photopolymers; Principles, Processes and Materials": Mid-Hudson Section, SPE: Ellenville, New York, Nov. 8-10, 1982, p. 185.
8. Pacansky, J.; Lyerla, J. R. *IBM J. Res. Develop.* 1979, *23*, 42.
9. Willson, C. G.; Miller, R.; McKean, D.; Clecak, N.; Tompkins, T.; Hofer, D.; Michl, J.; Downing, J. *Proc. Reg. Tech. Conf.* on "Photopolymers: Principles, Processes and Materials"; Mid-Hudson Section, SPE: Ellenville, New York, Nov. 8-10, 1982, p. 111.
10. Miller, R. D.; Willson, C. G.; McKean, D. R.; Tompkins, T.; Clecak, N.; Michl, J.; Downing, J. *Org. Coat. and Appl. Polym. Sci. Proc.*, 1983, *48*, p. 54.
11. MacDonald, S. A.; Miller, R. D.; Willson, C. G.; Feinberg, G. M.; Gleason, R. T.; Halverson, R. M.; MacIntyre, M. W.; Motsiff, W. T. "Kodak Microelectronics Seminar"; San Diego, 1982.
12. Dill, F. H.; Neureuther, A. R.; Tuttle, J. A.; Walker, E. J. *IEEE Trans. Electron Devices* 1975, *ED-22*, 456.
13. Oldham, W. G.; Nandagaonkar, S. N.; Neureuther, A. R.; O'Toole, M. M. *IEEE Trans. Electron Dev.* 1979, *ED-26*, 717.

14. Rubner, R.; Kuhn, E. *ACS Div. Org. Coatings and Plast. Chem. Preprints* 1977, *37(2)*, 118.
15. Babie, W. T.; Chow, M-F.; Moreau, W. M. *Org. Coatings and Appl. Polym. Sci. Proc.*, 1983, *48*, p. 53.
16. Dill, F. H.; Hornberger, W. P.; Hange, P. S.; Shaw, J. M. *IEEE Trans. Electron Dev.* 1975, *ED-22*, 445.
17. Jain, K.; Willson, C. G.; Lin, B. J. *IEEE Electron Device Letters* 1982, *EDL-3(3)*, 53.
18. Jain, K.; Willson, C. G.; Lin, B. J. *IBM J. Res. Develop.* 1982, *26(2)*, 151.
19. Jain, K.; Willson, C. G.; Rice, S.; Pederson, L.; Lin, B. J. *Proc. Reg. Tech. Conf.* on "Photopolymers: Principles, Processes and Materials"; Mid-Hudson Section, SPE: Ellenville, New York, Nov. 8-10, 1982, p. 173.
20. Lin, B. J. *J. Vac. Sci. Technol.* 1975, *12(6)*, 1317.
21. Mimura, Y.; Ohkubo, T.; Takeuchi, T.; Sekikawa, K. *Japan J. Appl. Phys.* 1978, *17(3)*, 541.
22. Chandross, E. A.; Reichmanis, E.; Wilkins. C. W., Jr.; Hartless, R. L. *Solid-State Technol.* 1981, *24(8)*, 81.
23. Chandross, E. A.; Reichmanis, E.; Wilkins, C. W., Jr.; Hartless, R. L. *Canadian J. Chem.*, in press.
24. Nakane, Y.; Tsumori, T.; Mifune, T. *Semiconductor Intl.* 1979, *2(1)*, 45.
25. Wilkins, C. W., Jr.; Reichmanis, E.; Chandross, E. A. *J. Electrochem. Soc.* 1980, *127(11)*, 2510.
26. Reichmanis, E.; Wilkins, C. W., Jr. *ACS Div. Organic Coatings and Plast. Chem. Preprints* 1980, *43*, 243.
27. Reichmanis, E.; Wilkins, C. W., Jr.; Chandross, E. A. *J. Electrochem. Soc.* 1980, *127(11)*, 2514.
28. Hartless, R. L.; Chandross, E. A. *J. Vac. Sci. Technol.* 1981, *19(4)*, 1333.
29. Nate K.; Kobayashi, T. *J. Electrochem. Soc.* 1981, *128*, 1394.
30. Applebaum, J.; Bowden, M. J.; Chandross, E. A.; Feldman, M.; White, D. L. In *Proc. Kodak Microelectronic Seminar – Interface*, Oct. 19-21, 1975, p. 40.
31. Bowden, M. J.; Chandross, E. A. *J. Electrochem. Soc.* 1975, *122(10)*, 1371.
32. Himics, R. J.; Ross, D. L. *Proc. Reg. Tech. Conf.* on "Photopolymers: Principles, Processes and Materials"; Mid-Hudson Section, SPE: Ellenville, New York, Oct. 13-15, 1976, p. 26.
33. Yamashita, Y.; Ogura, K.; Kunishi, M.; Kawazu, R.; Ohno, S.; Mizokami, Y. *J. Vac. Sci. Technol.* 1979, *16(6)*, 2026.
34. Grant, B. D.; Clecak, N. J.; Twieg, R. J.; Willson, C. G. *IEEE Trans. Electron Dev.* 1981, *ED-28*, 1300.
35. Reichmanis, E.; Wilkins, C. W., Jr.; Chandross, E. A. *J. Vac. Sci. Technol.* 1981, *19(4)*, 1338.
36. Wilkins, C. W., Jr.; Reichmanis, E.; Chandross, E. A. *J. Electrochem. Soc.* 1982, *129(11)*, 2552.

37. Iwayanagi, T.; Kohashi, T.; Nonogaki, S. *J. Electrochem. Soc.* 1980, *127*, 2759.
38. Iwayanagi, T.; Kohashi, T.; Nonogaki, S.; Matsuzawa, T.; Douta, K.; Yanazawa, H. *IEEE Trans. Electron Dev.* 1981, *ED-28*, 1306.
39. Matsuzawa, T.; Tomioka, H. *IEEE Trans. Electron Dev.* 1981, *ED-28*, 1284.
40. Iwayanagi, T.; Hashimoto, M.; Nonogaki, S.; Koibuchi, S.; Makino, D. *Proc. Reg. Tech. Conf.* on "Photopolymers: Principles, Processes and Materials"; Mid-Hudson Section, SPE: Ellenville, New York, Nov. 8-10, 1982, p. 21.
41. Nakane, H.; Yokota, A.; Yamamoto, S.; Kanai, W. *Proc. Reg. Tech. Conf.* on "Photopolymers: Principles, Processes and Materials"; Mid-Hudson Section, SPE: Ellenville, New York, Nov. 8-10, 1982, p. 43.
42. Imamura, S.; Sugawara, S. *Japan J. Appl. Phys.* 1982, *21(5)*, 776.
43. Liutkus, J.; Hatzakis, M.; Shaw, J.; Paraszczak, J. *Proc. Reg. Conf.* on "Photopolymers: Principles, Processes and Materials"; Mid-Hudson Section, SPE: Ellenville, New York, Nov. 8-10, 1982, p. 223.
44. Harita, Y.; Kamoshida, Y.; Tsutsumi, K.; Koshiba, M.; Yoshimoto, H.; Harada, K. "Preprints of Papers presented at 22nd Symposium of Unconventional Imaging Science and Technology SPSE"; Arlington, VA, Nov. 15-18, 1982, p. 34.
45. Moreau, W. M. *Proc. SPIE*, 1982, *333*; *Submicron Lithography* 1982, *2*.
46. Hatzakis, M. *J. Electrochem. Soc.* 1969, *116(7)*, 1033.
47. Kitakohji, T.; Yoneda, Y.; Kitamura, K. Okuyama, H.; Murakawa, K. *J. Electrochem. Soc.* 1979, *126(11)*, 1881.
48. Tada, T. *J. Electrochem. Soc.* 1981, *128(8)*, 1791.
49. Kakuchi, M.; Sugawara, S.; Murase, K.; Matsuyama, K. *J. Electrochem. Soc.* 1977, *124*, 1648.
50. Sugawara, S.; Kogure, O.; Harada, K.; Kakuchi, M.; Sukegawa, K.; Imamura, S.; Miyoshi, K. *Electrochem. Soc., Spring Meeting*, St. Louis, Missouri, May 11-16, 1980, Extended Abstracts p. 680 (presented at *9th Intl. Conf.* on "Electron and Ion Beam Sci. and Technol.").
51. Tada, T. *J. Electrochem. Soc.* 1979, *126(9)*, 1635.
52. Pittman, C. U., Jr.; Ueda, M.; Chen, C. Y.; Kwiatkowski, J. H.; Cook, C. F., Jr.; Helbert, J. N. *J. Electrochem. Soc.* 1981, *128(8)*, 1758.
53. Lai, J. H.; Helbert, J. N.; Cook, C. F., Jr.; Pittman, C. U., Jr. *J. Vac. Sci. Technol.* 1979, *16*, 1992.
54. Matsuda, S.; Tsuchiya, S.; Honma, M.; Nagamatsu, G. U.S. Patent 4 279 984, 1981.
55. Tada, T. *J. Electrochem. Soc.* 1979, *126(11)*, 1829.
56. Helbert, J. N.; Cook, C. F., Jr.; Chen, C. Y.; Pittman, C. U., Jr. *J. Electrochem. Soc.* 1979, *126(4)*, 695.
57. Helbert, J. N.; Chen, C. Y.; Pittman, C. U., Jr.; Hagnauer, G. L. *Macromolecules* 1978, *11*, 1104.
58. Chen, C. Y.; Pittman, C. U., Jr.; Helbert, J. N. *J. Polymer Sci., Polymer Chem. Ed.* 1980, *18*, 169.
59. Haller, I.; Feder, R.; Hatzakis, M.; Spiller, E. *J. Electrochem. Soc.* 1979, *126*, 154.

60. Moreau, W.; Merritt, D.; Moyer, W.; Hatzakis, M. *J. Vac. Sci. Technol.* 1979, *16(6)*, 1989.
61. Hatano, Y.; Shiraishi, H.; Taniguchi, Y.; Horigome, S.; Nonogaki, S.; Naraoka, K. *Proc. 8th Intl. Conf.* on "Electron and Ion Beam Sci. and Technol., Bakish, R., Ed.; Electrochem. Soc.: Princeton, 1978, p. 332.
62. Lai, J. H.; Shephard, L. T.; Ulmer, R.; Griep, C. *Proc. Reg. Tech. Conf.* on "Photopolymers: Principles, Processes and Materials"; Mid-Hudson Section, SPE: Ellenville, New York, Oct. 13-15, 1976, p. 259.
63. Stillwagon, L. E.; Doerries, E. M.; Thompson, L. F.; Bowden, M. J. *ACS Div. Org. Coatings and Plast. Chem. Preprints* 1977, *37(2)*, 38.
64. Gipstein, E.; Moreau, W.; Need, O. *J. Electrochem. Soc.* 1976, *123(7)*, 1105.
65. Nonogaki, S. In *Proc. Reg. Tech. Conf.* on "Photopolymers: Principles, Processes and Materials"; Mid-Hudson Section, SPE: Ellenville, New York, Nov. 8-10, 1982, p. 1.
66. Saeki, H.; Kohda, M. *Proc. 17th Symp.* on "Semiconductor and Integrated Circuit Technol."; Tokyo, 1979, p. 48
67. Lai, J. H.; Kwiatkowski, J. H.; Cook, C. F., Jr. *J. Electrochem. Soc.* 1982, *129(7)*, 1596.
68. Lai, J. H.; Shepherd, L. T. *Proc. 10th Intl. Symp.* on "Electron and Ion Beam Sci. and Technol."; 1983, *83-2*, p. 185.
69. Quano, A. C. *Polymer Eng. and Sci.* 1978, *78(4)*, 306.
70. Roberts, E. D. *ACS Div. Org. Coatings and Plast. Chem. Preprints* 1973, *33(1)*, 359.
71. Roberts, E. D. *ACS Div. Org. Coatings and Plast. Chem. Preprints* 1977, *37(2)*, 36.
72. Suzuki, M.; Ohnishi, Y. *J. Electrochem. Soc.* 1982, *129(2)*, 402.
73. Bowden, M. J.; Thompson, L. F.; Ballantyne, J. P. *J. Vac. Sci. Technol.* 1975, *12(6)*, 1294.
74. Ballantyne, J. P.; Bowden, M. J.; Frackoviak, J.; Pease, R. F. W.; Thompson, L. F.; Yau, L. D. In "Microcircuit Engineering"; Ahmed, H.; Nixon, W. C., Eds.; Cambridge University Press: Cambridge, 1980, p. 239.
75. Bowden, M. J.; Thompson, L. F.; Fahrenholtz, S. R.; Doerris, E. M. *J. Electrochem. Soc.* 1981, *128*, 1304.
76. Bowden, M. J.; Allara, D. L.; Vroom, W. I.; Frackoviak, J.; Kelley, L. C.; Falcone, D. R. *Org. Coatings and Applied Polym. Sci. Proc.*, 1983, *48*, p. 161.
77. Bowmer, T. N.; Bowden, M. J. *Org. Coatings and Applied Polym. Sci. Proc.*, 1983, *48*, p. 171.
78. Shiraishi, H.; Isobe, A.; Murai, F.; Nongaki, S. *Org. Coatings and Applied Polym Sci. Proc.*, 1983, *48*, p. 178.
79. Cheng, Y. Y.; Grant, B. D.; Pederson, L. A.; Willson, C. G. U.S. Patent 4 398 001, 1983.
80. Kilichowski, K. B.; Pampalone, T. R. U.S. Patent 4 357 369, 1982.
81. Yamazaki, T.; Watakabe, Y.; Suzuki, Y.; Nakata, H. *J. Electrochem. Soc.* 1980, *127(8)*, 1859.

82. Taniguchi, Y.; Hatano, Y.; Shiraishi, H.; Horigome, S.; Nonogaki, S.; Naraoka, K. *Japan J. Appl. Phys.* 1979, *28*, 1143.
83. Thompson, L. F.; Ballantyne, J. P.; Feit, E. D. *J. Vac. Sci. Technol.* 1975, *12(6)*, 1280.
84. Thompson, L. F.; Yau, L. D.; Doerries, E. M. *J. Electrochem. Soc.* 1979, *126(10)*, 1703.
85. (a) Imamura, S. *J. Electrochem. Soc.* 1979, *(9)*, 1628.

    (b) Imamura, S.; Tamamura, T.; Harada, K.; Sugawara, S. *J. Appl. Polym. Sci.* 1982, *27*, 937.
86. Shiraishi, H.; Taniguchi, Y.; Horigome, S.; Nanogaki, S. *Proc. Reg. Tech. Conf.* on "Photopolymers: Principles, Processes and Materials"; Mid-Hudson Section, SPE: Ellenville, New York, Oct. 10-12, 1979, p. 56.
87. (a) Tan, Z. C.; Petropoulos, C. C.; Rauner, F. J. *J. Vac. Sci. Technol.* 1981, *19(4)*, 1348.

    (b) Tan, Z.; Georgia, S. *Proc. Reg. Tech. Conf.* on "Photopolymers: Principles, Processes and Materials"; Mid-Hudson Section, SPE: Ellenville, New York, Nov. 8-10, 1982, p. 221.
88. Yoneda, Y.; Kitamura, K.; Naito, J.; Kitakohji, T.; Okuyama, H.; Murakawa, K. *Proc. Reg. Tech. Conf.* on "Photopolymers: Principles, Processes and Materials"; Mid-Hudson Section, SPE: Ellenville, New York, Oct. 10-12, 1979, p. 44.
89. Choong, H. S.; Kahn, F. J. *J. Vac. Sci. Technol.* 1981, *19(4)*, 1121.
90. Sukegawa, K.; Sugawara, S. *Japan J. Appl. Phys.* 1981, *20*, L583.
91. Sukegawa, K.; Tamamura, T.; Sugawara, S. *Proc. 10th Intl. Symp.* on "Electron and Ion Beam Sci. and Technol."; 1983, *83-2*, p. 193.
92. Kamoshida, Y.; Koshiba, M.; Yoshimoto, H.; Harita, Y.; Harada, K. *Abstracts, 1983 Internatl. Symp.* on "Electron, Ion and Photon Beams"; Los Angeles, May 31 - June 3, 1983, p. I-1.
93. G.A.F. Technical Literature.
94. Shaw, J. M.; Hatzakis, M.; Paraszczak, J.; Liutkus, J.; Babich, E. *Proc. Reg. Tech. Conf.* on "Photopolymers: Principles, Processes and Materials"; Mid-Hudson Section, SPE: Ellenville, New York, Nov. 8-10, 1982, p. 285.
95. Novembre, A.; Masakowski, L.; Frackoviak, J.; Bowden, M. *Proc. Reg. Tech. Conf.* on "Photopolymers: Principles, Processes and Materials"; Mid-Hudson Section, SPE: Ellenville, New York, Nov. 8-10, 1982, p. 285.
96. Oldham, W. G.; Hieke, E. *IEEE Electron Dev. Lett.* 1980, *EDL-1(10)*, 217.
97. Bruning, J. H., private communication.
98. Taylor, G. N. *Solid-State Technol.* 1980, *23(5)*, 73.
99. Taylor, G. N.; Wolf, T. M.; Moran, J. M. *J. Vac. Sci. Technol.* 1981, *19(4)*, 872.
100. Choong, H. S.; Kahn, F. J. *Abstracts, 1983 Intl. Symp.* on "Electron, Ion and Photon Beams"; Los Angeles, May 31 - June 3, 1983, p. F-2.
101. Taylor, G. N.; Hellman, M. Y.; Feather, M. D.; Willenbrock, W. E. *Proc. Reg. Tech. Conf.*, on "Photopolymers: Principles, Processes and Materials"; Mid-Hudson Section, SPE: Ellenville, New York, Nov. 8-10, 1982, p. 355.

102. Brown, W. L.; Venkatesan, T.; Wagner, A. *Solid-State Technol.* 1979, *24(6)*, 60.
103. Hall, T. M.; Wagner, A.; Thompson, L. F. *J. Vac. Sci. Technol.* 1979, *16(6)*, 1889.
104. Brault, R. G.; Kubena, R. L.; Jensen, J. E. *Proc. Reg. Tech. Conf.* on "Photopolymers: Principles, Processes and Materials"; Mid-Hudson Section, SPE: Ellenville, New York, Nov. 8-10, 1982, p. 235.
105. Wagner, A.; Barr, D.; Venkatesan, T.; Crane, W. S.; Lamberti, V. E.; Tai, K. L.; Vadimsky, R. G. *J. Vac. Sci. Technol.* 1981, *19(4)*, 1363.
106. Tai, K. L.; Vadimsky, R. G.; Kemmerer, C. T.; Wagner, J. S.; Lamberti, V. E.; Timko, A. G. *J. Vac. Sci. Technol.* 1980, *17*, 1169.
107. Nagai, H.; Yoshikawa, A.; Togoshima, Y.; Ochi, O.; Mizushima, Y. *Appl. Phys. Lett.* 1976, *28*, 145.
108. Yoshikawa, A.; Ochi, O.; Nagai, H.; Mizushima, Y. *Appl. Phys. Lett.* 1976, *29*, 667.
109. Yoshikawa, A.; Ochi, O.; Nagai, H.; Mizushima, Y. *Appl. Phys. Lett.* 1977, *31*, 167.
110. Tai, K. L.; Johnson, L. F.; Murphy, D. W.; Chung, M. S. C. *ECS Ext. Abstr.* 1979, *79-1*, 244.
111. Tai, K. L.; Sinclair, W. R.; Vadimsky, R. G.; Moran, J. M.; Rand, M. J. *J. Vac. Sci. Technol.* 1979, *16(6)*, 1977.
112. Vadimsky, R. G.; Tai, K. L.; Ong, E. *Proc. Symp. on Inorganic Resist Systems, Electrochem. Soc.*, 1982 *82-9*, p. 37.
113. Chang, M. S.; Chen, J. T. *Appl. Phys. Lett.* 1978, *33(10)*, 892.
114. Chang, M. S.; Hou, T. W. *Thin Solid Film* 1978, *55*, 463.
115. Chang, M. S.; Hou, T. W.; Chen, J. T.; Kolwicz, K. D.; Zemel, J. N. *J. Vac. Sci. Technol.* 1979, *16(6)*, 1973.
116. Wagner, A.; Barr, D. *Proc. Symp. on Inorganic Resist Systems, Electrochem. Soc.*, 1982, *82-9*, p. 281.
117. Tai, K. L.; Ong, E.; Vadimsky, R. G.; Kemmerer, C. T.; Bridenbaugh, P. M. *Proc. Symp. on Inorganic Resist Systems, Electrochem. Soc.*, 1982, *82-9*, p. 49.
118. (a) Lin, B. J. *Solid-State Technol.* 1983, *26(5)*, 105.

     (b) Burggraaf, P. S. *Semiconductor Intl.* 1983, *6(6)*, 48.
119. O'Toole, M. M.; Liu, E. D.; Chang, M. S. "SPIE"; Semiconductor Microlithography VI: 1981; Vol. 275, p. 128.
120. Hatzakis, M.; Canavello, B. J.; Shaw, J. M. *IBM J. Res. Develop.* 1980, *24*, 452.
121. Halverson, R. M.; MacIntyne, M. W.; Motsiff, W. T. *IBM J. Res. Develop.* 1982, *26(5)*, 590.
122. Hatzakis, M. U.S. Patent 4 024 293, 1977.
123. Hatzakis, M.; Hofer, D.; Chang, T. H. P. *J. Vac. Sci. Technol.* 1979, *16(6)*, 1631.
124. Lin, B. J. *J. Electrochem. Soc.* 1980, *202*, 127.
125. Lin, B. J.; Chang, T. H. P. *J. Vac. Sci. Technol.* 1979, *16(6)*, 1669.
126. MacDonald, S. A.; Steinmann, A. S.; Ito, H.; Hatzakis, M.; Lee, W.; Hiroaka, H.; Willson, C. G. *Abstracts, 1983 Intl. Symp.* on "Electron Ion and Photon Beams"; Los Angeles, May 31 - June 3, 1983, p. I-4.

127. Moran, J. M.; Maydan, D. *Bell System Tech. J.* 1979, *58(5)*, 1027.
128. Moran, J. M.; Maydan, D. *J. Vac. Sci. Technol.* 1979, *16(6)*, 1620.
129. Ray, G. W.; Peng, S.; Burriesci, D.; O'Toole, M. M.; Liu, E-D. *J. Electrochem. Soc.* 1982, *129(9)*, 2152.
130. Bowden, M. J.; Thompson, L. F. *Polymer Eng. and Sci.* 1974, *14*, 525.
131. Willson, C. G.; Ito, H.; Frechet, M. J.; Houlihan, F. *Proc. IUPAC 28th Macromolecular Symposium,* Amherst, Massachusetts, July 1982, p. 448.
132. Ito, H.; Willson, C. G. *Proc. Reg. Tech. Conf.* on "Photopolymers: Principles, Processes and Materials"; Mid-Hudson Section, SPE: Ellenville, New York, Nov. 8-10, 1982, p. 331.
133. Ito, H.; Willson, C. G. *Org. Coatings and Appl. Polymer Sci. Proc.*, 1983, *48*, p. 60.
134. Hiroaka, H. *J. Electrochem. Soc.* 1981, *128*, 1065.
135. Yamada, M.; Tamano, J.; Yoneda, K.; Morita, S.; Hattori, S. *Japan J. Appl. Phys.* 1982, *21(5)*, 768.
136. Taylor, G. N.; Wolf, T. M. *J. Electrochem. Soc.* 1980, *127(12)*, 2665.
137. Smith, J. N.; Hughes, H. G.; Keller, J. V.; Goodner, W. R.; Wood, T. E. *Semiconductor International* 1979, *2(10)*, 41.
138. Tsuda, M.; Oikawa, S.; Kanai, W.; Yokota, A.; Hijikata, I.; Vehara, A.; Nakane, H. *J. Vac. Sci. Technol.* 1981, *19(2)*, 259.
139. Tsuda, M.; Oikawa, S.; Yokota, A.; Yabuta, M.; Kanai, W.; Kashiwagi, K.; Hijikata, I.; Nakane, H. *Proc. Reg. Tech. Conf.* on "Photopolymers: Principles, Processes and Materials"; Mid-Hudson Section, SPE: Ellenville, New York, Nov. 8-10, 1982, p. 397.
140. Imamura, S.; Morita, M.; Tamamura, T.; Kogure, O.; Murase, K. *Abstracts, 1983 Intl. Conf.* on "Electron, Ion and Photon Beams"; Los Angeles, May 31 - June 3, 1983, p. I-5.
141. Lai, J. H. *Org. Coatings and Applied Polymer Sci. Proc.*, 1983, *48*, p. 189.
142. Schlessinger, S. I. *Proc. Reg. Tech. Conf.* on "Photopolymers: Principles, Processes and Materials"; Mid-Hudson Section, SPE: Ellenville, New York, Oct. 24-26, 1973, p. 85.
143. Smith, G. H. Belg. Patent 828 841, 1975.
144. Crivello, J. V. *Org. Coatings and Applied Polymer Sci. Proc.*, 1983, *48*, p. 65.
145. Hofer, D.; Kaufman, F.; Kramer, S. *Proc. Reg. Tech. Conf.* on "Photopolymers: Principles, Processes and Materials"; Mid-Hudson Section, SPE: Ellenville, New York, Nov. 8-10, 1982, p. 245.
146. Kaufman, F. B.; Hofer, D. C.; Kramer, S. R. *Preprints, ACS Div. Org. Coatings and Plastics Chem.* 1980, *43*, 375.
147. Kuwano, H.; Yoshida, K.; Yamazaki, S. *Japan J. Appl. Phys.* 1980, *19*, L615.
148. Venkatesan, T.; Taylor, G. N.; Wagner, A.; Wilkens, B.; Barr, D. *J. Vac. Sci. Technol.* 1981, *19(4)*, 1379.
149. Taylor, G. N.; Stillwagon, L. E.; Venkatesan, T. N. C. *Abstracts, 1983 International Symp.* on "Electron, Ion and Photon Beams"; Los Angeles, May 31 - June 3, 1983, p. F-5.

RECEIVED September 28, 1984

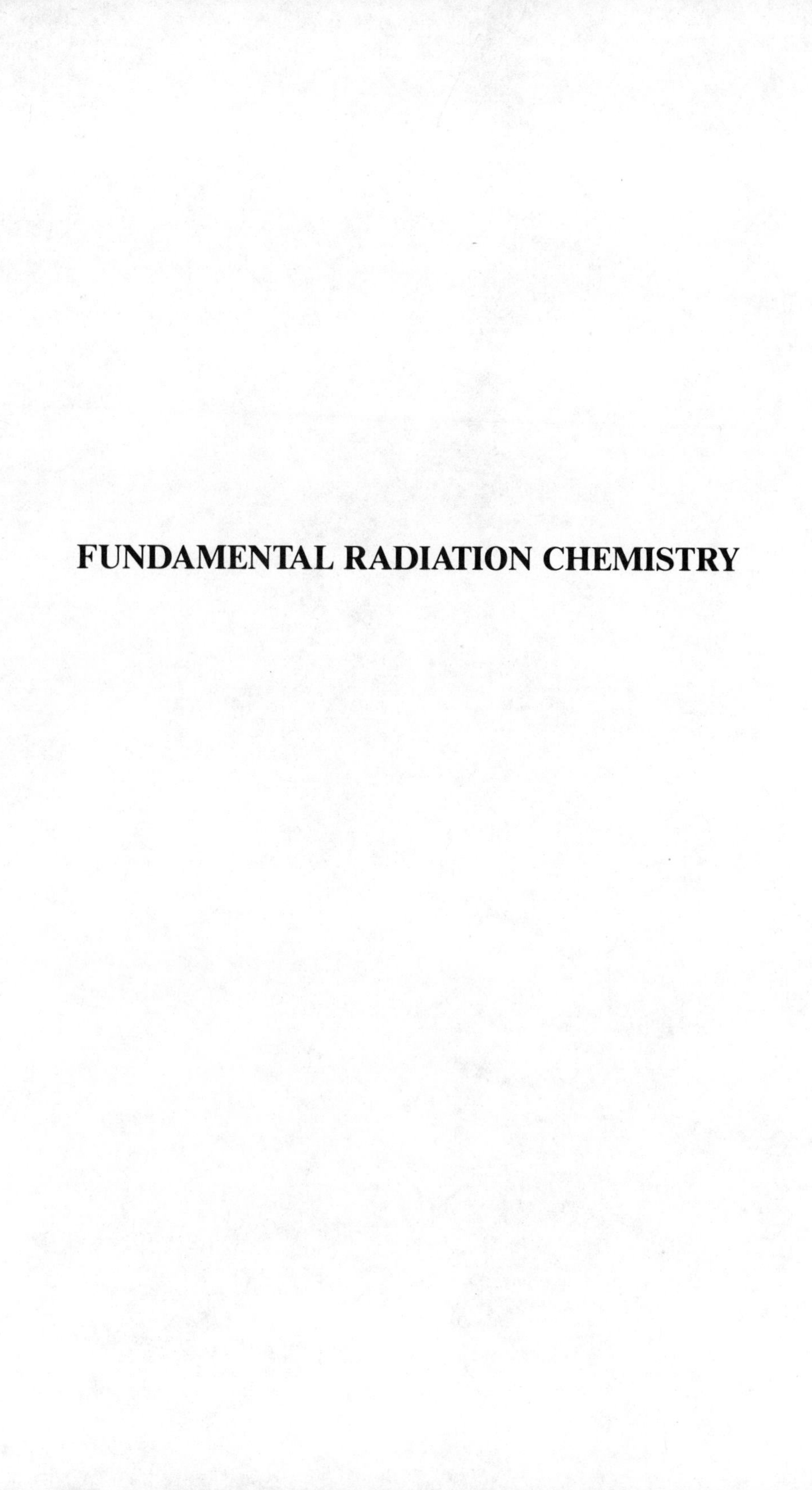

FUNDAMENTAL RADIATION CHEMISTRY

Fundamental Radiation Chemistry

The design of photoresists for microelectronics is a task which usually has succumbed to the pressures of the very competitive environment of high technology and has therefore progressed very rapidly in selected directions corresponding to the most acute needs of the microelectronics industry. Although a number of sophisticated resist systems were designed, thereby meeting the challenge of more and more demanding production needs, much of the fundamental studies which normally precede or accompany technological developments were not done. As the field matured and new technological advances required a better and more thorough understanding of the fundamental principles underlying resist methodology, more basic studies were undertaken and great progress was made with the development of new generations of resist materials. Future improvements in the field will likely depend more and more on the exploitation of the results of basic knowledge in the field of radiation sensitive polymeric materials and composites. Chemistry plays a central role in microlithography and significant contributions can be made by scientists active in areas as varied as polymer and organic synthesis, fundamental radiation chemistry, theoretical chemistry, etc. Of particular importance is a good understanding of the phenomena which occur when polymeric materials and/or small molecules are subjected to irradiation.

For example, numerous studies have sought to establish a definite correlation between polymer structure and radiation degradation behavior. This goal has essentially been accomplished in the case of vinyl polymers and the resist chemist is often able to predict accurately the radiation behavior of a new vinyl polymer. In contrast, a much smaller data base is available for many other types of polymers, particularly those containing a heteroatom in the main chain, and, in many cases, it is difficult to predict the result of radiation induced modifications and further studies are required to elucidate the mechanism of the interaction of the polymer with radiation. This "predictive" goal is well within reach as was shown elegantly by several researchers in their work with polysulfones. It is now well established that, even when high energy radiation is used, some bonds are cleaved selectively while others are very resistant. To simplify the task of characterization which is required to help gain an understanding of the mechanism of radiation modification, small molecules can be used efficiently in model studies; it must be remembered however, that polymers have unique properties directly related to their high molecular weights and thus model studies cannot replace direct experimentation with the polymers themselves but can only complement them.

Most of the current generation of resists consists of one of three types of polymeric materials which undergo radiation induced modifications:

- Polymers which exhibit intrinsic radiation sensitivity and undergo either main-chain cleavage or coupling upon irradiation.

- Polymers which exhibit intrinsic radiation sensitivity and undergo either side-chain cleavage or coupling upon irradiation.
- Polymers which are not modified themselves by irradiation but are used in blends with radiation sensitive additives; in this case the result of irradiation may be the cross-linking, depolymerization, side-chain modification, or change in physical properties of the polymer blend.

Another important factor which can affect significantly the properties of a resist material is the morphology of the polymer film as crystalline and amorphous areas of a film may not undergo the same radiation-induced changes. These changes are also affected by temperature since reductions in the mobility of polymer chains can prevent the occurrence of the molecular motions which are required for processes such as the formation of excited states, molecular rearrangements, disproportionation, etc.

In the case of blends containing small molecules dispersed in a polymer matrix the solubility and the diffusion constant of the small molecule in the polymer will be important factors, as will the absorptivity of both the starting and final components of the blend. Part II of this book will focus on a number of different aspects of the interaction of radiation with polymeric materials and includes many of the points discussed briefly above.

Chapter 4 is a detailed account of the techniques used and the results obtained in the search for a correlation between polymer structure and radiation degradation behavior for poly(olefin sulfones), poly(amino acids), poly(olefins), and aromatic polymers. Of particular interest is the suggestion that ESR measurements can be used to evaluate the radiation resistance of various polymers at low radiation doses.

Chapter 5 is a masterful account of the use of time-resolved absorption spectroscopy of radiation-induced intermediates to elucidate the mechanism responsible for the high sensitivity of chloromethylated polystyrene as a negative electron resist.

Chapter 6 reviews the photochemistry of ketone polymers in the solid phase and shows the usefulness of small molecules as model compounds to predict the photochemistry of polymeric materials, while pointing out that polymer chains need a certain free volume to allow for the motions necessary for formation of excited states or products. The importance of T_g and of diffusion of small molecules is also placed in proper focus.

Chapter 7 is a continuation of the previous chapter concerned with the application of basic principles of ketone photochemistry to the optimization of the radiolysis of poly(vinyl ketones). The approach which is chosen incorporates a synthetic design which maximizes the occurrence of the desired main chain cleavage.

Chapter 8 covers another important area with a look at photoinduced electron transfer reactions using polymer-bound sensitizers as probes. The chapter contains an in-depth look at interfacial electron-transfer sensitization using a solid-liquid boundary and compares the efficiency of polymer-bound sensitizers with their unbound counterparts.

Chapter 9 reviews laser-induced polymerization while pointing out the advantages of this approach over more conventional photopolymerization. Two types of UV-lasers are compared in this study which includes kinetic measurements on the photocross-linking of multifunctional acrylate oligomers.

Chapter 10 reports on the synthesis of polymers containing both photosensitive and photosensitizer pendant groups; the photochemical reactivity and the practical photosensitivity of the polymers is then studied with a detailed comparison of two different polymer systems.

Chapter 11 takes the reader to a different area of great current interest with the description of a new technique for the determination of radiation chemical yields of negative electron beam resists. The new methods compare favorably with the more conventional soxhlet extraction technique.

Finally, this section is closed in Chapter 12 by a fascinating report on the anomalous topochemical photoreaction of olefinic crystals. Although small molecules only are used in this study, the results can no doubt be extended to polymeric systems.

Clearly, this section cannot claim to be comprehensive but it includes some striking illustrations of the application of basic principles of radiation chemistry to the design and study of radiation-sensitive polymers with potential use in microelectronics.

4

Fundamental Aspects of Polymer Degradation By High-Energy Radiation

D. J. T. HILL, J. H. O'DONNELL[1] and P. J. POMERY

Department of Chemistry
University of Queensland
Brisbane, Australia, 4067

Understanding of the mechanism of radiation degradation of polymer molecules is essential for development of improved and new industrial processes, for radiation-induced modification of polymer properties, and for selection of polymers for use in radiation environments. This means that the detailed chemical reactions resulting from absorption of radiation must be known. This fundamental understanding must enable us to relate the chemical structure of a polymer to changes in its chemical, physical and material properties. Such structure-property relationships require a great deal of research work, but they are the key to further advancement on a scientific basis.

The morphology of solid polymers is also an important parameter. Thus, radiation-induced changes can be expected to differ in crystalline and amorphous regions — but in what way and to what extent? "Crystallinity" and "amorphous" are not absolute terms and as more becomes known about the solid-state structure of polymers this should be related to radiation degradation.

The aim of such fundamental research is to be able to predict the behavior of new polymer structures or new irradiation conditions. Understanding of the radiation degradation of *pure* polymers is only a starting point for industrial processes as few polymers in commerce have the high purity used in laboratory research experiments. Very small levels of residual solvents, water (which is strongly attached by hydrogen bonding to polar groups in polymers), oxygen and especially metal compounds, including residues of catalysts, may sensitize degradation. Stabilizers, lubricants, fillers, etc. may be present in larger quantities. Involvement of additives and impurities in the radiation degradation of polymers is a vast and largely unexplored field.

Temperature is an important parameter in radiation degradation, yet remarkably little experimental work has been reported for irradiation at other than ambient temperature. The transition temperatures, T_g and T_m, of the polymer are as important in the radiolysis as they are for the properties. The irradiation temperature can be deliberately varied to enhance the radiation-induced modifications of polymers and should not be regarded as just an environmental property. Similarly, once the sensitization to, or protection from, radiation degradation by small molecule additives is understood — as a function

[1] To whom correspondence should be addressed.

0097–6156/84/0266–0125$11.50/0

of the structures of the additive and the polymer — such additives can be used with increased effectiveness to modify the radiolysis process.

Areas of application of radiation degradation of polymers can be usefully classified into (1) modification of polymers, e.g. molar mass or structure, (2) radiation-sensitive polymers, and (3) radiation-resistant polymers. Within the constraints of the required material properties of the polymers, either maximization or minimization of the response to radiation is usually the objective.

A variety of techniques should be utilized to elucidate the mechanism of radiation degradation. One important aim should be to measure changes at small radiation doses so that (1) the substrate is not substantially changed, (2) secondary reactions of radiolysis products are not serious, and (3) results can be obtained in a reasonable time. Multiple experiments should be performed to enable (1) statistical evaluation of confidence limits from derived values, and (2) statistical testing of the significance of conclusions. Therefore, methods which utilize small amounts of polymer are desirable so that carefully prepared, purified and characterized polymers can be used. The major changes resulting from high-energy irradiation of polymers are (1) scission and cross-linking of the polymer molecules (2) formation of small molecule products, and (3) modification of the structure of the polymer. Increased attention should be given to the development of testing techniques using miniature samples for evaluation of all of these changes.

The degradation of polymer materials in service results in deterioration of mechanical properties, leading to failure of their practical function. Appropriate mechanical property tests are exceedingly sensitive to degradation, even when chemical changes are quite small. However, standard procedures for mechanical testing, e.g., ASTM, normally require samples of large dimensions, whereas for research in polymer degradation only small amounts of polymer may be available. New instrumental techniques requiring samples of small size, or the careful use of small samples for conventional measurements, such as tensile elongation, offer great scope in the study of radiation degradation of polymers.

The role of specific chemical groups in or on a polymer chain in radiation degradation can be usefully investigated using model compounds of low molar mass. Much valuable information can be deduced in this way, but the unique properties of polymers are a consequence of their high molar mass and this can also apply to radiation degradation, thereby limiting conclusions drawn from radiation studies on model compounds of low molar mass.

High-energy radiation causes ionization and excitation in polymers, as in small molecules (*1*). The primary ionization (Equation 1) produces an electron and the parent cation radical. The electron must be slowed down to thermal energies before it can undergo further reaction, such as capture by suitable receptive groups, e.g., carbonyl, to give an anion radical (Equation 2). Recombination of cation radicals and anion radicals, especially the geminate ions (electron and parent cation radical) will produce excited molecules. The energy of excitation, if sufficiently high, will lead to molecular decomposition, and this will frequently occur by homolytic bond scission, with the formation of neutral radicals.

$$P \rightarrow P^* + \overset{+}{P^{\cdot}} + e^- \tag{1}$$

$$P + e^- \rightarrow \overset{-}{P^{\cdot}} \tag{2}$$

Thus, there are three possible pathways for the radiation degradation of polymer molecules: neutral radical, cation-radical and/or anion-radical intermediates. Interest in the formation of these three types of reaction intermediates has fluctuated over the years with the utilization of different techniques and with the particular interests of different investigators. It is likely that all three species will be produced, but their relative importance in the degradation mechanism will depend on the chemical structure of the polymer. Evidence for their involvement will depend on the experimental methods used and the temperature and time scale of observation. In this paper we illustrate our investigations of many of the fundamental aspects of the radiation degradation of polymers through studies of series of polymers and copolymers.

Experimental

Materials. Poly(olefin sulfone)s were prepared by copolymerization of liquid mixtures of sulfur dioxide and the appropriate olefin using tert.-butyl hydroperoxide as initiator in the temperature range from −80 to 0°C. The poly(amino acid)s were obtained from Sigma Chemical Co. and used without further purification. The poly(olefin)s were provided by Mr. O. Delatycki and Dr. T. N. Bowmer and were prepared under controlled conditions. The aromatic polysulfones were prepared and purified by Mr. J. Hedrick. The purity of all polymers was checked by ^{1}H and ^{13}C NMR.

Irradiation. Samples of the polymers were weighed into small glass ampoules and evacuated at ca. 100 Pa for at least 48 h before sealing. During evacuation the samples of poly(amino acid)s and aromatic polysulfones were heated to assist in removal of occluded moisture.

The samples were irradiated with ^{60}Co gamma rays in an AECL Gammacell or in the pond facility of the AAEC, at dose rates from 0.1 to 2 Mrad h^{-1}, to total doses from 0.1 to several hundred Mrad. Constant temperatures from 77° to 448°K (-196° to 175°C) were maintained using liquid nitrogen, slush baths, solid CO_2 and aluminum block heaters.

Post-Irradiation Treatment. Irradiated solid polymers contain trapped radicals which eventually decay via reactions which can be expected to cause further chemical changes in the polymers. The concentrations of these radicals are highest at low temperatures, and ESR studies during progressive, post-irradiation warming can provide valuable information on the mechanisms of degradation. Radical concentrations are usually significant up to the glass transition or melting temperature of the polymer and the radicals usually decay with time at rates which increase with temperature. The trapped radicals are likely to react with oxygen if the samples are opened in air, which can lead to enhanced chain scission, whereas cross-linking usually occurs if the polymer radicals decay in vacuum. Since the concentration of radicals can be monitored by ESR, we have adopted a procedure of determining the temperature-time profile of the radical concentration, and utilizing this information to subject the irradiated samples to post-irradiation heating at the lowest temperature which will result in disappearance of most of the radicals within 18 hours before the sample tubes are opened.

Characterization. A major emphasis has been placed on the use of small samples of polymer. This enables 1) a variety of different types of experiments such as G value determination and 2) a sufficient number of experiments to enable statistical evaluation of derived values, to be performed on small amounts of polymer produced in the laboratory, as in copolymerizations to low conversions (necessary to avoid drifting in composition). Yields of small molecule products were obtained from 10-50 mg samples of polymer utilizing a technique of breaking an ampoule of the irradiated polymer in a specially constructed injection system in a gas chromatograph (*2*).

Changes in molar mass of irradiated polymers were measured by viscosity, gel permeation chromatography, osmometry and light scattering. G(scission) and G(cross-link) values can be derived from gel contents if cross-linking predominates (*3*) and a GPC technique requiring only 50 mg of polymer, which could be in the form of powder, was developed for this purpose (*4*).

Although average molar masses are convenient for evaluation of $G(S)$ and $G(X)$, we have shown that the molar mass distribution should be considered (*5*). In particular, the formation of a high molar mass tail can result in serious underestimation of the correct average molar masses, especially after high doses of radiation. Cross-linking causes changes in the hydrodynamic volume of the polymer molecules relative to linear molecules and this affects viscosity and GPC estimates of molar mass, which should be taken into consideration.

ESR Spectroscopy. Electron Spin Resonance spectroscopy is an important technique for investigating the role of radical intermediates in radiation chemistry. The technique has been used widely for many years in the study of radicals occurring in irradiated solid polymers (*6,7*). However, by their very nature, such species are reactive and may only exist in low concentration. The identification of these species can also be a problem since in the majority of polymers the environment of the radicals leads to broad, unresolved ESR spectra, which makes detailed spectral analysis difficult. In recent years, many of these problems of sensitivity and resolution have been reduced by more sensitive and stable ESR spectrometers and by development of new methods of data handling and manipulation.

All work reported in this paper has been performed on a Bruker 200D ESR Spectrometer interfaced with a Digital Equipment PDP11/34 computer with an AK/11 A/D converter. ESR studies using these facilities can provide information on

a) Concentration of total radical species
b) Identification of radicals present
c) Proportions of component radicals.

Thermal Trapping and Annealing. The primary radicals in the radiolysis of polymers usually cannot be studied by irradiation at ambient temperature as the radicals which are observed often result from secondary radical reactions or trapping in specific areas of the polymer matrix. Observation and identification of primary radicals is best achieved by thermal trapping of the radicals. The polymers are irradiated at low temperature and their reactions observed during subsequent warming. This technique should result in the same radicals as those observed on irradiation at higher temperature. Such an approach has been widely used in mechanistic studies of polymer degradation and is based on the implicit assumptions that 1) non-radical reactions are relatively unimportant, and 2) the sequence of radical reactions during warming is the same as that occurring on a much shorter time scale during (and following) irradiation at a higher temperature.

Radical Concentrations. Radical concentrations are measured by double integration of the observed ESR signal and subsequent comparison with a known standard (e.g. DPPH) run under similar instrumental conditions. A plot of radical concentration versus temperature for poly(methyl methacrylate) irradiated at 77°K is shown in Figure 1. In each region a number of radicals can be identified: the radical decay *at low temperatures* is normally associated with recombination of radicals in close proximity (i.e. spurs, etc.). At higher temperatures, H atom abstraction or "hopping" mechanisms occur. Chain motion near the glass transition (T_g) or melting (T_m) temperatures results in

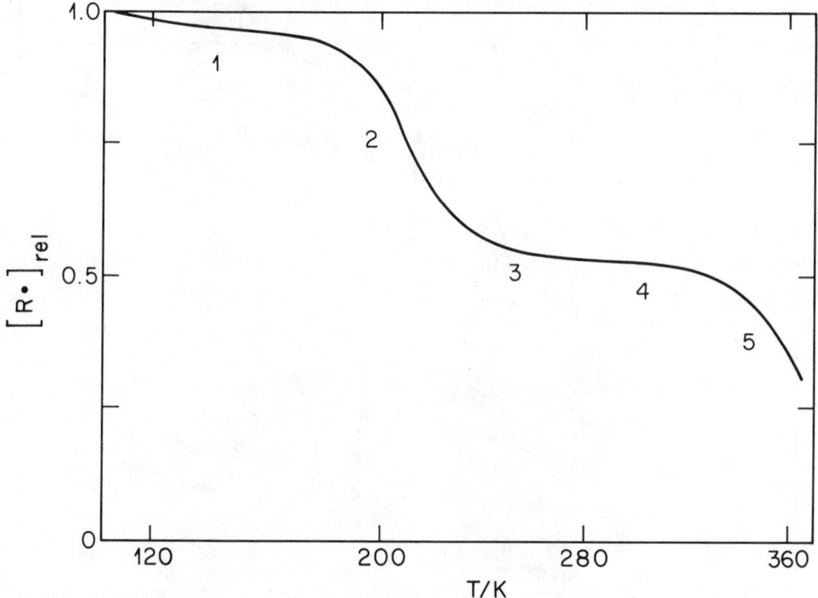

Figure 1. Variation in the concentration of radicals in poly(methyl methacrylate) irradiated at 77°K and warmed progressively to 370°K. The numbers indicate the temperature range where different radicals decay.

rapid decay. Such information may lead to useful information about the chemical reactions involved in the radiolysis.

Spectrum Accumulation. The importance of studies at low radiation doses has already been indicated. Although radical yields are small at such doses, the high stability of modern spectrometers allows spectrum accumulation to be performed on low concentrations of radicals. This is illustrated in Figure 2 for

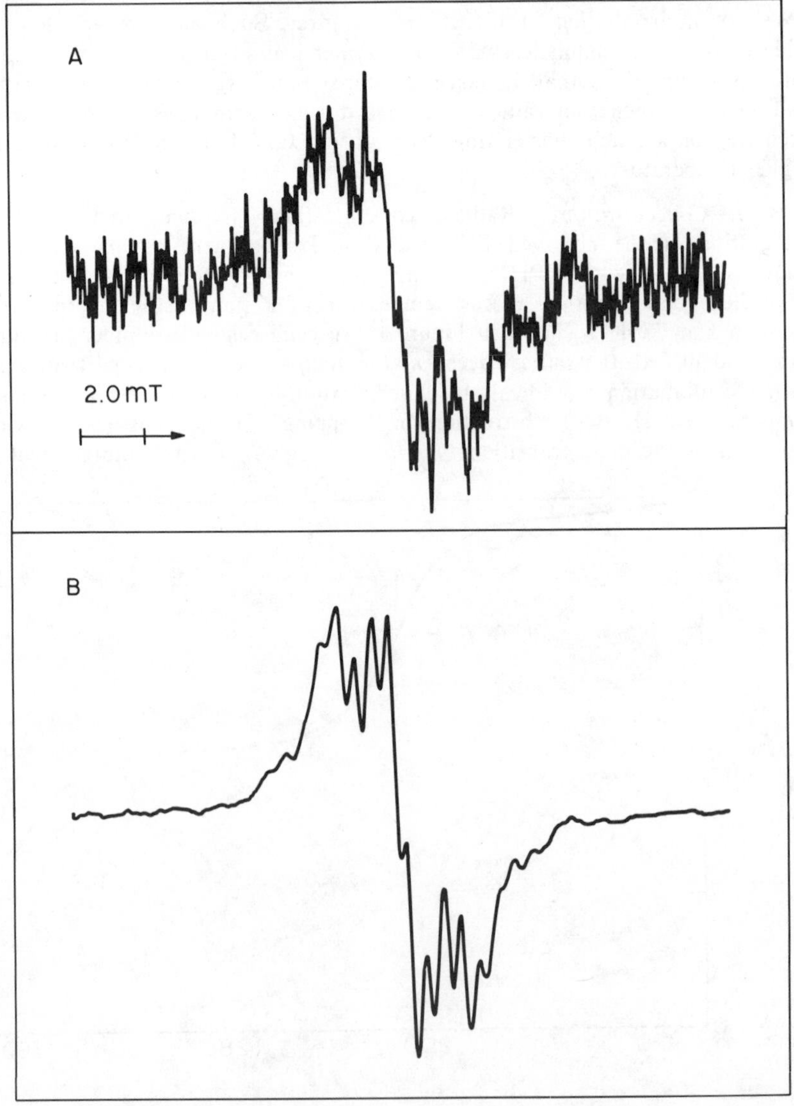

Figure 2. ESR spectrum of poly(α-methyl styrene) irradiated at 77°K to 0.6 Mrad. (A) a single 200 s scan, (B) accumulation of 150 scans.

a sample of poly(α-methyl styrene) irradiated at 25°C to a dose of 0.6 Mrad. Clearly, accumulation greatly improves the signal: noise ratio and the resolution of the spectrum and can assist substantially in analysis.

Greater concentrations of radicals can be produced at higher irradiation doses. However, at high doses, dose saturation effects may well become important. Normally, the radical concentration is proportional to the dose, but with dose saturation this proportionality becomes invalid as a result of secondary radical reactions occurring or a change in the physical nature of the polymer matrix.

Analysis of Spectra. An understanding of the mechanism of polymer degradation must involve identification of the radical intermediates. However, anisotropy due to spin lattice interactions in the solid state invariably results in broad, poorly resolved ESR spectra and together with the low concentration of radicals which is usually present, can result in major problems with analysis. We have developed two approaches to this problem: 1) increasing resolution and 2) sophisticated analysis routines.

Resolution. Resolution can normally be increased by lowering the concentration of radical species and/or by varying the temperature. The former technique can result in an inadequate signal to noise ratio, whilst charged radical species may be formed when irradiations are performed at low temperature. It has been found that radical cations are not stable at 77°K and act as the principal precursors for bond scission reactions. Radical anions are often stable at 77°K and result in broad singlets or doublets, which make spectrum analysis difficult. Such species can be removed by photo-bleaching, in which the sample is irradiated with light of wavelength above 500 nm – usually giving non-radical products. Selective removal of lines in the ESR spectra associated with different radicals can be achieved by variation of the microwave power incident on the sample in the spectrometer. A plot of spectrum area versus the square root of microwave power should give a linear relationship, but deviation from linearity can occur at high powers and this is known as power saturation. The power above which this nonlinearity can occur is known as the "critical" power level and has a unique value for each radical. It is therefore possible to increase the intensities of the resonances due to certain radical spectra by increasing the incident power whilst the spectra contributions from other radicals increase more slowly.

Analysis. A range of data manipulation and simulation programs can be used to analyze complex spectra. The best signal to noise ratio and resolution should be obtained by utilizing the techniques described above before analysis is attempted. Subtraction of ESR spectra after variation in temperature, time or microwave power will enable the spectra of species that have decayed or formed to be observed. However, in many cases the spectra observed are complex and composed of a variety of radicals with multiple hyperfine splitting of the electron energy levels. Analysis of such spectra can be achieved using a Fourier transformation (FT) technique. In this technique the ESR spectrum is subjected to a Fourier transformation in which the differing frequencies associated with the ESR spectra can be plotted as the modulus of the Fourier coefficients versus time. Distinct frequency bands can be observed, which are associated with different levels of hyperfine splitting. By application of suitable truncation functions, these distinct frequencies can be separated and, by suitable reverse transformation, specific hyperfine components of the ESR spectrum can be obtained.

When the ESR spectrum has been analyzed into spectral components assigned to particular radicals, the individual spectra are simulated, taking into consideration (1) g value, (2) hyperfine splitting, (3) line shape and (4) line width. These simulated spectra are then summed, corresponding to the percentages of individual radicals, to produce a spectrum similar to that obtained experimentally. The application of ESR to the study of radiation degradation of polymers is extremely valuable and the techniques described above have been used extensively in the present work.

Results and Discussion

Poly(olefin sulfone)s. The copolymers of sulfur dioxide with olefins show many interesting properties. The depropagation reaction is favored at quite low temperatures and consequently the ceiling temperature for the polymerization is in the range −100 to +100°C, depending on the nature of the olefin, rather than above the thermal degradation temperature of the polymer, as is usual (*8*). All olefins which copolymerize with SO_2, except ethylene in the gas phase, form alternating copolymers; this may be due to polymerization proceeding via the known 1:1 comonomer complex, but this question is still unresolved. A limited number of substituted olefins form copolymers with SO_2 and the composition may vary considerably from 1:1; examples are vinyl chloride and styrene.

A low ceiling temperature means that:

(1) polymer of high molar mass cannot be formed above this temperatures (the molar mass decreases rapidly approaching T_c — and this can be utilized to produce polymers of low molar mass).

(2) the polymer is unstable with respect to formation of monomer if active sites, for example free radical centers corresponding to propagating/depropagating radicals, are produced in the polymer, as will occur on irradiation.

Under these conditions there is a theoretical equilibrium vapor pressure of monomer above solid polymer which varies with temperature. This pressure is substantial at ambient temperature for poly(olefin sulfone)s and will increase exponentially with temperature. Thus, there will be a critical temperature above which the depropagation equilibrium vapor pressure will be higher than the condensation pressure and the polymer will be thermodynamically completely unstable.

The unusual properties of poly(olefin sulfone)s can be attributed to the weak C-S bonds in the polymer, which also has interesting implications in the degradation of these polymers by high-energy radiation. Firstly, there is highly specific cleavage of C-S bonds, illustrating that radiation degradation can be highly selective and not occur at random. This is confirmed by the distribution of volatile products — there are only small yields of products resulting from scission of any other bonds (*9*).

Secondly, there is a high yield, $G(S)$, for permanent main-chain scission and an accompanying rapid and substantial decrease in molar mass on irradiation. This implies not only that there is a high number of main-chain scissions, but that a large proportion of the initial scissions must result in permanent cleavage. This is apparently due to the elimination of SO_2 from the polymer chain. The C-S bond in RSO_2 is known to be particularly weak and the C-S bond adjacent to the site of the original cleavage can be expected to have a high possibility of secondary scission. SO_2 elimination could stabilize main-chain scission by formation of a 'hole', which would counteract the normal recombination of the chain ends due to the cage effect which operates very effectively in the solid state.

The G values for SO_2 and butene production from the irradiation of poly(1-butene sulfone) are shown in Figure 3 as a function of temperature from

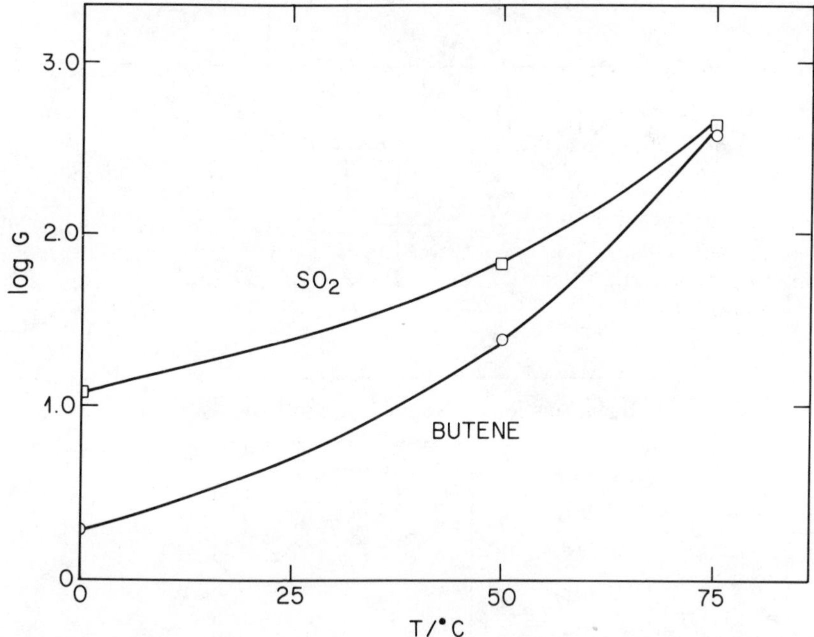

Figure 3. Temperature dependence of the yields (G values) of SO_2 *and butene from* γ *irradiation of poly(1-butene sulfone).*

0 to 150°C. They support the argument above as follows:

(1) the SO_2 yield is high even at 0°C, indicating that the two adjacent C-S scission reactions and loss of the SO_2 molecule occur at low temperatures, and hence preferably to depropagation, which would also yield butene.

(2) the rate of depropagation increases markedly with temperature, with the increase accelerating rapidly above about 50°C, (note that the scale for the G values is logarithmic).

(3) the G values for SO_2 and butene approach equality at high temperatures, indicating the predominance of depolymerization.

Further evidence for specificity of C-S scission as the primary degradation reaction is provided by ESR. For example, Table I shows possible bond cleavages (A-E) in poly(isobutene sulfone) and the expected number of hyperfine lines in the ESR spectrum of the resultant radicals. The ESR spectrum observed after irradiation of poly(isobutene sulfone) at 77°K is shown

Table I. Possible Radiation Induced Bond Scissions in Poly(Isobutene Sulfone) and Expected Number of Lines in the ESR Spectra of the Corresponding Radicals

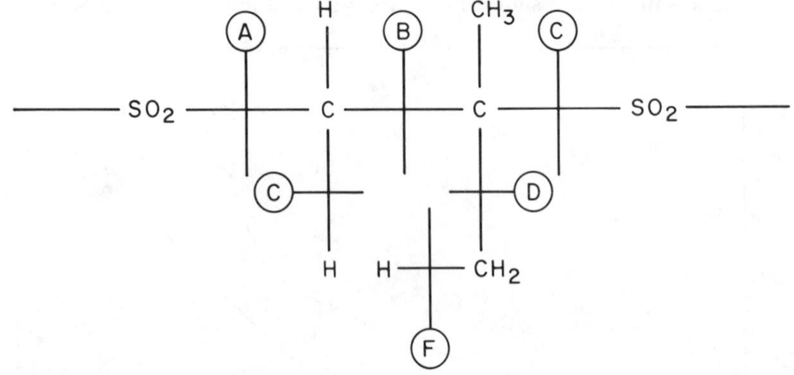

Bond Scission	No. of Lines in ESR
A	3 or 4
B	3 or 4, 7
C	2
D	4
E	⩾ 8
F	3 or 4

in Figure 4. It contains two components — a multi-line spectrum with hyperfine splitting (hfs) of appox. 2.3 mT containing 8 or more lines, and a central singlet with a shifted center position.

These two components of the ESR spectrum may be attributed to an alkyl radical and a sulfonyl radical. The presence of two radicals can be confirmed by power saturation experiments. The spectrum at high microwave power shows increased intensities for the lines due to the alkyl radical and

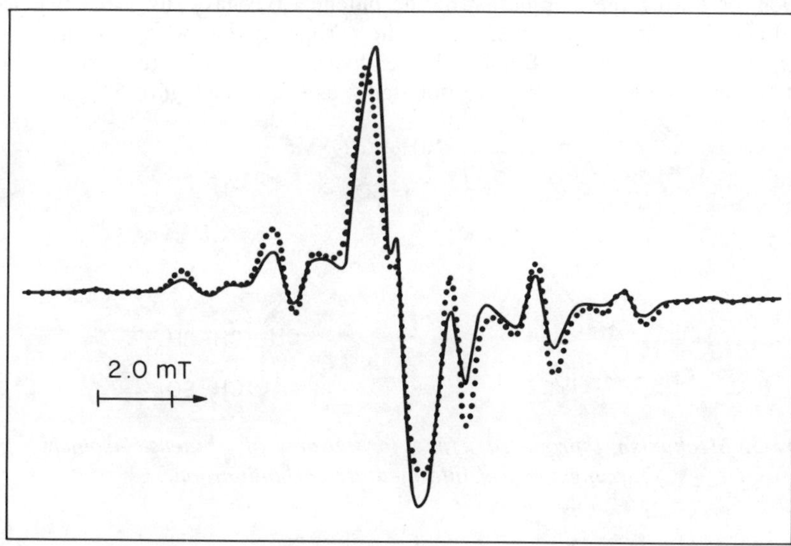

Figure 4. ESR spectrum of poly(isobutene sulfone) after γ irradiation at 77°K. (–) low and (· · ·) high microwave power.

decreased intensity of the singlet from the sulfonyl radical, in agreement with expectation. The number of lines in the alkyl radical spectrum is consistent with scission of bond E. It is quite clear that C-H scission (A) cannot explain the spectrum. It can also be deduced that the C-S scission occurs preferentially adjacent to the more heavily substituted carbon atom. However, there must be smaller amounts of other radicals as indicated by the analyses of volatile products.

There is evidence that both ionic and free radical species are involved in the degradation and depolymerization of poly(olefin sulfone)s by high energy radiation (*10*). Thus, the yields of olefins from poly(1-butene sulfone) at 30°C (the sample was heated to 70°C during removal of the gaseous products) are shown in Table II. The butene is not solely 1-butene, but comprises significant proportions of all three isomers, 1-butene, 2-butene and isobutene.

Isomerization must occur during the radiation-induced degradation, either of the initially produced 1-butene as a radiation-induced secondary

Table II. Yields of Butene Isomers from the γ Radiolysis of Poly(1-Butene Sulfone) at 30°C

Radiolysis Product	*G* value
1-butene	11.8
iso-butene	2.8
2-butene	2.2

reaction, or during the elimination of the butene. We have suggested that the most likely process is rearrangement of the carbonium ion A to give the more favorable secondary species B and C by 1,3 hydride shift (with the formation of 2-butene) or 1,3 alkyl shift (to give isobutene) as shown in Figure 5.

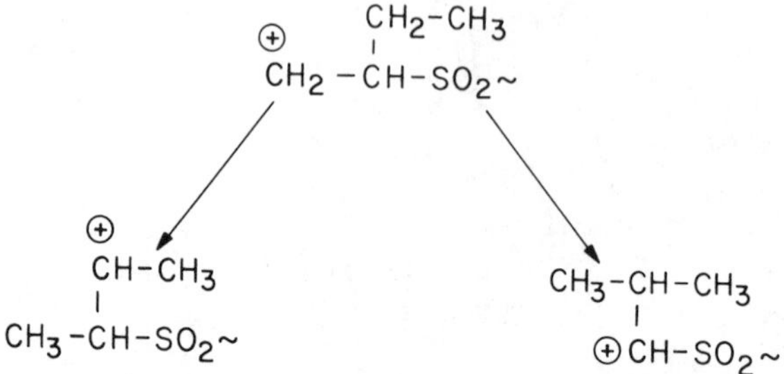

Figure 5. Mechanism proposed for formation of butene isomers by rearrangement of intermediate carbonium ion.

Poly(amino acids).

Homopolymers. The synthetic poly(amino acid)s serve as useful models for investigation of the radiation damage to biological polypeptides because, while they have the same basic backbone structure as the proteins, they do not have the complicating variety of amino acid residues, and hence side chains. The homopolymers also offer the advantage over small molecule models, such as the N-acetyl amino acids and oligomeric peptides, that end-group effects do not play a significant role in determining the major degradation pathways (*11*).

Relatively little quantative work has been carried out on the radiation degradation of synthetic amino acid homopolymers or copolymers, either in the solid state or in aqueous solutions. We have investigated the solid state degradation of a series of amino acid polymers by gamma radiation, (*12*) including glycine (R=H), L-alanine (R=CH_3), L-valine (R=$CH(CH_3)_2$), L-glutamic acid (R=$(CH_2)_2COOH$), L-phenylalanine (R=$CH_2\phi$) and L-tyrosine (R=$CH_2\phi OH$), in which the polymers have the general structure I.

$$-\!\!\left(NH - \underset{\displaystyle R}{\underset{|}{CH}} - CO \right)\!\!-_n$$

I

In each case we have identified the major radical intermediates that are formed on radiolysis at ambient temperatures, and measured their yields. We have also identified the major low molecular weight products and measured their yields.

Radicals. The major radical species found at room temperature for amino acids with aliphatic side chains are secondary radicals formed by hydrogen abstraction by the primary radicals that result from main-chain cleavage. Abstraction occurs usually at the α-carbon in the backbone chain, although in poly(valine) some abstraction also occurs at the tertiary carbon atom in the side chain, and in poly(glutamic acid) abstraction at the carbon atom adjacent to the side-chain carboxylic acid group also takes place. The carboxylic acid group in poly(glutamic acid) is radiation sensitive, with decarboxylation occurring readily, leading to formation of side chain alkyl radicals which are stable at room temperature.

The poly(amino acid)s with aromatic side chains behave somewhat differently. In poly(phenylalanine) the α-carbon radical is the major radical species observed, but radicals formed by hydrogen atom addition to the ring are also found. Benzyl radicals formed by side-chain cleavage are present, but only in very low yield. In poly(tyrosine) the only radical species observed is the tyrosyl phenoxyl radical formed by loss of the hydroxyl hydrogen. There is no evidence for formation of significant concentrations of α-carbon radicals. Thus, the nature of the substituents can strongly influence the radiation sensitivity of the backbone chain.

The radical intermediates and their yields are presented in Table III. It is noticeable that the G-values for production of α-carbon radicals are remarkably similar for the amino acids glycine, alanine, valine and glutamic acid, and are considerably higher than those for phenylalanine and tyrosine, indicative of the protective effect of the aromatic side chains.

Table III. Radicals Observed in Poly(Amino Acids) After γ Irradiation at 25°C and G Values for Their Production.
(Abbreviations for Amino Acids: Glycine = gly, Alanine = ala, Valine = val, Glutamic Acid = glu, Phenylalanine = phe, Tyrosine = tyr)

Polymer	Radicals	$G(R\cdot)$
gly	$\text{-NH-}\dot{\text{C}}\text{H-CO-}$	2.2
ala	$\text{-NH-}\dot{\text{C}}(\text{CH}_3)\text{-CO-}$	2.9
val	$\text{-NH-}\dot{\text{C}}(\text{CH}_2(\text{CH}_3)_2)\text{-CO}$	2.4
	$\text{-NH-CH}(\dot{\text{C}}\text{H}(\text{CH}_3)_2)\text{-CO-}$	0.6
glu	$\text{-NH-}\dot{\text{C}}(\text{CH}_2)_2\text{COOH})\text{-CO-}$	2.1
	$\text{-NH-CH}(\text{CH}_2\ \dot{\text{C}}\text{H}_2)\text{-CO-}$	0.8
	$\text{-NH-CH}(\text{CH}_2\ \dot{\text{C}}\text{H COOH})\text{-CO-}$	0.3
phe	$\text{-NH-}\dot{\text{C}}(\text{CH}_2\phi)\text{-CO-}$	1.0
	$\text{-NH-CH}(\text{CH}_2\ \dot{\phi}\text{H})\text{-CO}$	0.25
	$\dot{\text{C}}\text{H}_2\phi$	0.05
tyr	$\text{-NH-CH}(\text{CH}_2\phi\dot{\text{O}})\text{-CO-}$	0.4

Volatile Products. The volatile products which result from breakdown of the polymer backbone include carbon monoxide, ammonia and acetamide, which was the major product from the aliphatic amino acids. Side-chain fragments are also found in significant yields. The presence of acetamide as the major product suggests that two related chain scission reactions coupled with elimination of the side chain must be highly favourable.

The presence of volatile nitrogen-containing products is indicative of main-chain scission. A plot of the total yield of these low molar mass products versus the yield of hydrogen abstraction radicals (which are formed from the primary radicals) is shown in Figure 6. The linear correlation with a slope close to one suggests that the formation of main-chain radicals is accompanied by the elimination of segments from the backbone polymer chain.

Copolymers. Studies of the radiation sensitivities of amino acid homopolymers highlight the protective effect of aromatic substituents in the amino-acid side-

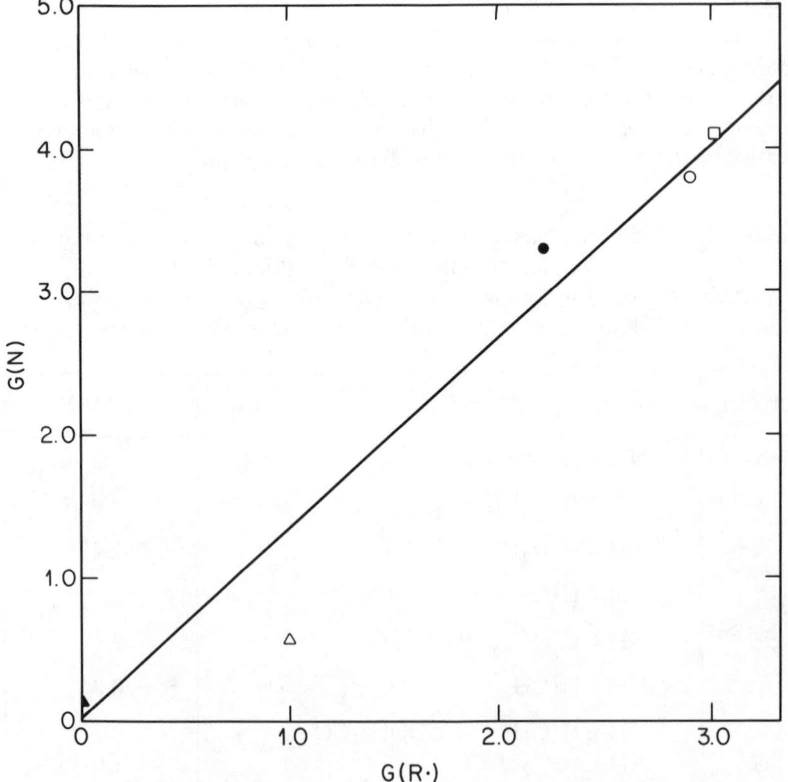

Figure 6. Relationship between yields of low molar mass nitrogen-containing products resulting from elimination of a section of the main chain and hydrogen abstraction radical from the γ irradiation of poly(amino acid)s at 25°C. (Δ) tyr, (Δ) phe, (●) gly, (○) ala, (□) val.

chain. In order to examine the range of this protective effect, we have examined two random copolymers of glutamic acid and tyrosine. The variation in the total radical yield with copolymer composition, and the dependence of the yield of carbon dioxide (resulting from decarboxylation of the side-chain) upon composition are illustrated in Figures 7A and B respectively. These figures demonstrate that incorporation of small amounts of tyrosine in a glutamic acid copolymer decreases considerably the radiation sensitivity of the copolymer. This indicates that the protection afforded by the tyrosine extends over several glutamic acid residues.

Poly(olefins)s. Radiation modification of the properties of poly(olefin)s has provided the basis for an industrial process of considerable magnitude. Polyethylene has superficially a particularly simple chemical structure and is therefore attractive to study. However, in practice, the molecular and morphological structures of PE can be exceedingly complex. In particular, short-chain branching has a major influence on the properties of LDPE. The complexities resulting from short-chain branching, unsaturation, chain ends and especially morphology have meant that there is still a great deal unknown and unquantified in the radiation degradation of PE.

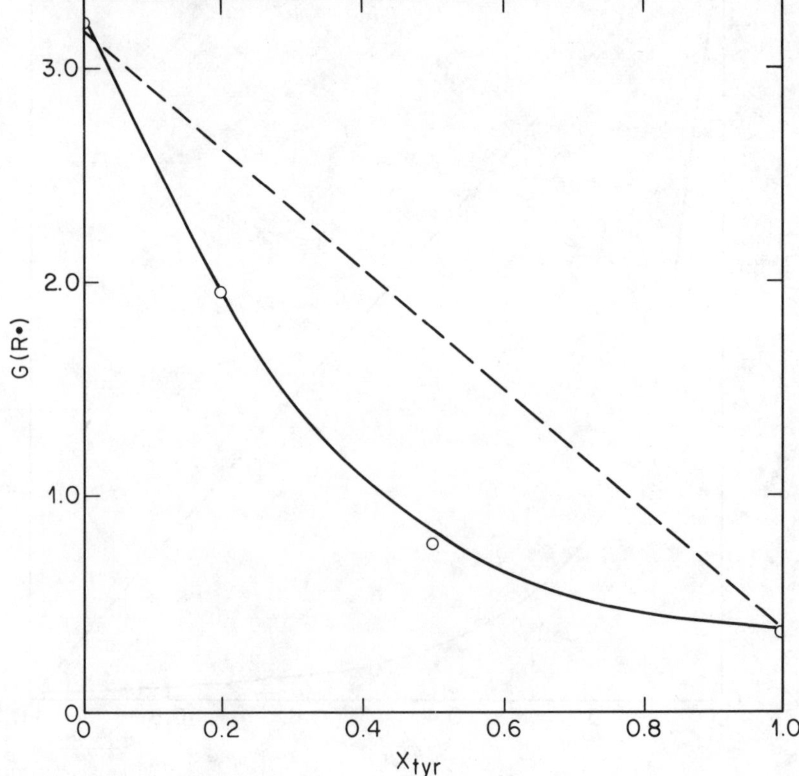

Figure 7a. Effect of tyrosine residues in copolymers with glutamic acid on the yields of total radicals.

The distribution of volatile products of low molar mass from the irradiation of poly(olefin)s is strongly dependent on the nature of substituents (short-chain branches) on the backbone chain. Hydrogen is the main volatile product with smaller quantities of alkanes and alkenes.

We have studied the alkane and alkene yields from the radiolysis of copolymers of ethylene with small amounts of propylene, butene and hexene. These are examples of linear low density polyethenes (LLDPE) and models for LDPE. Alkanes from C_1 to C_6 are readily observed after irradiation of all the polymers in vacuum. The distribution of alkanes shows a maximum corresponding to elimination of the short-chain branch. This is illustrated in Figure 8 for the irradiation of poly (ethylene-co-1-butene) containing 0.5 branches per 1,000 carbon atoms at 20°C.

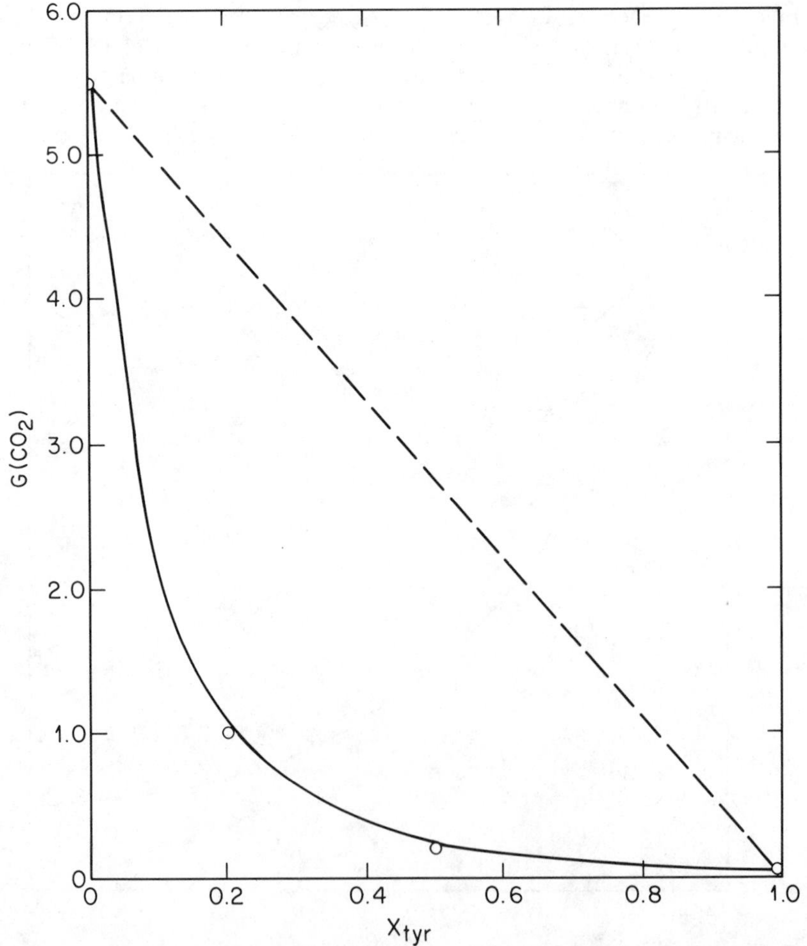

Figure 7B. Effect of tyrosine residues in copolymers with glutamic acid on the yield of CO_2 from γ irradiation at 25°C.

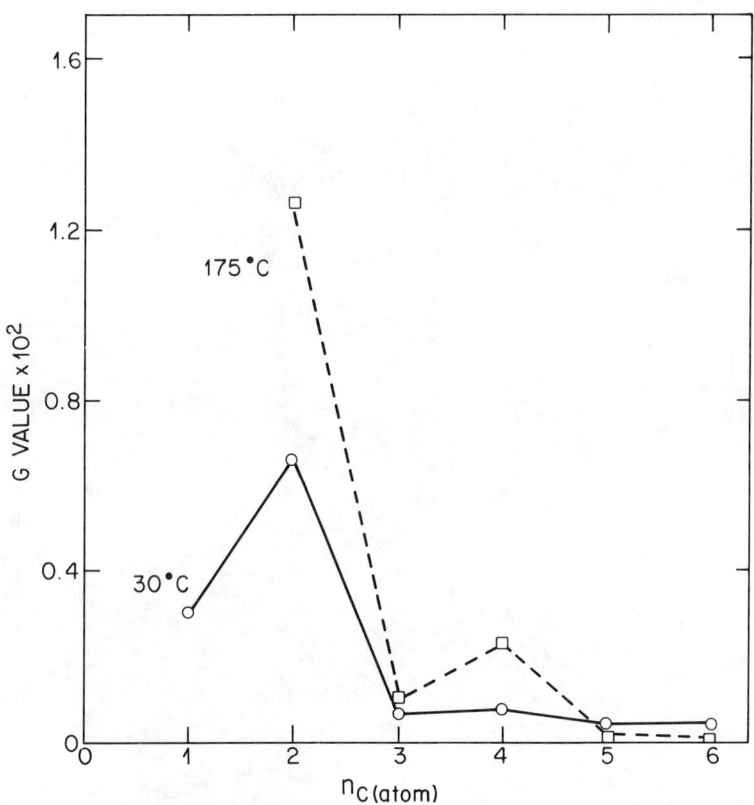

Figure 8. Distribution of alkane and alkene products from the γ irradiation of poly(ethylene-co-1-butene). (○) alkanes at 30°C, (□) alkenes at 175°C.

The relative sensitivity of short-chain alkyl branches of different sizes to elimination on irradiation with formation of the corresponding alkane has been variously reported as being constant or varying (*13,14*). Figure 9 compares *G* values for formation of the alkane corresponding to the short-chain branch from samples of these three polymers with branch frequencies from 0.5 to 6 per 1,000 C atoms. There is a notably higher scission efficiency for ethyl branches.

This technique has been used to investigate the distribution of short-chain branches in LDPE and the relatively higher scission efficiency of ethyl branches would rationalize the yields with ^{13}C NMR measurements of branch frequencies (*2*).

Alkenes are only produced in significant amounts above ca. 80°C. Ethylene is produced with the highest yield, which may be comparable to that for alkanes from short-chain branches after irradiation above 150°C (*15*). Typical results for the increasing yields of alkanes and alkenes with irradiation temperature are shown in Figure 10. Closer examination of the butane and butene produced has shown that they include considerable proportions of isobutane and isobutene. Typical *G* values for the formation of the butenes at

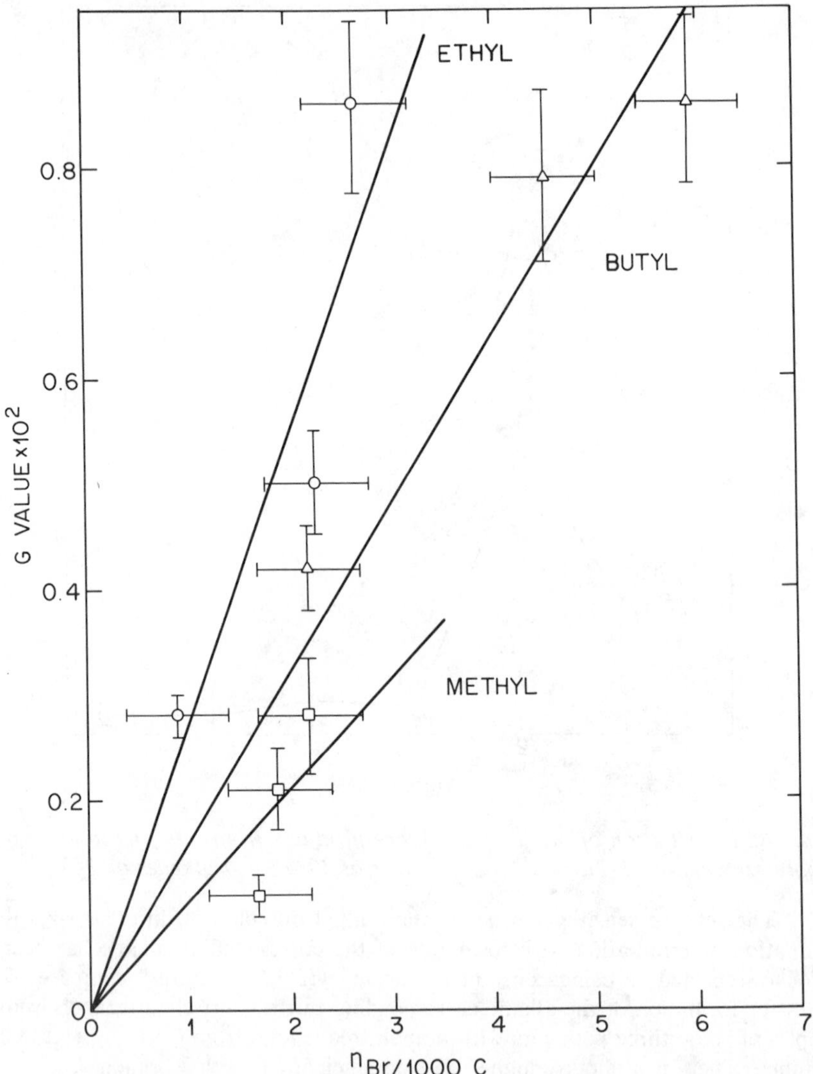

Figure 9. Relative scission efficiencies of short-chain branches in the γ radiolysis of ethylene-α-olefin copolymers. Branch types are indicated.

150°C are given in Table IV. The relatively greater proportion of 1-butene from ethylene-hexene copolymers indicates that there are two mechanisms for the formation of the butenes, one involving the butyl branches and that this pathway yields a much higher proportion (perhaps 100%) of 1-butene. The C_4 hydrocarbons are apparently also formed by fragmentation of chain ends in the polymers. These are probably formed mainly by radiation-induced scission.

Irradiation of polymer samples containing added n-butane or 1-butene, produced no marked increases in the yields of the isomers. Therefore, isobutane

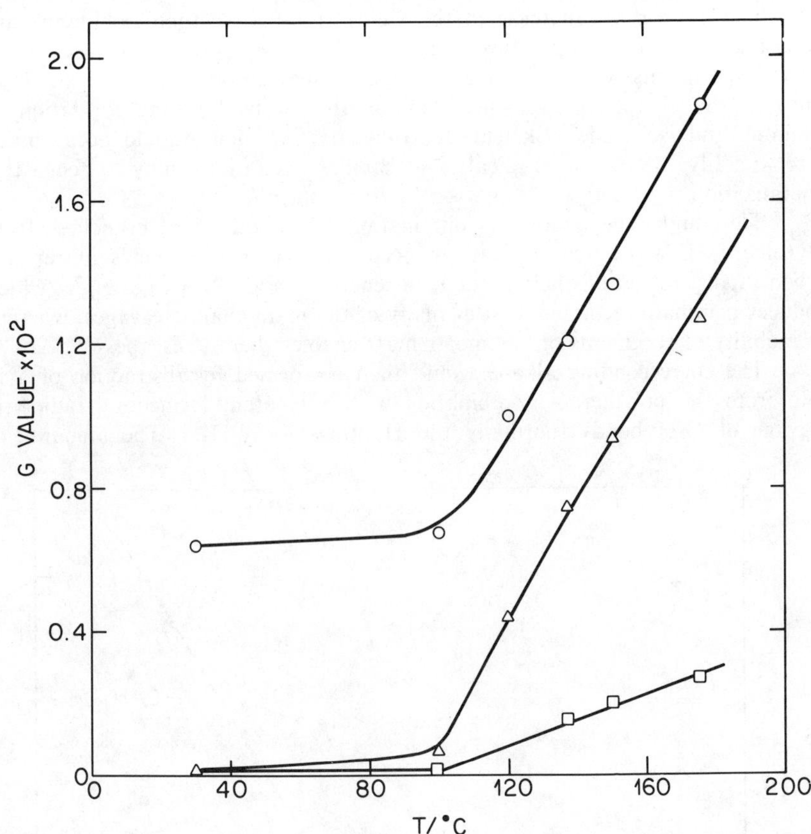

Figure 10. Temperature dependence of the yields of (○) butane, (△) ethylene and (□) butene from γ irradiation of poly(ethylene-co-1-hexene).

Table IV. Yields of Isobutene and 1-Butene from Irradiation of Ethylene-α-Olefin Copolymers at 150°C

	G Value ×10^2	
Polymer	Isobutene	1-Butene
EP	0.35	0.09
EB	0.44	0.12
EH	0.41	0.51

and isobutene are not formed significantly by radiation-induced isomerization of n-butane and 1-butene. However, efficient scavenging of H atoms by 1-butene *was* observed, highlighting the need to use low doses of radiation to avoid secondary reactions of unsaturated radiolysis products.

The more rapid increase in the yields of n-butane than isobutane with increasing irradiation temperature from 30 to 175°C is shown in Figure 11.

Experiments with selective scavengers for cations and radicals have indicated that both types of intermediate are involved in the formation of isobutane and isobutene. Skeletal rearrangements of ions should occur much more readily than of radicals, although excitation may affect this generalization.

The high specificity of elimination of alkyl side branches from poly(olefin)s is attributed to (1) enhanced scission of C-C bonds at tertiary carbon atoms, reflecting their lower bond energies, and (2) a cage effect which produces geminate recombination of many of the main-chain cleavages, whereas the mobility of fragments of low molar mass enables them to escape.

The corresponding alkane would then be formed by abstraction of a H atom from the polymer or combination with a H atom formed by radiolytic cleavage of C-H bonds (normally the H atoms form H_2). The amounts of

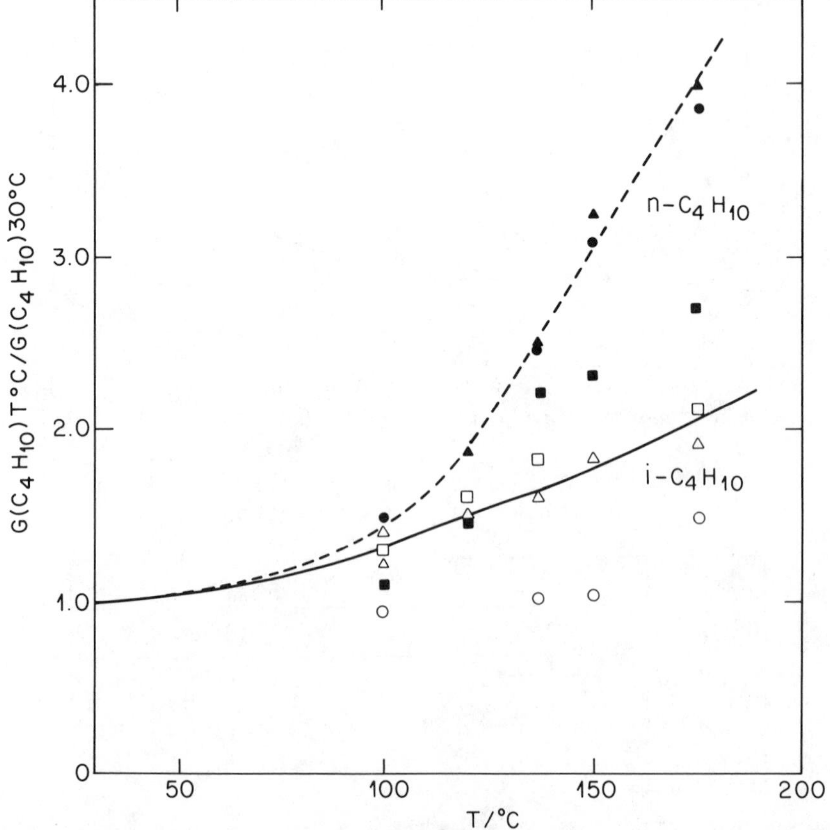

Figure 11. Variation of the relative yields of C_4 *alkane and alkene isomers with irradiation temperature.*

alkane dimer and alkene formed by combination and disproportionation of the eliminated alkyl branches is small apparently due to their low concentrations.

This radiolysis technique can also be used to investigate the nature and frequency of short-chain branches in other polymers. For example, poly(vinyl chloride) can be reduced to the corresponding polyethylene by treatment with $LiAlH_4$ or $(n\text{-}Bu)_3SnH$. The distribution of volatile hydrocarbon products can then be compared with the results obtained for ethylene-α-olefin copolymers. Thus, the relatively high yield of methane indicates methyl branches in the reduced polymer. This is consistent with the presence of CH_2Cl groups in the original PVC, which would result from head-head addition of monomer, followed by a 1,2 Cl shift reaction. Corrections to the yields of other hydrocarbons for chain end fragmentation are substantial, but the net results suggest the presence of small concentrations of ethyl and butyl branches.

Aromatic Polymers. Recent ESR investigations at 77°K utilizing dose-saturation, microwave power saturation, photobleaching and progressive warming techniques have provided strong evidence for the role of ionic as well as neutral radical intermediates in the radiation degradation of polymers containing aromatic groups in the backbone chain. These observations have been made on polysulfones, poly(phenylene oxide), poly(phenylene sulfide), polyamides and polyimides. For example, the radical concentration versus radiation dose relationship of an irradiated polysulfone (Figure 12) shows a marked reduction in the rate of radical formation above approx. 0.5 Mrad. The spectrum comprises mainly a sharp singlet, which shows dose saturation, and several other broader singlets.

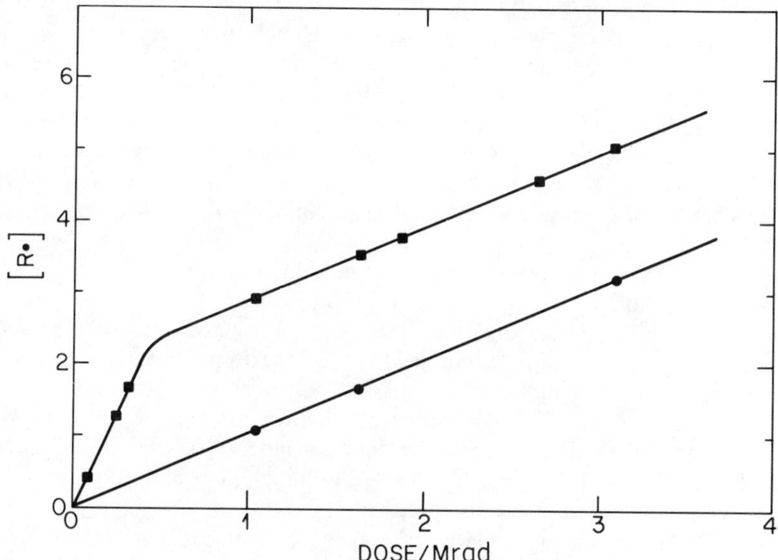

Figure 12. Variation of the yield of radicals determined by ESR with radiation dose for bisphenol-A aromatic polysulfone showing (□) dose saturation and (●) effect of photobleaching.

If ESR spectra are recorded at different microwave powers and a plot of spectrum area versus the square root of power constructed, then the spectrum shows increasing deviation from linearity above a critical "saturation" power. This is normally the maximum microwave power which should be used to record spectra. However, observation of the ESR spectra from irradiated aromatic polymers at higher power levels has shown selective saturation of the singlet produced preferentially at low doses. Thus the dose saturation behavior is mirrored by the power saturation.

We have photolysed the irradiated polymers at 77°K with visible light above a minimum wavelength, which has been varied from the blue to the red region of the spectrum. Complete bleaching of the sharp singlet in the ESR spectrum was observed. The combined evidence from the ESR spectra indicates that the saturating, photobleachable, singlet species can be attributed, at least in part, to a radical ion which is likely to be associated with the aromatic rings in the polymer. The other singlets are probably due to phenoxy and sulfonyl radicals. It is interesting that the G value for the formation of the ionic species is considerably greater than the G values for chemical change, suggesting that it does not continue to be produced in such high yields at higher doses of radiation, or that it has mainly a non-chemical fate.

The presence of aromatic groups in polymers greatly reduces their radiation sensitivity. Aromatic polysulfones are commercially important engineering plastics with high temperature resistance and also show good radiation resistance (*16*). Development of polymers with improved radiation resistance should be possible by copolymerization of other aromatic structures into the chain.

The radiation degradation of a series of aromatic polysulfone homopolymers and copolymers with different molecular structures has been investigated. The techniques of (i) gaseous product analysis at moderate doses, (ii) gel fraction measurements by extraction at high doses, and (iii) ESR studies of radical intermediates formed during degradation at low temperatures and at low doses were used to investigate their radiation degradation behavior. This work provides an excellent example of the utilization of techniques which we have developed for use when only small amounts of research material are available.

The GPC technique was used to determine gel formation. Figure 13 shows the soluble fraction of aromatic polysulfone I measured with this technique after irradiation at 30 and 150°C. The relative radiation resistance of different polymers can be obtained by comparison of the gel doses (the highest dose for complete solubility of the polymer) provided that the initial molar masses of the of the polymers are known, or from $G(S)$ and $G(X)$ values; these can be derived from the dose dependence of the soluble fractions beyond the gel dose, using a Charlesby-Pinner, or Saito-type plot with allowance for the molar mass distribution.

ESR provides a rapid method for assessing the sensitivity of a particular polymer structure to radiation. We have shown in several systems that there is a good correlation between $G(R\cdot)$ and the G values for permanent change in

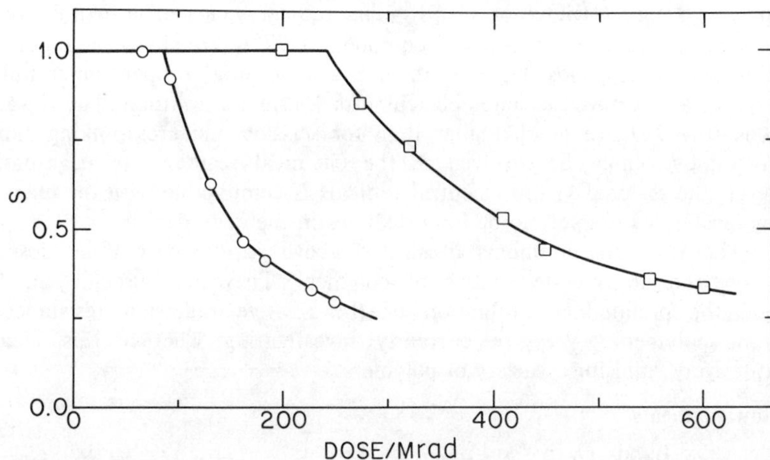

Figure 13. Dependence of the soluble fraction of bisphenol-A aromatic polysulfone on irradiation dose for irradiation at (□) 30°C, (○) 150°C, determined by GPC using miniature sample technique.

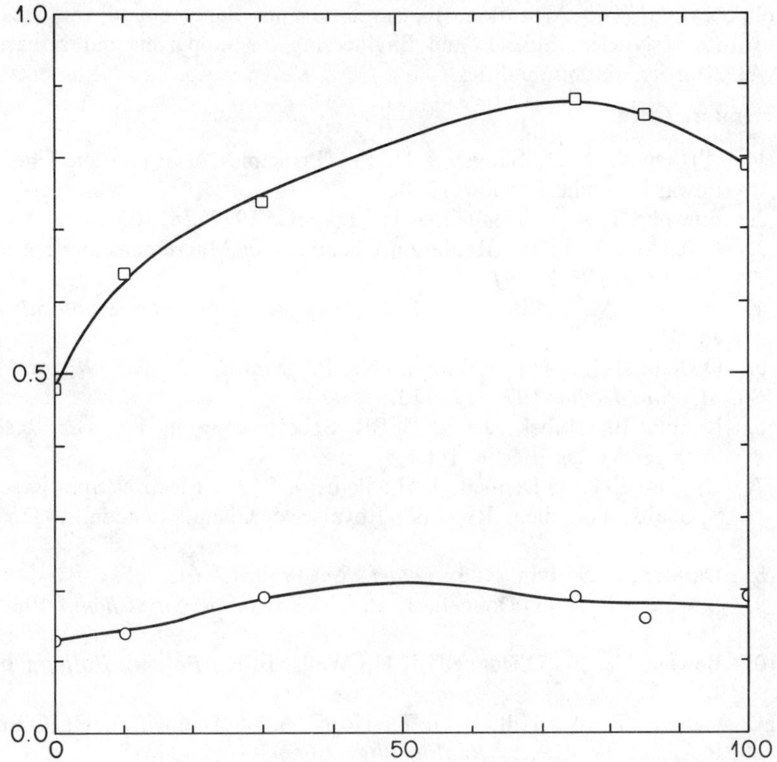

Figure 14. Variation of radical yield with copolymer composition for bisphenol-A — hydroquinone aromatic polysulfone copolymers after γ irradiation at 77°K. (□) initial (low dose) G values (○) high dose G values.

the polymer, e.g., $G(SO_2)$ or $G(S)$. The radical yields after irradiation of copolymers of bisphenol A and hydroquinone at 77°K are shown in Figure 14. The initial radical yields at 77°K include a substantial proportion of radical ions, which are removed by photobleaching or thermal annealing. The G values for ions (0.4-0.8) are much higher than for scission and cross-linking (about 0.1), so they cannot be involved in the chemical reactions of degradation. However, the residual yield of neutral radicals is comparable with the chemical change and they are likely to be intermediates in the degradation.

The ESR measurements illustrated above require only small doses of radiation and are non-destructive of polymer. They may provide the best method for preliminary evaluation of the relative radiation resistance of different polymers. We are currently investigating whether this idea is quantitatively valid for a variety of polymers.

Acknowledgments

We wish to thank Dr. T. N. Bowmer, Miss S. Y. Ho, Mr. R. W. Garrett, Mr. P. W. O'Sullivan, Mr. D.A. Lewis, Mr. A. K. Whittaker and Miss E. Grespos for experimental work; Mr. O. Delatycki and Mr. J. Hedrick for preparation of polymers; Prof. J. E. McGrath and Mr. G. Sykes for collaboration; The Australian Research Grants Scheme and the Australian Institute of Nuclear Science and Engineering for supporting our research; the AAEC for irradiation facilities.

Literature Cited

1. O'Donnell, J. H; Sangster, D. F. "Principles of Radiation Chemistry", Edward Arnold: London, 1970.
2. Bowmer, T. N.; O'Donnell, J. H. *Polymer* 1977, *18*, 1032.
3. Dole, M. (Ed.) "The Radiation Chemistry of Macromolecules"; Academic Press: New York, 1972.
4. Hellman, M. Y.; Bowmer, T. N.; Taylor, G. N. *Macromolecules* 1983, *16*, 34.
5. O'Donnell, J. H.; Rahman, N. P.; Smith, C. A.; Winzor, D. J. *Macromolecules* 1979, *12*, 113.
6. Ranby, B.; Rabek, J. F. "ESR Spectroscopy in Polymer Research"; Springer-Verlag: Berlin, 1964.
7. Hill, D. J. T.; O'Donnell, J. H.; Pomery, P. J. "Electron Spin Resonance, Specialist Periodical Reports"; Royal Soc. Chem.: London, 1982, Vol. 7, Ch. 1.
8. Dainton, F. S.; Ivin, K. J. *Quart. Rev* 1958, *12*, 61.
9. Bowmer, T. N.; O'Donnell, J. H. *J. Macromol. Sci. Chem.* 1982, *A17*, 243.
10. Bowmer, T. N.; O'Donnell, J. H.; Wells, P. R. *Polym. Bulletin* 1980, *2*, 103.
11. Garrett, R. W.; Hill, D. J. T.; Ho, S. Y.; O'Donnell, J. H.; O'Sullivan, P. W.; Pomery, P. J. *Radiat. Phys. Chem.* 1982, *20*, 351.
12. Hill, D. J. T.; Garrett, R. W.; Ho, S. Y.; O'Donnell, J. H.; O'Sullivan, P. W.; Pomery, P. J. *Radiat. Phys. Chem.* 1981, *17*, 163.

13. Boyle, D. A.; Simpson, W.; Waldron, J. D. *Polymer* 1961, *2*, 323.
14. Kamath, P. M.; Barlow, A. *J. Polym. Sci.*, A-1, 1967, *5*, 2023.
15. Bowmer, T. N.; Ho, S. Y.; O'Donnell, J. H. *Radiation Degrad. Stab.* 1983, *5*, 449.
16. Brown, J. R.; O'Donnell, J. H. *J. Appl. Polym. Sci.* 1979, *23*, 2763.

RECEIVED August 6, 1984

5

Pulse Radiolysis Studies on the Mechanism of the High Sensitivity of Chloromethylated Polystyrene as an Electron Negative Resist

Y. TABATA*, S. TAGAWA*,**, and M. WASHIO*

*Nuclear Engineering Research Laboratory
Faculty of Engineering
University of Tokyo
22-2 Shirane Shirakata
Tokai-mura, Naka-gun
Ibaraki-ken, Japan

**Research Center for Nuclear Science and Technology
University of Tokyo
22-2 Shirane Shirakata
Tokai-mura, Naka-gun
Ibaraki-ken, Japan

The progress of technology for the high-resolution fabrication of semiconductor and magnetic bubble devices has required sub-micron exposure techniques such as electron beam x-ray and deep UV. Although a number of papers have been published on electron beam resists, reaction mechanisms of electron resists are still largely unknown since few studies on reactive intermediates by means of direct measurements have been done in order to elucidate the reaction mechanisms.

Recently chloromethylated polystyrene (CMS), a highly sensitive, high resolution electron resist with excellent dry etching durability, was developed. Very recently reactive intermediates in irradiated polystyrene, which is a starting material of CMS, have been studied and the transient absorption spectra of excimer (*2-4*), triplet states (*2,5*), charge-transfer complexes, and radical cations (*6*) of polystyrene have been measured. The present paper describes the cross-linking mechanism of the high sensitivity CMS resist and compares it to that of polystyrene on the basis of data on reactive intermediates of polystyrene and CMS.

Experimental

Details of the picosecond pulse radiolysis system for emission (*7*) and absorption (*8*) spectroscopies with response time of 20 and 60 ps, respectively, including a specially designed linear accelerator (*9*) and very fast response optical detection system have been reported previously. The typical pulse radiolysis systems are shown in Figures 1 and 2. The detection system for emission spectroscopy is composed of a streak camera (C979, HTV), a SIT

0097–6156/84/0266–0151$06.00/0

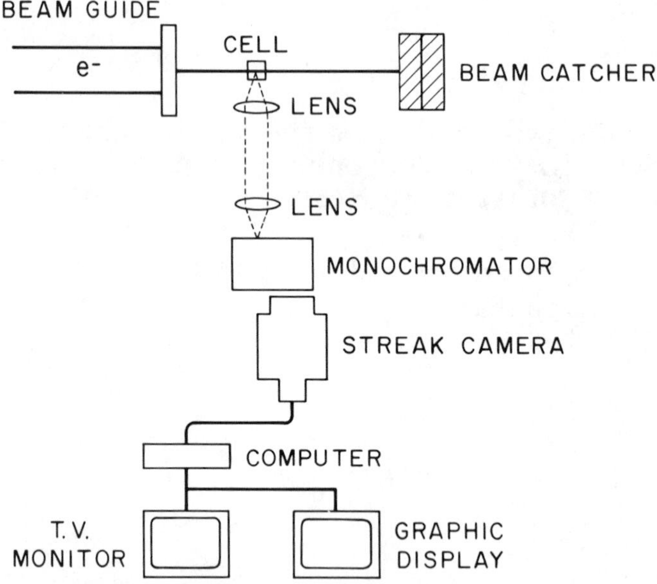

Figure 1. The schematic diagram of the picosecond pulse radiolysis system for emission spectroscopy.

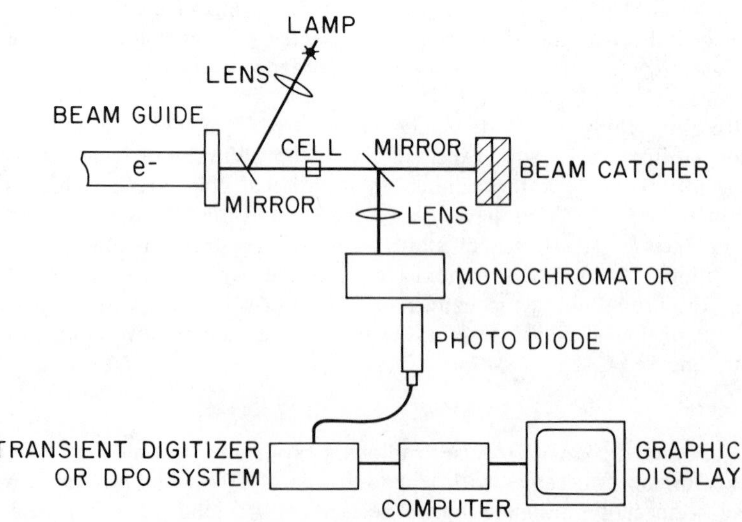

Figure 2. The schematic diagram of the picosecond pulse radiolysis system for absorption spectroscopy.

camera (C1000-12, HTV), an analyzer (C1098, HTV), a computer (PDP 11/34), a display system (HTV and Tektronix) (see Figure 1). The detection system for absorption spectroscopy is composed of a very fast response photodiode (R1328U, HTV), transient digitizer (R7912, Tektronix) or DPO system (R7704, Tektronix), a computer (PDP 11/34), and a display unit (Tektronix) (see Figure 2). CMS was prepared from polystyrene and chloromethyl-methylether using $SnCl_4$ as the catalyst (*1,10,11*). The structure of CMS is shown in Figure 3.

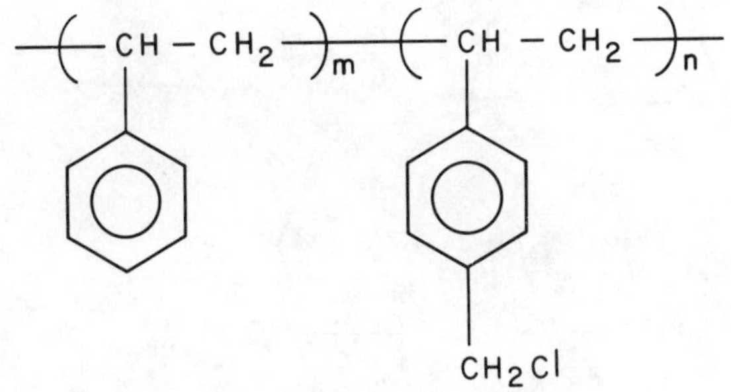

$$-\!\left(CH - CH_2\right)_m\!-\!\left(CH - CH_2\right)_n\!-$$

CH_2Cl

Figure 3. The structure of CMS: n/(m+n) chloromethylation ratio.

Results and Discussions

Solid Films. The excimer fluorescence of solid films of polystyrene was observed using pulse radiolysis. The decay curves of the excimer fluorescence observed at 340 nm for solid films of polystyrene and CMS are shown in Figures 4(a) and (b), respectively. The lifetime of the excimer fluorescence of polystyrene agree with the reference data (*12*). In CMS, the initial yield decreases and the decay rate of excimer fluorescence increases with increasing chloromethylation ratio of CMS. These experimental results indicate that the chloromethyl part of CMS quenches the excimer of CMS and scavenges the precursors of the excimer as described below.

precursors of the excimer ⟶ the excimer

| scavenged by $-CH_2Cl$ | quenched by $-CH_2Cl$ |

The absorption spectrum observed in the pulse radiolysis of solid films of polystyrene is shown in Figure 5. The absorption spectrum around 540 nm is also very similar to the absorption spectrum of polystyrene excimer observed in irradiated polystyrene solutions in cyclohexane as reported previously (*2,3*). The absorption with the maximum at 410 nm was reported previously and was assigned to anionic species (*13,14*). The longer life absorptions, attributed to triplet excited polystyrene repeat units and nonidentifiable free radicals, were observed in a wave length region <400 nm. The absorption spectrum of CMS films obtained in pulse radiolysis showed a peak around 320 nm and a very broad absorption around 500 nm as shown in Figure 6.

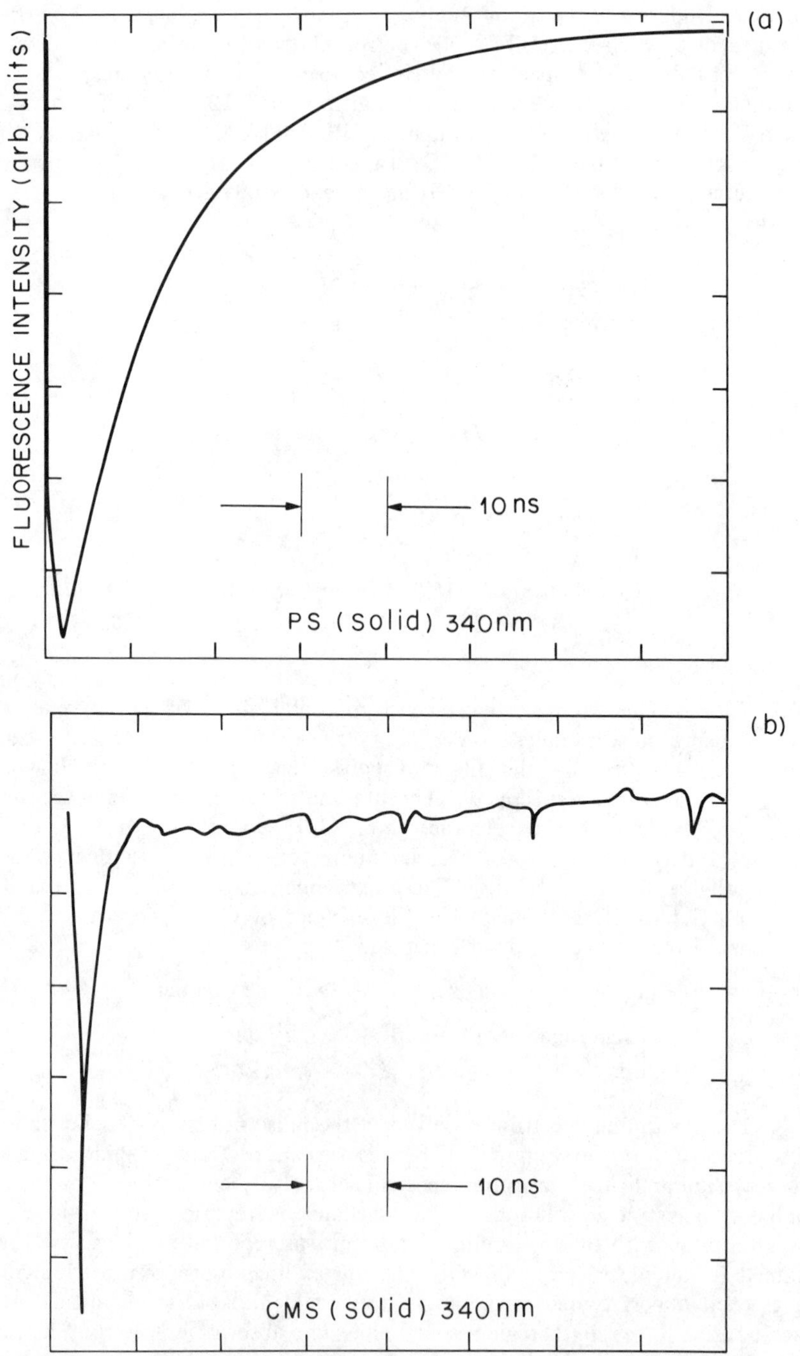

Figure 4. Typical oscilloscope trace of emission behavior obtained from pulse radiolysis of (a) polystyrene solid and (b) CMS solid at 340 nm.

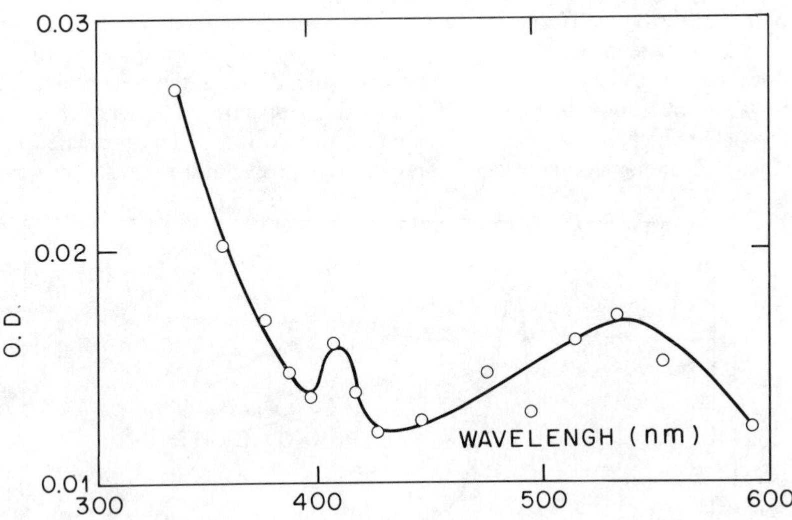

Figure 5. The absorption spectrum obtained from the pulse radiolysis of polystyrene solid.

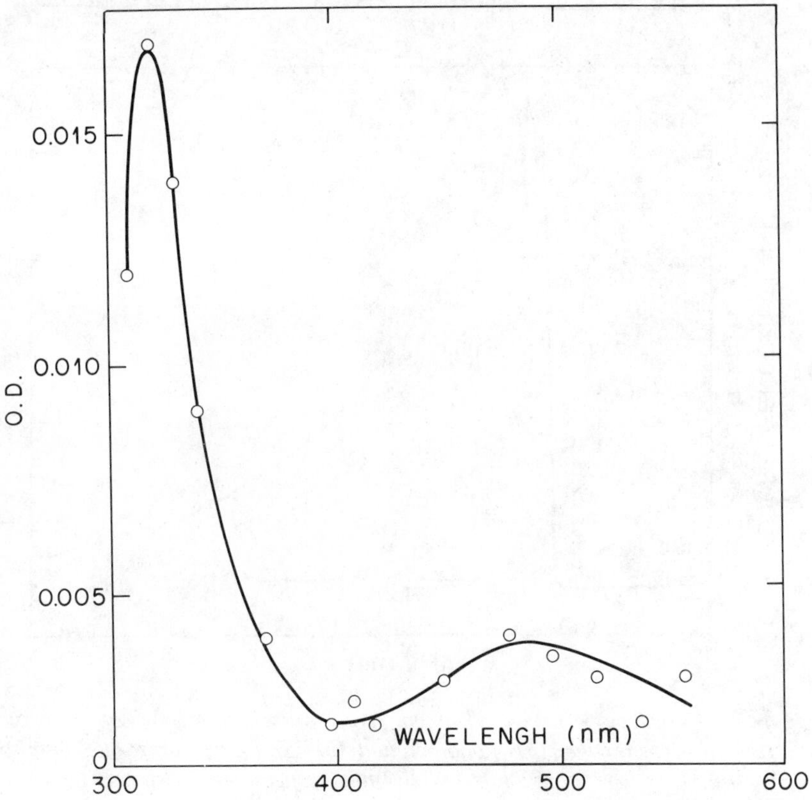

Figure 6. The absorption spectrum obtained from pulse radiolysis of CMS solid.

CMS and Polystyrene Solutions in Cyclohexane. Both monomer and excimer fluorescences were observed in the pulse radiolysis of polystyrene solution in cyclohexane. The decay curves of monomer and excimer fluorescences at 287 and 360 nm are shown in Figures 7(a) and (b), respectively. Energy migration on the polymer chain has been discussed elsewhere (*15*). The dependences of the decay of monomer fluorescence and the rise of excimer fluorescence on the

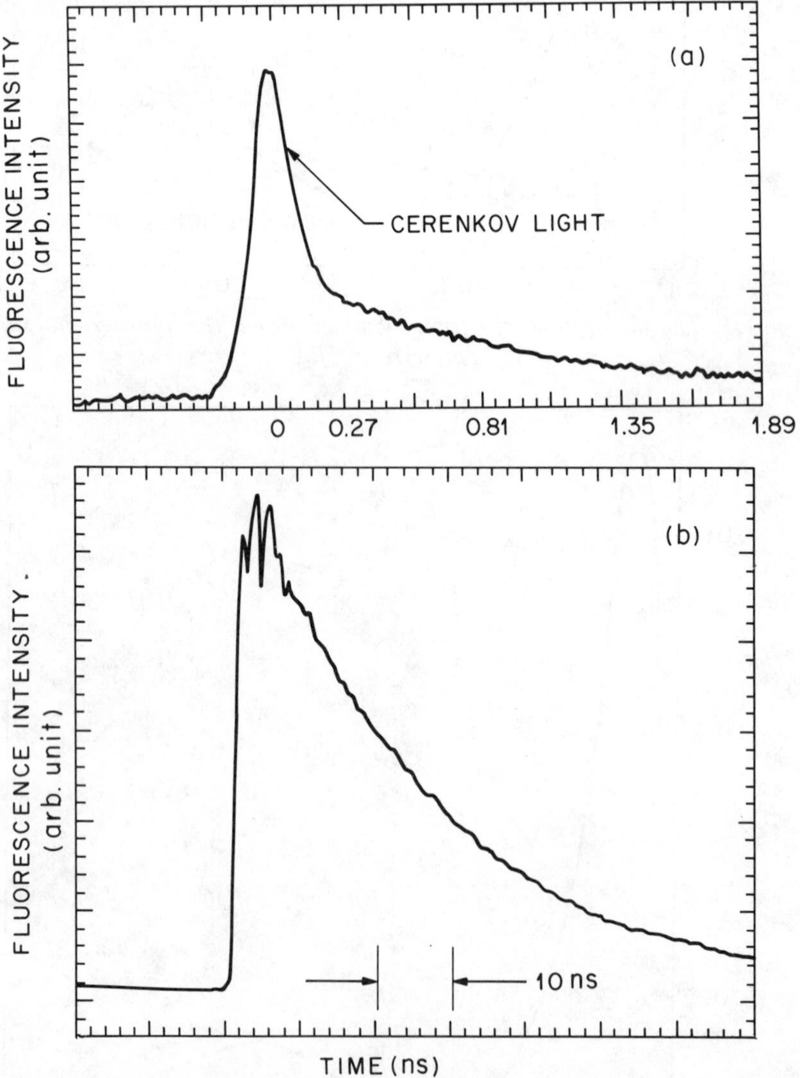

Figure 7. The decay curves obtained from pulse radiolysis of polystyrene solution in cyclohexane; (a) monomer and (b) excimer fluorescence monitored at 287 nm and 360 nm, respectively.

chain length of oligomers have also been reported elsewhere (*15*). The lifetime of monomer fluorescence and the formation rate of excimer of polystyrene solution in cyclohexane are almost constant regardless of the chain length of the polymer. Figure 8 shows the transient absorption spectrum obtained in the pulse radiolysis of polystyrene solution in cyclohexane.

The absorption band around 520 nm is very similar to that of polystyrene excimer (*2,3,5*). The decay follows first order kinetics with a lifetime of 20 ns. The decay rate agrees with that of the excimer fluorescence and excimer absorption. The longer life absorptions, attributed to the triplet states and free radicals (*2,5*), were observed at wave lengths <400 nm, although the anionic species of polystyrene with the absorption maximum at 410 nm as seen in solid films (cf. Figure 5) was not observed. Figure 9 shows the absorption spectrum observed in the pulse radiolysis of CMS solution in cyclohexane.

Figure 10 shows the absorption observed in the pulse radiolysis of benzyl chloride solutions in cyclohexane using the same radiolysis system. The absorption spectrum agrees very well with the absorption spectrum of benzyl radicals observed in cyclohexane solutions reported previously (*16*). In irradiated cyclohexane solutions, the mobility of electrons (so-called quasi-free electrons) is very high ($\mu = 0.23$ cm $V^{-1}s^{-1}$) and electrons react with benzyl chloride very effectively. This allows the effective dissociative electron attachment reaction. The similar reaction (reaction (4)) may occur in irradiated CMS solutions in cyclohexane and the benzyl type radical, P_1, is

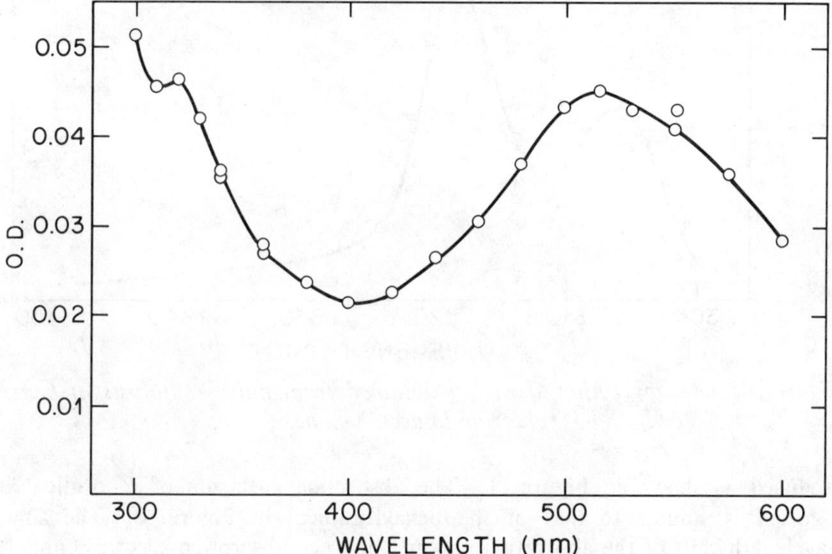

Figure 8. Transient absorption spectrum obtained by pulse radiolysis of 200 mM polystyrene solution in cyclohexane.

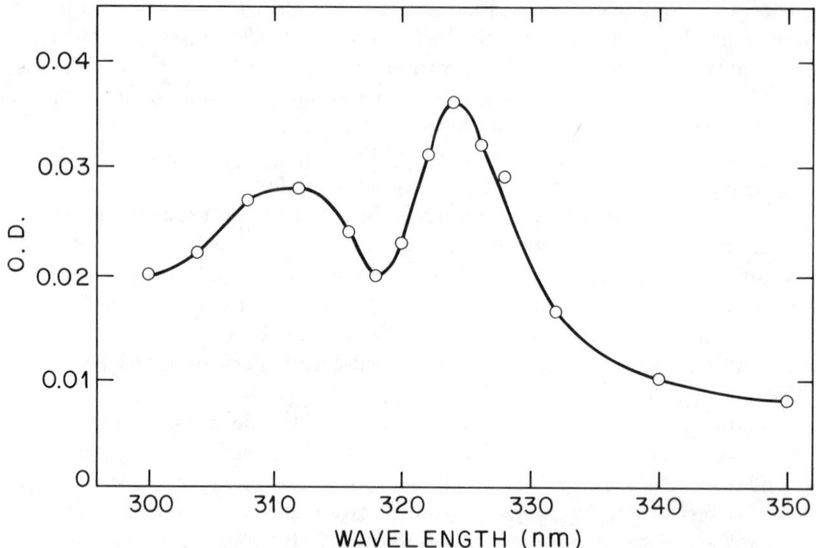

Figure 9. The absorption spectrum obtained from pulse radiolysis of CMS in cyclohexane.

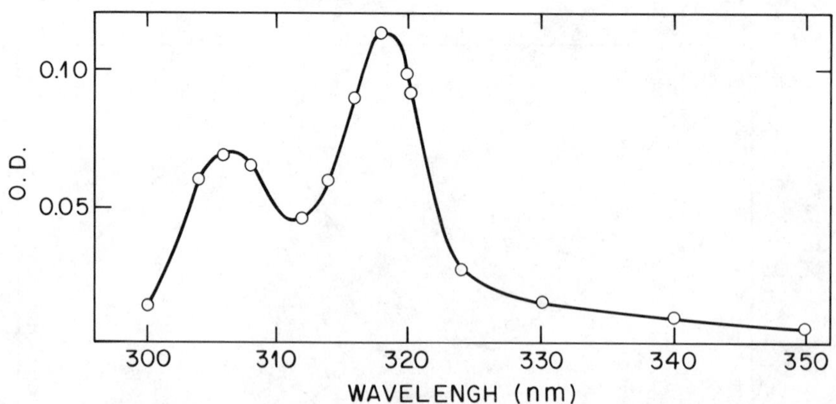

Figure 10. The absorption spectrum obtained from pulse radiolysis of benzyl chloride in cyclohexane.

produced as shown in Figure 11. The absorption spectrum of P_1 radical in Figure 9 is similar to that of the benzyl radical in Figure 10. The small wavelength shift of the absorption maxima between absorption spectra 9 and 10 is generally observed among substituted benzyl type radicals.

Polystyrene Solution in $CHCl_3$ and CCl_4. In the laser photolysis and pulse radiolysis of polystyrene solution in $CHCl_3$ one observes an absorption spectra with maxima at 320 nm and around 500 nm (*2,4,17*). as shown in Figure 12.

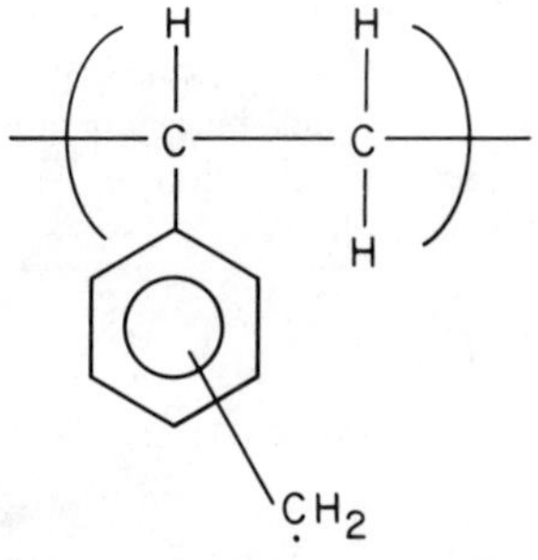

Figure 11. The structure of the P_1 radical.

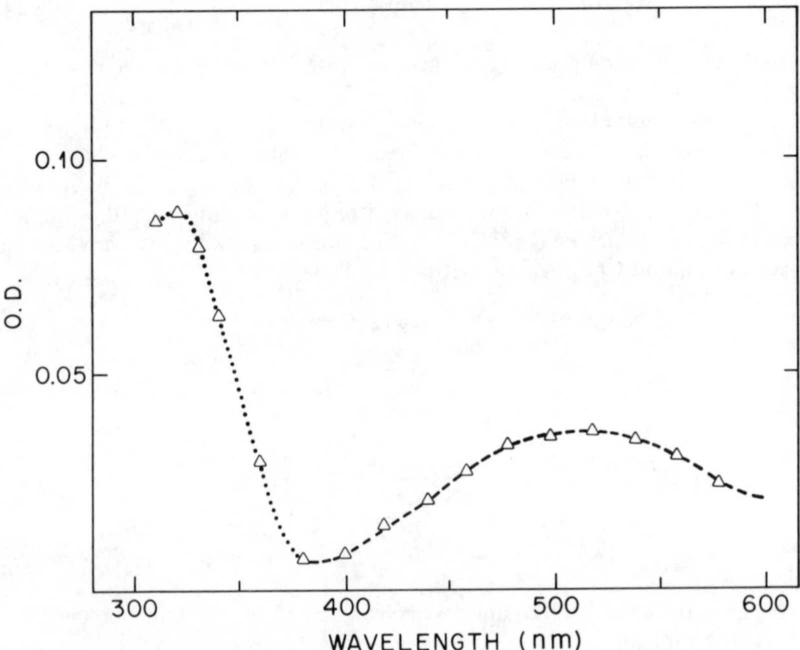

Figure 12. Transient absorption spectra obtained by pulse radiolysis of 200 mM polystyrene in $CHCl_3$. 170 ns after the pulse. (Absorbed dose is about 6 krad.)

The absorptions at both 500 nm and 320 nm follow first order kinetics with a lifetime of 420 ns. This absorption species is neither the excimer of polystyrene nor free cationic species of polystyrene. Although the excimer of polystyrene has an absorption band around 500 nm, the lifetime is only 20 ns. Further the free cationic species of polystyrene should live for a longer time in this solution, and the absorption band should exist in a longer wavelength region (*6*). These considerations of lifetime and absorption spectrum lead us to conclude that the absorption spectrum shown in Figure 12 is due to the charge transfer-radical complex between polystyrene and Cl radical (*2,4,17*). A very similar absorption spectrum is obtained by the pulse radiolysis of benzene in $CHCl_3$ solution. Therefore, the structure of the species shown in Figure 12 is most likely the charge transfer radical complex between phenyl rings and Cl radical. The possible formation processes of this charge transfer radical complex in liquid $CHCl_3$ can be considered as follows.

$$CHCl_3 \rightsquigarrow CHCl_3^{+\cdot} + e^- \quad (1)$$

$$CHCl_3 + e^- \longrightarrow \cdot CHCl_2 + Cl^- \quad (2)$$

$$CHCl_3^{+\cdot} + PS \longrightarrow PS^{+\cdot} + CHCl_3 \quad (3)$$

$$PS^{+\cdot} + Cl^- \longrightarrow (PS^{\delta+}\ Cl^{\delta-})\cdot \quad (4)$$

The last reaction should be very fast, because it represents geminate ion recombination.

The transient absorption spectrum obtained in the pulse radiolysis of polystyrene solution in CCl is shown in Figure 13. The spectrum is very similar to the charge transfer radical complex $(PS^{\delta+}Cl^{\delta-})\cdot$species. The lifetime is about 200 ns. Consideration of the absorption spectrum and the lifetime suggest that this species is $(PS^{\delta+}Cl^{\delta-})\cdot$. The processes leading to formation of this species in liquid CCl_4 can be written as follows (*4*,*7*).

$$CCl_4 \rightsquigarrow CCl_4^{+\cdot} + e^- \quad (5)$$

$$CCl_4 + e^- \longrightarrow \cdot CCl_3 + Cl^- \quad (6)$$

$$CCl_4^{+\cdot} + PS \longrightarrow PS^{+\cdot} + CCl_4 \quad (7)$$

$$PS^{+\cdot} + Cl^- \longrightarrow (PS^{\delta+}\ Cl^{\delta-})\cdot, \quad (8)$$

The charge transfer radical complex is considered to be the main precursor of the polystyryl radical.

Reaction Scheme of CMS Resists. The transient absorption spectrum shown in Figure 6 and observed for irradiated CMS films is mainly composed of two components as based on pulse radiolysis data of solid films of CMS and polystyrene, and CMS and polystyrene solutions in cyclohexane, chloroform, and carbon tetrachloride. An absorption with a maxima at 320 nm and 500 nm as due to the charge transfer radical-complex of the phenyl ring of CMS and chlorine atom (see Figure 14) and an absorption with maxima at 312 and 324 nm is due to benzyl type radicals (see Figure 11).

The scheme of radiation-induced reactions of CMS, negative electron resist, is proposed as follows on the basis of the present pulse radiolysis data.

$$CMS \rightarrow CMS^{+\cdot} + e^- \quad (9)$$

$$CMS \rightarrow CMS^* \quad (10)$$

$$CMS^{\dot{+}} + e^- \rightarrow CMS^* \quad (11)$$

$$CMS + e^- \rightarrow P_{\dot{1}} + Cl^- \quad (12)$$

$$Cl^- + CMS^{\dot{+}} \rightarrow \text{complex} \quad (13)$$

$$\text{complex} \rightarrow P_{\dot{2}} + HCl \quad (14)$$

$$CMS^* \rightarrow P_{\dot{1}} + Cl \quad (15)$$

$$CMS + Cl \rightarrow P_{\dot{2}} + HCl \quad (16)$$

$$CMS + Cl \rightarrow \text{complex} \quad (17)$$

$$CMS^* + CMS \rightarrow P_{\dot{1}} + \text{complex} \quad (18)$$

$$Cl + Cl \rightarrow Cl_2 \quad (19)$$

$$P_{\dot{1}} + P_{\dot{1}} \rightarrow P_1{-}P_1 \quad (20)$$

$$P_{\dot{1}} + P_{\dot{2}} \rightarrow P_1{-}P_2 \quad (21)$$

$$P_{\dot{2}} + P_{\dot{2}} \rightarrow P_2{-}P_2 \quad (22)$$

Cl reacts with hydrogen atoms on the main chain (reaction (16)) and can also interact with phenyl rings through reaction (17). CMS corresponds to

$$(C_6H_4{-}CH_2Cl)^*$$

in reaction (15) and

$$-(CH(C_6H_5)^*-CH_2)-$$

in reaction (18). The excited phenyl rings of CMS react with chloromethyl groups of CMS through reaction (10).

The high cross-linking sensitivity of CMS can be explained by the pair production of P_1 and P_2 radicals. The structure of the P_2 radical is shown in Figure 15. The formation of the P_2 radical from the charge transfer radical

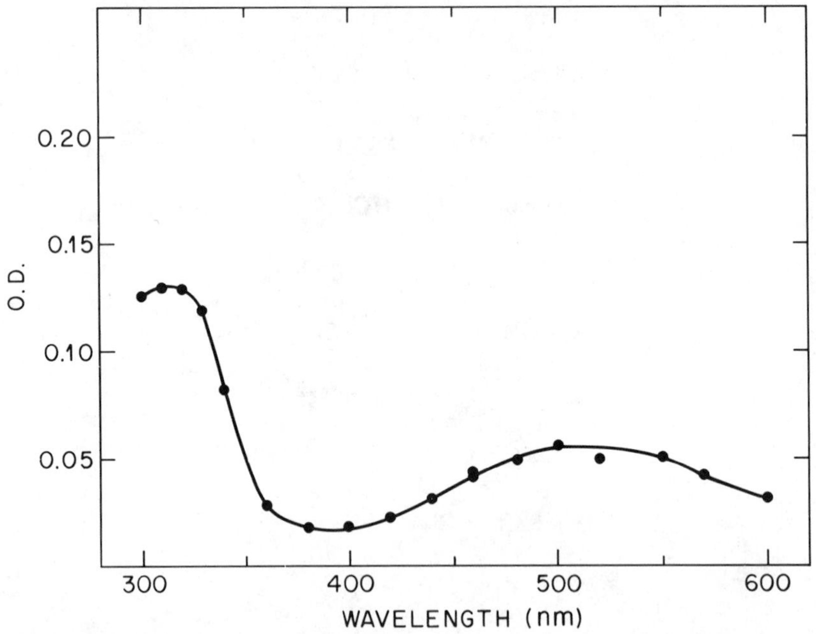

Figure 13. Transient absorption spectra obtained by pulse radiolysis of 200 mM polystyrene in CCl_4. 70 ns after the pulse. (Absorbed dose is about 6 krad.)

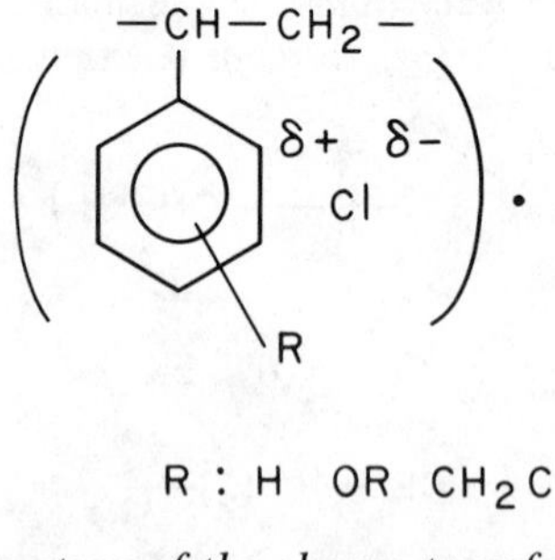

R : H OR CH_2Cl

Figure 14. The structure of the charge-transfer radical complex.

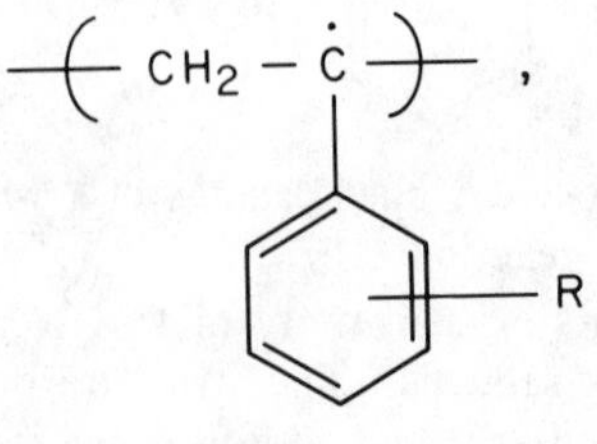

Figure 15. The structure of the P_2 radical.

complex has been reported elsewhere (*2,4,17*). There are three types of the formation of radical pairs of P_1 and P_2 radicals: (a) reactions (9) (12) (13) and (14), (b) reactions (10) (11) (15) and (16), (c) reactions (10),(11) (18) and (14).

Acknowledgments

The authors are grateful to S. Imamura, O. Kogure, and K. Murase of Ibaraki Electrical Communication Laboratory for helpful discussions.

Literature Cited

1. Imamura, S. *J. Electrochem. Soc.* 1979, *126*, 1628.
2. (a) Tagawa, S. and Schnabel, W. *Makromol. Chem., Rapid Commun.* 1980, *1*, 345. (b) Tagawa S. and Schnabel, W. *Polymer Photochemistry*, 1983, *3*, 203.
3. Tagawa, S. and Schnabel, W. *Chem. Phys. Lett.* 1980, *75*, 120.
4. Tagawa, S.; Schnabel, W.; Washio, M. and Tabata, Y. *Radiat. Phys. Chem.* 1981, *18*, 1087.
5. Tagawa, S.; Nakashima, N. and Yoshiwara, K. *Macromolecules*, in press.
6. Tagawa, S.; Beck, G. and Schanbel, W. *Z. Naturforsch* 1982 *37a*, p. 982 and unpublished data.
7. (a) Tagawa, S.; Katsumura, Y.; Ueda, T. and Tabata, Y. *Radiat. Phys. Chem.* 1980, *15*, 287. (b) Tabata, Y.; Katsumura, H.; Kobayashi, H.; Washio, M. and Tagawa, S. In "Picosecond Rhenomena II", Hochstrasser, R. M.; Kaiser, W.; Shank, C. V., Eds.; 266 Springer-Verlag, Berlin 1980; p. 266.
8. Kobayashi, H.; Ueda, T.; Kobayashi, T.; Washio, M.; Tabata, Y. and Tagawa, S. *Radiat. Phys. Chem.* 1983, *21*, 13.
9. (a) Tabata, Y.; Tanaka, J.; Tagawa, S.; Katsumura, Y. Ueda, T. and Hasegawa, K. *J. Fac. Eng. Univ. Tokyo* 1978 *34B*, 619. (b) Kobayashi, H.; Ueda, T.; Kobayashi, T.; Tagawa, S. and Tabata, Y. *Nucl Instrum. Meth.* 1982 *179*, 223. (c) Kobayashi, H.; Ueda, T.; Kobayashi, T.; Tagawa, and Tabata, Y. *J. Fac. Eng. Univ. Tokyo* 1981 *36B*, 85.
10. Imamura, S.; Tamamura, T.; Harada, L. and Sugawara, S. *J. Appl. Polym. Sci.* 1982, *27*, 937.
11. Lehman, H. W. and Widmer, R. *Appl. Phys. Lett.* 1978, *32*, 163.
12. Basile, L. J. *Trans. Faraday Soc.* 1964, *60*, 1702.
13. (a) Ho, S. K.; Siegel, S. and Schwarz, H. A. *J. Phys. Chem.* 1967, *71*, 4527. (b) Ho, S. K. and Siegel, S. *J. Chem. Phys.* 1968, *50*, 1142.
14. Thomas, J. K. *J. Chem. Phys.* 1969, *51*, 770.
15. Itagaki, H.; Horie, K.; Mita, I. Washio, M.; Tagawa, S. and Tabata, Y. *J. Chem. Phys.* 1983, *79*, 3996.
16. Hagemann, R. J. and Schwarz, H. A. *J. Phys. Chem.* 1965, *71*, 1960.
17. Washio, M.; Tagawa, S. and Tabata, Y. *Radiat. Phys. Chem.* 1983, *21*, 239.

RECEIVED August 6, 1984

6

Photochemistry of Ketone Polymers in the Solid Phase: A Review

J. E. GUILLET, S.-K. L. LI[1], and H. C. Ng[2]

Department of Chemistry, University of Toronto
Toronto, Canada M5S 1A1

The increasing importance of photopolymeric materials for lithography and the manufacture of microcircuits has led to an increased understanding of the photochemical processes in solid polymers. Although polymers are known to react readily in solution, the fact that they may be equally reactive in the solid phase has not been generally appreciated. In this paper we will discuss the characteristics of photochemical reactions of the ketone chromophore, both in solid and liquid systems. Since it is generally accepted that chemical reactions of small molecules in solution and in the gas phase occur as a result of activated collisions, it was expected that reactions would not occur in solid matrices because of the restrictions on molecular motion imposed by the solid. However, as we will see, organic polymeric materials, even in the glassy state, retain substantial freedom of motion of individual groups down to very low temperatures and these are often sufficient to permit chemical reactions to occur, particularly when the excitation is provided by a photon. This work will deal primarily with the photochemical and photophysical processes associated with the ketone group contained in a variety of polymeric macromolecules.

The ketone group is a useful model for other types of chromophores because it can be selectively excited in the presence of other groups in polymer chains such as the phenyl rings in polystyrene and so the locus of excitation is well defined. Furthermore there is a great deal known about the photochemistry of aromatic and aliphatic ketones and one can draw on this information in interpreting the results. A further advantage of the ketone chromophore is that it exhibits at least three photochemical processes from the same excited state and thus one has a probe of the effects of the polymer matrix on these different processes by determination of the quantum yields for the following photophysical or photochemical steps: 1) fluorescence, 2) phosphorescence, 3) the Norrish type I reaction, 4) the Norrish type II reaction, 5) photoreduction, and 6) the formation of cyclobutanol.

Current address: [1]Facelle Co. Ltd., 1551 Weston Road, Toronto, Ontario M6M 4Y4.
[2]Department of Chemistry, Chinese University of Hong Kong, Shatin, Hong Kong.

0097-6156/84/0266-0165$06.00/0

$$\text{>C(=O)-R} \xrightarrow{h\nu} \text{>C(=O}^{*}\text{)-R}\ (S^1) \longrightarrow \text{>C(=O)-R}\ (S^0) + h\nu_F \qquad (1)$$

$$\text{>C(=O)-R} \xrightarrow{h\nu} (S^1) \longrightarrow (T^1) \longrightarrow \text{>C(=O)-R}\ (S^0) + h\nu_P \qquad (2)$$

$$\text{>C(=O)-R} \xrightarrow{h\nu} \text{>}\cdot + \cdot\text{C(=O)-R} \qquad (3)$$

$$\text{>C(=O)-R} \xrightarrow{h\nu} \text{>=} + \text{CH}_3\text{-C(=O)-R} \qquad (4)$$

$$\text{>C(=O)-R} \xrightarrow[RH]{h\nu} \text{>C(OH)(H)-R} \qquad (5)$$

$$\text{>C(=O)-R} \xrightarrow{h\nu} \text{cyclobutane-C(OH)-R} \qquad (6)$$

$h\nu_F$ = FLUORESCENCE $h\nu_P$ = PHOSPHORESCENCE

The polymers discussed in this paper will generally be polymers of ethylene or styrene which are copolymerized with monomers capable of introducing ketone groups with a variety of structural features. These copolymer structures are summarized in Table I.

When a thin film of solid polymer containing a ketone function is cooled to 0°K no motion of the groups or constituents of the polymer is possible. If the film is gradually warmed to room temperature in the presence of oxygen and irradiated with UV light, phosphorescence emission is observed. Since this is a strictly electronic transition, no molecular motion is required and the intensity of the emission remains relatively high until certain transition points are observed. If one plots the phosphorescence intensity from thin films of polystyrene or polyethylene in the form of an Arrhenius plot, these transitions are identified by intersections of straight line portions of the response curve shown in Figure 1. As the temperature is increased from 0°K the specific volume of the polymer increases in a manner shown schematically in Figure 2. Since there is very little change in the bond lengths with temperature the observed increases in the specific volume must be due to the formation of small holes or voids in the system which collectively increase in size and/or number as the temperature is raised. These holes or voids constitute the "free volume" in the polymer which is indicated in Figure 2 by the area lying above the straight line parallel to the temperature axis.

Table I. Structures of Ketone Copolymers Studied

	COPOLYMER	STRUCTURE
I	POLY (ETHYLENE - co - CARBON MONOXIDE) (PE - CO)	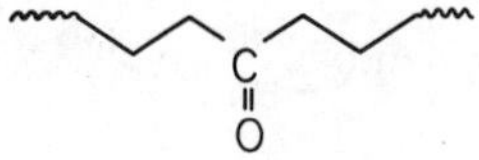 C ‖ O
II	POLY (ETHYLENE - co - METHYL VINYL KETONE) (PE - MVK)	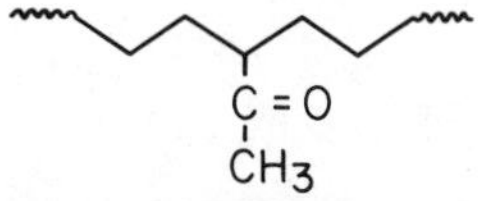 C = O CH_3
III	POLY (ETHYLENE - co - METHYL ISOPROPENYL KETONE) (PE - MIPK)	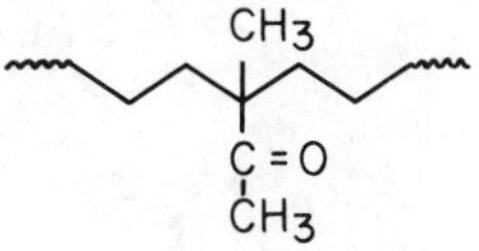 CH_3 C = O CH_3
IV	POLY (STYRENE - co - METHYL VINYL KETONE) (PS - MVK)	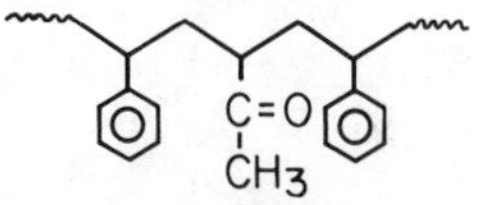 C=O CH_3
V	POLY (STYRENE - co - METHYL ISOPROPENYL KETONE) (PS - MIPK)	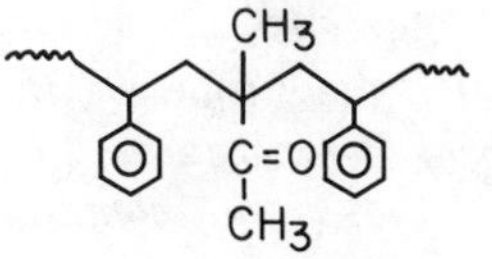 CH_3 C=O CH_3
VI	POLY (STYRENE - co - tert-BUTYL VINYL KETONE) (PS - tBVK)	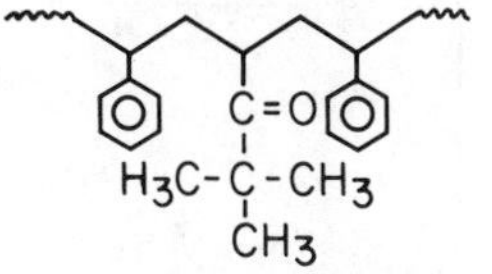 C=O H_3C-C-CH_3 CH_3
VII	POLY (STYRENE - co - PHENYL VINYL KETONE) (PS - PVK)	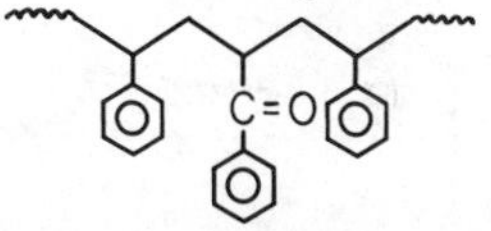 C=O
VIII	POLY (STYRENE - co - PHENYL ISOPROPENYL KETONE) (PS - PIPK)	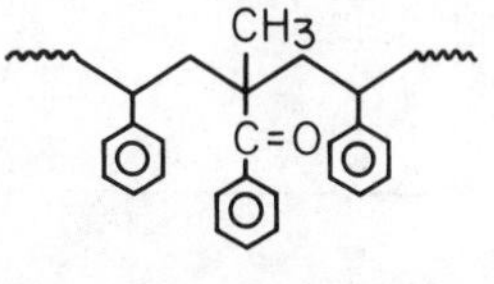 CH_3 C=O

As the free volume in the polymer increases, various types of molecular motion can begin to occur and these are identified by the transitions shown in Figure 1 with the motion of various subgroups in the polymer. The important ones observable by this experiment are the crankshaft motion of polyethylene observable at about -85°C, and the transition associated with movement of the phenyl ring polystyrene at about -80°C. The straight line portion of the Arrhenius curve above about 100°K observed in both cases is attributed to quenching of the phosphorescence emission by oxygen, and the slope of this curve accurately reflects the activation energy associated with the diffusion of the oxygen quencher. This effect can be used (*1, 2*) as a means of measuring the rates of oxygen diffusion in a variety of polymers. As a general rule, the

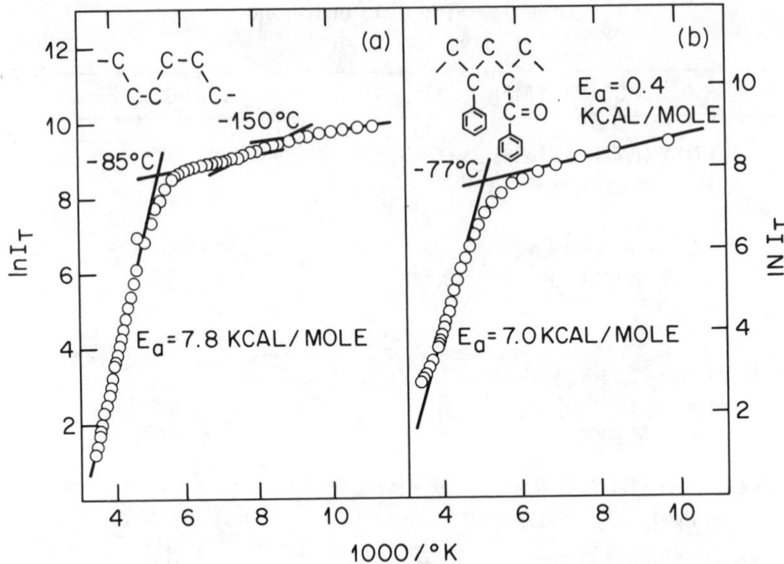

Figure 1. Logarithm of the phosphorescence intensity I_T as a function of reciprocal temperature for films of (a) polyethylene and (b) polystyrene containing traces of carbonyl groups.

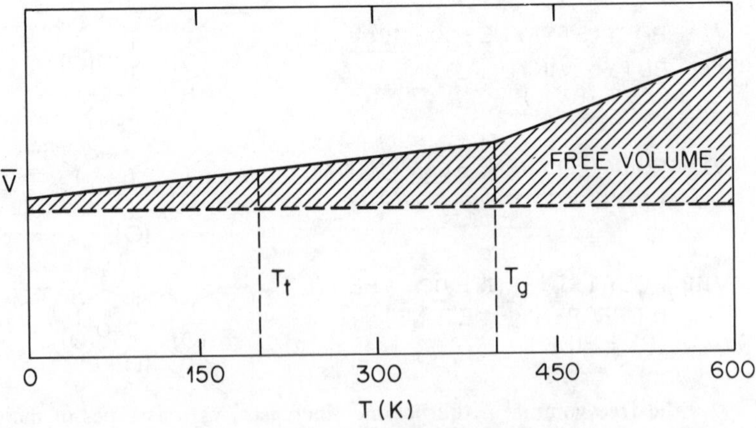

Figure 2. Specific volume and "free volume" of polymeric material.

activation energy for a given process in a solid matrix can be related to the amount of free volume required for that process, so that small scale motions requiring relatively small amounts of free volume adjacent to the moving group will have low activation energies, whereas those requiring larger packets of free volume will have large activation energies. Typically, the activation energies for diffusion of oxygen in polymers range from 7 to 10 kcal/mol, whereas those for small scale rotations of methyl and methoxy groups are of the order of 1 to 3 kcal/mol.

Before looking at the effect of the polymeric matrix on quantum yields and efficiencies of photochemical processes it is important to look first at variations which are due to the structure of the ketone chromophore itself which are observable regardless of whether the chromophore is in the solid, liquid, or gaseous state. The first of these is illustrated in Table II which illustrates the quantum yields for esters of dimethyl keto azelate (*3*).

Table II. Quantum Yields for Photolysis of Esters of δ Keto Azelaic Acid[a]

Ester	ϕ_{total}
Dimethyl	0.41
Copolyester with ethylene glycol containing 10% δ keto azelate groups	0.021

a. Ref. (*3*).

The dimethyl ester of this acid in solution shows a quantum efficiency ϕ_{total} of about 0.41 for all photochemical products. On the other hand, when the same acid is copolymerized with a glycol to form a polymeric compound with molecular weight 10,000 the quantum yield drops by about two orders of magnitude, $\phi \sim 0.012$. The reason for this behavior appears to be that when the chromophore is in the backbone of a long polymer chain the mobility of the two fragments formed in the photochemical process is severely restricted and as a result the photochemical reactions are much reduced. If radicals are formed the chances are very good that they will recombine within the solvent cage before they can escape and form further products. Presumably the Norrish type II process also is restricted by a mechanism which will be discussed below.

A similar variation in the quantum yield of the Norrish type I process is illustrated in Figure 3 for solid copolymers of ethylene containing three different ketone structures. The ketone groups in the backbone of the polymer chain in ethylene-CO copolymers show much lower quantum yields than those from the secondary or tertiary structures induced by copolymerization of methyl vinyl ketone and methyl isopropenyl ketone with ethylene. (See Table I, structures I, II and III.) In the latter two cases, the Norrish type I cleavage produces a small radical and a polymer radical, and it seems likely that the small radical has a much greater probability of escaping the cage than when the radicals produced are both polymeric, as in the case of structure I.

Effect of Crystallinity

Most polymers used as photoresists are amorphous. If, however, the polymer is capable of crystallization, the free volume relationship described in Figure 2 will be obeyed only by the amorphous regions. The crystalline regions will have very little free volume at temperatures below the melting point, T_m. The

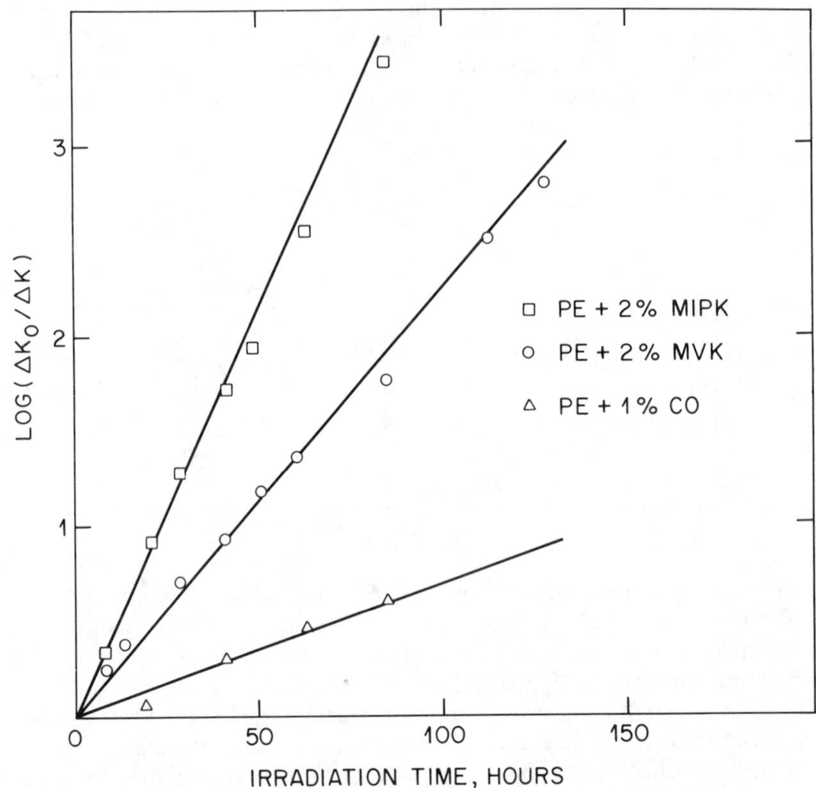

Figure 3. Plot of ln (ΔKo/ΔK) with time for irradiation of ketone copolymers in nitrogen.

rigidity of the crystalline lattice is such that most photochemical reactions cannot occur.

To illustrate the effects of a crystal lattice on photochemical reaction Slivinskas and Guillet (*4*) carried out a number of experiments with a model ketone 7-tridecanone. This ketone has a melting point of 33°C. When irradiated with ultraviolet light in the solid crystalline phase at 10°C the quantum yield for both type I and type II processes was so low as to be undetectable by gas chromatographic procedures. However, when irradiated a few degrees above its melting point, the quantum yield for type II was 0.013, the same as was observed in solution at the same temperature. The low quantum yields in the crystalline regions can be attributed to one of two effects: 1) the possible delocalization of the excitation throughout the crystal or 2) the lack of free volume imposed by the rigidity of the crystal lattice. Unless the photon energy is sufficient to cause local melting of the lattice to provide the necessary free volume, it seems unlikely that photochemical processes, particularly of the type requiring large rearrangements of molecular structure, can occur efficiently in crystalline regions of either a monomer or a polymer.

An elegant exception to this rule has been the recent work of Wegner on the polymerization of diacetylene compounds where the crystal form of the monomer and the polymer are nearly identical. In this case, photochemical reaction can and does take place (*5*).

The Effect of the Glass Transition Temperature

The work described previously for the ethylene copolymers was carried out in solid polymer films at 25°C and very little change in these yields is observed on raising the temperature of the film to about 100°C where it becomes molten. However, if the polymer is cooled, the quantum yields are relatively independent of temperature down to -40°C, after which a progressive decrease in the quantum yield for the type I process is observed at temperatures down to -100°C, below which no observable photochemical reaction occurs. As pointed out by Hartley and Guillet (*6*) it seems likely that the reason for this phenomenon is that the Norrish type II process requires a cyclic six-membered intermediate which requires at least two rotations about carbon-carbon bonds adjacent to the carbonyl group in order to be produced from the most stable planar zig-zag conformation of the polymer chain. Below the glass transition of polyethylene at -40°C the only motions observable by dielectric and other test procedures is the Shatzki crankshaft motion which is just the type of coordinated bond motion which would give rise to the necessary configuration for chemical reaction to occur. However, the frequency of these rotations will decrease with temperature and thus the quantum yields will also decrease as is observed experimentally. Below the transition at -100°C further motion of the polymer chain is restricted and even short segments of the polymer chain can no longer move. As a result, photochemical reaction is completely inhibited. On the other hand, the photophysical processes of fluorescence and phosphorescence may be quite efficient.

Further confirmation of the important effect of solid-phase transitions in polymer photochemistry was reported by Dan and Guillet (*7*). They studied the quantum yields of chain scission, ϕ_s, as a function of temperature in thin solid films of vinyl ketone homo- and copolymers. For polymers where the Norrish type II mechanism was possible, large increases in ϕ_{II} were observed at and above the glass transition, T_g. Figure 4 illustrates the effect in a styrene copolymer containing minor amounts of phenyl vinyl ketone (PVK). Below T_g, ϕ_s is about 0.07, but at T_g it rises to about 0.3, a value very similar to that observed for photolysis of this polymer in solution at 25°C. A similar effect was observed with poly(methyl methacrylate-methyl vinyl ketone) (PMMA-MVK) and PVK itself. The generality of this effect suggests that above T_g, the conformational freedom of the polymer chain is comparable to that in solution and consequently one should expect to find equivalent chemical reactivity. When this effect was first discovered it seemed important to determine whether or not it was a general phenomenon applicable to all photochemical reactions in solid systems. Fortunately, as it turned out, this is definitely not the case and only those reactions requiring relatively large amounts of free volume such as the Norrish type II, will show this marked effect at the glass transition temperature. For example, Table III shows data on the quantum yield for the loss of carbonyl function in films of PS-MIPK irradiated at 313 nm in nitrogen. There is a small but continuous increase in

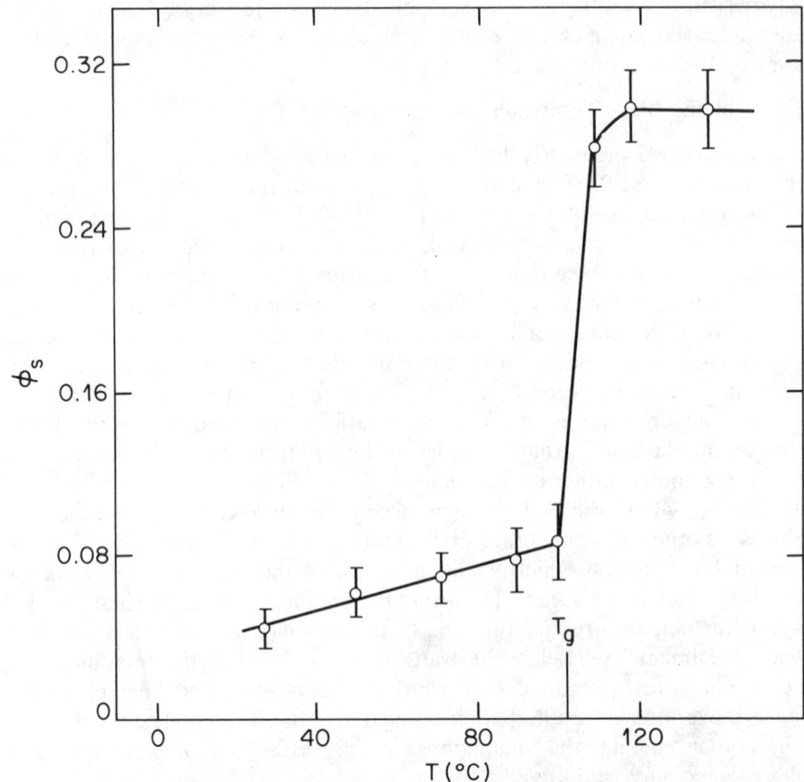

Figure 4. Quantum yield of Norrish type II reaction as a function of temperature for 0.5 mm PS-PVK films irradiated with monochromatic radiation of 313 mμ wavelength as measured by GPC. Error bars represent errors (±20%) in reproducibility of results.

Table III. Temperature Effects in the Irradiation of PS-MIPK Copolymer Films at 313 nm in N_2

Temperature °C	ϕ_{-CO}
23	0.26
64	0.26
89	0.30
104	0.32

the quantum yield from 23 to 104°C but no discontinuous change is observed in the region of the glass transition, around 98°C for these polymers. The total loss of carbonyl is considered to be the sum of the type I and photoreduction process, both of which presumably require very small volumes of activation in order to occur.

It is interesting to note that the efficiency of radical reactions in solid glassy matrices appears to be uninhibited by the presence of the polymer matrix and in fact in many cases we have observed that the efficiency of radical escape when at least one of the radicals is a small molecule is just as great in polymeric glasses as it is in solution. One can deduce from this that polymeric glasses are not particularly good at trapping radical species unless they are cooled to very low temperatures.

As shown in Table IV, the photoreduction process can be quite efficient in PVK copolymers with styrene. In the case of these polymers, because of the rapid intersystem crossing in the phenyl ketone chromophore it seems likely that in most of these reactions both the reduction and chain scission occur via the intermediacy of the triplet state. An interesting feature shown in Table IV is that, with PVK copolymers (structure VII) ϕ_{-CO} corresponds almost exactly to ϕ_{OH}, suggesting that photoreduction (or cyclization) is the major route for loss of carbonyl in these polymers. However, with PIPK copolymers, ϕ_{-CO} is four times as great as ϕ_{OH}. We suggest that in this case, because of the greater stability of the tertiary radical formed, the major loss of carbonyl in these polymers (structure VIII) is by the Norrish type I reaction. Because of mobility restrictions for the type II process in the solid phase, the quantum yield ϕ_I appears to be significantly higher than in solution.

Table IV. Various Quantum Yields in the Irradiation of Poly(styrene-*co*-vinyl aromatic ketone)s at 313 nm in N_2

Comonomer	Mol %[a]	Temp. °C	ϕ_{-CO} film	ϕ_{OH} film	ϕ_S film	ϕ'_S solution
PVK	1.1	23	0.20	0.19	0.043	0.13
PVK	3.6	23	0.21	0.18	0.054	0.15
PVK	7.6	23	0.22	0.20	0.072	0.42
PVK	7.6	55	0.23	—	—	0.53
PIPK	6.1	23	0.32	0.08	—	0.36

[a] Determined by UV spectrophotometry with the homopolymers as standards.

CH3 C=O hν CH3 IX + ·C(=O)–C6H5 (7)

The β-scission of the tertiary radical IX so produced provides another, potentially efficient, method of causing main-chain scission in the polymeric solid phase. Similar high quantum yields for the Norrish type I process were

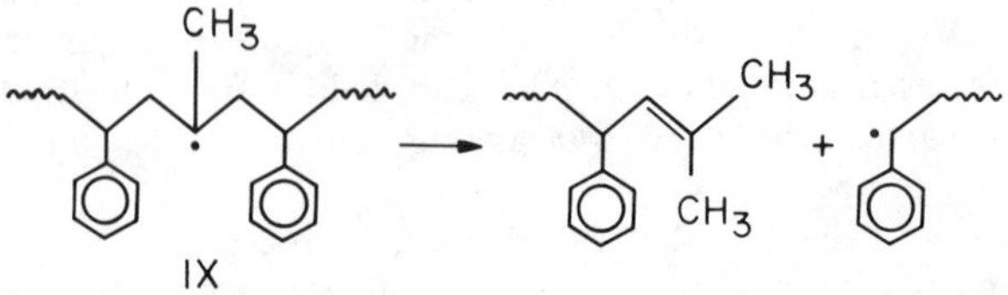

(8)

observed in the solid phase photolysis of PS-tBVK copolymer (structure VI):

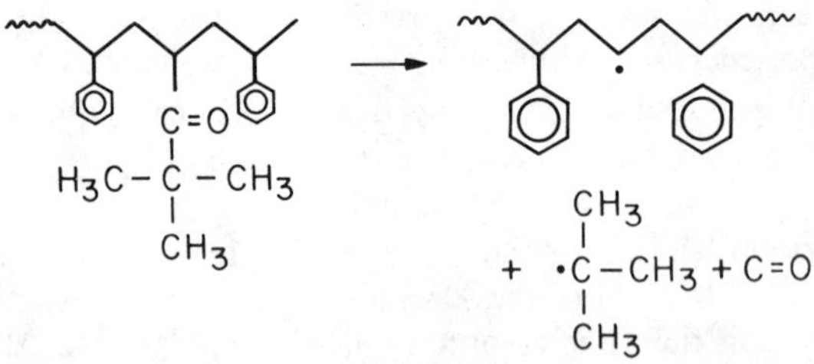

(9)

Depending on the fate of the secondary polymer radical produced, this could lead to either chain scission or cross-linking in the solid phase.

The Photo-Fries Reaction

Another reaction which occurs readily in solid polymers is the Fries reaction which can be observed in phenyl esters, particularly in phenyl acrylate and phenyl methacrylate polymers. The course of the reaction can be followed very

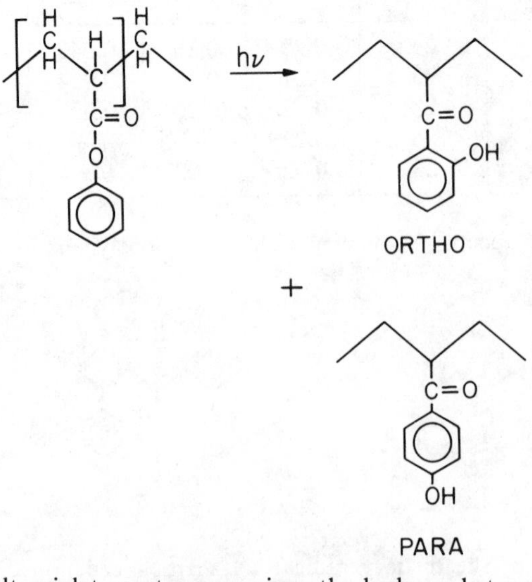

(10)

easily by ultraviolet spectroscopy since the hydroxy ketone products have strong absorbance at 260 and 320 nm (Figure 5). Reaction occurs with equal

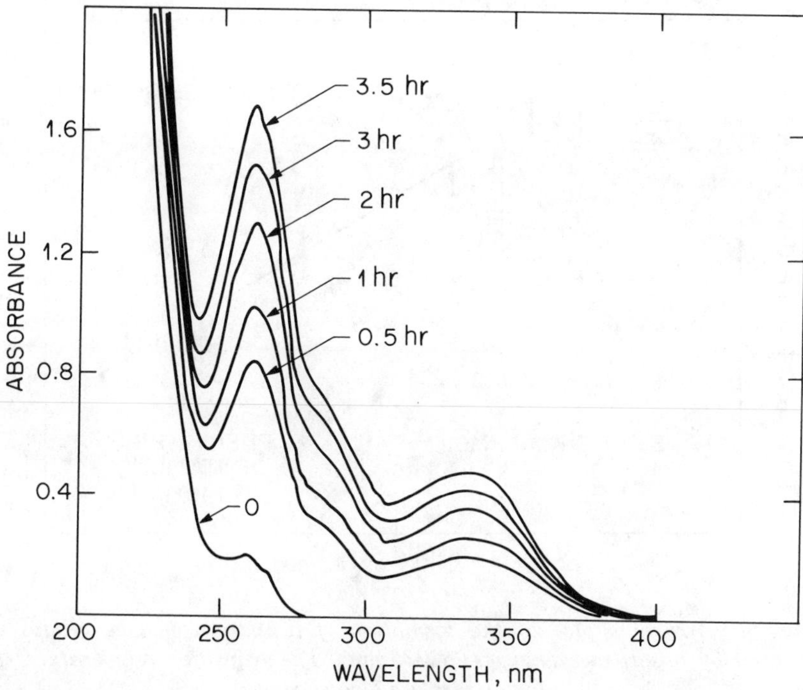

Figure 5. Absorption spectra of a PPA film after different periods of irradiation at room temperature, using light of wavelengths between 220 and 340 nm from an A.E.I. medium-pressure mercury lamp.

efficiency in small model compounds in solution and in the polymers in the solid phase (*8*). An Arrhenius plot of the quantum yield for the ortho product versus temperature (Figure 6) shows a linear increase up to 294°K, above which no further change in quantum efficiency is observed above or below the glass transition temperature. The linear portion of the curve has an activation energy of about 2 kcal/mol and is believed to be associated with the activated process involving small motions of the phenyl ring on the ester group. The position of the transitions is readily determinable by the phosphorescence procedures outlined earlier and are shown in Figure 7. The small value of the activation energy is presumably associated with the very small volume required for the rotation of the phenoxy radical before it recombines to form the hydroxy ketone.

Conclusions

What can be seen from the foregoing examples is that one can use the photochemistry of small model compounds to predict the photochemistry of a polymeric material, provided that certain structural features are included and that one has some idea of the free volume required for the conformational or other motions necessary for the formation of the excited state and rearrangements or disproportionation into products. It can be concluded that

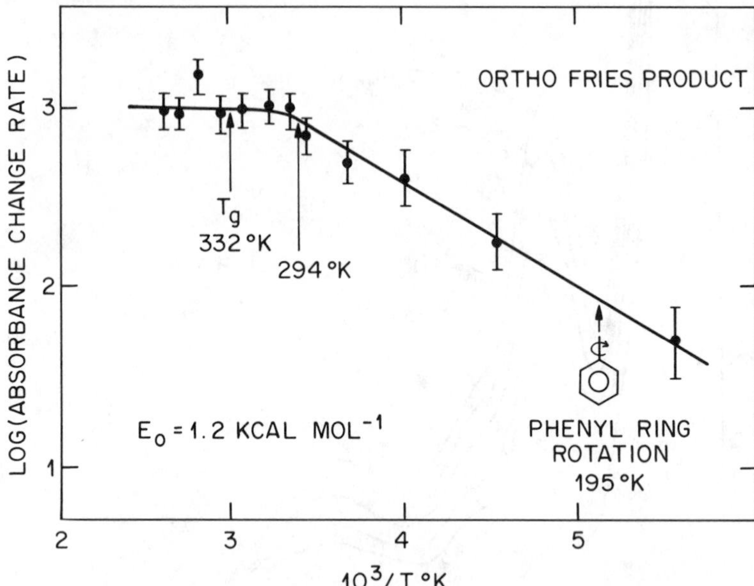

Figure 6. Arrhenius plot of the formation of o-hydroxyphenone groups as measured by absorbance changes at 335 nm. The transition temperatures of the polymer are also shown.

reactions which require very little change in the geometry of the excited state and reactants should proceed equally well in solid glassy polymer matrices as in solution. Dissociation of free radical pairs will be relatively efficient in the solid-state if one of the components is a small free radical but will be significantly inhibited if both components are polymer radicals. Reactions which can be considered to be associated with caged radicals, such as the photo-Fries, will require very little free volume and can be expected to be quite efficient in solid polymers below the glass transition, whereas photochemical processes like the Norrish type II process will be expected to be substantially reduced in glassy polymers below T_g unless the geometry of the cyclic six-membered ring is particularly favored by steric factors in the chain, so that the most stable conformation corresponds to that required for reaction. And finally, bimolecular reactions which require the diffusion of a small molecule reagent to a species in a polymer matrix will depend on both the diffusion constant and the solubility of the material in the matrix. However, it is worth pointing out that diffusion in solid glassy matrices, particularly of small molecules, is much higher than would be predicted from the bulk viscosity of the medium. Solid polymers generally have internal viscosities only two to three orders of magnitude less than those for simple liquids such as benzene or hexane so that under suitable conditions quite efficient bimolecular reaction can be induced to occur by diffusional processes in polymeric materials. The feasibility of such processes has been adequately demonstrated by the success of the technology of instant photography.

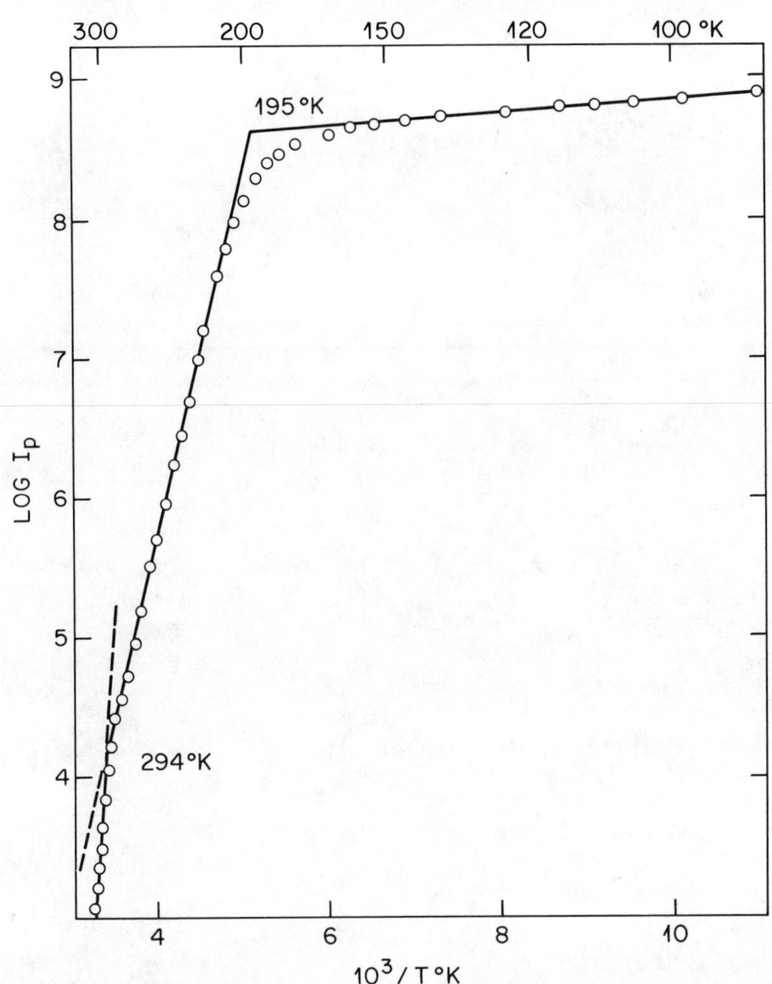

Figure 7. Arrhenius plot of PPA phosphorescence.

Literature Cited

1. Somersall, A. C.; Dan, E.; Guillet, J. E. *Macromolecules* 1974, *7*, 233.
2. Andrews, M.; Guillet, J. E. unpublished work.
3. Plooard, P. I.; Guillet, J. E. *Macromolecules* 1972, *5*, 405.
4. Slivinskas, J. A.; Guillet, J. E. *J. Polym. Sci., Polym. Chem. Ed.* 1973, *11*, 3043.
5. Wegner, G. *Pure Appl. Chem.* 1977, *49*, 443.
6. Hartley, G. H.; Guillet, J. E. *Macromolecules* 1968, *1*, 165.
7. Dan, E.; Guillet, J. E. *Macromolecules* 1973, *6*, 230.
8. Li, S.-K. L.; Guillet, J. E. *Macromolecules* 1977, *10*, 840.

RECEIVED October 3, 1984

7

Radiolysis of Poly(isopropenyl *t*-butyl ketone)

S. A. MACDONALD, H. ITO, and C. G. WILLSON

IBM Research (K42-282)
5600 Cottle Road
San Jose, CA 95193

J.W. MOORE, H. M. GHARAPETIAN, and J. E. GUILLET

Department of Chemistry
University of Toronto
Toronto, Canada M5S 1A1

The chemistry associated with molecular weight reduction of ketone containing polymers has been the subject of several fundamental studies and has been used by the lithographic community in the design of radiation sensitive resist materials. Poly(methyl isopropenyl ketone) (PMIPK) is a positive tone resist that undergoes molecular weight reduction upon exposure to either electron beam or UV radiation (*1,2*). While PMIPK does undergo photodegradation, this resist is not very sensitive to UV radiation, although the lithographic sensitivity to deep UV light can be improved by adding compounds such as 3,4-dimethoybenzoic acid (*2*). Another approach towards increasing the sensitivity of PMIPK is to prepare structural analogs that will undergo radiolysis by a more efficient pathway. In the previous paper, Guillet thoroughly reviewed the photochemistry associated with the degradation of macromolecules containing a ketone moiety. This paper will use those concepts in the design and synthesis of a poly(vinyl ketone) that has a high quantum yield for chain scission in the solid state.

Guillet and his colleagues have shown that molecular weight changes which occur when poly(vinyl ketones) are subjected to radiation arise primarily from Norrish Type I and Norrish Type II cleavage routes. They have also shown that in the solid state (below the glass transition temperature) the Type I cleavage predominates, while in solution or above the glass transition temperature the Type II pathway is favored (*3*). Since most lithographically useful resists are irradiated at temperatures below T_g we have prepared and studied an analogue of PMIPK in which the propensity for Norrish Type I cleavage has been increased. In poly(isopropenyl t-butyl ketone), (PIPTBK), the carbonyl is located between two quaternary centers; thus α-cleavage on either side of the carbonyl carbon will generate a stable tertiary radical. This should be contrasted with the situation in PMIPK, where α-cleavage can yield either a tertiary butyl or a methyl radical. As the stability of these two species is significantly different, cleavage only occurs to generate the tertiary radical. The structure of these two polymers is shown in Figure 1. Figure 2 shows that both alpha cleavage routes in PIPTBK, followed by rapid decarbonylation,

0097-6156/84/0266-0179$06.00/0

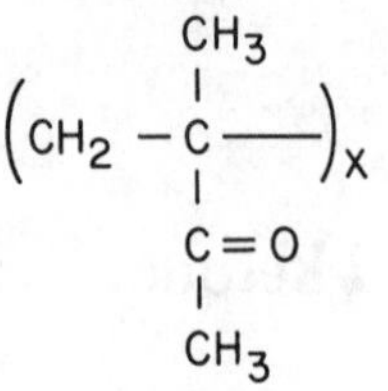

PMIPK 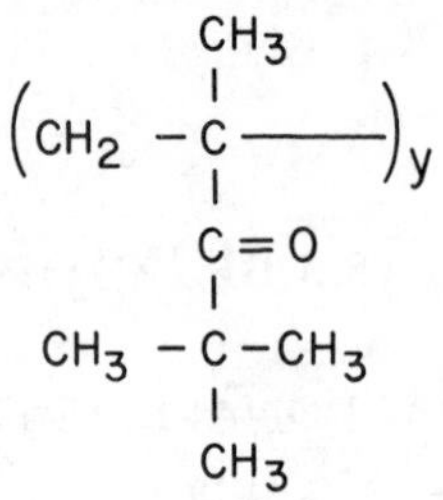 PIPTBK

Figure 1. Structure of poly(methyl isopropenyl ketone), (PMIPK), and poly(isopropenyl t-butyl ketone), (PIPTBK).

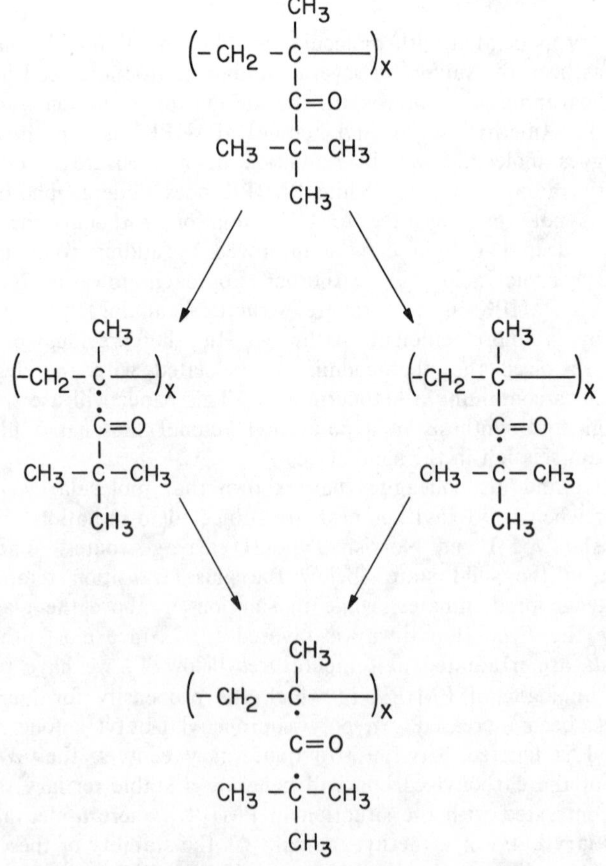

Figure 2. Norrish Type I cleavage in PIPTBK.

produce a radical on the main chain of the polymer. This species is now free to undergo the reactions outlined in the previous paper. It should also be noted that PIPTBK does not contain a hydrogen atom that is located alpha to the carbonyl carbon. As a result, the six-membered cyclic transition state of the Norrish Type II pathway is not feasible. This lack of an alpha hydrogen, also eliminates the seven-membered cyclic transition state that was originally proposed as a degradative mechanism in copolymers of methyl vinyl ketone and methyl methacrylate (*4*). This same seven-membered transition state has been proposed by Levine as the major pathway for chain-scission in the electron beam radiolysis of PMIPK (*1*).

This paper will describe the synthesis of PIPTBK and report the chain scission quantum yield, $\Phi(s)$ for this material both in thin films and in solution. For comparison, the chain scission quantum yield for PMIPK was measured under similar conditions.

Results

Preparation of Polymer. The monomer, 2,4,4-trimethylpentene-3-one, required to obtain the desired polymer has been reported in the literature (*5*). However, the scheme used by these workers requires preparative gas chromatography and does not conveniently yield gram quantities of this monomer. The synthetic procedure used in our study is outlined in Scheme I and has been successfully

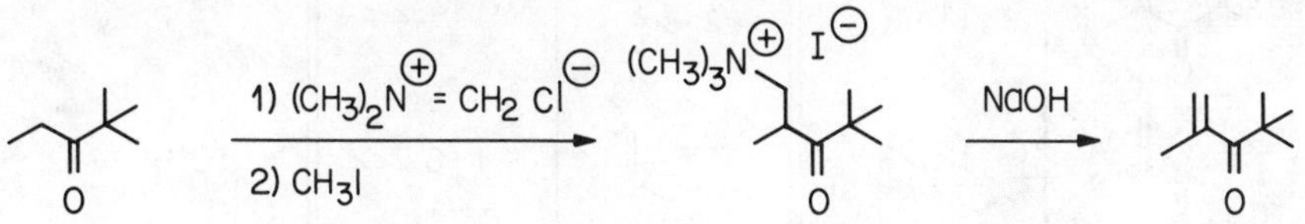

1) $(CH_3)_2N^{\oplus}=CH_2\ Cl^{\ominus}$ 2) CH_3I → $(CH_3)_3N^{\oplus}\ I^{\ominus}$ → NaOH →

Scheme I Synthesis of 2,4,4-trimethylpentene-3-one.

run on a 0.2 mole scale. This enone was obtained by allowing 2,2-dimethyl-3-pentanone to react with a 20% excess of N,N-dimethylmethyleneimmonium chloride in refluxing acetonitrile. The amine was alkylated with an excess of methyl iodide in methanol at room temperature to yield the quanternary ammonium salt. Elimination of trimethylamine in a 5% sodium hydroxide solution followed by distillation under reduced pressure (70°/80 mm Hg) afforded the desired monomer in 61% overall yield. The monomer was polymerized in THF (1/1 by volume) at −78° using n-butyllithium (0.05 equivalents) and 18-crown-6. Polymerization was terminated with cold methanol after three days at −78°, and the polymer was taken up into dichloromethane. The polymer was purified by successive precipitations into methanol and dried under reduced pressure at 38°. The number-average molecular weight of PIPTBK obtained in this fashion was found to be 23,700, as determined by membrane osmometry in toluene. The weight-average molecular weight was determined by light scattering in THF, and found to be 48,000.

While most vinyl ketones readily undergo radical polymerization, and can only be stored in the monomeric state if an inhibitor is present, this enone failed to polymerize with either benzoyl peroxide or azobisisobutyronitrile under a variety of conditions. Examining the C-13 NMR spectrum of the monomer provides some insight into the lack of reactivity displayed by this unsaturated ketone.

The β-methylene of an enone is typically 10-15 ppm downfield from the corresponding alkene (*6*). This general observation is attributed to a reduction of electron density at the β-methylene carbon due to conjugation with the carbonyl. In the case of 2,4,4-trimethylpentene-3-one, the β-methylene carbon of the enone resonates only 3.5 ppm downfield from the β-methylene carbon of the analogous alkene. Table I contains the carbon-13 chemical shift data for a series of enones and their corresponding alkenes that were measured under the same conditions. These C-13 NMR data indicate that 2,4,4-trimethylpentene-3-one is not a typical, delocalized system and suggest that the monomer will not polymerize in the manner anticipated for an α-β unsaturated ketone. Previously reported spectroscopic studies also support the argument that this enone does not adopt the planer conformation required to obtain a conjugated system (*7*).

Table I. Carbon-13 NMR Data (in ppm downfield from TMS) for the β-Carbon of an Enone and the Corresponding Alkene. Samples Were Run as Neat Liquids Using a Varian CFT-20 Spectrometer.

ENONE	δ β-CH_2	ALKENE	δ β-CH_2	Δ ppm
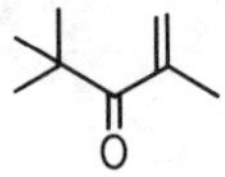 O	117.7	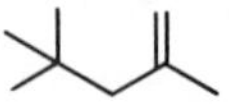	114.2	3.5
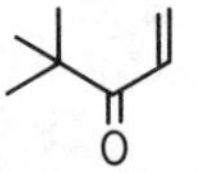 O	127.6	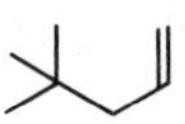	116.7	10.9
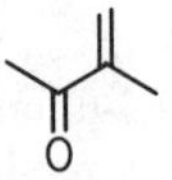 O	125.3	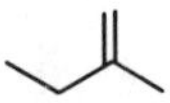	109.1	16.2
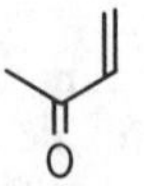 O	128.7	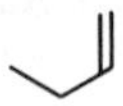	113.3	15.4

Irradiation Studies in Thin Films. The quantum yield for chain scission of PIPTBK in the solid state was determined at 313 nm (narrow band pass filter) using 30 μ thick films of the polymer that were deposited onto quartz substrates. The solid films were coated from a 8.0 wt-% solution of polymer in toluene. The coated substrates were placed into a heated chamber that was saturated with toluene vapor and the vapor concentration was gradually lowered allowing the films to dry slowly and uniformly. Finally, the films were baked for 2 days at 60° *in vacuo*. In this determination, coated discs were exposed to eight different doses of 313 nm radiation and the irradiated polymer was removed from the quartz substrate by rinsing with THF. The light intensity was determined by potassium ferrioxylate actinometry. The weight-average molecular weight of each sample was determined by low angle light scattering

(LALS) in THF and the intrinsic viscosity of five of the samples was measured by the single point method. Combining the LALS data with the intrinsic viscosity work allows one to obtain the Mark-Houwink constants, K and α, by plotting $\log[\eta]$ vs $\log M_w$. For PIPTBK, K and α in THF at 25° were found to be 5×10^{-3} ml/g and 0.75 respectively.

The quantum yield for the chain-breaking process in a polymer molecule is given by

$$\Phi(s) = \frac{W}{(M_n)_o} \frac{d\left(\frac{(M_n)_o}{M_n} - 1\right)}{d(It)} \tag{1}$$

In this equation, $\Phi(s)$ is the quantum yield, which is the number of events occuring per quantum absorbed; $(M_n)_o$ is the initial number-average molecular weight; M_n is the molecular weight at time t; I is the intensity of the light absorbed; the quantity $\{[(M_n)_o/M_n] - 1\}$ is the number of scissions per original polymer; and W is the weight of the polymer in the sample (3). For most degrading polymer systems in which the initial polydispersity is between 1.5 and 2.5, Equation 1 can be used in the following form

$$\Phi(s) = \frac{W}{(M_n)_o} \frac{d\left(\frac{(M_w)_o}{M_w} - 1\right)}{d(It)} \tag{2}$$

A plot of chain scissions per original polymer of PIPTBK, $\{[(M_w)_o/M_w] - 1\}$, vs absorbed light (at 313 nm) is shown in Figure 3. From the slope of the line shown in Figure 3, $\Phi(s)$, is calculated to be 0.29.

In a similar fashion, 3.0 μ thick films of PMIPK were coated on quartz substrates, irradiated with 313 nm light, and dissolved in dioxane. The intrinsic viscosity and weight-average molecular weight of each sample was determined by the single point method and LALS. From these data the K and α parameters in dioxane at 25° were found to be 7.5×10^{-3} ml/gram and 0.73 respectively (*8*). A plot of $\{[M_w)_o/M_w] - 1\}$ vs absorbed light was generated for the irradiated samples and $\Phi(s)$ for PMIPK at 313 nm was found to be 0.024. This study was also carried out at 254 nm, and the quantum yield for chain scission at this wavelength was the same as that observed at 313 nm. The quantum yield of chain scission for PMIPK at 254 nm was investigated twenty years ago with thicker films than those used in this study and reported to be 0.22 (*9*).

Irradiation Studies in Solution. The degradation studies on both PIPTBK and PMIPK in solution were carried out in an automatic UV irradiation-viscometer apparatus that has been described elsewhere (*10*). Each solution was irradiated at 313 nm for a fixed time period and then automatically transferred to a viscometer, where nine repetitive measurements of the efflux time were performed. The solution was then returned to the UV cell for further irradiation, and the cycle was repeated seven times. The intrinsic viscosity of

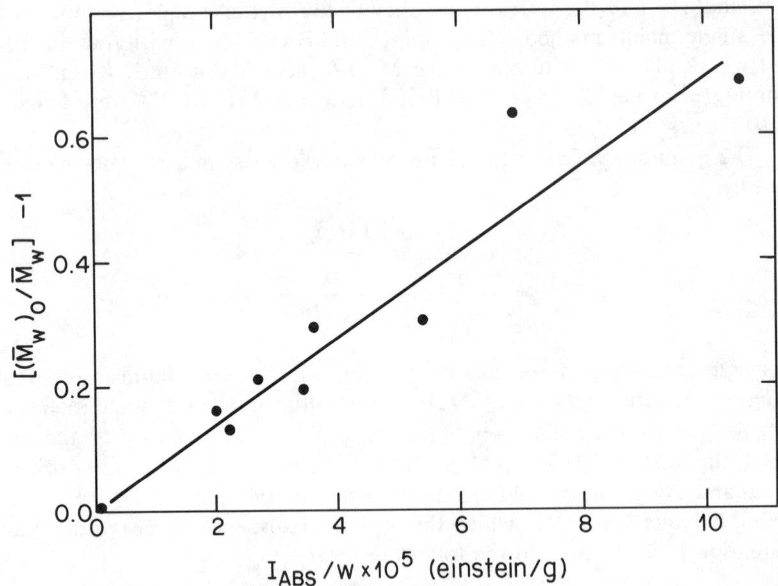

Figure 3. Determination of Φ(s) for PTBIPK films at 313 nm. Plot shows number of chain breaks per original molecule vs moles of photons absorbed per gram of film.

each sample was calculated from the measure efflux times by the single point method. With the Mark-Houwink constants that were determined previously and the known relationship between $[\eta]$ and absorbed dose, the quantum yield for chain scission was calculated from

$$\Phi(s) = \frac{W}{(M_n)_o} \frac{d[([\eta]_o/[\eta])^{1/a} - 1]}{d(It)} \tag{3}$$

Using these values we have found $\Phi(s)$ in solution to be 0.45 for both PMIPK and PIPTBK.

Cobalt-60 γ-Radiolysis Studies. Samples of PIPTBK powder were sealed under vacuum and exposed to ^{60}Co γ-radiation at the National Bureau of Standards in Washington D.C. The samples received 0, 6, 12, and 20 Mrads and the number-average and weight-average molecular weight of each polymer, relative to polystyrene, were determined in THF by GPC. The scission efficiency and the cross-linking efficiency were determined from

$$1/M_n = 1/(M_n)_o + [G(s)-G(x)](D/100N_A) \tag{4}$$

$$1/M_w = 1/(M_w)_o + [G(s)-4G(x)](D/200N_A) \tag{5}$$

In Equations 4 and 5, $G(s)$ is the scission efficiency; $G(x)$ is the cross-linking efficiency; M_n and M_w are the number-average and weight-average molecular weights after exposure to a dose of D; $(M_n)_o$ and $(M_w)_o$ are the initial molecular weight values; N_A is Avogadro's number (*11,12*). The plots of $1/M_n$ and $1/M_w$ vs exposure dose for this experiment are shown in Figure 4. From this graph, $G(s)$ and $G(x)$ were determined to be 1.5 and −0.02 respectively. These values should be compared with those reported for PMIPK of 1.95 and 0 (*13*).

Table II summarizes the quantum yield of chain-scission in the solid state and in solution for both PMIPK and PIPTBK.

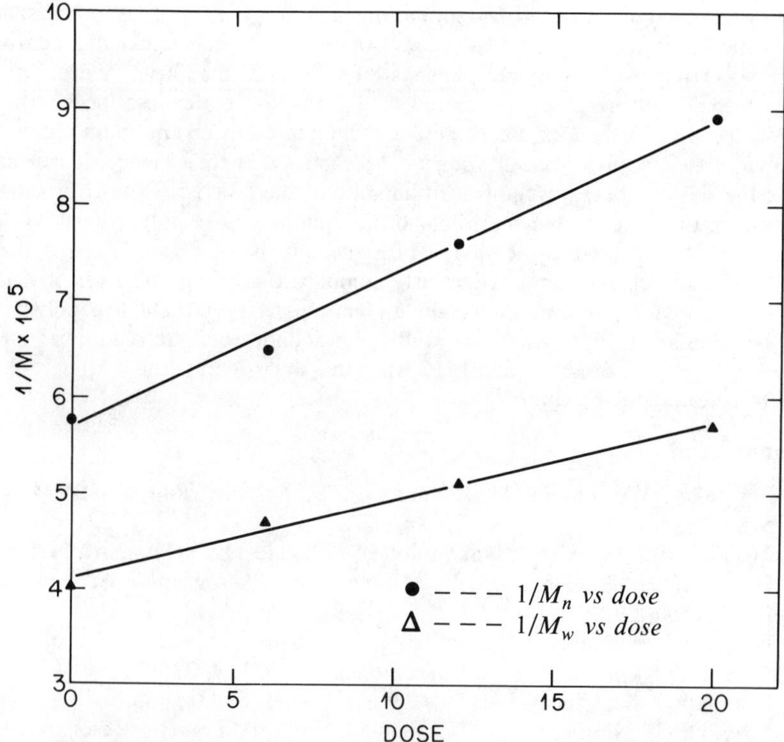

Figure 4. Determination of G(s) and G(x) for PTBIPK.

Table II. Summary of Quantum Yield Data

Polymer	$\Phi(s)$ Solid State	$\Phi(s)$ Solution	$G(s)$
PIPTBK	0.29	0.45	1.5
PMIPK	0.024	0.45	1.95 (*13*)

Discussion

An examination of the $\Phi(s)$ data in Table II shows several significant trends. In the solid state, the quantum yield for chain scission of PIPTBK is a factor of ten larger than that of PMIPK. This observation confirms the proposal put forth in the introduction that replacement of the methyl in MIPK by a t-butyl group will increase the propensity for photochemical chain degradation. Furthermore, the low Φ(solution)/Φ(solid state) ratio for PIPTBK argues for a degradative pathway, such as Norrish Type I, which is not strongly dependent on conformational mobility. In a similar fashion the high Φ(solution)/Φ(solid state) ratio exhibited by PMIPK suggest that the photochemical degradative process occurring in this polymer depends on segmental mobility. A process of this type could be the seven-membered ring transition state described earlier. PMIPK, like PIPTBK, does not possess a hydrogen atom on the main chain of the polymer that is in a γ orientation to the carbonyl carbon; hence, it can not assume the six-membered cyclic transition state of the Norrish Type II process.

Since the lithographic sensitivity of any photoresist is proportional to the quantum yield for that system, PIPTBK should be a "faster" resist than PMIPK. However, in order to rigorously compare the lithographic sensitivities of these systems, the molecular weight and dispersivity of these two polymers must be similar (*12*). Synthetic studies are underway to obtain a high molecular weight sample of PIPTBK to investigate the lithographic characteristics of this polymer.

Literature Cited

1. Levine, A. W.; Kaplin, M.; Poliniak, E. S. *Polymer Eng. Sci.* 1974, *14*, 518.
2. Tusuda, M.; Oikawa, S.; Nakamura, Y.; Nagata H.; Yokota, A.; Nakane, H.; Tsumori, T.; Nakane, Y.; Mifune, T. *Photographic Science and Engineering* 1979, *23*, 290.
3. Dan, E.; Somerstall, A. C.; Guillet, J. E. *Macromolecules* 1973, *6*, 228.
4. Amerik, Y.; Guillet, J. E. *Macromolecules* 1971, *4*, 375.
5. Crandall, J. K.; Chang, L-H. *J. Org. Chem.* 1967, *32*, 435.
6. Levy, G. C.; Nelson, G. L. "Carbon-13 Nuclear Magnetic Resonance for Organic Chemists"; Wiley: New York, 1972.
7. Bienvenue, A.; Duchatellier, B. *Tetrahedron* 1972, *28*, 833.
8. Li, X. B.; Redpath, A. E.; Gharapetian, H. M. "22nd Canadian High Poster Forum"; University of Waterloo, 1983.
9. Schultz, A. R. *J. Poly. Sci.* 1960, *47*, 267.
10. Kilp, T.; Guillet, J. E. *Macromolecules* 1977, *10*, 90.
11. O'Donnell, J. H.; Rahman, N. P.; Smith, C. A.; Winzor, D. J. *Macromolecules* 1979, *12*, 113.
12. Willson, C. G. In "Introduction to Microlithography"; Thompson, L. F.; Willson, C. G.; Bowden, M. J., Eds.; ACS SYMPOSIUM SERIES No. 219, American Chemical Society: Washington, D. C., 1983; Chapter 3.
13. Harada, K. J. *Applied Poly. Sci.* 1981, *86*, 3395.

RECEIVED August 6, 1984

8

Polymer-Bonded Electron-Transfer Sensitizers

S. TAZUKE, R. TAKASAKI, Y. IWAYA,[1] and Y. SUZUKI[2]

Research Laboratory of Resources Utilization
Tokyo Institute of Technology
4259 Nagatsuta, Midori-ku, Yokohama 227, Japan

The use of photoenergy for either synthetic processes or chemical conversion of solar energy requires efficient sensitizers (photocatalysts). Of particular interest are a group of sensitizers called electron transfer sensitizers which exchange electrons with substrates rather than transfer electronic excitation energy. They can induce useful photoredox reactions such as water splitting (*1*), reductive fixation of carbon dioxide (*2,3*), polymerization (*4*), dye formation (*5*), and so forth. The overall sensitization processes are shown in Figure 1. To achieve highly efficient photosensitization, the following conditions must be satisfied: (a) efficient photoabsorption over a wide range of wavelength (step 1), (b) efficient quenching (either oxidative or reductive) of the excited sensitizer by electron relay components (step 2), (c) efficient charge separation after the initial electron transfer (i.e., prevention of back electron transfer) (step 3), (d) efficient turn over of photosensitizer, and (e) conversion of active species into useful compounds (step 4).

The best example of electron transfer sensitization is found in photosynthesis in which photoabsorption is achieved by means of energy transfer along the array of antenna pigments to the reaction center, quenching is made possible by electron transfer across thylacoid membrane, and the active species are eventually consumed to reduce carbon dioxide and to oxidize water. In trying to mimic the function of photosynthesis, we succeeded in the following: mimicking the antenna pigments by means of molecular aggregate sensitizers (*6*), achieving complete charge separation (*7,8*), and reducing carbon dioxide to formic acid in a fair yield (*3*). However, the most difficult part is the mimicry of thylacoid membrane which enables product separation. In recent years, a number of examples of cross-membrane electron transfer reactions were demonstrated using a bilayer membrane or vesicle (*9,10*). Although these examples are encouraging, the practical system should be constructed from a polymer membrane which is mechanically strong enough for

Current address: [1]Research and Development Center, Unitika Ltd., 23 Kozakura, Uji, Japan.
[2]Research Institute for Polymers and Textiles, 1-1 Yatabe, Tsukuba, Japan.

0097–6156/84/0266–0187$06.00/0

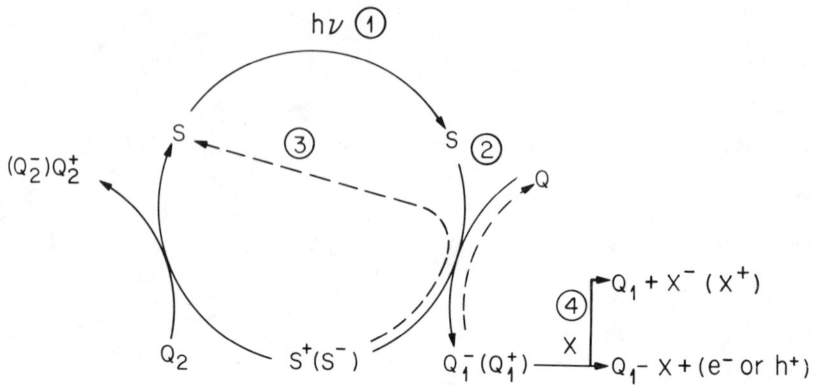

Figure 1. Schematic representation of electron transfer sensitization. 1: photo-oxidation of sensitizer 2: forward electron transfer (fluorescence quenching) 3: back electron transfer 4: product formation

large scale product separation. A reasonable approach to this problem is to study the behavior of electron transfer sensitizers bonded to polymers, to determine factors that control heterogeneous electron transfer at the polymer-solution interface, and finally to couple redox reactions between the surface and the reverse side of a polymer membrane. In this article we discuss the first two points. The possibility of the cross polymer membrane thermal redox reaction was recently demonstrated with a low efficiency (*11*).

Our extensive past studies on the excited state behavior of polymer-bound chromophores (*12,13*) indicate that the segment mobility of the chromophores is reduced and excimer/exciplex formation enhanced. Energy migration among chromophores is effective in increasing the fraction of photoenergy used for the subsequent reactions. The first two factors are negative to sensitizer efficiency of the polymer-bound chromophores, while the last factor is positive. In addition, the polymer chain causes so-called polymer-environmental effects so that it is difficult to decide whether the overall polymer effect on electron transfer sensitization is positive or negative. However, we know for sure that the Coulombic effect plays a decisive role in quenching and charge separation (Figure 1) (*7,14*) and therefore the use of a polyelectrolyte as a polymer support must be a promising approach.

Experimental

Materials. Polymer-bonded sensitizers and their model compounds are shown in Figure 2. Samples *1-5* were prepared by radical copolymerization of corresponding ethylenic monomers using azobisisobutyronitrile as initiator at 70°C in DMF(5) or DMF/H_2O (9/1 for *1*, *2* and *4*; 8/1 for *3*) (*15*). Sample *7* was prepared by the same procedure as *6* (*16*). The copolymer compositions were determined by elemental analysis. Pyrenylated polymer films were prepared by heterogeneous esterification of poly(ethylene-g-acrylic acid) with 1-hydroxymethylpyrene using dicyclohexyl-carbodiimide (DCC) as the coupling agent in either THF or acetonitrile (*17*). The grafted polyethylene film was donated by Dr. J. Okamoto, Japan Atomic Energy Agency in Takasaki. It was prepared by grafting acrylic acid onto LDPE (low density polyethylene) film by the post-irradiation method. Esterification in THF brought about deep penetration of the pyrenyl group (Film 1), whereas acetonitrile confined the esterification to a thin surface layer (Film 2).

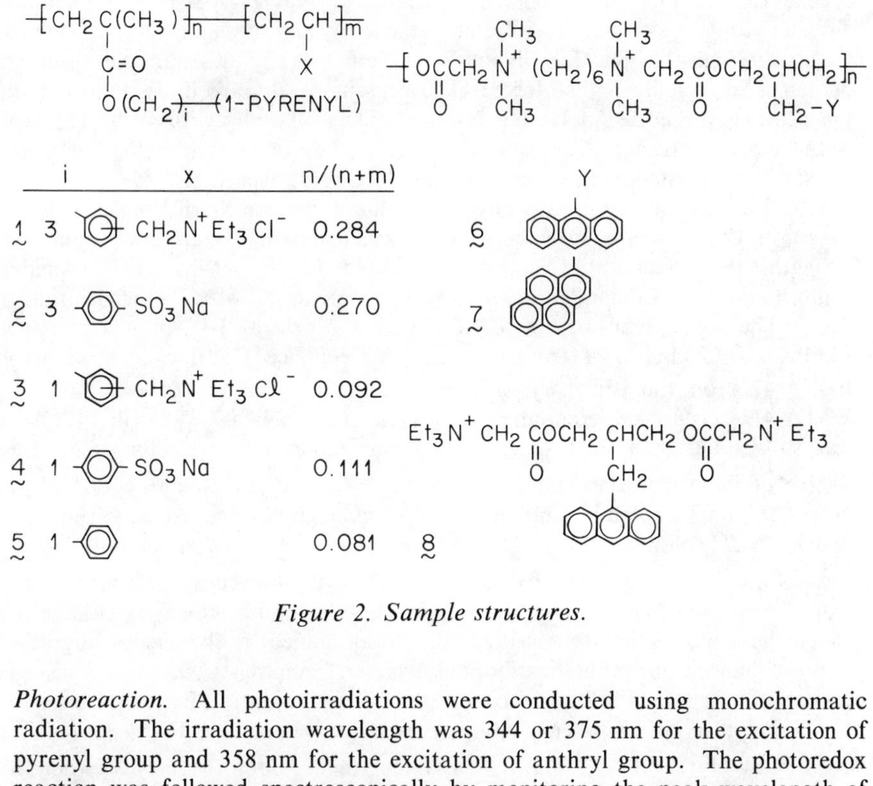

$$-[CH_2C(CH_3)]_n-[CH_2CH]_m-$$ with side groups C=O, O(CH$_2$)$_i$-(1-PYRENYL) and X

$$-[OCCH_2N^+(CH_3)_2(CH_2)_6N^+(CH_3)_2CH_2COCH_2CHCH_2]_n-$$ with side group CH$_2$-Y

	i	X	n/(n+m)
1	3	phenylene-$CH_2N^+Et_3Cl^-$	0.284
2	3	phenylene-SO_3Na	0.270
3	1	phenylene-$CH_2N^+Et_3Cl^-$	0.092
4	1	phenylene-SO_3Na	0.111
5	1	phenyl	0.081

Y: 6 (anthryl); 7 (pyrenyl)

8: $Et_3N^+CH_2COCH_2CHCH_2OCCH_2N^+Et_3$ (C=O groups; CH$_2$-anthryl side group)

Figure 2. Sample structures.

Photoreaction. All photoirradiations were conducted using monochromatic radiation. The irradiation wavelength was 344 or 375 nm for the excitation of pyrenyl group and 358 nm for the excitation of anthryl group. The photoredox reaction was followed spectroscopically by monitoring the peak wavelength of methyl viologen cation radical ($MV^{+\cdot}$) (605 ± 5 nm depending upon solvent) and that of crystal violet cation (CV^+, 595 nm).

Results and Discussion

(A) Polyelectrolyte-Bound Electron Transfer Sensitizers in Homogeneous Solution

Anthracene or Pyrene Bonded to Polyionene. When anthracene (An) or pyrene (Py) is irradiated in the presence of MV^{2+} and ethylenediaminetetraacetic acid tetrasodium salt (EDTA 4Na) in water, MV^{2+} is readily reduced to its cation radical ($MV^{\dot{+}}$). Formation of $MV^{\dot{+}}$ was not observed in the absence of the aromatic hydrocarbons. The reaction was followed spectroscopically, and examples are shown in Figure 3. The effect of the cationic environment is prominent. The enhanced quantum yield ($\phi_{MV^{\dot{+}}}$) is attributed to Coulombic repulsion between the sensitizer cation radical ($S^{\dot{+}}$) and $MV^{\dot{+}}$ assisted by the polyionene environment. From the fluorescence quenching experiment, anthracene fluorescence was shown to be hardly quenched by EDTA 4Na but efficiently quenched by MV^{2+}, indicating that $MV^{\dot{+}}$ is produced by the reaction of S^{*1} with MV^{2+}. Furthermore, the quenching efficiency is higher for polymer *6* than for the monomer model *8* as shown in Figure 4, despite the smaller diffusion constant of *6*. Possibly singlet energy migration along the polymer chain is responsible for this. Namely, the efficiency of utilizing the absorbed photon is expectedly higher for *6*. The present results are in agreement with the previously reported polyester systems. The polyesters with anthryl groups were better electron transfer sensitizers than their monomer models, whereas the similar polyesters with pyrenyl groups were poorer sensitizers (*6*).

We have no immediate answer for the nonlinear Stern-Volmer plots in Figure 4. Since the measurements were made under the condition of high ionic strength, the effect could not be attributed to the change of molecular conformation and Coulombic interaction as a function of MV^{2+} concentration.

The higher efficiency of *6* and *8* than the neutral 3-(9-anthryl)propanol (AnPrOH) (Figure 3) is caused in part by enhanced local concentration of EDTA 4Na on the periphery of An so that the reaction between $An^{\dot{+}}$ and EDTA 4Na occurs preferentially to back electron transfer. When EDTA 4Na was substituted with EDTA, $\phi_{MV^{\dot{+}}}$ of *8* was reduced by a factor of 4. The degree of ionic dissociation of EDTA($pK_a = 6.27$ at 25°C) at neutral pH is only 2.5%; thus Coulombic interaction is of minor importance. Comparing *6*, *8*, and AnPrOH, positive and negative effects influencing $\phi_{MV^{\dot{+}}}$ are summarized in Table I. Considering AnPrOH as a neutral small molecule reference, both *6* and *8* are more reactive mainly because of Coulombic effects. Although *6* should be a more effective sensitizer than *8* as judged by the energy migration effect, enhanced photodimerization and excimer formation (*18*), which consume excitation energy and nearly cancel out the energy migration (antenna) effect.

Polymer *7* with pyrenyl group is less effective than *6*. The $\phi_{MV^{\dot{+}}}$ for *7* is about half of that of *6*. This is the same trend as for the polyester-bonded sensitizers used in photo-oxidation of LCV. Efficient excimer formation by pyrenyl groups accounts for the results.

Copolymers of 3-(1-Pyrenyl)propyl Methacrylate. The pyrene-containing methacrylate was copolymerized with vinylbenzyltriethylammonium chloride or sodium *p*-styrenesulfonate. These copolymers were expected to behave very differently from each other as electron transfer sensitizers. Photoreduction of MV^{2+} sensitized by the copolymers in the presence of EDTA is shown in Figure 5. The photoabsorbing species is exclusively pyrenyl groups. It is noteworthy that polycations which repel MV^{2+} are more effective sensitizers

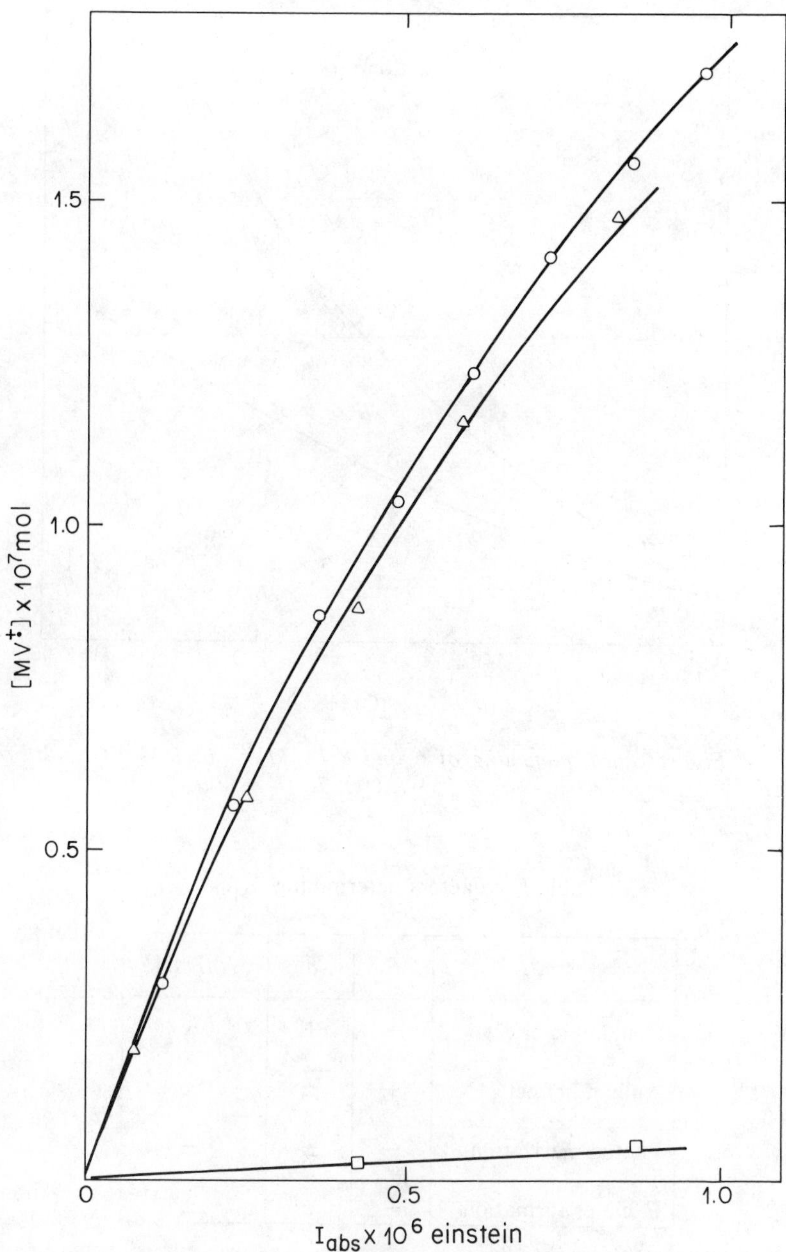

Figure 3. Photoreduction of MV^{2+} by 6, 8 and AnPrOH in the presence of EDTA 4Na. $[MV^{2+}] = 1.0 \times 10^{-4}M$, $[EDTA\ 4Na] = 1.0 \times 10^{-3}M$, $[6\ or\ 8] = 1.2 \times 10^{-4}M$, $[AnPrOH] = 1.7 \times 10^{-4}M$, irradiation at 358 nm. 6 (○), 8 (Δ), ANPrOH (□)

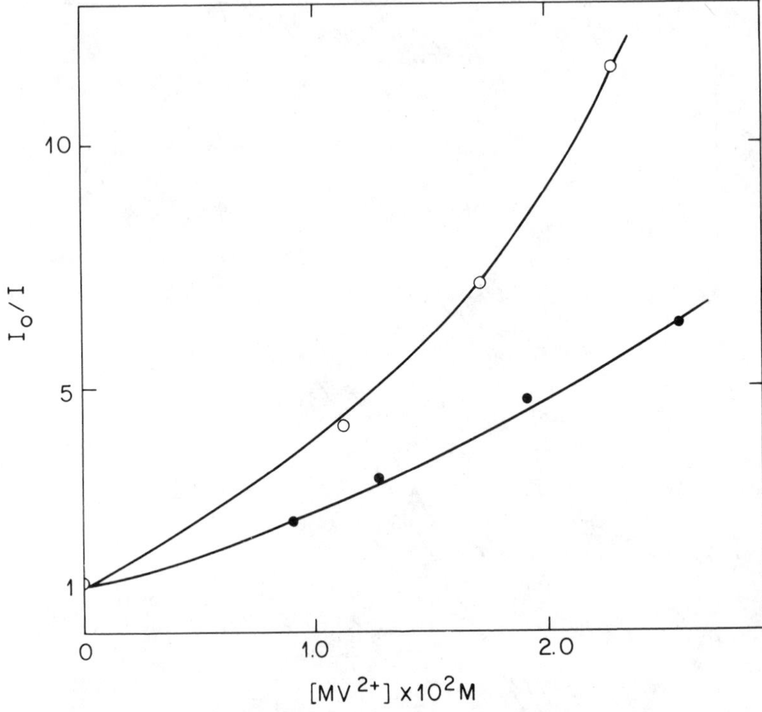

Figure 4. Fluorescence quenching of 6 and 8 by MV^{2+} in 1 M KCl solution. 6 (○), 8 (●).

Table I. Factors determining ϕ_{MV^+}

	6	*8*	AnPrOH
Coulombic effect	+	+	±
Antenna effect	+	±	±
Photodimerization	−	±	±
Excimer formation	−	±	±

+: positive effect on ϕ_{MV^+}

±: very weak or no effect on ϕ_{MV^+}

−: negative effect on ϕ_{MV^+}

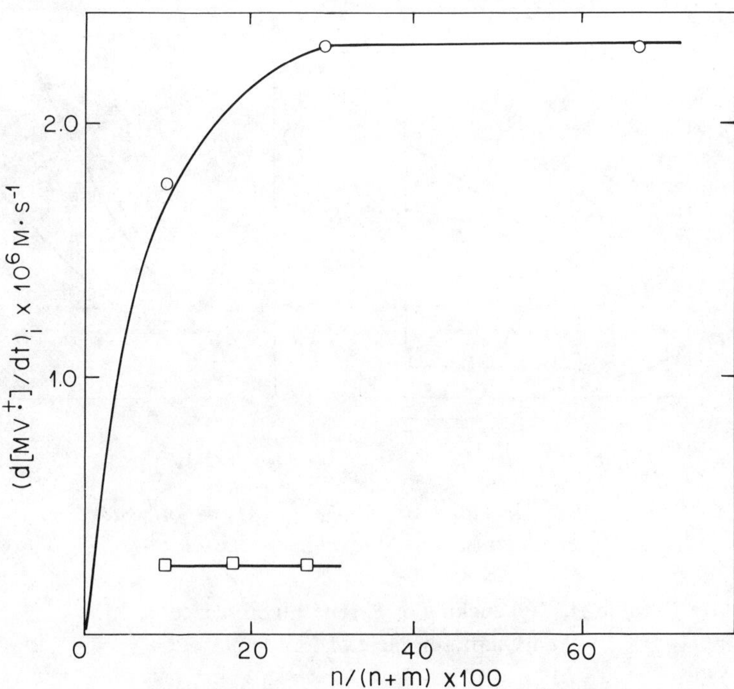

Figure 5. Photoreduction of MV^{2+} by pyrene containing polymers. [Py] = 5 × 10^{-5}M, [EDTA] = 1 × 10^{-3}M, [MV^{2+}] = 5 × 10^{4}M in DMF/water (85/15). 1 (○), 2 (□) (The pyrene content in copolymer is variable).

than polyanions. By increasing the molar fraction of the pyrene-containing methacrylate in the cationic copolymer, the sensitizer efficiency increases to a certain level and then remains constant.

These results suggest that the suppression of back electron transfer is again important in increasing the quantum efficiency. Fluorescence by pyrenyl groups in a polycation is scarcely quenched by MV^{2+}, whereas that in a polyanion is very effectively quenched (Figure 6). Furthermore, the mode of quenching does not obey the Stern-Volmer equation. The nonlinear, concaved-upward plots would suggest the participation of static quenching in the polyanion. The Stern-Volmer quenching constants (K_q) calculated from the initial slope and the initial quantum yield of MV^+ formation ($\phi_{MV^{\dot{+}}}$) are given in Table II. The general features of the reaction are shown in Figure 7. The role of EDTA is to reduce the pyrene cation radical ($Py^{\dot{+}}$); and consequently the expected attraction between the polyanion and EDTA is again a favorable condition to suppress the back electron transfer from $MV^{\dot{+}}$ to $Py^{\dot{+}}$.

Although the general picture can be drawn, several problems must be solved that arise from the interaction of polyelectrolytes (polycation and polyanion) with ions (MV^{2+} and EDTA). These interactions bring about changes in polymer chain configuration. This effect is manifested by the

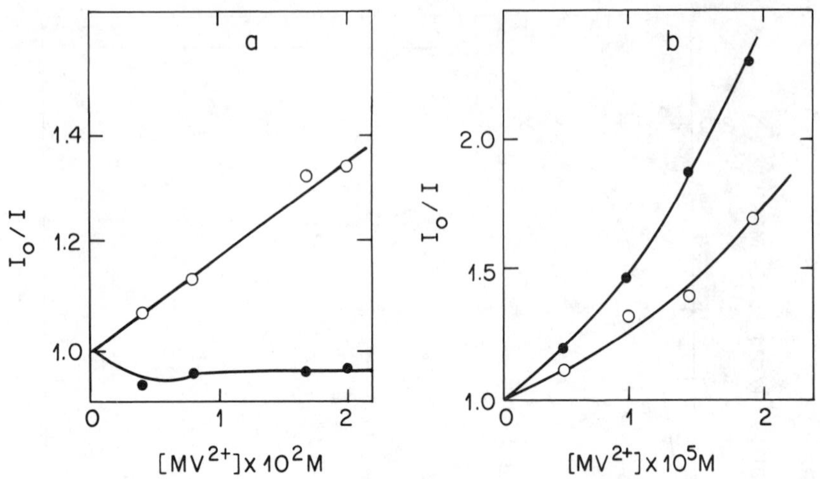

Figure 6. Fluorescence quenching of pyrene by MV^{2+} in water. a) 1, b) 2. Monomer emission (○), excimer emission (●).

Table II. Quenching of Pyrene Fluorescence by MV^{2+} and Quantum Yield of $MV^{\dot{+}}$ Formation

S	$K_1(M^{-1})$	$(\phi_{MV^+})_i$ *1
1	17.5	0.061
2	2×10^4	0.00

*1 $[S] = 5 \times 10^{-5} M$,
$[EDTA] = 1 \times 10^{-3} M$,
$[MV^{2+}] = 5 \times 10^{-4} M$ in water.

variation of excimer intensity as a function of $[MV^{2+}]$ and [EDTA]. At the moment we do not understand these phenomena adequately.

Instead of MV^{2+} (in the photo-oxidation of leuco crystal violet (LCV)), a neutral species is sensitized by pyrene containing polymers and the Coulombic effect is not as drastic as in the case of MV^{2+}. As shown in Figure 8, the cationic polymer is more effective than the neutral or anionic polymer. This is attributed to the Coulombic repulsion between $LCV^{\dot{+}}$ and $Py^{\dot{-}}$ assisted by the cationic environment of the polycation. However, the Coulombic effect occurs only after forward electron transfer.

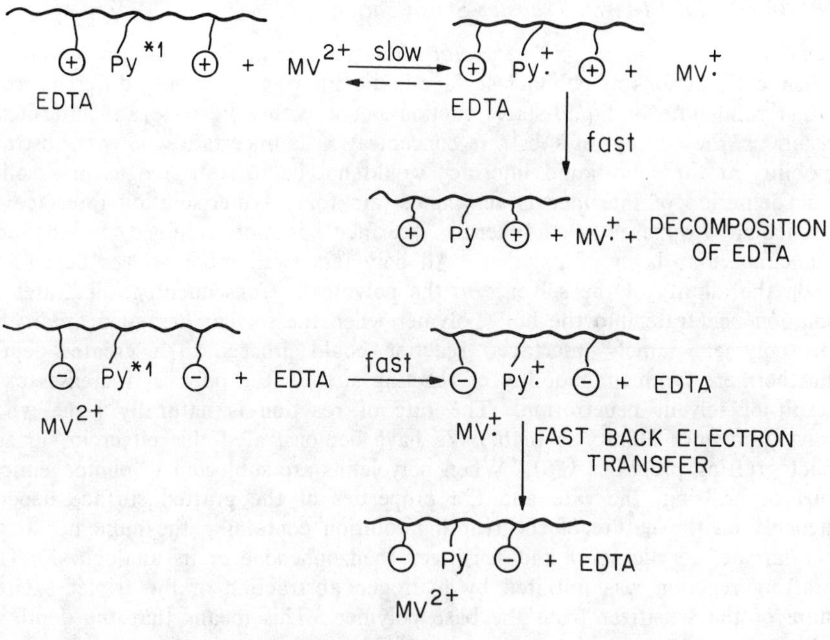

⊕ Py*1 ⊕ + MV^{2+} slow → ⊕ Py^{+}· ⊕ + MV^{+}·
EDTA EDTA
fast
⊕ Py ⊕ + MV^{+}· + DECOMPOSITION OF EDTA

⊖ Py*1 ⊖ + EDTA fast → ⊖ Py^{+}· ⊖ + EDTA
MV^{2+} MV^{+}·
FAST BACK ELECTRON TRANSFER
⊖ Py ⊖ + EDTA
MV^{2+}

Figure 7. Coulombic effects on photoreduction of MV^{2+}.

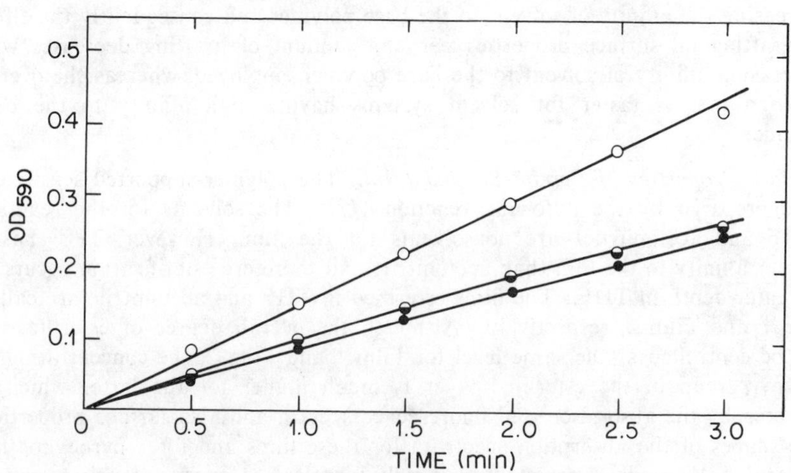

Figure 8. Photo-oxidation of LCV in homogeneous solution. Sensitizer 3 (○), 4 (●), 5 (○), solvent: DMF/H_2O(9/1), [LCV] = 4×10^{-3}M, irradiation at 344 nm (OD_{344} 1.5). "Reproduced with permission from Ref. 15. Copyright 1983, 'John Wiley & Sons, Inc.'"

(B) Interfacial Electron Transfer Sensitization

General Considerations on Interfacial Photoreaction at Solid-Liquid Boundary. Chemical reactions at polymer surface-liquid interfaces are very different from either solid-state or liquid-state reactions. For both solid-state and interfacial reactions, the concept of substrate concentration is uncertain whereas substrate mobility at the solid-liquid interface would not be so restricted as in a solid. The definition of interface is very uncertain for polymer-solution interface in particular. Apparently the term "interface" is not confined to the top monomolecular layer of polymer. All polymers swell more or less depending upon the affinity of the solvents to the polymers. Consequently, substrates in solution penetrate into the bulk polymer when the solvent has good affinity to the polymer, namely interfacial reaction could proceed to a greater depth. Furthermore, segment mobility of reacting sites in the polymer increases as a result of solvent penetration. The rate of reaction is naturally large when solvent-polymer affinity is high. We have demonstrated this effect in surface photografting processes (*20*). When polyolefins are subjected to photochemical surface grafting, the rate and the properties of the grafted surface depend strongly on the nature of the reacting solution containing the monomer (e.g., acrylamide, acrylic acid) and sensitizer (benzophenone or its analogues). The grafting reaction was initiated by hydrogen abstraction of the triplet excited state of the sensitizer from the base polymer. This means that the depth to which the monomer and sensitizer can diffuse is the main controlling factor for the grafting rate and for the degree of surface modification. Even if the same monomer and sensitizer are used, the depth of the grafted layer increases by increasing the affinity of solvent to the base polymer and consequently the effect of grafting on surface properties per unit amount of grafting decreases with decreasing affinity of solvent to the base polymer employed, whereas the overall grafting rate is faster for solvent systems having high affinity to the base polymer.

Surface Properties of Pyrene-Bonded Film. The polymer-supported sensitizers were prepared by the following reaction (*17*). The solvents for the reaction (THF and acetonitrile) are nonsolvents for the film. However, THF has a greater affinity to the film than acetonitrile and therefore esterification occurs to a greater depth in THF. The films esterified in THF and acetonitrile are called Film 1 and Film 2, respectively. Although the overall degree of esterification can be controlled at the same level for Film 1 and Film 2, the concentration of pyrenyl group in the esterified layer is much higher for the latter which is reflected by the absorption and fluorescence spectra and also surface properties. The shapes of the absorption spectrum for these films and for 1-pyrenylmethyl acetate as the model compound are nearly identical. However the sharpness of the ^1L_b band expressed by the ratio of OD at 375 nm (peak) to OD at 372 nm (valley) is less for Film 2. Furthermore, the spectral sharpness decreases with increasing degree of esterification as shown in Figure 9. The spectrum

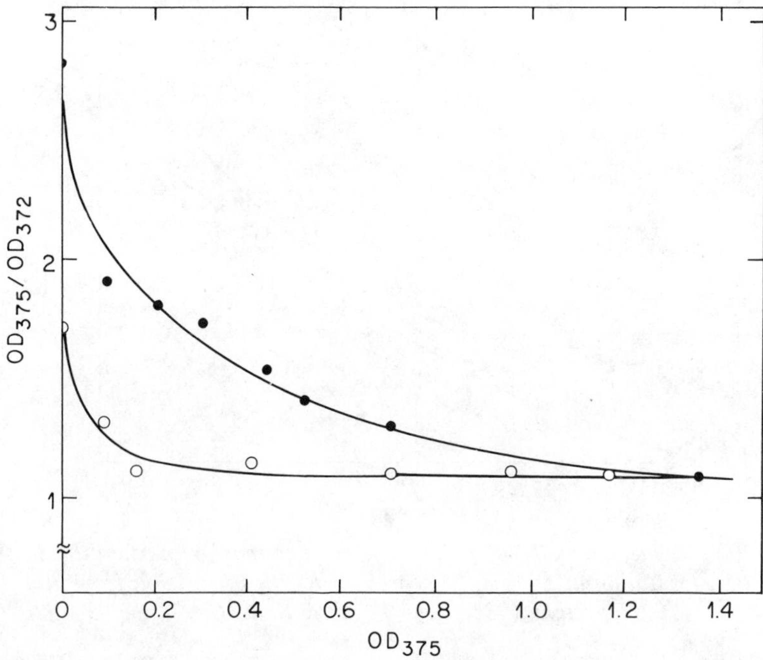

Figure 9. Broadening of absorption spectra of pyrenyl groups bonded to polymers. Film 1 (●), Film 2 (○). "Reproduced with permission from Ref. 17. Copyright 1983, 'John Wiley & Sons, Inc.'"

broadening is a common trend when aromatic hydrocarbons are attached to a polymer in particular when they are congested (*21,22*). The peak-to-valley ratio drops suddenly and remains constant for Film 2 whereas the absorption spectrum of Film 1 changes gradually with the degree of esterification. This shows that the esterification of Film 1 occurs gradually in a thick layer, but that esterification proceeds densely only in a thin surface layer in Film 2. This picture is well supported by the excimer intensity (F_e) of the pyrenyl group relative to the monomer fluorescence intensity (F_m). Excimer intensity studies are a sensitive method of indicating the local chromophore concentration provided the chromophore mobility is not frozen. The excimer intensity at 465 nm increases with an increase in the degree of esterification for both films (Figure 10). When they are compared at the same level of esterification, the intensity is very much higher for Film 2 than for Film 1 indicating the layer thickness of esterification must be deeper for Film 1. The maximum wavelength of excimer emission is 465 nm for Film 1, which is shorter by about 20 nm than that of (1-pyrenyl)methyl acetate solution. The formation of high energy excimers in Film 1 indicates restricted excimer conformation in the polymer film. In the case of intramolecular excimer formation in α,ω-bispyrenylalkanes, a blue shift was reported when the alkane chain length was unfavorable for face-to-face encounter of the two pyrenyl end groups (*23*).

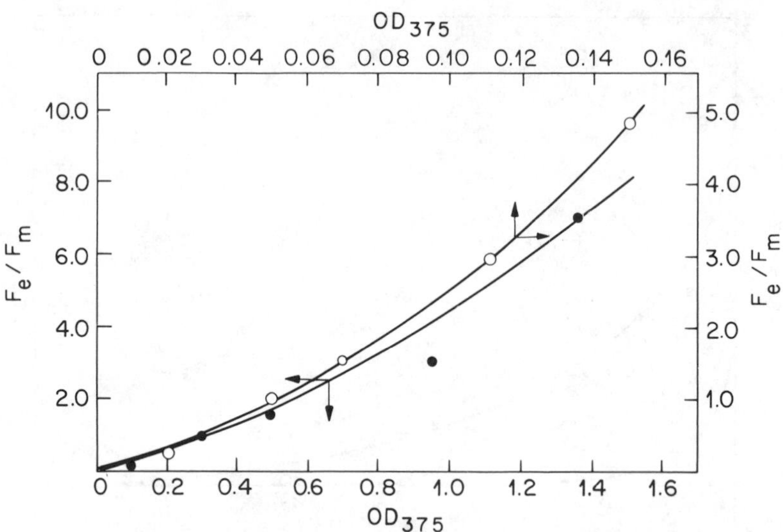

Figure 10. Excimer intensity of polymer bonded pyrenyl groups. Film 1 (●), Film 2 (○), excitation at 344 nm. "Reproduced with permission from Ref. 17. Copyright 1983, 'John Wiley & Sons, Inc.'"

These results agree with the surface property change as a function of the degree of esterification. The polyethylene film grafted with acrylic acid is hydrophilic before esterification. With increasing the degree of esterification, the surface becomes hydrophobic. This change, expressed by the change in contact angle measured with water, is shown in Figure 11. Interestingly, when the esterification is limited to the thin surface layer (Film 2) the contact angle reaches a limiting value at an early stage of esterification whereas a deeper esterification layer of Film 1 is reflected by the more gradual change. The esterification reaction does not alter the surface morphology. As examined by scanning electron microscopy Film 1 and Film 2 are identical.

Photo-oxidation of LCV by the Pyrene-Bonded Film. The surface properties and structure of these systems should be related to the sensitizer efficiency of pyrenyl groups. Pyrene is a good sensitizer for the oxidative color formation of LCV. The singlet excited state of Py acts as an electron acceptor bringing about one electron oxidation of LCV^+. The unit processes are considered as

$$Py^{*1} + LCV \longrightarrow Py^{\overline{\cdot}} + LCV^{\dot{+}}$$

$$LCV^{\dot{+}} \rightleftharpoons H^+ + CV\cdot$$

$$LCV^{\dot{+}} + CV\cdot \longrightarrow CV^+ + LCV$$

follows: 1) The time-conversion relation of CV^+ formation in acetonitrile/acetone (1/1) mixture is linear during the initial period. 2) The quantum efficiency of CV^+ formation (ϕ_{CV^+}) depends strongly on the film structure, as shown in Figure 12. To understand the variation of ϕ_{CV^+}, all results are summarized in Table III. Subtle differences in the sensitizer

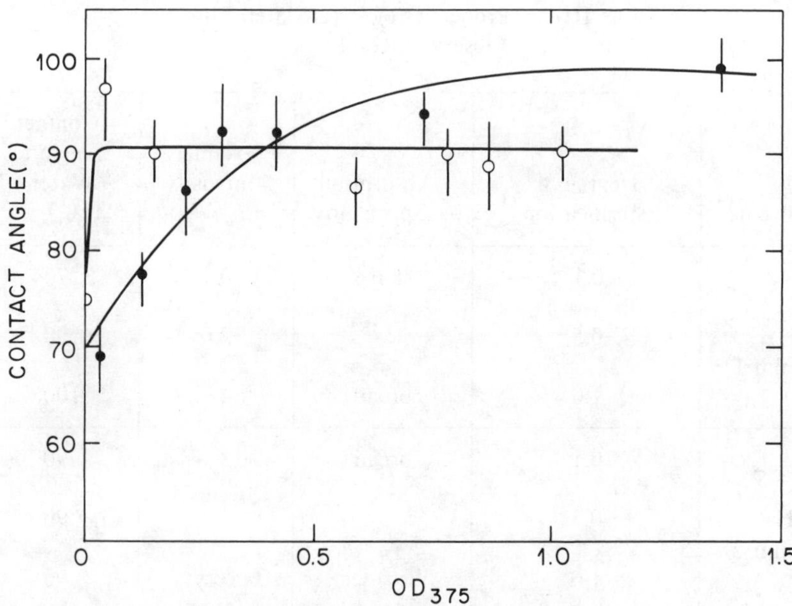

Figure 11. Contact angle of film to water versus the degree of esterification. Film 1 (●), Film 2 (○).

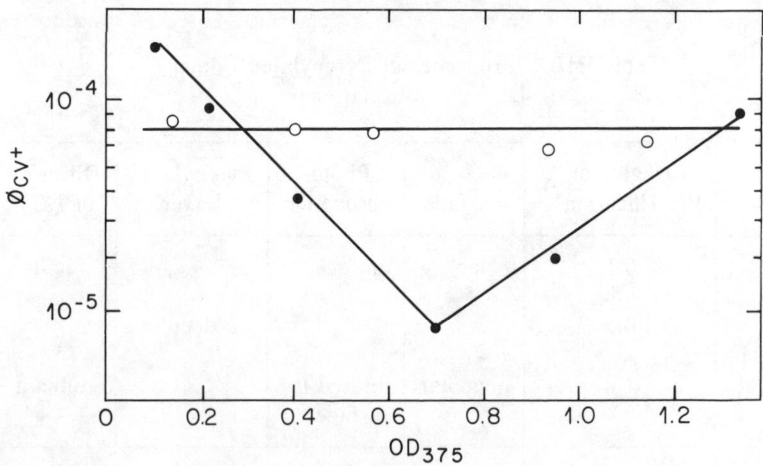

Figure 12. The quantum yield of CV^+ formation as a function of pyrene concentration in film. Film 1 (●), Film 2 (○). "Reproduced with permission from Ref. 17. Copyright 1983, 'John Wiley & Sons, Inc.'"

Table IIIA. Properties of Pyrenylated Films: Observed Results

Solvent[1)]	Degree of Esterification[2)]	Absorption Spectrum	Excimer Intensity F_e/F_m	Contact Angle to Water (°)
THF (Film 1)	0.1	sharp	0.1	70
	0.5		2.0	90
	1.0	broad	4	100
CH_3CN (Film 2)	0.1	broad	3	90
	0.5			90
	1.0	broader	very large	90

1) Used in the esterification reaction

2) Expressed by OD at 375 nm (the ^1L_b band of the pyrenyl group)

Table IIIB. Properties of Pyrenylated Films: Derived Information

Solvent[1)]	Degree of Esterification[2)]	Surface	Photo-absorption	Pyrenylated Layer	Diffusion of LCV
THF (Film 1)	0.1	polar	uniform		facille
	0.5			deep	
	1.0	nonpolar	limited to surface		difficult
CH_3CN (Film 2)	0.1	nonpolar			
	0.5		limited to surface	shallow	difficult
	1.0	nonpolar			

1) Used in the esterification reaction

2) Expressed by OD at 375 nm (the ^1L_b band of pyrenyl group)

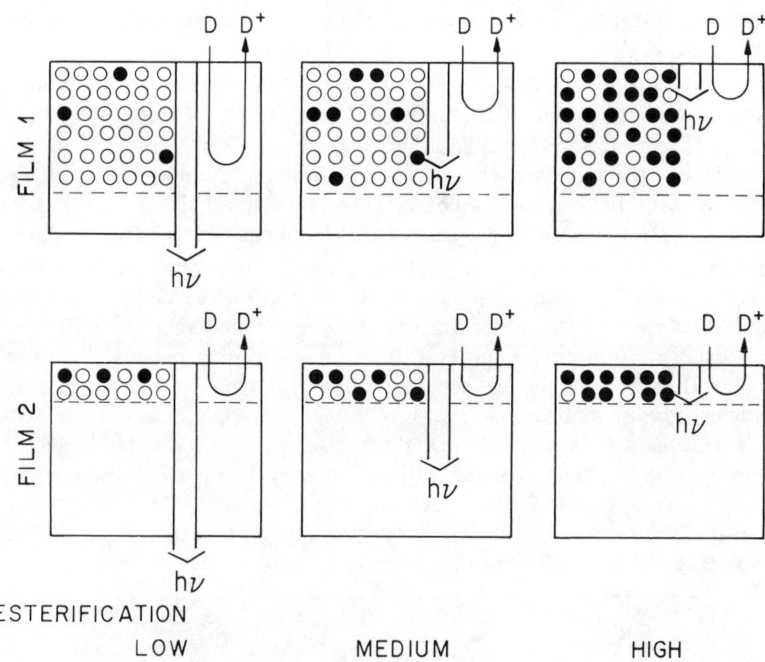

Figure 13. Sketches of the sensitizer-containing films which explain the differences between Films 1 and 2 in photochemical processes. Acrylic acid site (○), pyrenylated site (●), LCV (D).

structure have a large influence on ϕ_{CV^+}. For Film 1, the value of ϕ_{CV^+} differs over an order of magnitude when the pyrene content changes. This can be explained schematically (Figure 13). When the degree of esterification is low, the surface remains polar and the solvent for the photoreaction (acetonitrile/acetone=1/1) has a good affinity for the film. The results of the study on surface photografting suggest that when the solvent for the grafting reaction wets the polymer surface well the monomer and the sensitizer diffuse into the polymer to produce a deeper grafted layer on which the degree of surface modification per unit graft yield is small (*20*). The same argument is applicable to the present system. For low degrees of esterification the condition of homogeneous irradiation is fulfilled and the diffusion of LCV into Film 1 is rapid and deep. Consequently the use of absorbed photoenergy is efficient. With an increasing degree of esterification, Film 1 surface becomes nonpolar and the diffusion of LCV is limited. The result is a decrease in the reaction volume for the reduction of ϕ_{CV^+}. A further increase in the degree of esterification above OD 0.7 does not alter the surface polarity (see Figure 11), and the depth of LCV diffusion remains constant. However, the increase in local pyrene concentration at the surface enhances surface photoabsorption, and ϕ_{CV^+} increases again.

The results for Film 2 reveal no dependence of ϕ_{CV^+} on the degree of esterification. The wettability of Film 2 is low over a wide range of esterification, and the diffusion of LCV is confined to a shallow surface layer.

Also, the esterification proceeds only on the film surface and photoabsorption is limited to this region. The result is a constant value for ϕ_{CV^+}.

Comparing ϕ_{CV^+} for Films 1 and 2, we found that the maximum value for Film 1 was higher than that for Film 2 at $OD_{375} \approx 0.1$ (Figure 12). This may be attributed to the higher excimer intensity of Film 2 (Figure 10). Being a stabilized excited state, the excimer state exhibits a lower sensitizer efficiency. The results with the model compound ((1-pyrenyl)methyl acetate) indicate a decrease in ϕ_{CV^+} and an increase in the excimer intensity (Figure 14). Furthermore, the previous work on mimicking the antenna pigments in photosynthesis by aggregating sensitizer molecules revealed that pyrene containing polymers were poorer sensitizers than the monomer model because of efficient excimer formation whereas anthracene containing polymers were better sensitizers than the monomer model because of inefficient excimer formation and efficient energy migration (*6*).

Consequently, high local concentration of pyrene is disadvantageous. In addition, it brings about the low surface polarity (compare Film 1 and 2 in Figure 11 at OD $\approx$ 0.1). Affinity of LCV in the solvent (acetonitrile/acetone=1/1) to Film 1 is certainly higher than that to Film 2, thus resulting a higher efficiency for Film 1.

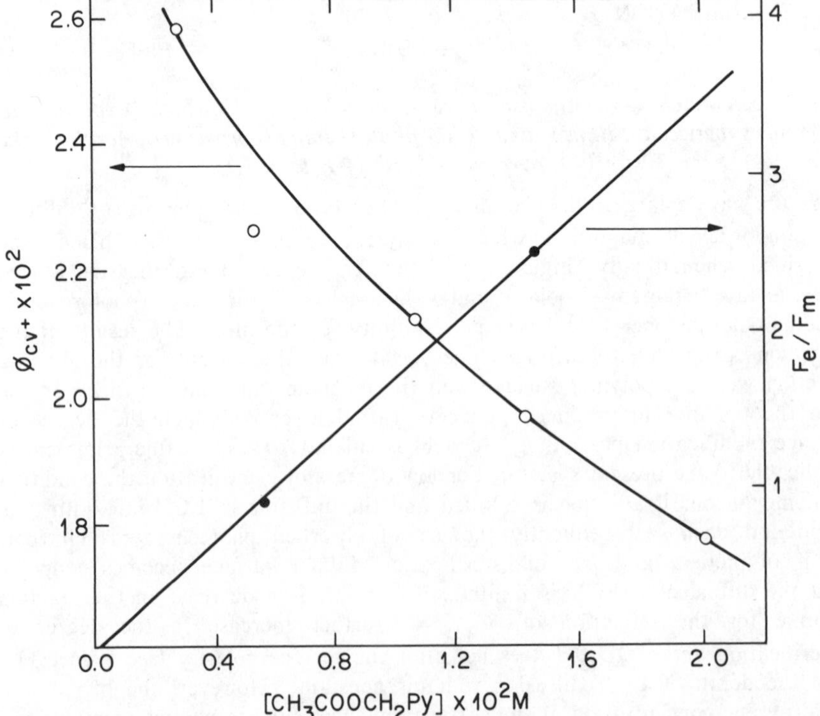

Figure 14. Plots of ϕ_{CV^+} and excimer intensity versus the concentration of (1-pyrenyl)methyl acetate. ϕ_{CV^+} (○) (The reaction conditions are the same as Table IV, run 4), F_e/F_m in the absence of LCV (●). "Reproduced with permission from Ref. 17. Copyright 1983, 'John Wiley & Sons, Inc.'"

Ionic Effects on Interfacial Photoinduced Electron Transfer. In the previous section we discussed the controlling factors for interfacial photoinduced electron transfer. However, the absolute quantum efficiency was rather low. Because we know that the introduction of Coulombic interaction is effective in improving ϕ_{CV^+}, we prepared three types of polymers bearing positive, negative, and no charge. These polymers are soluble in aqueous DMF but insoluble in acetonitrile. We then compared the ionic effects in homogeneous and heterogeneous systems as shown in Figure 15 and Table IV. Ionic effects are shown to be even more prominent in interfacial electron transfer systems (compare Figures 8 and 15). The strikingly large ionic effects in Figure 15 are in part attributed to different wettability of polymers *3*, *4*, and *5*. The wettability of *3*, *4*, and *5* with acetonitrile expressed by contact angle is 3°, 32.2 ± 5.2°, and 16.4 ± 2.5°, respectively. This sequence is consistent with the reactivity profile.

Conclusions

When electron transfer sensitizers are bonded to polymers the sensitizer efficiency is in general reduced. This is caused by (a) loss of segment mobility, (b) enhanced excimer formation (energy trap), (c) enhanced side reactions, and (d) reduction of solvent polarity in the microenvironment of polymer chain. Laser flash photolysis studies of carbazole or pyrene-containing polymers in the presence of dimethyl terephthalate showed negative polymer effects on transient

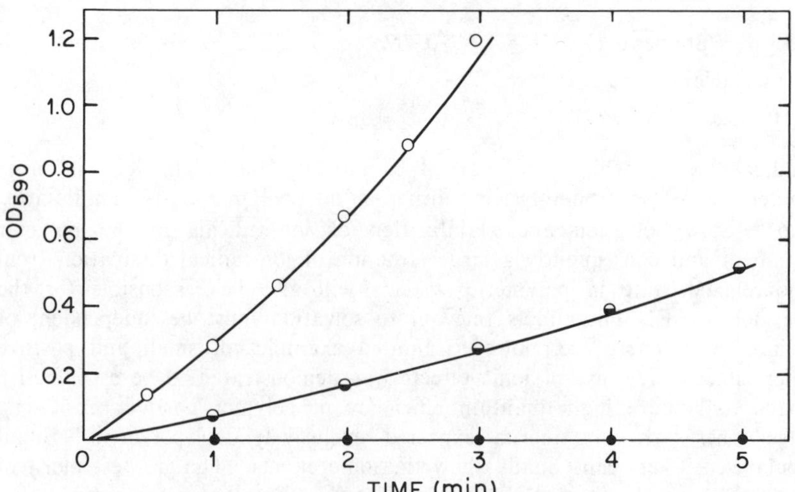

Figure 15. Photo-oxidation of LCV in the heterogeneous system. Sensitizer 3 (○), 4 (●), 5 (○). Solution [LCV] = 4 × 10^{-3}M in acetonitrile, irradiation at 375 nm, OD_{375} of the films > 1.5 nm. "Reproduced with permission from Ref. 15. Copyright 1983, 'John Wiley & Sons, Inc.'"

Table IV. Quantum Efficiency of Crystal Violet Formation

Run	Sensitizer	Solvent	Wavelength of Irradiation (nm)	ϕ_{CV^+}
1[a]	5	CH_3CN	375	2.97×10^{-3}
2[a]	3	CH_3CN	375	1.41×10^{-2}
3[a]	4	CH_3CN	375	~ 0
4[a]	Film 2	CH_3CN/acetone(1/1)	375	$10^{-4} - 10^{-5}$
5[b]	5	DMF/H_2O (9/1)	344	2.19×10^{-3}
6[b]	3	DMF/H_2O (9/1)	344	3.63×10^{-3}
7[b]	4	DMF/H_2O (9/1)	344	1.73×10^{-3}
8[b]	(1-pyrenyl)methyl	CH_3CN acetate	375	1.80×10^{-2}

For all measurements [LCV] = 4×10^{-3}M.

a = Interfacial

b = Homogeneous

ion-radical yields, which was caused mainly by the high light intensity characteristic of laser photolysis resulting in intrapolymer S_1-S_1 annihilation (*24-26*). Slow solvation and stabilization of ion-radicals in the polymer environment and consequently a larger fraction of ion-radical dissipation from the nonrelaxed state in polymeric systems will also be responsible for the smaller ion yields. The effects relevant to solvation must be independent of light intensity. There are, however, limited examples of small and positive polymer effects. The use of ionic effects was demonstrated to be an effective approach to achieve high quantum efficiency in polymer bonded sensitizers. However these are not merits expected exclusively for polymers. Small molecular sensitizers can equally show the ionic effects. The greatest merit of the polymer-bonded sensitizer is the interfacial electron transfer sensitization and more probably, the photoredox coupling across the polymer membrane in view of facile product separation. If the interfacial properties are well analyzed and the solution-solid interface is properly designed, the efficiency of the polymer-bonded sensitizer could approach the value of small molecular models in homogeneous systems (compare runs 1 and 8 in Table IV).

Literature Cited

1. "Photochemical Conversion and Storage of Solar Energy — Part A and B"; Rabani, J., Ed.; Weizmann Science Press: Israel, 1982.
2. Tazuke, S.; Kitamura, N. *Nature* 1978, *275*, 301-2.
3. Kitamura, N.; Tazuke, S. *Chem. Lett.* 1983, 1109-13.
4. Kitamura, N.; Tazuke, S. *Bull. Chem. Soc. Jpn.* 1980, *53*, 2594-7.
5. Kitamura, N.; Tazuke, S. *Bull. Chem. Soc. Jpn.* 1980, *53*, 2598-604.
6. Tazuke, S.; Tomono, N.; Kitamura, N.; Sato, K.; Hayashi, N. *Chem. Lett.* 1979, 85-8.
7. Tazuke, S.; Kawasaki, N.; Kitamura, N.; Inoue, T. *Chem. Lett.* 1980, 251-4.
8. Kitamura, N.; Kawanishi, Y.; Tazuke, S. *Chem. Lett.* 1983, 1185-8.
9. Calvin, M. *Acc. Chem. Res.* 1978, *11*, 369-74.
10. Ford, W. F.; Otvos, J. W.; Calvin, M. *Proc. Natl. Acad. Sci., USA.*, 1979, *76*, p. 3590-3.
11. Ageishi, K.; Endo, T.; Okawara, M. *Macromolecules* 1983, *16*, 884-7.
12. Tazuke, S. *J. Syn. Org. Chem. Jpn.* 1982, *40*, 806-23.
13. Yuan, H. L.; Tazuke, S. *Polym. J.* 1983, *15*, 125-33 and many references therein.
14. Kitamura, N.; Okano, S.; Tazuke, S. *Chem. Phys. Lett.* 1982, *90*, 13-6.
15. Tazuke, S.; Takasaki, R. *J. Polym. Sci., Polym. Chem. Ed.* 1983, *21*, 1529-34.
16. Tazuke, S.; Suzuki, Y. *J. Polym. Sci., Polym. Lett. Ed.* 1978, *16*, 223-8.
17. Tazuke, S.; Takasaki, R. *J. Polym. Sci., Polym. Chem. Ed.* 1983, *21*, 1517-27.
18. Suzuki, Y.; Tazuke, S. *Macromolecules* 1980, *13*, 25-30.
19. Suzuki, Y.; Tazuke, S. *Macromolecules* 1981, *14*, 1742-7.
20. Tazuke, S.; Matoba, T.; Kimura, H.; Okada, T. "Modification of Polymers"; ACS SYMPOSIUM SERIES, 1980, *121*, p. 217-41.
21. Tazuke, S.; Hayashi, N. *Polym. J.* 1978, *10*, 443-50.
22. Tazuke, S.; Ooki, H.; Sato, K. *Macromolecules* 1982, *15*, 400-6.
23. Zachariasse, K.; Kühnle, W. *Z. Phys. Chem. NF* 1976, *101*, 267-76.
24. Masuhara, H.; Ohwada, S.; Mataga, N.; Itaya, A.; Okamoto, K.; Kusabayashi, S. *J. Phys. Chem.* 1980, *84*, 2363-8.
25. Masuhara, H.; Ohwada, S.; Seki, Y.; Mataga, N.; Sato, K.; Tazuke, S. *Photochem. Photobiol.* 1980, *32*, 9-15.
26. Masuhara, H.; Shioyama, H.; Mataga, N.; Inoue, T.; Kitamura, N.; Tanabe, T.; Tazuke, S. *Macromolecules* 1981, *14*, 1738-42.

RECEIVED August 6, 1984

9

Laser-Induced Polymerization

C. DECKER

Laboratoire de Photochimie Générale associé au CNRS
Ecole Nationale Supérieure de Chimie
3 rue A. Werner
68200 Mulhouse, France

Ultra-violet radiation is being increasingly used to initiate polymerization reactions, mainly because of the high efficiency and the rapidity characteristic of this type of initiation. The limited penetration of UV light into organic materials has restricted the field of applications of this technique, mostly to surface treatment processes; the highly cross-linked photopolymer films obtained give very resistant protective coatings as well as the high-resolution relief images required in microprinting and microlithography (*1,2*). Such photoresist materials often consist of multifunctional systems that polymerize under UV light within a fraction of a second leading to a totally insoluble polymer network.

In the continuing search for faster cure rates, one would naturally conclude that lasers should be the ultimate light source that would provide a quasi-instantaneous polymerization and allow operation at extremely high scanning speeds. Several investigations on laser-induced polymerization have been reported in the last few years (*3-11*) but the actual development of this technology has apparently not met the efficiency, reliability and economic requirements essential for industrial applications. Still laser-initiated radical production offers some remarkable advantages over conventional UV initiation that result mainly from the large power output available, the narrow bandwidth of the emission and the spatial coherence of the laser beam which can be finely focused. Rapid curing of thick sections (up to 2 inches) of polymers can be achieved by visible-laser irradiation, as shown recently by Castle (*12*). Full use of the high power of the laser beam can only be made if the irradiated system obeys the reciprocity law, i.e., if the product of the light intensity and the required exposure time is independent of the intensity. This was shown to be the case for several positive working photoresists that undergo degradation under laser exposure (*13*). This law was not expected to hold true for conventional photopolymerization reactions where the kinetic chain length and thus the quantum efficiency is known to decrease as the light-intensity is increased.

Most of the work reported so far on laser-induced photopolymerization deals with near UV and visible radiation ranging from blue (~400 nm) (*3,12*) to red light (~700 nm) (*4*). We present here a kinetic investigation on the

0097-6156/84/0266-0207$06.00/0

photopolymerization of multifunctional acrylate monomers initiated by UV laser irradiation. A pulsed UV laser has been used for large area processing whereas a continuous wave (C.W.) UV laser has been employed to perform spatially localized polymerizations, suitable for direct writing of micrometer-size structures. One of the advantages of light-induced reactions, from a pure scientific point of view, is that the rate of initiation can be easily varied over a large range by varying the light-intensity. From the relationship between the rate of polymerization (R_p) and the rate of initiation (r_i), one can infer some basic information about the mechanism of this process, especially the termination step. By using the powerful UV lasers which are now on the market, the light-intensity range available can be greatly extended, primarily towards the very high values. The close to first-order dependence of R_p on the light-intensity that we observed in these multifunctional systems, over an 8 order of magnitude range, suggests that the growing polymer chains terminate mostly by a unimolecular process, probably because of the limited mobility of the reactive species in the rigid network. This means, that the reciprocity law is still obeyed so that the extent of the polymerization will not be restricted by the high rates of initiation provided by the laser beam, as would be the case for monofunctional systems.

Background

Laser-initiated Radical Production. Although there are different physical mechanisms involved in laser chemistry, we are concerned here with the photodissociation, i.e., the breaking of molecular bonds directly by UV photons. The laser emission is used to produce electronically excited molecules which split into reactive radicals, with the highest possible quantum yield. Since the substrate usually behaves as a poor photoinitiator, an additional molecule must be introduced in order to enhance the radical production, much in the same way as in conventional photoinitiated reactions. In this work, 2,2dimethoxy-2-phenylacetophenone (DMPA) was chosen as photoinitiator for two main reasons: (i) its absorption spectrum extends into the near UV region where both the C.W. and pulsed lasers used exhibit their emission lines (337.1 nm and 363.8 nm) and (ii) its quantum yield of radical production is high (*14*). The laser-initiation reaction can be written formally as:

(A)

$$\underset{\text{DMPA}}{C_6H_5-\underset{\parallel}{\overset{}{C}}(O)-C(OCH_3)_2-C_6H_5} \xrightarrow[\text{LASER}]{h\nu} \left[C_6H_5-C(=O)-C(OCH_3)_2-C_6H_5\right]^{*}$$

$$\downarrow$$

$$C_6H_5-\dot{C}(=O) + \dot{C}H_3 + CH_3O-C(=O)-C_6H_5$$

Both the benzoyl and the methyl radicals react with the double bond of the monomer and thus initiate the polymerization. Other types of photoinitiators, like 1-benzoylcyclohexanol or 2,2-dimethyl-2-hydroxyacetophenone, were shown to be as efficient as DMPA in initiating the polymerization of acrylate monomers (*15,16*); however, their absorption in the near UV is less pronounced so that the overall rate of the laser-induced polymerization is substantially lower when compared to DMPA (*8*). By contrast, thioxanthone derivatives (2CTX or 2DTX) exhibit large absorptions in the 330-380 nm wavelength region but have low initiation quantum yields. Actually, the highest rate of laser-initiated radical production was obtained using a photoinitiator system consisting of a mixture of DMPA, 2CTX and a tertiary amine (*17*).

Photocross-linking Polymerization. It was previously shown that, when a monomer like acrylamide is exposed to laser irradiation in the presence of an adequate photoinitiator, polymerization occurs rapidly (*4*). If the monomer contains more than one reactive double bond, the process develops in the three dimensions, leading to a highly cross-linked and insoluble material that can be used as protective coating or negative working photoresist. Thanks to the large power output available in the laser emission, the exposure time can be considerably shortened, down into the millisecond or even microsecond range. Most of these UV curable systems consist of three basic components: a photoinitiator that cleaves into radicals under laser irradiation, a multifunctional prepolymer that will constitute the backbone of the network and a reactive diluent that also participate to the cross-linking polymerization. The cure rate and the final properties of the cross-linked polymer will primarily depend on the functionality and the chemical nature of the prepolymer that usually consists of a polyester, polyurethane or epoxy chain. When fast speed is required, acrylic systems are commonly employed because of the high reactivity of the polymerizable double bond.

Figure 1 shows a schematic representation of the polymer network obtained by laser curing of a photoresist based on a bis-phenol *A* epoxy-

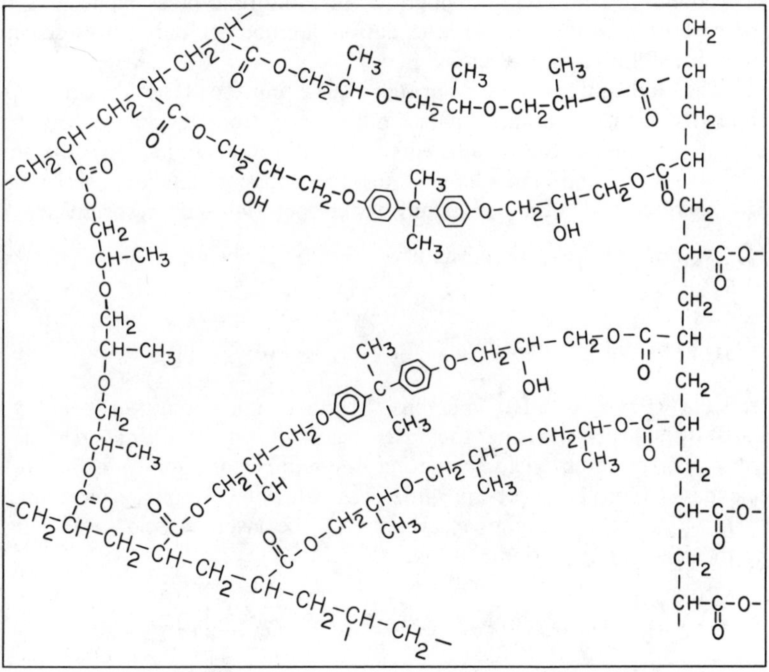

Figure 1. Schematic representation of a TPGDA-epoxy diacrylate network.

diacrylate and tripropyleneglycol diacrylate, assuming equal reactivity of the acrylate end group of both the epoxy-oligomer and the ether monomer. The cross-link density of this material reaches very high values (up to 8 mol L^{-1}) since for each acrylate group that polymerizes one cross-link unit will be formed. This leads to a complete insolubilization of the thoroughly cured polymer which shows no swelling at all in any organic solvent and exhibits remarkable optical and mechanical properties.

Experimental

Irradiation. Two types of lasers were used that both emit in the UV range:

(i) An argon-ion laser (Spectra Physics — model 170) tuned to its emission line at 363.8 nm; the maximum power available in the continuous mode was 100 mW for a cross section of the laser beam of 2 mm^2. The short exposure times required (10^{-3} to 10^{-6} s) were obtained by a fast scanning of the sample in front of the laser beam that could be focused down to a 10 μm spot by means of a microscope objective lens. For some experiments, the other UV emission line, located at 351.1 nm, was used and proved to be equally effective (*8*), the somewhat lower power output being compensated by a larger absorption of the polymer film at this wavelength.

(ii) A pulsed nitrogen laser (Sopra 804-C) that emits at 337.1 nm; each pulse, 8 ns wide, delivers an energy of 5 mJ, which corresponds to an instantaneous power output of 0.6 mW over an exposed area of 1.2 cm^2. The power of the laser emission was tunable in the range 0.05 to 0.6 mW. The laser was operated in a multiple pulse mode at a typical repetition rate of 20 Hz; the desired number of pulses was selected by means of a camera shutter.

The light-intensity (or photon flux) of the laser beam (I_0) was calculated from the measured power output and from the known value of the energy of the photons at the desired wavelength: 328 kJ.mol^{-1} for the emission line at 363.8 nm and 354 kJ.mol^{-1} for the emission line at 337.1 nm. The maximum value of I_0 obtained at full power operation were respectively:

$(I)_{363.8}$ nm = 1.5×10^{-5} einstein* s^{-1} cm^{-2} for the argon–ion laser beam

and

$(I)_{337.1}$ nm = 1.45 einstein s^{-1} cm^{-2} for the nitrogen laser beam.

(1 einstein (E) = 6×10^{23} photons.) Higher light-intensities could still be reached by sharply focusing the laser beam, down to a tiny spot of a few microns. For quantum yield evaluation, the number of photons absorbed by the laser-exposed film had to be known; it was calculated from the incident photon flux (I_0) and from the film absorbance at the laser emission wavelength by using the following equation:

$$I_{a_{(\mathrm{E.L^{-1}\,s^{-1}})}} = I_{0_{(\mathrm{E\,s^{-1}\,cm^{-2}})}} \left[1 - \exp(-2.3\epsilon e c)\right] \frac{10^7 p}{e_{(\mu m)}} \qquad (1)$$

where I_a is the absorbed photon flux, e the thickness of the film, c the concentration of the photoinitiator, ϵ its molar extinction coefficient at the laser emission wavelength and p the ratio of the measured power output of the laser to the maximum power available. Values of I_a are typically in the order of 0.5 to 3 E.L^{-1} s^{-1} for the unfocused argon-ion laser beam and 10^5 to 5×10^5 E.L^{-1} s^{-1} for the pulsed nitrogen laser beam. The initiation rate of the polymerization (r_i) is then given by the simple equation: $r_i = \phi_i I_a$ where ϕ_i is the quantum yield of initiating radicals; for DMPA, ϕ_i was evaluated to be 0.1 (*18,19*) which leads to maximum r_i values of 0.3 and 5×10^4 radicals.L^{-1} s^{-1} for the CW and pulsed laser, respectively.

It must be pointed out that, at the laser power and scanning rates used, the surface temperature of the sample did not rise more than a few degrees above ambient, as shown recently by Tsao and Ehrlich (*10*). Since the laser exposure time was five orders of magnitude shorter in our experiments as in Tsao's work, thermal initiation can be neglected. Further support of an exclusive photoeffect in the laser curing of these resins came from the absence of any detectable polymerization when the photoinitiator was not introduced in the formulation.

Photoresist Material. Two types of multifunctional oligomers were used in these experiments, both containing acrylate end groups as reactive functions: (i) a polyester chain terminated at each extremity by 3 acrylate functions, with a molecular weight of 1400 (Ebecryl 830 from UCB), and (ii) a diacrylate epoxy resin derived from the glycidyl ether of bisphenol A, with a molecular weight of 500 (Ebecryl 605 A from UCB). The reactive diluent common to all the formulations was a triacrylate monomer, trimethylolpropanetriacrylate (TMPTA from UCB), at a concentration of 50%. As photoinitiator, DMPA (Irgacure 651 from Ciba Geigy) was employed at a concentration of 5 to 10%.

The formulation was applied with a calibrated wire wound applicator onto a quartz or sodium chloride plate; the thickness of the uniform layer was typically between 5 and 10 μm. Samples were exposed to the laser beam for various durations up to a maximum absorbed dose of 100 mJ cm^{-2}. Except for some specific experiments carried out in a nitrogen saturated reactor, all the irradiations were performed in the presence of air, at room temperature.

The kinetics of the laser-induced polymerization was followed either by measuring the thickness of the insoluble polymer film formed on the quartz plate after laser exposure and solvent development, using UV spectroscopy or by monitoring the decrease of the IR absorption of the coating at 810 cm^{-1} which corresponds to the twisting vibration of the acrylate CH_2=CH double bond. This last method permits accurate evaluation of the rate of polymerization (R_p) by observing the variation of the 810 cm^{-1} band, and using Equation 2:

$$R_p = \left(\frac{A_1 - A_2}{A_0}\right)\left(\frac{[M]_0}{t_1 - t_2}\right) \qquad (2)$$

where $[M]_0$ is the initial concentration in acrylate function of the formulation and A_0, A_1 and A_2 the IR absorbance at 810 cm^{-1} before and after laser exposure during time t_1 and t_2, respectively.

Kinetics of the Laser-induced Photopolymerization

Under UV-laser irradiation, photosensitive multifunctional acrylate resins become rapidly cross-linked and completely insoluble. The extent of the reaction was followed continuously by both UV and IR spectroscopy in order to evaluate the rate and quantum yield of the laser-induced polymerization of these photoresist systems. Two basic types of lasers emitting in the UV range were employed, either a continuous wave (C.W.) argon-ion laser, or a pulsed nitrogen laser.

Argon-ion Laser. Both the polyester and the epoxy-based multiacrylate photoresists were found to polymerize within a few milliseconds when exposed in the presence of air, to the argon-ion laser beam tuned to its C.W. emission line at 363.8 nm and operated at an incident photon flux of 7.5×10^{-6} E s^{-1} cm^{-2}. Characteristic *S* shape kinetic curves were obtained by plotting the normalized thickness of the insoluble polymer film formed against the duration of the laser exposure, in a semi-logarithmic scale (Figure 2). As expected, the induction period observed in the early stages of the irradiation is primarily due to the presence of atmospheric oxygen which is known to strongly inhibit the photopolymerization of acrylic systems by reacting with the initiating radicals as well as with the growing polymer radicals (*20,21*). As shown by Figure 2, the induction period disappeared almost completely when the laser irradiation was carried out under a dry nitrogen purge; total insolubilization of the photoresist was then reached within 2 milliseconds, compared to about 10 milliseconds in the presence of air.

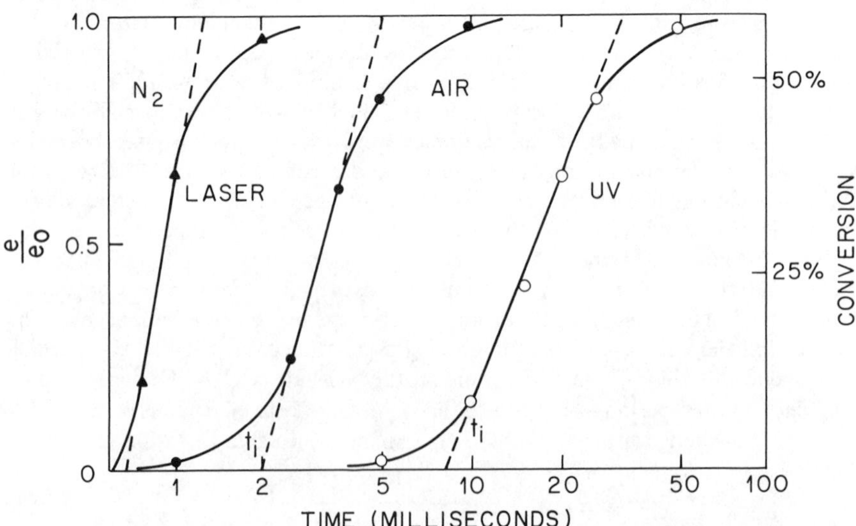

Figure 2. Normalized thickness of the polymerized epoxy-acrylate film versus duration of the exposure to the argon-ion laser beam (Δ in N_2, ● in air) or to conventional UV radiation (○ in air).

As the polymerization proceeds, the viscosity of the polymer system increases steadily to finally yield a solid and rigid highly cross-linked polymer in which the segmental mobility is quite restricted. Consequently, the encounter probability of the polymer radicals with the reactive acrylate double bonds is then sharply reduced, which accounts for the rate slowing down observed in the latter stages of the irradiation (Figure 2).

In order to compare conventional and laser UV initiation, some polymerizations were carried out by replacing the laser beam by a powerful and focused medium-pressure mercury lamp, all the other experimental conditions remained unchanged. The incident photon flux, emitted in the 250-400 nm UV range was measured to be 1.5×10^{-6} E s^{-1} cm^{-2}, i.e., five times less than the laser photon flux. The reaction proceeds then at about one fifth of the rate of the laser polymerization, as expected from the concomitant decrease in light-intensity; 50% insolubilization of the polymer film was obtained after 3.2 and 16 ms of laser and UV irradiation, respectively (Figure 2). This means that the intrinsic sensitivity (S) of the photoresist is essentially the same, whatever the type of light used, either conventional UV or laser radiations. In photoimaging technology, the sensitivity of a negative working photoresist is usually defined as the amount of incident energy required to transform 50% of the initial film into a totally insoluble material. By taking into account the corresponding photon fluxes, S values were calculated from exposure times to be 7.5 and 8 mJ cm^{-1} for the laser and UV polymerization, respectively.

The second important parameter of a photoresist material is the contrast (γ) that characterizes the sharpness of the high-resolution relief image needed in microlithography. It was evaluated from the slope of the linear portion of the kinetic curves shown in Figure 2, by using Equation 3:

$$\gamma = \frac{e}{e_0} (\log t - \log t_i)^{-1} \tag{3}$$

where e_0 and e represent the thickness of the coating before and after laser exposure during time t, respectively, and t_i is the induction period due to oxygen inhibition. Values of γ range from 1.8 to 2.5, thus revealing a fairly sharp contrast for these negative acrylate photoresists. This remarkable feature may result from the very high cross-link density of the cured polymer (> 6 mol L^{-1}) that prevents it from swelling during solvent development of the patterns.

Laser-induced photopolymerizations can be greatly accelerated either by increasing the power output or by finely focusing the laser beam, down into the micrometer range. For instance, by using a 20 μm laser spot and a corresponding light intensity of 10^{-2} E s^{-1} cm^{-2}, the rate of polymerization was increased by 3 orders of magnitude for both the polyester- and the epoxy-acrylate photoresists. Scanning speeds of up to 10m s^{-1} were then reached (*22*), which corresponds to exposure times in the microsecond range. If necessary, faster scanning rates can still be reached either by increasing the laser power output or by focusing more sharply the laser beam down into the micrometer range. Although one would expect ablation to become the limiting step beyond a certain power density, it can hardly occur under the present conditions because the laser radiation is absorbed exclusively by the photoinitiator and not by the substrate. Such a scanning spot technique has already been reported for highly focused beams of lasers emitting in the visible (*23,24*) or UV (*25*) range, but operated at much lower scanning rates (a few mm/s); it offers the advantage of successfully eliminating the detrimental speckle effect due to interference patterns in laser lithography (*26*). Direct maskless writing on the wafer by a computer directed laser beam becomes feasible, much in the same way as in electron beam lithography.

Since the argon-ion laser has its most intense emission lines in the visible range at 488 and 514.5 nm, it was tempting to see if polymerization could also be induced at these wavelengths in order to take advantage of the huge power output available, up to 5 W compared to only 100 mW in the UV range. In a recent study, Castle (*12*) has demonstrated that the 488 nm laser emission can indeed initiate the polymerization of multiacrylate monomers if benzoquinone is used as photoinitiator; the quantum efficiency was found to be much lower than with UV laser initiation (*22*). Recent polymerization experiments carried out with camphoroquinone were very promising since they show a 10 fold increase of the rate of polymerization of epoxy-acrylate photoresists, as compared to benzoquinone-based systems. Such a large improvement is assumed to result primarily from two favorable factors: (i) a higher absorbance of the 488 nm emission band since camphoroquinone exhibits its maximum absorption in the 460-480 nm range, compared to 430-450 nm for benzoquinone, and (ii) a larger initiation efficiency of camphoroquinone, especially when it is complexed with a tertiary amine like dimethylethanolamine. Scanning rates of a few meters per second were thus obtained for the curing of acrylate photoresists by visible laser radiation.

Pulsed Nitrogen Laser. Although the scanning spot technique appears to be well suited for the direct writing of patterns with a focused laser beam, one of the major problems that plague C.W. mode operated lasers is the lack of power available in the UV range. This is not the case for pulsed lasers, such as nitrogen or excimer lasers, that can deliver in the near or deep UV region extremely high power densities, up to 10 W cm^{-2}, during the few nanoseconds duration of the pulse.

In this study, a pulsed nitrogen laser was selected as excitation source first because its emission at 337.1 nm is best suited to the photoresists in this study which exhibit their highest sensitivity in the 310-350 nm wavelength region ($n\pi^*$ transition of the photoinitiator), and secondly because this type of pulsed laser is less expensive to operate than excimer lasers, while still delivering large output energies. For irradiations carried out in the presence of air, a few laser shots, 8 ns wide each, proved to be sufficient for complete insolubilization of the resist, whereas in N_2-saturated systems, one single pulse was sufficient to give 100% polymer insolubility.

In a more quantitative approach, the decrease of the IR absorption band at 810 cm^{-1}, characteristic of the acrylate double bond, was followed as a function of the exposure time (Figure 3). This permits the precise evaluation how many acrylate functions have polymerized as a function of dose and then to deduce the degree of conversion of the polymer formed. Figure 4 shows the

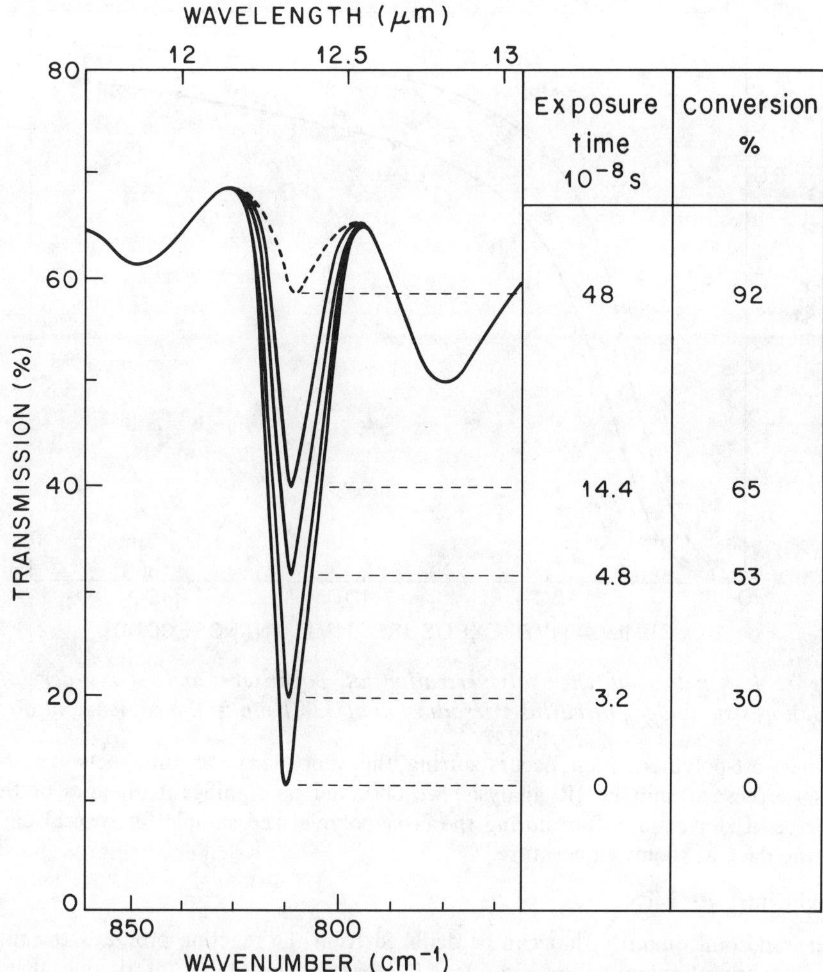

Figure 3. IR spectra of a polyester-acrylate photoresist exposed for various times to nitrogen-laser pulses in the presence of air. (--- polyurethane-acrylate).

reaction profile of the laser-induced polymerization that was obtained by plotting the degree of conversion against the cumulative exposure time. In the presence of air, typical *S* shape kinetic curves were again observed for both the epoxy- and the polyester-based photoresists which show comparable reactivities, with a slight advantage for the latter. It is interesting to note that half of the original photoresist layer was recovered as an insoluble polymer when the overall degree of conversion reached 30%, and that complete insolubilization required about 60% conversion (Figure 2). Upon extended laser exposure, the degree of conversion levelled off at about 70% for the epoxy- and polyester oligomers, while it exceeded 90% using the more flexible polyurethane chain, despite a lower rate of polymerization (*27*). Although we cannot exclude that

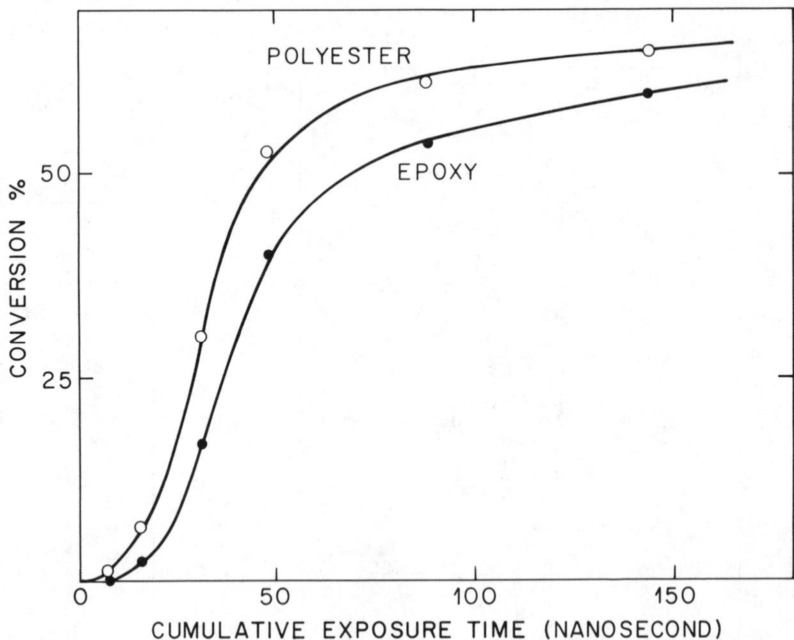

Figure 4. Kinetics of the polymerization of polyester- and epoxy-acrylate photoresists under pulsed laser irradiation at 337.1 nm in the presence of air.

some post-polymerization occurs during the short lapse of time between the laser exposure and the IR analysis, we observed no significant changes of the degree of conversion after storing the laser-polymerized sample for several days in the dark at room temperature.

Light Intensity Effect

An important quantity that can be deduced from the reaction profile is the rate of the cross-linking polymerization (R_p), i.e., the number of double bonds polymerized or of cross-links formed per second. R_p values were determined from the maximum slope of the kinetic curves (usually reached for conversion degrees between 20 and 40%). Table I summarizes the R_p values for the two photoresists tested under various conditions, namely conventional UV and continuous or pulsed laser irradiation at different light intensities. According to these kinetic data, R_p increases almost as fast as the light-intensity; the ratio I_0/R_p which is directly related to the product of the light-intensity and the required exposure time was found to vary only in the range 10^{-8} to 2.6×10^{-7} E cm^{-2}L mol^{-1} when the light-intensity value was increased by 8 orders of magnitude. Such a small departure from the reciprocity law was quite unexpected, especially for a polymerization which is basically a chain reaction process and thus strongly dependent on the initiation rate. This surprising light-intensity effect becomes clearly apparent from a plot of R_p

Table I. Rates and Quantum Yields of Polymerization of an Epoxy-diacrylate Photoresist Exposed to Conventional or Laser UV-irradiation

UV-Light Source	Irradiation Wavelength (nm)	Light Intensity I_0 E s^{-1} cm^{-2}	Polymerization Rate R_p (mol L^{-1} s^1)	Quantum Yield (ϕ_p) (mol E^{-1})
Mercury Lamp (Med. Press.)	> 250	7×10^{-8}	29	3200
		1.5×10^{-6}	5.8×10^2	2900
C.W.-Ar Ion Laser	363.8	4×10^{-6}	2×10^3	1700
		8×10^{-4}	1.2×10^5	600
		1.2×10^{-2}	1.6×10^6	540
Pulsed Nitrogen Laser	337.1	1.1	1×10^8	420
		12	$\sim 10^9$	400

against I_0, in a logarithmic scale (Figure 5). The value of the slope of the straight line obtained, 0.85, reveals a close to first-order process for the polymerization of these multifunctional acrylate systems instead of the expected half-order relationship between R_p and I_0. The important consequences of this kinetic law on the cross-linking polymerization mechanism are discussed below.

For an accurate evaluation of the efficiency of UV photons in initiating the polymerization of acrylate photoresists, it is necessary to determine the polymerization quantum yield, ϕ_p, that corresponds to the number of acrylate functions which have polymerized per photon absorbed. ϕ_p can be expressed as the ratio of the rate of polymerization to the absorbed light-intensity and calculated from the following expression, after introducing Equations 1 and 2:

$$\phi_p = \frac{R_p}{I_a} = \frac{e\ (A_1 - A_2)\ [M]_0}{10^7\ p\ (t_2 - t_1)\ A_0\ I_0[1 - \exp(-2.3\epsilon ec)]} \tag{4}$$

The high values of ϕ_p reported in Table I indicate that, despite the high rate of initiation provided by the laser irradiation, the propagation chain reaction still develops effectively in these multifunctional systems; each photon induces the polymerization of up to 1700 monomer units in air-saturated systems and up to 10,000 in the absence of oxygen. Such a large amplification factor is necessary in this system when one considers the high cost of laser photons (*28*). When using the pulsed N_2 laser as excitation source, the polymerization quantum yield value was found to be in the order of 400, instead of less than 1 if the expected half-order rate equation ($R_p \sim I_0^{1/2}$) would have prevailed (Figure 6).

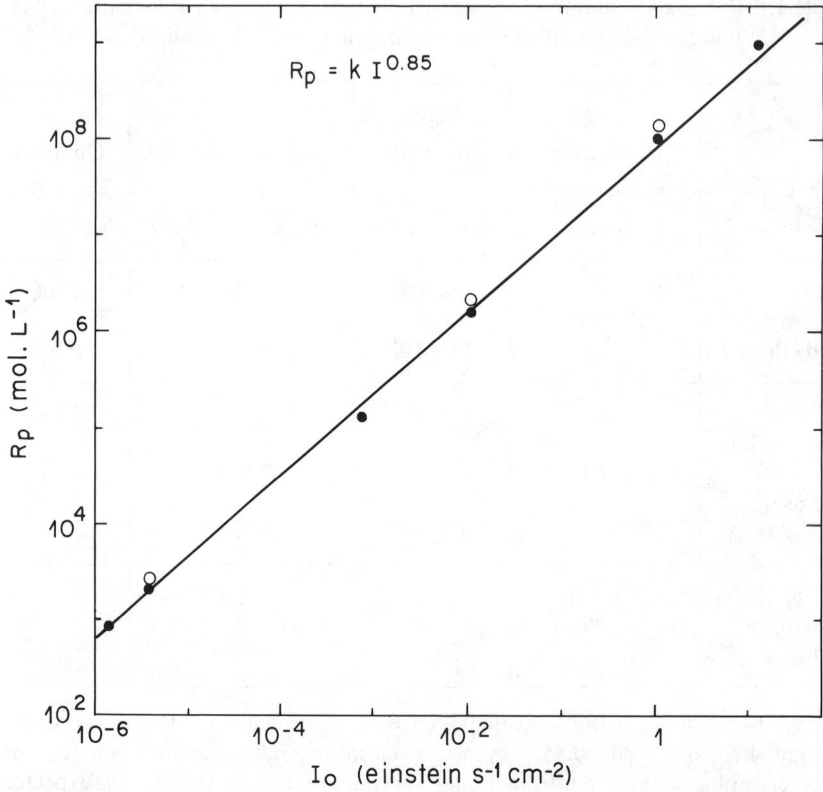

Figure 5. Dependence of the rate of polymerization (R_p) on the light-intensity (I_0) upon UV-laser irradiation of polyester- (○) and epoxy- (●) multiacrylate photoresists in the presence of air.

By taking a quantum yield value of 0.1 (*18,19*) for the production of initiating radical in DMPA photolysis, it was then possible to evaluate from ϕ_p values the kinetic chain length (*kcl*) or the degree of polymerization. Over the whole range of light intensity investigated, *kcl* values range between 4000 and 32,000 acrylate functions polymerized or cross-links formed per initiating radical; it corresponds to a calculated average molecular weight of the polymer network of over 3×10^6 for the laser curing in air and up to 10^7 in N_2-saturated systems.

It is important to note that ϕ_p values are of the same order of magnitude whether conventional UV or laser radiation were used (Table I). This result is in contrast with the recent work of Sadhir et al. (*7*) on the photopolymerization of maleic anhydride and styrene, induced by the 363.8 nm emission of an argon ion laser. That work showed to be the laser initiation 1000 times more energy efficient than the UV-induced polymerization. This discrepancy may arise from two factors: (i) the lower light-intensity used in the laser irradiation that should favor chain propagation and (ii) differences in the way of comparing the

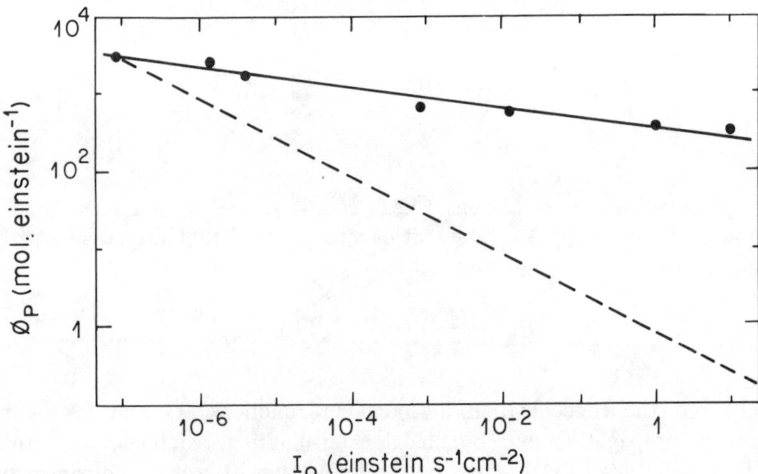

Figure 6. Dependence of the polymerization quantum yield (ϕ_p) on the light-intensity (I_0) in the laser-induced polymerization of epoxy-acrylate photoresists (---: expected variation of ϕ_p on I_0 for a half-order kinetic law).

efficiency from the energy output (7) rather than from the energy actually absorbed by the polymer. Moreover, the two systems undergoing polymerization, multiacrylates in one case and vinyl monomers in the other, were quite different, as well as different initiation processes.

Discussion

Overall the aforementioned kinetic results clearly demonstrate that multiacrylate photoresists can be polymerized efficiently by a short exposure to near-UV laser radiation. Using this technique it becomes possible to achieve extremely fast polymerizations that are not possible with conventional UV light sources. One of the most remarkable features is that the huge concentration of initiating radicals generated by the intense continuous or pulsed laser irradiation does not prevent the chain propagation from developing extensively, as indicated by the surprisingly large values of the polymerization quantum yield. Although unexpected, this favorable result permits consideration of this laser technique for applications where extremely fast curing is required. Further, it has also direct implications which shed new light on the basic mechanism of the cross-linking polymerization of these photoresist materials.

In consideration of the kinetic law obtained, $R_p \sim I_0^{0.85}$ over a 8 orders of magnitude range, one can conclude that the common polymerization mechanism, based on bimolecular termination reactions, is no longer valid for these multifunctional systems when irradiated in condensed phase. Indeed, for conventional radical-induced polymerizations, the termination step consists of the interaction of a growing polymer radical with another radical from the initiator ($R\cdot$), monomer ($M\cdot$) or polymer ($P\cdot$) through recombination or disproportionation reactions:

$$P^{\cdot} + P^{\cdot} \longrightarrow P - P \text{ or PH + P-CH=CH-R}'$$

$$P^{\cdot} + M^{\cdot} \longrightarrow P - M$$

$$P^{\cdot} + R^{\cdot} \longrightarrow P - R$$

Assuming steady-state conditions, i.e., equal rates of initiation and of termination ($r_i = \phi_i\ I_a = 2\ k_t[P^{\cdot}]^2$), it can be inferred that the polymerization rate must be half-order in light-intensity:

$$R_p = \frac{k_p}{2\ k_t^{1/2}}\ \phi_i^{1/2}\ I_a^{1/2}\ [M] \tag{5}$$

where $[M]$ is the concentration in monomer functions, k_p and k_t the rate constants of propagation and bimolecular termination reactions, respectively. Actually, it was found that increasing the light-intensity was accompanied by a more rapid increase of R_p than that which would be expected on the basis of Equation 5.

In order to account for the close to first-order kinetic law observed, one has to consider the contribution of a unimolecular termination process that would involve only one reactive radical. Various reactions can be postulated:

- *a radical transfer* from the growing polymer chain to an acceptor site that would yield inactive species.
- *the so-called wall-effect* that consists in the deactivation or scavenging of polymer radicals by electron traps located on the support; if possible, such a reaction is likely to play some role in the present case because of the large surface to volume ratios involved: 1000 cm^{-1} for a 10 μm thick film, as compared to less than 0.1 cm^{-1} for conventional photochemical reactors.
- *a degradative reaction* between the propagating radicals and the monomer or oligomer molecules; such a reaction was already suggested to explain a similar first-order kinetic law observed in the photopolymerization of vinylimidazole (*29*); it seems difficult to consider in our acrylic systems.
- *the scavenging by oxygen* of polymer radicals to yield peroxy radicals that would be inactive; the fact that the close to first-order rate equation was also observed in N_2-saturated systems eliminates this possibility.
- *a radical occlusion phenomenon* due to the formation of a tight network which strongly restricts the molecular mobility and thus prevents the monomer from diffusing towards the radical sites.

This last explanation appears to be the most feasible since we did not observe deviations from the expected half-order kinetic law when those multiacrylate monomers were polymerized in dilute solution where no rigid network is formed (*30*). A further feature which corroborates this conclusion is that neat monofunctional acrylic monomers do polymerize according to Equation 5 kinetics. Strong arguments in favor of the radical isolation hypothesis have been provided together by Tryson and Shultz (*31*) through their calorimetric study of the photopolymerization of multifunctional acrylates in condensed phase and by Julien (*32*) who detected long living polymer radicals by e.s.r. spectroscopy in UV cured multifunctional acrylate photoresists.

Based on the assumption that termination of the growing polymer chains occurs by both mono- and bimolecular pathways, the overall rate of polymerization can then be expressed, at a first approximate, as the sum of two terms which depend on the first power and on the square root of the absorbed light-intensity, respectively:

$$R_p = \alpha \frac{k_p}{k'_t} [M]\phi_i I_a + (1 - \alpha) \frac{k_p}{2 k_t^{1/2}} [M]\phi_i^{1/2} I_a^{1/2} \qquad (6)$$

where k'_t is the rate constant of the unimolecular termination process, i.e., the reciprocal of the polymer radical lifetime, and α a coefficient that reflects the relative contribution of the unimolecular pathway in the termination step, i.e., the probability for a given polymer radical to become occluded. From our kinetic data, α was calculated to be close to 0.8 for both the epoxy and polyester multiacrylate photoresists investigated; it means that, under the given experimental conditions, one fifth of the polymer radicals are likely to terminate through bimolecular reactions, the remaining radicals becoming occluded in the polymer matrix. The value of α is yet assumed to strongly depend on the structure and cross-link density of the network; for instance, it is expected to be substantially lower in polyurethane based photoresists that are known to give more flexible UV-cured materials and thus exhibit a higher segmental mobility of the network chains (*33*).

Moreover, it should be noticed that polymerization rates were determined from the maximum slope of the kinetic curves, namely at degrees of conversion between 20 and 40%. At that time, the large increase in viscosity of the photoresist may already have reduced the chain mobility, thus favoring radical isolation and first-order termination. It is therefore very likely that the intensity exponent of the photopolymerization rate equation will be less than 0.85 in the early stages and that it increases with conversion to reach almost unity in the solid network. Such a kinetic behavior was indeed observed for the photopolymerization of neat hexanedioldiacrylate (*31*).

Finally, it must be pointed out that the close to first-order kinetic law observed in this study is by no means specific to polymerizations induced by intense laser irradiation; a similar kinetic law was obtained by exposing these multiacrylic photoresists to conventional UV light sources that were operated at much lower light-intensities (*27,34*). This indicates that the unimolecular termination process does not depend so much on the rate and type of initiation used but rather on the monomer functionality and on the cross-link density which appear as the decisive factors.

Conclusion

As a result of this study, it should be apparent that both C.W. and pulsed laser beams are capable of very efficiently initiating the cross-linking polymerization of multifunctional acrylate photoresists, provided adequate initiators are used to absorb the laser photons and generate the reactive species. The main advantages of this technology arise from the specific properties of the laser emission:

- Spatial and temporal control of the radical production.
- A large power output allowing instantaneous and deep section polymerizations to be carried out.
- The spatial coherence that permits to draw high-resolution images by means of sharply focused laser beams.
- The narrow bandwidth that can be matched with the maximum absorption of the initiator and thus eliminates undesirable secondary photochemical reactions; non-linear processes may however become important at high power densities.

The multifunctional acrylate systems investigated appear to be particularly appropriate for UV-laser initiation because kinetic chain length of the radicals is very long, giving DP's of up to 10^4 and thus making the economics attractive. Potential applications of laser curing are expected to concentrate mostly in microprinting and microelectronics for the production of polymer relief images by a lithographic process. While pulsed lasers will serve mainly as powerful projection light-sources for the processing of entire wafers, C.W. laser beams are more suited to the direct writing of small size patterns.

Other possible uses of high-intensity UV laser beams include the curing of thick sections of polymers, the high-speed surface treatment of optical fibers by UV curable coatings and the direct etching of photodegradable polymers working then as positive photoresists. The economic prospects of this laser technology applied to polymer processing will primarily depend on further progress in the scientific and technological understanding and control of the basic phenomena involved, together with the development of reliable and low cost lasers, such as those used in this work, that can be safely operated in the chemical industry environment.

Literature Cited

1. Pappas, S. P. "UV Curing Science and Technology" – Technology Marketing Corporation, Stamford, Connecticut 1978.
2. Green, G. E.; Stark, B. P.; Zahir, S. A. *J. Macromol. Sci., Rev. Macro. Chem.* 1982, *C21*, 187.
3. Parts, L. P.; Feairheller, W. R., Jr. U.S. Patent 3 477 932, 1969.
4. Frigerio, G. E.; Stefanini, A. *Lett. Nuovo Cimento Soc. Ital. Fis.* 1971, *2*, 810.
5. Decker, C. *Microcircuit Engineering* 1982, *82*, 299. Intern. Conf. Microlithography, Grenoble, France 1982.

6. Williamson, M.A.; Smith, J. D. B.; Castle, P. M.; Kauffman, R. N. *J. Polym. Sci., Polym. Chem. Ed.* 1982, *20*, 1875.
7. Sadhir, R. K.; Smith, J. D. B.; Castle, P. M. *J. Polym. Sci., Polym. Chem. Ed.* 1983 *21*, 1315.
8. Decker, C. *Polym. Photochem.* 1983, *3*, 131.
9. Decker, C. *J. Polym. Sci., Polym. Chem. Ed.* 1983, *21*, 2451.
10. Tsao, J. Y.; Ehrlich, D. J. *Appl. Phys. Lett.* 1983, *42*, 997.
11. Hoyle, C. E.; Hensel, R. D.; Grubb, M. B.; *Polym. Photochem.*, (in press).
12. Castle, P. M. IUPAC 28th Macromol. Symp. Amherst 1982 — Preprints p. 282.
13. Jain, K.; Willson, C. G.; Rice, S.; Pederson, L.; Lin, B. J. *Proc. Int. Conf. Microcircuit Engineering*, Grenoble, France, 1982, p. 69.
14. Kirchmayr, R.; Berner, G.; Rist, G. *Farbe and Lack* 1980, *86*, 224.
15. Decker, C.; Fizet, M. *Makromol Chem. Rapid. Commun.* 1980, *1*, 637.
16. Lieberman, R. A. *Radiation Curing* 1981, *8*, 13.
17. Decker, C. SME Techn. Paper, FC 83-265 (1983); Int. Conf. Rad. Curing, Lausanne 1983.
18. Carlblom, L. H.; Pappas, P. *J. Polym. Sci., Polym. Chem. Ed.* 1977, *15*, 381.
19. Merlin, A.; Fouassier, J. P. *J. Chim. Phys.* 1981, *78*, 267.
20. Wight, F. R. *J. Polym. Sci., Polym. Lett.* 1978, *16*, 1121.
21. Decker, C. *J. Appl. Polym. Sci.* 1983, *28*, 97.
22. Decker, C. *J. Coat. Technol.* (in press), *Polym. Mat. Sci. Eng.* 1983, *49*, 32.
23. Biedermann, K.; Holmgren, O. *Appl. Opt.* 1977, *16*, 2014.
24. Becker, R. A.; Sopori, B. L.; Chang, S. C. *Appl. Opt.* 1978, *17*, 1069.
25. Loh, I. H.; Martin, G. C.; Kowel, S. T.; Kornreich, P. *Polymer Preprints* 1982, *23*, 195.
26. Jain, K. *Lasers and Applications* September 1983, 49.
27. Decker, C.; Bendaikha, T. *Europ. Polym. J.*, (in press).
28. Kaldor, A.; Woodin, R. *Proc. IEEE*, 1982, *70*, 565.
29. Bamford, C. H.; Schofield, E. *Polymer* 1981, *22*, 1227.
30. Fizet, M., Ph.D. Thesis, University of Haute Alsace, Mulhouse 1981.
31. Tryson, G. R.; Shultz, A. R. *J. Polym. Sci., Polym. Chem. Ed.* 1979, *17*, 2059.
32. Julien, private communication.
33. Roffey, C. G. "Photopolymerization of Surface Coatings"; J. Wiley Sons: New York, 1982.
34. Decker, C.; Bendaikha, T. GFP Conference on Tridimensional Polymer Systems, Strasbourg 1983, Preprints p. 121.

RECEIVED October 3, 1984

10

Novel Synthesis and Photochemical Reaction of the Polymers with Pendant Photosensitive and Photosensitizer Groups

T. NISHIKUBO, T. IIZAWA and E. TAKAHASHI

Department of Applied Chemistry
Faculty of Engineering
Kanagawa University
Rokkakubashi, Kanagawa-ku
Yokohama-shi, 221 Japan

Polymers with pendant cinnamoyl groups are well known as photosensitive polymers (*1*) and have been used as photoresists in the fabrication of printed circuits, integrated circuits or printing plates. Their excellent thermal stability, resolving power, high tensile strength, good resistance to solvents, and photosensitivity are all properties that have led to their acceptance. The use of cinnamic acid as a starting material for the syntheses of polymers is advantageous because of its commercial availability. Polymers derived from cinnamic acid are frequently used with suitable, low molecular weight photosensitizers, because the polymers alone lack satisfactory photosensitivity. However, some of the photosensitizers used have vaporized from the polymer film during use of the photosensitive polymer. This phenomenon is sometimes responsible for the lack of reproducibility of the photosensitivity and resolving power, and corrosion of the working environment.

Polymers with other pendant photosensitive moieties such as β-furylacrylic ester (*2*) or β-styrylacrylic ester (*3*) are highly photosensitive and have even higher photosensitivity after the addition of photosensitizers. However, the thermal stability of these polymers is inferior to that of the polymer with pendant cinnamic esters (*4*). Polymers with pendant benzalacetophenone (*5*), styrylpyridinium (*6*), α-cyanocinnamic ester (*7*) or α-phenylmaleimide (*8*) have high photosensitivity but they can not be sensitized. In addition, the photosensitive moieties that are used in the syntheses of these polymers are not commercially available, in contrast to cinnamic acid.

Accordingly, the synthesis of novel cinnamate polymers with high functionality and performance is very important from the viewpoint of both polymer chemistry and practical use. Recently, we have reported the synthesis of polymers with pendant photosensitive moieties such as cinnamic ester and suitable photosensitizer groups by radical copolymerizations of 2-(cinnamoyloxy)ethyl methacrylate with photosensitizer monomers (*9*), by copolymerizations of chloromethylated styrene with the photosensitizer monomers followed by the reactions of the copolymers with salts of

0097–6156/84/0266–0225$06.00/0

photosensitive compounds (*10*), and by substitution reactions of poly(chloromethylstyrene) with salts of photosensitizer compounds (*11*).

This article reports on the synthesis of photosensitive polymers with pendant cinnamic ester moieties and suitable photosensitizer groups by cationic copolymerizations of 2-(cinnamoyloxy)ethyl vinyl ether (CEVE) (*12*) with other vinyl ethers containing photosensitizer groups, and by cationic polymerization of 2-chloroethyl vinyl ether (CVE) followed by substitution reactions of the resulting poly(2-chloroethyl vinyl ether) (PCVE) with salts of photosensitizer compounds and potassium cinnamate using a phase transfer catalyst in an aprotic polar solvent. The photochemical reactivity of the obtained polymers was also investigated.

Experimental

Materials. CEVE and 4-nitrophenyl vinyl ether (VNP) were synthesized and purified as reported earlier (*12,13*), respectively. 2-(4-Nitrophenoxy)ethyl vinyl ether (NPVE) (m.p. 72-73°C) was prepared by reacting of potassium 4-nitrophenoxide (PNP) (142 g; 0.8 mol) and CVE (842 g; 7.9 mol) using tetra-n-butylammonium bromide (TBAB) (4.0 g; 12 mmol) as a phase transfer catalyst at the boiling temperature of CVE for 12 h. The potassium chloride produced was filtered off, the filtrate washed with water, excess CVE evaporated, and then the crude product was recrystallized twice from n-hexane. (Yield: 61.7%. IR (KBr): 1630 (C=C), 1520 ($-NO_2$), and 1340 cm^{-1} ($-NO_2$).) Elemental analysis on the product provided the following data: Calculated for $C_{10}H_{11}NO_4$: C, 57.40%, H, 5.30%, N, 6.69%. Found: C, 57.49%, H, 5.3%, N, 6.72%.

2-(4-Nitro-1-napthoxy)ethyl vinyl ether (NNVE) (m.p. 79-80°C) was obtained by the reaction of CVE with potassium 4-nitro-l-naphthoxide (PNN) under the similar conditions as the synthesis of NPVE, and then it was recrystallized twice from methanol. (Yield: 72.0%. IR (KBr): 1620 (C=C), 1500 ($-NO_2$), and 1320^{-1} ($-NO_2$).) Elemental analysis on the product provided the following data: Calculated for $C_{14}H_{13}NO_4$: C, 64.85%, H, 5.05%, N, 5.40%. Found: C, 64.64%, H, 4.97%, N, 5.29%.

2-[2-(4-Nitrophenoxy)ethoxy]ethyl vinyl ether (NPEVE) (m.p. 49-50°C) was synthesized with 30% yield by the reaction of CVE (86.3 g; 0.80 mol) with 2-(4-nitrophenoxy)ethanol (16.3 g; 0.09 mol) and potassium hydroxide (9.3 g; 0.14 mol) using TBAB (2.9 g; 0.01 mol) at the boiling temperature of CVE for 8 h and treated in the similar way for NPVE. (IR (KBr): 1630 (C=C), 1520 ($-NO_2$), and 1340 cm^{-1} ($-NO_2$).) Elemental analysis on the product provided the following data: Calculated for $C_{16}H_{17}NO_5$: C, 56.91%, H, 5.97%, N, 5.53%. Found: C, 56.70%, H, 5.99%, N, 5.59%.

PCVE (reduced viscosity: 0.33, measured at 0.5 g/dl in DMF at 30°C) was prepared quantitatively by cationic polymerization of CVE using trifluoroboron ether complex (TFB) in toluene at −65°C for 3 h.

Typical Procedure for Cationic Copolymerization of CEVE with Photosensitizer Monomer. CEVE (9.81 g; 45 mmol) and 0.86 g (5 mmol) of NPVE were dissolved in 50 ml of toluene and the solution cooled with dry ice/methanol. Next 0.13 ml (1 mmol) of TFB was dissolved in 5 ml of toluene and added to the solution, and then polymerization was carried out at −65°C for 3 h in a stream of dry nitrogen. The polymerization was terminated by the addition of a small amount of 2-aminoethanol, then the solution was poured into methanol. The obtained copolymer was reprecipitated from THF into methanol, and dried in vacuum at 50°C. The yield of polymer was 96.8%. The amount of photosensitizer unit was 9.2 mol-% (determined by elemental analysis). The reduced viscosity was: 0.32 (0.5 g/dl in DMF at 30°C). (IR (film): 1710 (C=O), 1640 (C=C), 1520 (-NO_2), and 1340 cm^{-1} (-NO_2). ^1H NMR (100 MHz, in $CDCl_3$): $\delta = 1.8$ (C-CH_2-C), 3.7 and 4.3 (C-CHO-C and O-CH_2-CH_2-O), 6.4 and 7.6 (-CH=CH- doublet), and 6.8-8.0 (aromatic proton).)

Typical Synthesis of Polymeric Photosensitizer. PCVE (1.07 g; 10 mmol) was dissolved in 10 ml of DMF, and then 0.18 g (1 mmol) of PNP and 0.32 g (1 mmol) of TBAB were added to the polymer solution. The reaction mixture was stirred at 80°C for 24 h and then poured into methanol. The resulting polymer was reprecipitated twice from THF into water and then from THF into methanol, and dried in vacuum at 50°C. The yield of polymer was 1.02 g. The conversion of chlorine in PCVE was 10.1 mol-% (calculated from the halogen analysis) (*14*). (Reduced viscosity: 0.38 (0.5 g/dl in DMF at 30°C). IR (film): 1520 (-NO_2) and 1340 cm^{-1} (-NO_2).)

Typical Reaction of Polymeric Photosensitizer with Potassium Cinnamate. The polymeric photosensitizer containing 90 mol-% of pendant 2-chloroethoxy group and 10 mol-% of the 2-(4-nitrophenoxy)ethoxy group (0.59 g; 4.5 mmol of chlorine in the polymer) was dissolved in 10 ml of DMF, and then 1.01 g (5.4 mmol) of potassium cinnamate and 0.15 g (0.45 mmol) of TBAB were added to the polymer solution. The reaction mixture was stirred at 100°C for 24 h and then poured into methanol. The resultant polymer was reprecipitated twice from THF into water and then from THF into methanol, and dried in vacuum at 50°C. The yield of polymer was 1.01 g. The conversion of chlorine was 100 mol-% (calculated from the halogen analysis). (Reduced viscosity: 0.33 (0.5 g/dl in DMF at 30°C). IR (film): 1710 (C=O), 1640 (C=C), 1520 (-NO_2), and 1340 cm^{-1} (-NO_2). ^1H NMR (100 MHz, in $CDCl_3$): $\delta = 1.8$ (C-CH_2-C), 3.7 and 4.3 (C-CHO-C and O-CH_2-CH_2-O), 6.4 and 7.6 (-CH=CH- doublet), and 6.8-8.0 aromatic proton).)

Measurement of Photochemical Reactivity. The polymer solution in THF was cast on a KRS plate and dried. The film obtained on the plate was irradiated by a high-pressure mercury lamp (Ushio Electric Co: USH-250D) without a filter at a distance of 30 cm in air. The rate of disappearance of the C=C bonds at 1640 cm^{-1} was measured by IR spectrometry (JASCO A-202 model).

Measurement of Practical Photosensitivity. The photosensitivity of the polymer was measured by a gray-scale method (*15*) as follows. The polymer solution (10%) in cyclohexanone was cast on a copper plate by using a rotary applicator and dried. The Kodak step tablet No. 2 (Eastman Kodak Co.) was placed upon the polymer film cast on the plate, exposed on a chemical lamp (15w × 7) from 3 cm for 1 min., and then the exposed film was developed by the solvent for 2 min.

Measurement of Glass Transition Temperature (T_g). The T_g values of the obtained polymers were measured by DSC analysis (Du Pont Inc., 910 model) at a heating rate of 20°C/min.

Results and Discussion

Syntheses of Self-Sensitized Polymers by Cationic Copolymerizations. The cationic polymerizations of several vinyl ethers containing pendant ester groups such as cinnamic ester (*12*), methacrylic ester (*16*), acrylic ester (*17*), and crotonic ester (*18*) have been reported. Based on these reports, cationic copolymerizations of CEVE with photosensitizer monomers such as NPVE, NNVE, VNP and NPEVE were carried out using TFB as a catalyst in toluene at −65°C. Each copolymer was obtained with high yield except in the case of copolymerization of CEVE with VNP as summarized in Table I.

The cationic copolymerizations of CEVE (M_1) with NPVE (M_2) and NNVE (M_2) gave the copolymers with the photosensitizer monomer in proportion to the photosensitizer monomers in the charge when the molar content of M_2 was lower than 30 mol-%. However, copolymers with equal amounts of photosensitizer units were not obtained by the copolymerization of 60 or 70 mol-% of CEVE with 40 mol-% of NPVE and 30 mol-% of NNVE in toluene, because a portion of the photosensitizer monomer in each case was precipitated during the copolymerization under similar reaction conditions.

It was reported (*19*) that a low T_g value is required from the viewpoint of the photochemical reactivity of pendant cinnamic ester in the polymer film. As summarized in Table I, the T_g of all the copolymers of PCEVE-NPVE and PCEVE-NNVE were in the vicinity of room temperature as expected from the flexible ethoxy chain in the polymer structure.

(1)

$CH_2=CH-O-CH_2-CH_2-O-C(=O)-CH=CH-C_6H_5$ (CEVE) + $CH_2=CH-O-CH_2-CH_2-O-R$ —(TFB AT −65°C IN TOLUENE)→ $-(CH_2-CH(O-CH_2-CH_2-O-C(=O)-CH=CH-C_6H_5))_n-(CH_2-CH(O-CH_2-CH_2-O-R))_m-$ (PCEVE-NPVE, PCEVE-NNVE, OR PCEVE-NPEVE)

R: $-C_6H_4-NO_2$ (NPVE)

$-C_{10}H_6-NO_2$ (NNVE)

$-CH_2-CH_2-O-C_6H_4-NO_2$ (NPEVE)

Table I. Reaction Conditions and Results of Cationic Copolymerizations of CEVE with Photosensitizer Monomers[a]

No.	Molar ratio of monomer (M_1/M_2)	Yield (%)	M_2 in copolymer (mol-%)	$\eta_{sp/c}$[b]	T_g (°C)	Sensitivity[c]
1	(9.5/0.5)[e]	98.1	4.4	0.38	29	4
2	(9.0/1.0)[e]	96.5	9.2	0.37	24	6
3	(.5/1.5)[e]	80.2	15.3	0.68	—	9
4	(8.0/2.0)[e]	97.4	19.2	0.35	27	5
5	(7.0/3.0)[e]	94.9	28.1	0.25	27	—
6	(6.0/4.0)[e]	83.8	28.4	0.22	—	—
7	(9.5/0.5)[e]	93.6	5.4	0.33	33	8
8	(9.0/1.0)[e]	96.9	9.7	0.28	30	9
9	(8.5/1.5)[e]	96.6	14.3	0.51	—	12
10	(8.0/2.0)[e]	92.9	20.0	0.17	32	12
11	(7.0/3.0)[e]	89.8	17.8	0.23	—	8
12	(9..0/1.0)[f]	80.7	0	0.33	—	3[d]
13	(9.0/1.0)[g]	99.2	10.2	0.49	—	8
14	(10.0)[h]	98.7	0	0.80	—	4[d]

[a] Copolymerization was carried out with 50 mmol of monomer in 50 ml of toluene using 1 mmol of TFB as a cationic catalyst at −65°C for 3 h.

[b] Measured at 0.5 g/dl in DMF at 30°C.

[c] Exposed with a chemical lamp (15w × 7) for 1 min., and then developed in toluene for 2 min. The numbers correspond to the step number from the Kodak step table number 2.

[d] Exposed for 3 min.

[e] CEVE/NNVE

[f] CEVE/VNP

[g] CEVE/NPEVE

[h] CEVE

PCEVE-NPEVE with 10 mol-% of pendant photosensitizer unit was also prepared by the cationic copolymerization of 90 mol-% of CEVE with 10 mol-% of NPEVE under similar reaction conditions used for the copolymerization of CEVE with NPVE. In contrast, a copolymer with pendant photosensitizer group could be obtained from the cationic copolymerization of CEVE with

VNP, and only homopolymer of poly[2-(cinnamoyloxy)ethyl vinyl ether] (PCEVE) was obtained. It seems that the cationic reactivity of the vinyl ether group in VNP was decreased because of the electron attracting nitro group attached at the 4-position on the phenoxide.

Syntheses of Polymeric Photosensitizers and Self-sensitized Polymers by the Reactions of PCVE. Photosensitizer monomers such as NPVE, NNVE and NPEVE were synthesized by the reaction of excess CVE with potassium salts of the corresponding photosensitizing compounds using TBAB as a phase transfer catalyst as described in the experimental part. Substitution reactions of chloroethyl groups in PCVE with PNP and PNN were also carried out using TBAB as a phase transfer catalyst in DMF at 80°C for 24 h.

The reactions of PCVE with PNP are quantitative under these reaction conditions, and polymeric photosensitizers such as poly[2-chloro-ethyl vinyl ether-co-2-(4-nitrophenoxy)ethyl vinyl ether] (PCVE-NPVE) containing about 4, 10, 19, 28, 37, and 46 mol-% of pendant 4-nitrophenoxy (NP) groups were prepared (Table II). Although the reactions of PCVE with PNN were carried

Table II. Reaction Condition and Results of the Syntheses of Polymeric Photosensitizers[a]

No.	PCVE (mmol)	Photosensitizer compound (mmol)	Yield (g)	Degree of substitution (mol-%)	$\eta_{sp/c}$[b]
15	10	PNP (0.5)	0.93	4.4	0.64
16	10	PNP (1.0)	1.02	10.1	0.38
17	10	PNP (2.0)	0.96	19.0	0.42
18	10	PNP (3.0)	1.03	28.2	0.42
19	10	PNP (4.0)	1.36	37.0	0.35
20	10	PNP (5.0)	1.46	46.0	0.36
21	10	PNN (0.5)	1.04	3.5	0.39
22	10	PNN (1.0)	1.08	7.9	0.56
23	10	PNN (1.5)	1.16	12.8	0.49
24	10	PNN (2.0)	1.21	15.8	0.49
25	10	PNN (3.0)	1.42	24.1	0.40
26	10	PNN (5.0)	1.45	27.0	0.48
27	10	PNN (5.0)	1.49	30.7	0.34

[a] The reaction was carried out using 1.0 mmol of TBAB in 10 ml of DMF at 80°C for 24 h.

[b] Measured at 0.5 g/dl in DMF at 30°C.

out under the similar reaction conditions which were used for the reactions of PNP, the resulting poly[2-chloroethyl vinyl ether-co-2-(4-nitro-l-naphthoxy)ethyl vinyl ether] (PCVE-NNVE) containing pendant 4-nitro-l-naphthoxy (NN) groups did not contain the expected amounts of the NN groups when more than 20 mol-% of PNN were charged in the reaction system.

The degree of substitution of chlorine measured by halogen analysis (*14*) agrees with the contents of the photosensitizer groups in the polymer determined by the UV absorptions. From these results, it can be concluded that no side reactions occur during the substitution reactions of PCVE with PNP and PNN using a phase transfer catalyst. The IR spectrum of PCVE-NPVE showed absorptions at 1530 and 1340 cm^{-1} due to the $-NO_2$ stretching of the photosensitizer units. The IR spectrum of PCVE-NNVE also showed the corresponding characteristic absorptions.

(2)

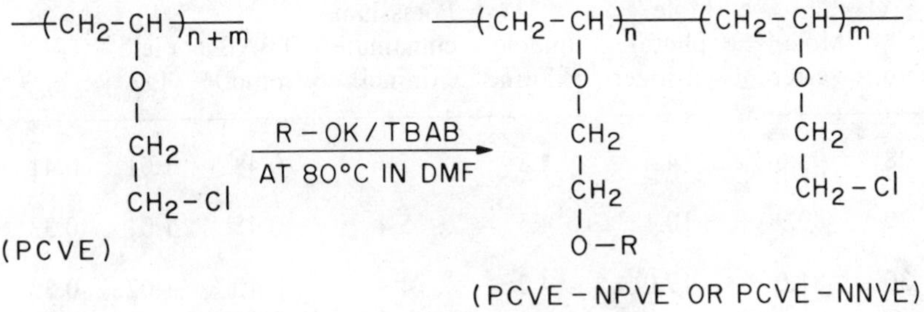

$$\text{(PCVE)}\ -\!\left(CH_2-\underset{\underset{\underset{\underset{\underset{CH_2-Cl}{|}}{CH_2}}{|}}{O}}{CH}\right)_{n+m}\xrightarrow[\text{AT 80°C IN DMF}]{\text{R}-\text{OK/TBAB}}-\!\left(CH_2-\underset{\underset{\underset{\underset{\underset{O-R}{|}}{CH_2}}{|}}{CH_2}}{\underset{|}{\overset{}{CH}}}\right)_{n}\!\!\left(CH_2-CH\right)_{m}\text{ with } -O-CH_2-CH_2-Cl$$

(PCVE − NPVE OR PCVE − NNVE)

(3)

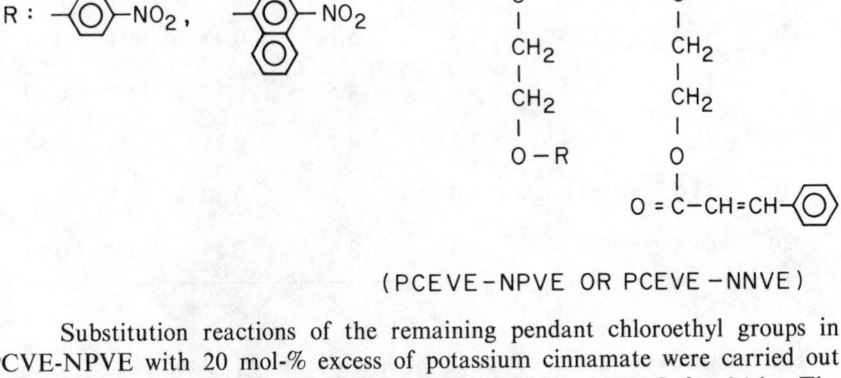

$$\text{PCVE-NPVE OR PCVE-NNVE}\xrightarrow[\text{AT 100°C IN DMF}]{C_6H_5-CH=CH-COOK/\text{TBAB}}-\!\left(CH_2-\underset{O-CH_2-CH_2-O-R}{CH}\right)_{n}\!\left(CH_2-\underset{O-CH_2-CH_2-O-C(=O)-CH=CH-C_6H_5}{CH}\right)_{m}$$

R : $-C_6H_4-NO_2$, $-C_{10}H_6-NO_2$ (4-nitro-1-naphthyl)

(PCEVE − NPVE OR PCEVE − NNVE)

Substitution reactions of the remaining pendant chloroethyl groups in PCVE-NPVE with 20 mol-% excess of potassium cinnamate were carried out using TBAB as a phase transfer catalyst in DMF at 100°C for 24 h. The reaction conditions and results are summarized in Table III. It should be noted that 100 mole % substitution occurred in all cases.

The reactions of PCVE-NPVE and PCVE-NNVE with excess potassium cinnamate proceeded quantitatively under these reaction conditions leading to PCEVE-NPVE and PCEVE-NNVE containing cinnamic ester as the photosensitive moiety, suitable photosensitizer groups such as NP and NN groups and no unreacted chloroethyl groups. The IR and ^1H NMR spectra of PCEVE-NPVE and PCEVE-NNVE obtained from the reaction of the polymeric photosensitizers with potassium cinnamate are the same as the spectra of the corresponding copolymers prepared from the cationic copolymerizations of CEVE with NPVE or NNVE.

Table III. Reaction Conditions and Results of the Reactions of Polymeric Photosensitizers with Potassium Cinnamate[a]

No.	Mole % ester	Mole % photo-sensitizer	mmole chlorine	Potassium cinnamate (mmol)	TBAB (mmol)	Yield (g)	$\eta_{sp/c}$[b]
28	95.6	4.4	4.8	5.7	0.48	1.01	0.41
29	89.9	10.1	4.5	5.4	0.45	1.01	0.33
30	81.0	19.0	4.0	4.8	0.40	1.02	0.33
31	72.8	28.0	3.5	4.2	0.35	0.94	0.35
32	63.0	37.0	3.0	3.6	0.30	1.00	0.32
33	54.0	46.0	2.5	3.0	0.25	1.02	0.32
34	96.5	3.5	4.8	5.7	0.48	1.03	0.34
35	92.1	7.9	4.5	5.4	0.45	1.03	0.39
36	87.2	12.8	4.3	5.1	0.43	1.04	0.33
37	84.5	15.5	4.0	4.8	0.40	1.00	0.39
38	75.9	24.1	3.5	4.2	0.35	1.05	0.34
39	73.0	27.0	3.0	3.6	0.30	1.15	0.39
40	69.3	30.7	2.5	3.0	0.25	1.00	0.34

[a] The reaction was carried out in 10 ml of DMF at 100°C for 24 h.

[b] Measured at 0.5 g/dl in DMF at 30°C.

These results suggest that the reaction conditions for the syntheses of PCEVE-NPVE and PCEVE-NNVE can be accomplished by the reactions of PCVE with any ratio of potassium cinnamate and PNP or PNN in one pot using a phase transfer catalyst. In addition, it is to be expected that PCEVE-NPVE and PCEVE-NNVE prepared from the reactions of PCVE have the same degree of polymerization if no side reactions occur during the substitution reactions. It is also expected that these copolymers are more random compared to the copolymers prepared from the cationic copolymerizations of the monomers, because the former is not affected by the monomer reactivity ratios.

Therefore, we believe that PCEVE-NPVE and PCEVE-NNVE prepared from the reactions of PCVE are better model polymers for the determination of the relationship between the photochemical reactivity and the contents of the photosensitizer units in the copolymers than the copolymers prepared by copolymerization. On the other hand, PCEVE-NPVE and PCEVE-NNVE prepared from the cationic copolymerization seem to be more photosensitive than PCEVE-NPVE and PCEVE-NNVE polymers prepared from the reactions of PCVE and have better properties practical applications since the former copolymers have higher purity than the latter copolymers.

Photochemical Reaction of PCEVE-NPVE and PCEVE-NNVE. Photochemical reactions of polymer films (3 μm thick) containing the cinnamic ester moiety and suitable photosensitizer groups were carried out by irradiation with a high-pressure mercury lamp on a KRS plate under identical reaction condition. As shown in Figure 1, the rates of disappearance of the C=C bonds in PCEVE-NPVE, PCEVE-NNVE, and PCEVE-NPEVE, which were prepared by the cationic copolymerizations of CEVE with the corresponding photosensitizer monomers, containing about 10 mol-% of pendant photosensitizer groups and 90 mol-% of the cinnamic ester moieties are much higher than that of PCEVE, which has only pendant cinnamic ester moieties. In addition, it was found that the photochemical reactivity of this PCEVE-NNVE was about two times higher than those of PCEVE-NPVE, PCEVE-NPEVE, and PCEVE with 4-nitroanisole as the corresponding low molecular weight photosensitizer, however, the photochemical reactivity of PCEVE-NNVE was slightly lower than that of PCEVE with the corresponding low molecular weight photosensitizer such as 1-methoxy-4-nitronaphthalene.

As shown in Figure 2, PCEVE-NPVE and PCEVE-NNVE containing about 10 mol-% of pendant sensitizer groups prepared from the reactions of PCVE with potassium cinnamate and PNP or PNN have photochemical reactivity equal to PCEVE-NPVE and PCEVE-NNVE prepared from the cationic copolymerizations, however, their photochemical reactivities were changed by the contents of photosensitizer units in the copolymers.

The relationship between the content of the photosensitizer unit in PCEVE-NPVE and the conversions of the C=C bonds after 10 min. irradiation is shown in Figure 3. The photochemical reactivities of PCEVE-NPVE prepared from the cationic copolymerizations were higher than those of PCEVE-NPVE prepared from the reactions of PCVE at photosensitizer contents lower than 15 mol-%. On the other hand, PCEVE-NPVE prepared from the reactions of PCVE has its highest rate of disappearance of the C=C

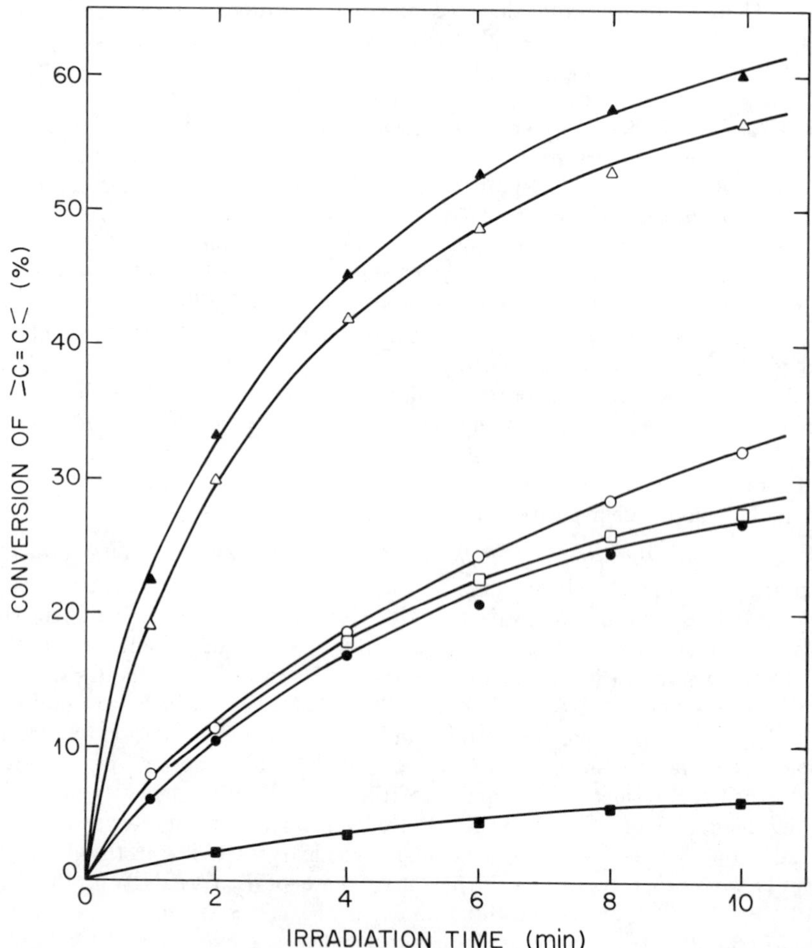

Figure 1. The rate of disappearance of the C=C group in cinnamic ester in the polymer (Δ) PCEVE-NNVE containing 9.7 mol-% of pendant NN group; (○) PCEVE-NPVE containing 9.2 mol-% of pendant NP group; (□) PCEVE-NPEVE containing 10.2 mol-% of pendant NP group (■); (▲) PCEVE + 9.7 mol-% of 4-nitro-l-methoxynaphthalene; (●) PCEVE + 9.2 mol-% of 4-nitroanisole.

bond at about 30 mol-% of pendant 4-nitrophenoxy (NP) groups. Furthermore, the photochemical reactivity of PCEVE with 4-nitroanisole gave linear plots from 5 to 45 mol-% of photosensitizer content, because they are highly compatible and the photosensitizers are relatively mobile in the polymer film.

It seems that PCEVE-NPVE prepared from the reactions of PCVE have a random distribution of substituents, and the maximum in the photochemical reactivity of these copolymers results from the balance of the positive effect of

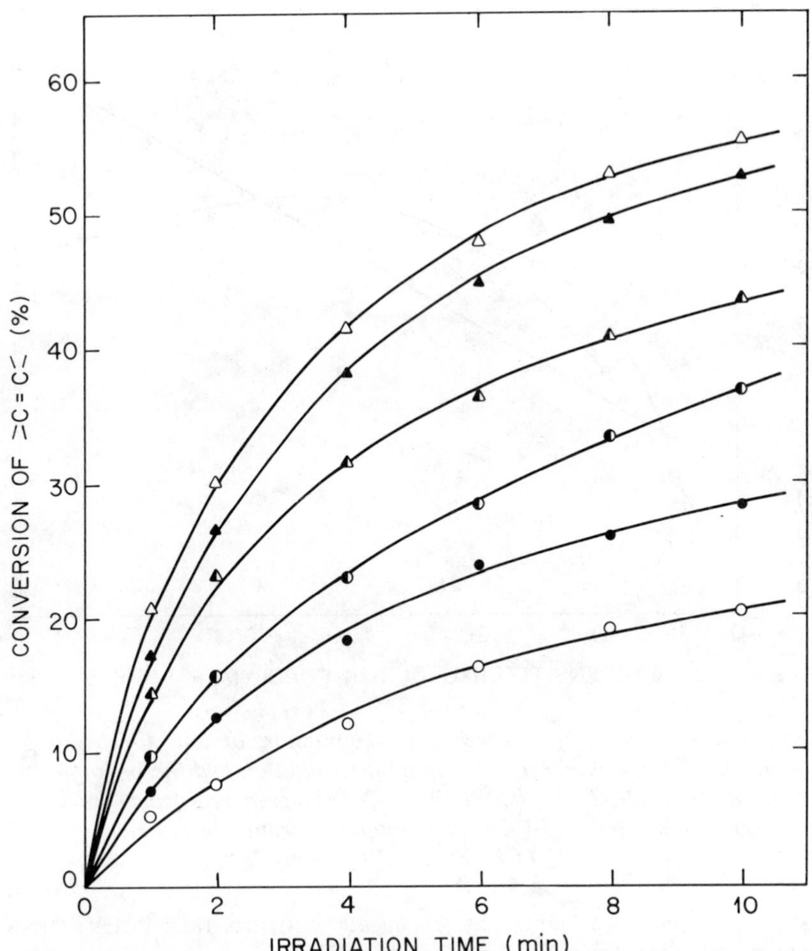

Figure 2. The rate of disappearance of the C=C group in cinnamic ester in the polymer prepared by the reaction of PCVE (Δ) PVEVE-NNVE containing 7.9 mol-% of pendant NN group; (Δ) PCEVE-NNVE containing 12.8 mol-% of the NN (Δ) PCEVE-NNVE containing 27.0 mol-% of the NN group; (○) PCEVE-NPVE containing 4.4 mol-% of the NP group; (●) PCEVE-NPVE containing 10.1 mol-% of the NP group; (○) PCEVE-NPVE containing 19.0 mol-% of the NP group.

pendant photosensitizer groups and the negative effect such as screening of the sensitizer on the surface of polymer film and steric interference of the sensitizer groups with the reaction of pendant photosensitive moieties (*11*).

The relationship between the content of the photosensitizer unit in PCEVE-NNVE and the conversions of the C=C bonds after 10 min. irradiation is shown in Figure 4. PCEVE-NNVE prepared from cationic

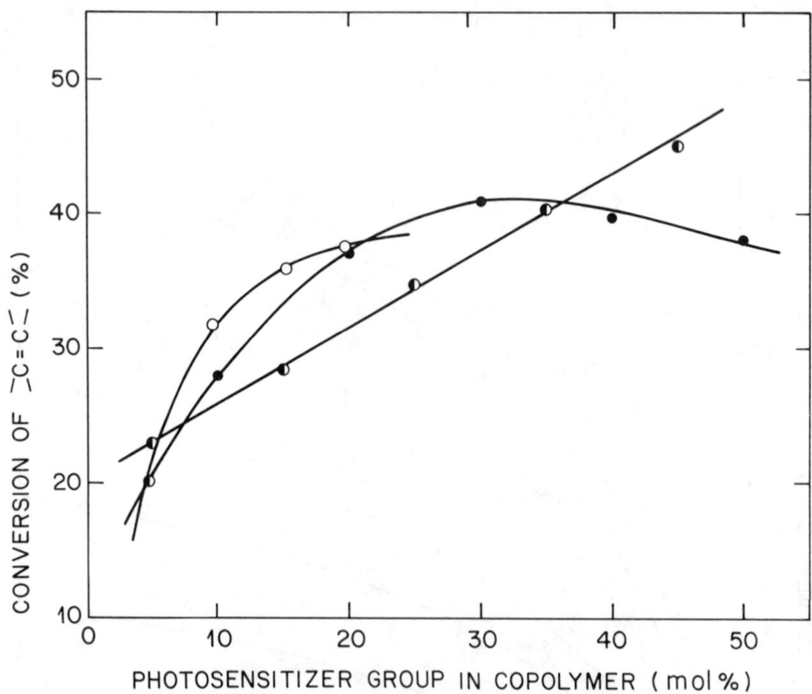

Figure 3. The relationship between the conversion of the C=C group in cinnamic ester in the polymer by 10 min. of irradiation and the photosensitizer content in PCEVE-NPVE: (○) PCEVE-NPVE prepared from the cationic copolymerization; (●) PCEVE-NPVE prepared from the reaction of PCVE; (◐) PCEVE with 4-nitroanisole.

copolymerizations have higher photochemical reactivities than PCEVE-NNVE polymers prepared from the reactions of PCVE at the low photosensitizer contents. The former copolymer has its highest rate of disappearance of the C=C bonds when 10 mol-% of pendant NN groups are incorporated. On the other hand, PCEVE-NNVE prepared from the reactions of PCVE has its highest rate at about 15 mol-% of pendant NN groups. PCEVE with 4-methoxy-l-nitronaphthalene has its highest rate with 10 mol-% of the low molecular weight photosensitizer present. However, the rate decreased rapidly with increasing content of the sensitizer, because they do not have good compatibility and the photosensitizer separates and forms the clusters in the polymer film.

The relationship between the thickness of polymer films containing about 10 mol-% of pendant photosensitizer groups and the conversions of the C=C bonds after 10 min. irradiation is shown in Figure 5. The photochemical reactivity of PCEVE-NNVE decreased slightly with increasing thickness of the film. Interestingly, the reactivity of PCEVE-NPVE was not influenced by the thickness of the film in contrast to the self-sensitized polymers previously

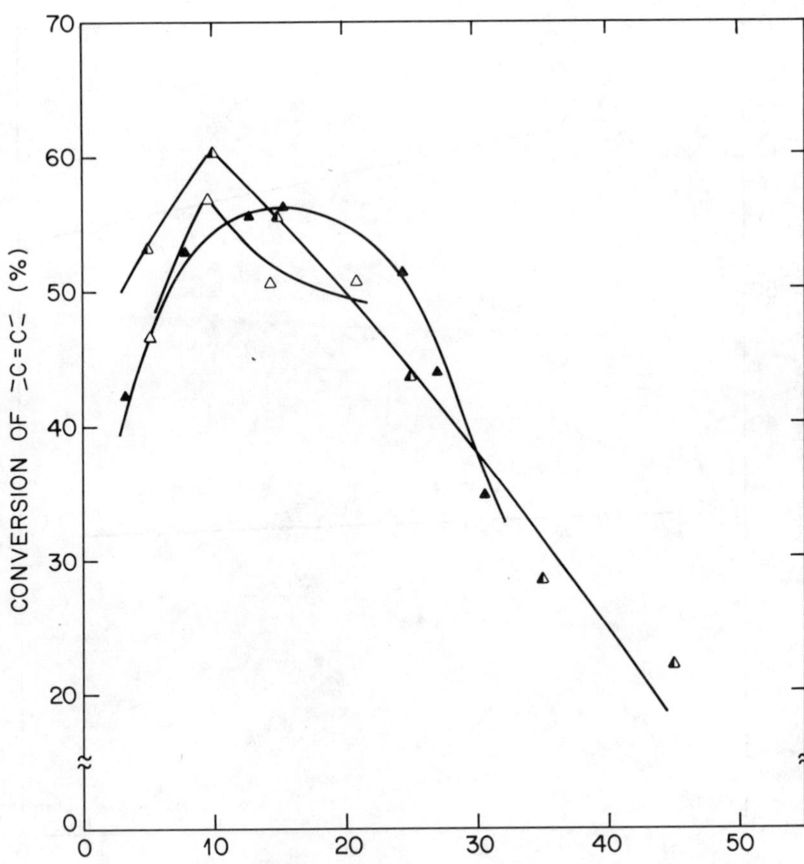

Figure 4. The relationship between the conversion of the C=C group in cinnamic ester in the polymer by 10 min. of irradiation and the photosensitizer content in PCEVE-NNVE: (Δ) PCEVE-NNVE prepared from the cationic copolymerization; (▲) PCEVE-NNVE prepared from the reaction of PCVE; (Δ) PCEVE with 4-nitro-1-methoxynaphthalene.

reported (*20*). In addition, these copolymers have the T_g near room temperature as summarized in Table I. These results raise the possibility of PCEVE-NPVE and PCEVE-NNVE use as the polymer for the preparation of dry film resists (*21*) containing cinnamic ester as a photosensitive moiety.

Practical photosensitivities, which were measured by a gray-scale method (*15*) of PCEVE-NPVE, PCEVE-NNVE and PCEVE-NPEVE prepared from the cationic copolymerization are summarized in Table I. The photosensitivities of PCEVE-NPVE, PCEVE-NNVE and PCEVE-NPEVE were much higher than that of PCEVE. In addition, it was found that PCEVE-NNVE containing pendant NN groups (λ_{max}: 360 nm) as a photosensitizer has a relatively higher photosensitivity than PCEVE-NPVE containing pendant NP groups (λ_{max}:

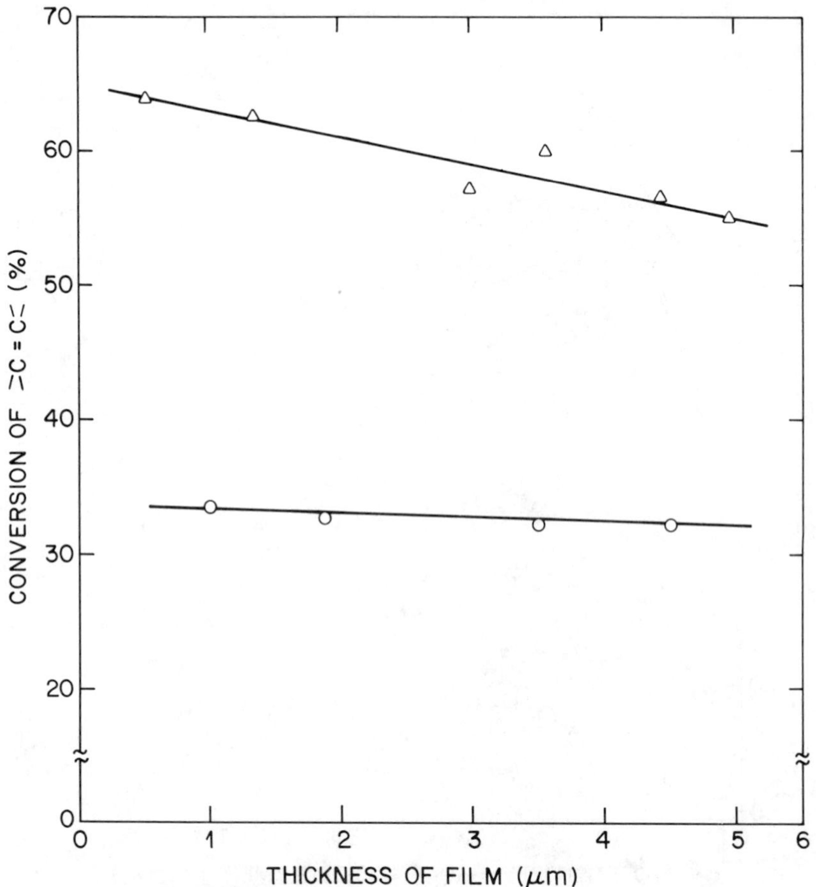

Figure 5. The relationship between the conversion of the C=C group in cinnamic ester in the polymer by 10 min. of irradiation and the thickness of the polymer film: (Δ) PCEVE-NNVE containing 9.7 mol-% of pendant NN group; (○) PCEVE-NPVE containing 9.2 mol-% of pendant NP group.

305 nm) as a photosensitizer. The rate of disappearance of the C=C bonds in the polymer indicates that photochemical reactivity of the system includes intra-molecular photo-dimerization reaction of pendant cinnamic ester in a polymer chain. However, the photosensitivity of the polymer depends on the rate of gelation of the polymer based on the inter-molecular photo-dimerization of pendant cinnamic ester in two polymer chains. In addition, the practical photosensitivity of the polymer is strongly affected by the contents of the photosensitizer in the polymer and by the degree of polymerization of the photosensitive polymer (*22*).

Accordingly, as shown in Figure 6, the practical photosensitivities of PCEVE-NPVE and PCEVE-NNVE prepared from the reactions of PCVE, which have the same degree of polymerization, were measured. The

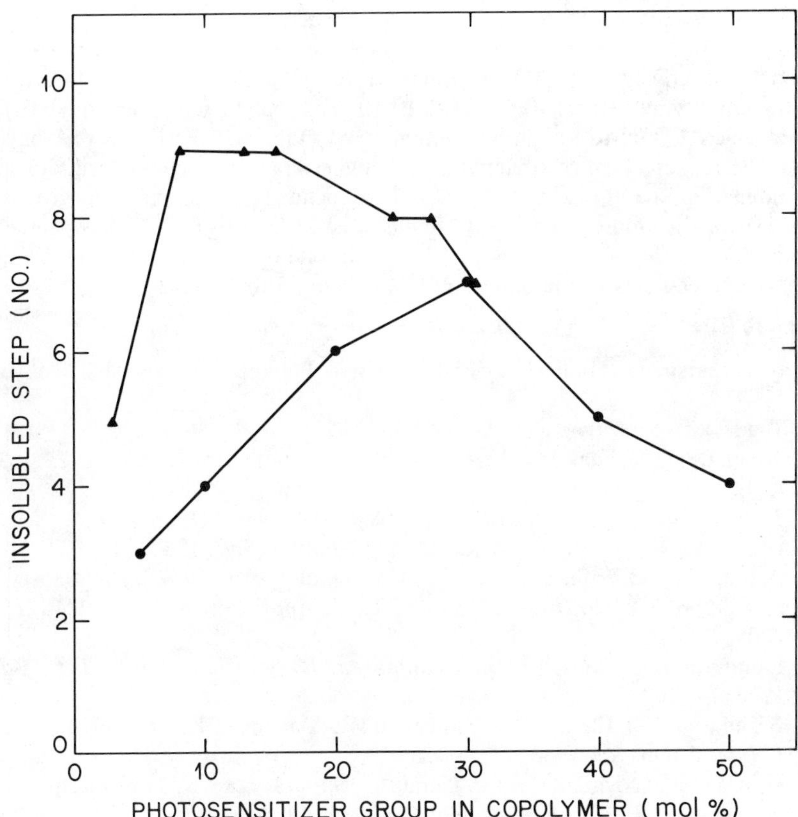

Figure 6. The relationship between the practical photosensitivity and the photosensitizer content: (Δ) PCEVE-NNVE; (○) PCEVE-NPVE.

development of the irradiated polymer film was carried out using ethyl methyl ketone, because polymers containing larger amounts of the photosensitizer groups were insoluble in toluene. PCEVE-NPVE and PCEVE-NNVE have their highest photosensitivity at the contents of about 30 mol-% of pendant NP groups and 15 mol-% of pendant NN groups, respectively. These maximum values agree with the maximum values of the rates of disappearance of the C=C bonds in these copolymers. Furthermore, PCEVE-NNVE has higher photosensitivity than PCEVE-NPVE at the contents of the photosensitizers from 25 to 5 mol-%. This result suggests that the pendant NN group is more effective photosensitizer than the pendant NP group in the self-sensitized photosensitive polymer in practice.

Conclusions

From all these results, it was concluded that polymers have high photochemical reactivity and high practical photosensitivity when synthesized by the cationic copolymerizations of CEVE with NNVE or NPVE, or by the reactions of PCVE with potassium cinnamate and PNN or PNP using phase transfer catalyst in DMF.

The photochemical reactivity and the practical photosensitivity of the resulting polymers were measured by IR spectroscopy and by the gray-scale method, respectively. Poly[2-(cinnamoyloxy)ethyl vinyl ether-co-2-(4-nitrophenoxy)ethyl vinyl ether] (PCEVE-NPVE) and poly[2-(cinnamoyloxy)ethyl vinyl ether-co-2-(4-nitro-l-naphthoxy)ethyl vinyl ether] (PCEVE-NNVE) have their highest photochemical reactivity and highest practical photosensitivity at the contents of about 30 and 15 mol-% of pendant photosensitizer groups, respectively. In addition, it was found that PCEVE-NNVE has higher photochemical reactivity and higher practical photosensitivity than PCEVE-NPVE at the contents of the photosensitizers from 25 to 5 mol-%.

Literature Cited

1. Nagamatsu, G.; Inui, H. "Photosensitive Polymers", Kodansha, Tokyo, 1977.
2. Tsuda, M. *J. Polym. Sci.* A-1, 1969, *7*, 259.
3. Tanaka, H.; Tsuda, M.; Nakanishi, H. *J. Polym. Sci.* A-1, 1972, *10*, 1729.
4. Tanaka, H.; Sato, Y. *J. Polym. Sci.* A-1, 1972, *10*, 3279.
5. Unruh, C. C.; Smith, Jr., A. C. *J. Appl. Polym. Sci.* 1960, *3*, 310.
6. Borden, D. G.; Williams, J. L. R. *Makromol. Chem.* 1977, *178*, 3035.
7. Nishikubo, T.; Ichijyo, T.; Takaoka, T. *J. Appl. Polym. Sci.* 1974, *218*, 2009.
8. Ichimura, K.; Watanabe, S.; Ochi, H. *J. Polym. Sci., Polym. Lett. Ed.* 1976, *14*, 207.
9. Nishikubo, T.; Iizawa, T.; Yamada, M.; Tsuchiya, K. *J. Polym. Sci., Polym. Chem. Ed.* 1983, *21*, 2025.
10. Iizawa, T.; Nishikubo, T.; Uemura, S.; Kakuta, K. Takahashi, E.; Hasegawa, M. *Kobunshi Ronbunshu* 1983, *40*, 425.
11. Iizawa, T.; Nishikubo, T.; Takahashi, E.; Hasegawa, M. *Makromol. Chem.* 1983, *184*, 2297.
12. Kato, M.; Ichijyo, T.; Ishii, K.; Hasegawa, M. *J. Polym. Sci.* A-1, 1971, *9*, 2109.
13. Montanari, F. *Bull. Sci. Fac. Chem. Ind. Balogan* 1956, *14*, 55; *Chem. Abstr.* 1957, *51*, 5723d.
14. Nara, A. "Micro Quantitative Analysis", Nankodo, Tokyo, 1968, p. 283.
15. Minsk, L. M.; Smith, J. G.; Van Deusen, W. P.; Wright, J. F. *J. Appl. Polym. Sci.* 1959, *2*, 302.
16. Haas, H. C.; Simon, M. S. *J. Polym. Sci.* 1955, *17*, 421.
17. Nishikubo, T.; Kishida, M.; Ichijyo, T.; Takaoka, T. *Makromol. Chem.* 1974, *175*, 3357.
18. Nishikubo, T.; Kishida, M.; Ichijyo, T.; Takaoka, T. *Nippon Kagaku Kaishi* 1974, *1581*; *Chem. Abstr.* 1974, *83*, 4370q.
19. Nishikubo, T.; Ichijyo, T.; Takaoka, T. *Nippon Kagaku Kaishi* 1973, *35*; *Chem. Abstr.* 1973, *78*, 150375q.
20. Nishikubo, T.; Takahashi, E.; Iizawa, T.; Hasegawa, M. *Nippon Kagaku Kaishi* 1984, (2), 306.
21. Celeste, J. R. U.S. Patent No. 3, 526, 504, 1970; *Chem. Abstr.* 1970, *73*, 104324g.

RECEIVED October 3, 1984

11

A Novel Technique for Determining Radiation Chemical Yields of Negative Electron-Beam Resists

A. NOVEMBRE and T. N. BOWMER[1]

AT&T Bell Laboratories, Murray Hill, NJ 07974

The susceptibility to undergo a structural change upon exposure to radiation is inherent in the choice of an organic polymer as a potential electron beam resist. The radiation chemical yields of a polymer are a measure of the intrinsic radiation sensitivity of the polymer (i.e., how many bonds break or cross-links form per unit of absorbed energy). The lithographic sensitivity and contrast are measures of the efficiency with which a relief image is generated and are related to the intrinsic radiation sensitivity of a material. Resist sensitometry is a measure of the intrinsic lithographic sensitivity (*1*) of a polymer and is related to the two events; chain scission of the polymer backbone and the formation of cross-links between polymer chains.

The radiation chemical yields are expressed in terms of G-values. G(scission), $G(s)$, equals the number of main chain scissions produced per 100 eV of energy absorbed and G(cross-linking), $G(x)$, the number of cross-links formed per 100 eV absorbed. The G-value is a structure dependent constant similar to quantum efficiency in photochemistry.

For resist applications, $G(s)/G(x)$ values greater than 4 indicate that chain scission reactions are predominating. The resultant effect is a decrease in the molecular weight of the exposed vs. the unexposed regions of the film. The difference in dissolution rates between the regions is then used to form a positive relief image. $G(s)/G(x)$ ratios less than 4 indicate cross-linking reactions will predominate resulting in an increase in molecular weight leading to an insoluble gel and negative tone formation. Other lithographic properties such as plasma etch resistance and dissolution rate have been correlated to G-values. Taylor and Wolf, (*2*) have shown for positive acting resists, the higher the $G(s)$ value, the greater the O_2 plasma etch rate, independent of the mechanism by which main chain scission occurs. Greenwich (*3*) related the scission efficiency of PMMA to the ratio of solubility rates of the exposed and unexposed areas.

We report here a novel technique used to measure the cross-linking and scission yields for negative acting resist systems. The three systems investigated were poly(glycidyl methacrylate-co-3-chlorostyrene) (GMC), poly(3-chlorostyrene) (PCLS), and polystyrene (PS). The technique requires generating an

[1] Current address: Bell Communications Research, Inc., Murray Hill, NJ 07974

0097-6156/84/0266-0241$06.00/0

electron exposure response curve, which measures the thickness of the resist film after exposure and development as a function of incident dose. The film thickness remaining is correlated to the gel fraction of the sample and this data is used to calculate the G-values of interest. The results of this technique are compared to a more standard technique in which the radiation chemical yield values are determined from gel data obtained by the Soxhlet extraction method (*4*).

Experimental

Materials. GMC and PCLS were synthesized by free radical solution polymerization initiated by benzoyl peroxide as described previously (*5,6*). Nearly mono and polydisperse polystyrenes were obtained from Pressure Chemical Co. and the National Bureau of Standards respectively. Molecular weight and polydispersity were determined by gel permeation chromatography (GPC) using a Water Model 244 GPC, equipped with a set (10^2-10^6 Å) of μ–Styragel columns using THF as the elution solvent. The molecular parameters of the above three polymers are listed in Table I. The copolymer, poly(GMA-co-3-CLS), contained 53.5 mole % 3-CLS and 46.5 mole % GMA, as determined by chlorine elemental analysis. The structure of the copolymer is shown in Figure 1.

Table I. Molecular Parameters

Polymer	$M_n \times 10^{-5}$	$M_w \times 10^{-5}$	R^*
GMC	1.05	1.94	1.85
PCLS	0.76	1.49	1.95
PS	1.23	2.58	2.10
PS†	2.20	2.33	1.06

$^*R = M_w/M_n$

†Data supplied by Pressure Chemical Co.

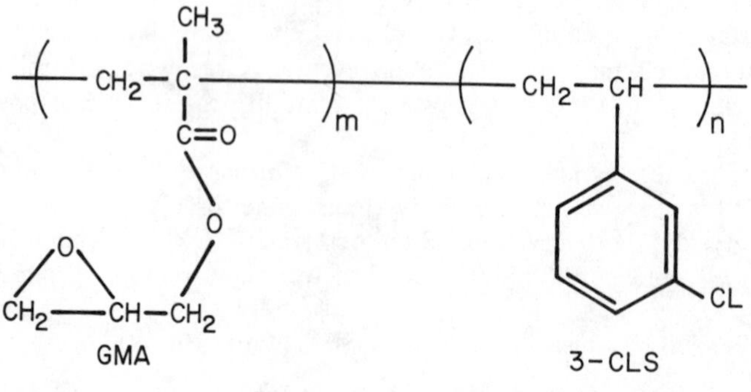

CH_3 — CH_2—C — C=O — O — CH_2—CH—CH_2 (epoxide O) ; m ; CH_2—CH — n ; CL

GMA 3-CLS

Figure 1. Structure of GMC.

Lithographic Analysis. Approximately 0.5 μm thick films were spun from solutions of polymers in chlorobenzene (7-10% w/v) onto 3" silicon wafers. All coated wafers were baked at 90°C for 30 minutes in air before exposure and after development. The baking steps were required to remove residual solvent and/or developer. Exposures were carried out on EBES-I @ 20 kV using a LaB_6 electron gun. Both the beam address and spot size were 0.25 μm. The exposed wafers were spray developed in an 8.5:1 ratio mixture of methylethyl ketone/ethanol for 25 sec. and rinsed in 2-propanol for 20 sec. Film thicknesses were measured optically using a Nanometries Nanospec/AFT microarea gauge. Characteristic electron exposure curves were plotted as % film thickness remaining (normalized to the initial film thickness) vs. log incident dose (expressed in $\mu C/cm^2$). The soluble fraction at a given dose was calculated from the appropriate exposure response curve. The absorbed dose was calculated from the incident dose using the method outlined by Bowden (*7*) (see Appendix). $G(x)$ and $G(s)$ values were determined using a Charlesby-Pinner plot.

Soxhlet Extraction Analysis. Powdered samples of GMC were degassed and sealed under vacuum into pyrex tubes, and irradiated at room temperature (25-30°C) in a Cobalt-60 Gammacell. The dose rate determined by Fricke dosimetry was 0.24 Mrad/hr. Doses up to 20 Mrad were utilized. After irradiation, the tubes were opened and the samples placed in cellulose thimbles. A Soxhlet extraction was performed with methylethyl ketone at 76 ± 1°C for 18-20 hrs. The insoluble material was dried to constant weight and the gel fraction calculated from initial weight.

Results and Discussion

The exposure response curves of the three negative acting resists are shown in Figure 2. GMC, containing the epoxide moiety, exhibited the highest sensitivity, followed by the meta-chlorinated derivative of polystyrene and polystyrene. The incipient gel dose ($D_g^{\,i}$) can be found from Figure 2 by an extrapolation of the linear region of the curve to 0% film thickness remaining. The values for GMC, PCLS, and polydisperse PS are 1.0, 2.25, and 22 $\mu C/cm^2$. $D_g^{\,i}$ for the nearly monodisperse PS (not shown in Figure 2) is 20 $\mu C/cm^2$. Feit and Stillwagon (*8*) have reported a value of 17 $\mu C/cm^2$ for nearly monodisperse polystyrene of similar molecular weight.

The soluble fraction for a given incident dose can be determined from Figure 2 provided that (1) all the soluble fraction was removed from the film during the development step, and (2) there was no volume contraction due to gel formation upon irradiation. If soluble material remained in the film, the measured soluble fraction will be systematically lower than the actual value. However, a volume contraction would result in a systematically higher than actual value for the soluble fraction.

To investigate these two possibilities, a wafer coated with GMC was exposed and then divided into a number of sections. Different sections were spray developed with varying strengths of the developer. Spray developing was found to be a more efficient and reproducible way to dissolve the soluble

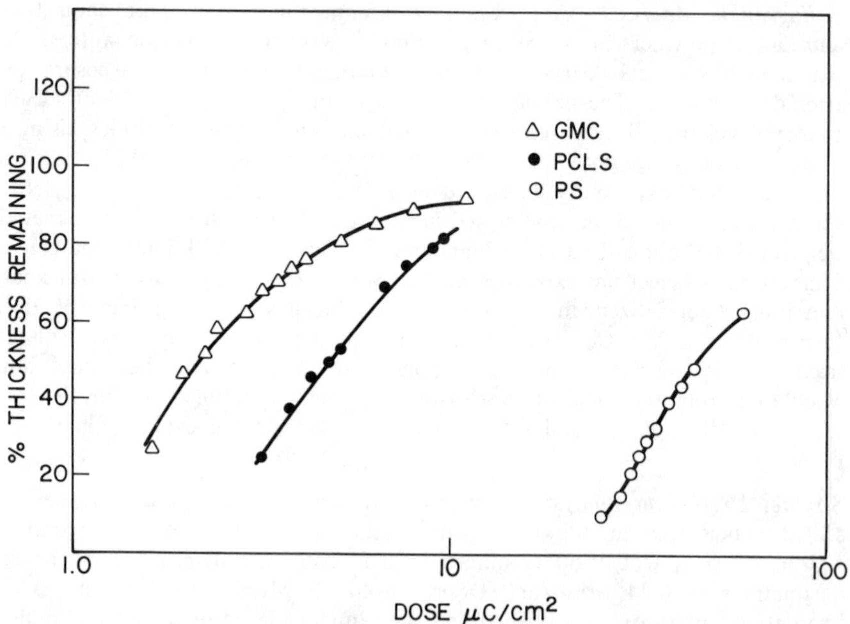

Figure 2. Exposure response curve of GMC, PCLS, and polydisperse PS.

fraction from the exposed areas of the film vs. tray or dip development. Figure 3 shows the exposure response curve for development with (a) the standard methylethyl ketone/ethanol (8.5/1.0) mixture, and (b) pure methylethyl ketone with the development time *twice* that of the standard developing time. There was no significant difference between the thickness

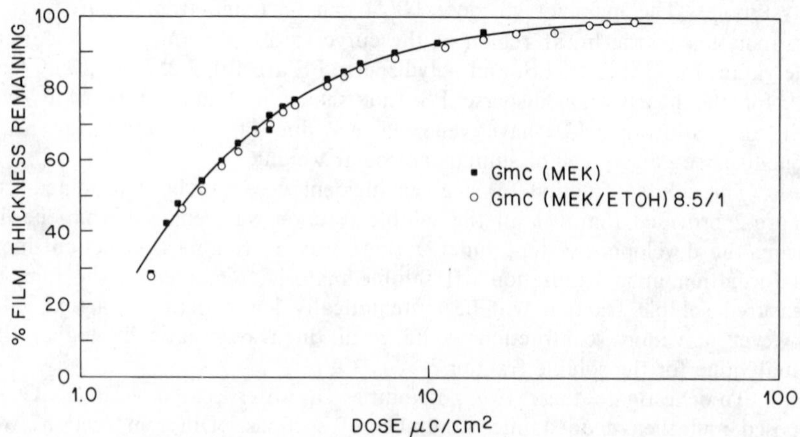

Figure 3. Exposure response curve of GMC at high gel fractions and varying developer strengths.

remaining versus dose curve for the two developers, indicating that all of the soluble material had been extracted from the exposed film. In addition, the exposure curves in Figure 3 extended from 1.6 to 40 $\mu C/cm^2$; (i.e., to high gel fractions). At doses > 30 $\mu C/cm^2$, the film thickness remaining after development approached the original thickness of the film, showing negligible volume contraction as a result of gel formation.

Calculation of G-values from the results shown in Figures 2 and 3 requires determining the amount of dose absorbed in the film for a given incident dose. Bowden (*7*) has shown, using the depth-dose model of Heidenreich, (*9*) an accurate measure of the energy absorbed in a polymer film can be determined. An example of such a calculation for polystyrene is given in the appendix.

Knowledge of the soluble fraction as a function of absorbed dose allows one to make use of the Charlesby-Pinner relationship (*4,10*) which states for a polymer with a *random distribution* (i.e., polydispersity equal to 2):

$$S + S^{1/2} = G(s)/2G(x) + \left(\frac{9.65\times10^5}{M_w}\right)\left(\frac{1}{G(x)}\right)\left(\frac{1}{r}\right)$$

where

S = soluble fraction

$G(s)$ = G-value for scission

$G(x)$ = G-value for cross-linking

M_w = weight average molecular weight (initial) in grams/mole

r = absorbed dose in Mrads

The probability a chain unit (e.g. -GMA-) undergoing scission or cross-linking upon absorption of unit dose is assumed to be independent of (1) the unit's position in the polymer molecule and (2) the energy previously absorbed by the molecule (*10a*). That is, the total number of scissions (or cross-links) is (a) small with respect to the total number of units in the molecule and (b) proportional to the total absorbed dose. Using these assumptions of random scission and cross-linking occurring in a polymer, the above analytical solution is found (*4,10*). Figure 4 shows the Charlesby-Pinner relationship for the data in Figure 2. The slope of each line is used to calculate the value $G(x)$, the intercept yielding the ratio of $G(s)/G(x)$. Calculated G-values for the polymers studied are listed in Table II.

Gel fractions as a function of absorbed dose were measured for GMC using the conventional Soxhlet extraction method and these results compared with values found from the above method. Figure 5 is a plot of gel fraction (soxhlet method) versus absorbed dose for GMC. (r_{gel}) equals 2.5 ± 0.1 Mrad as compared to 2.1 Mrads found from the new technique. Above 2.5 Mrads an insoluble gel was rapidly formed with the gel fraction approaching a limiting asymptotic value of 1.0 at doses greater than 25 Mrad. The Charlesby-Pinner

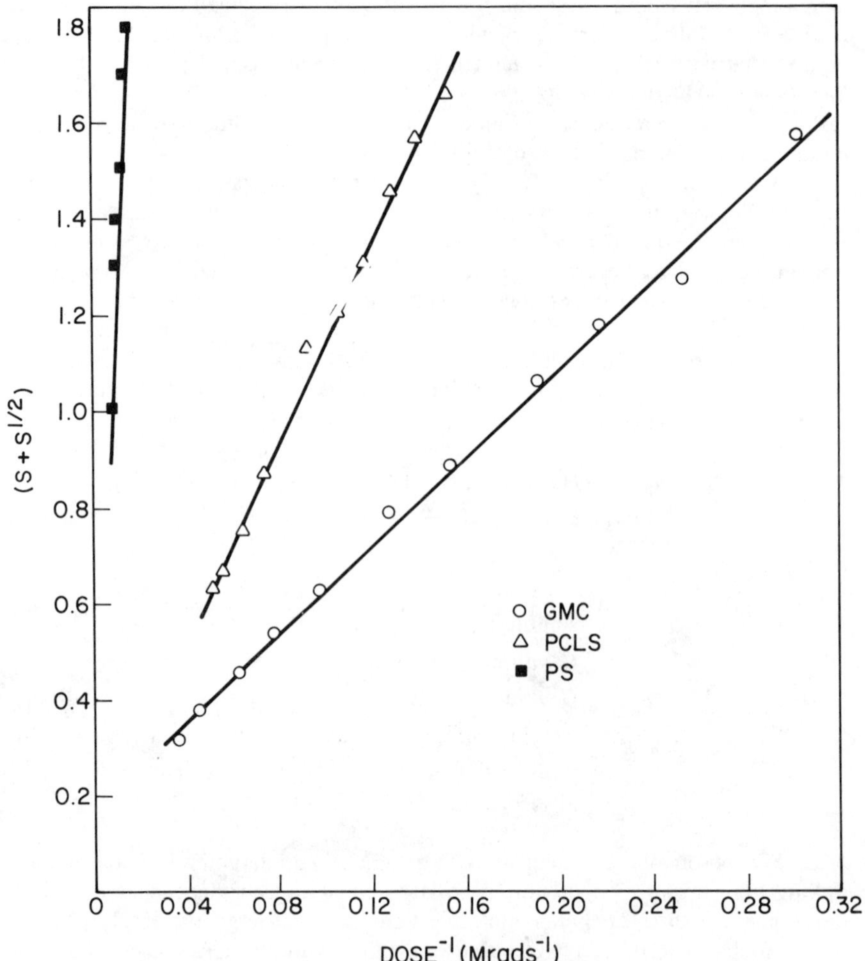

Figure 4. Charlesby-Pinner plot of GMC, PCLS, and polydisperse PS using data taken from Figure 2.

plot of the data in Figure 5 is given in Figure 6 from which a $G(x)$ value of 1.13 ± 0.07 and a $G(s)$ value of 0.65 ± 0.1 were calculated. Both values were slightly higher than those obtained using the exposure response curve method (see Table II). This difference may be attributed to not taking into account 1) the contribution of backscatter dose to the total incident dose, and 2) inefficient extraction by the soxhlet technique. For the highest dose measured, the $S + S^{1/2}$ value was lower than expected. In this region the highly

Table II. G-Values Calculated From Figure 3

Polymer	$G(x)$	$G(s)$	$G(s)/G(x)$
GMC	1.02	0.42	0.43
PCLS	0.61	0.16	0.26
PS	0.038	0.020	0.54

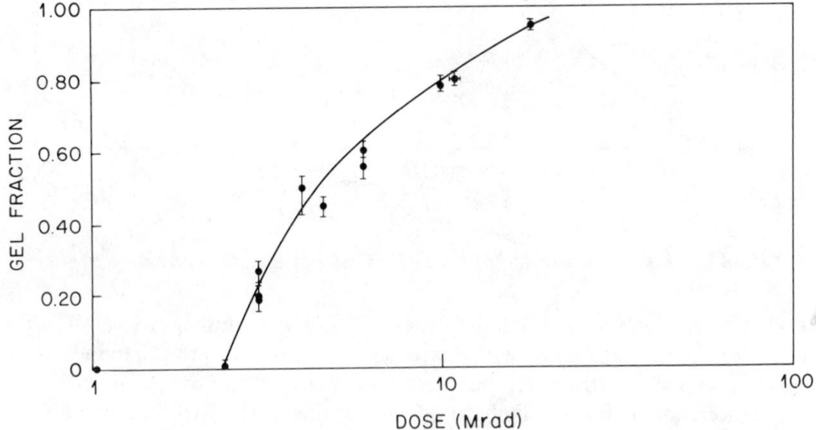

Figure 5. Plot of gel fraction (soxhlet extraction method) vs. absorbed dose for GMC using ^{60}Co radiation.

cross-linked network may inhibit the extraction of soluble molecules by entangling them and this will increase the measured gel content and therefore decrease the $S + S^{1/2}$ values (*31*).

There have been numerous radiolysis studies (in vacuum) of polystyrene reported (*4,10-20*) with $G(x)$ ranging from 0.02 (*16*) to 0.05 (*11*) and $G(s)$ ranging from 0.0 (*10*) to 0.02 (*15*). These are low values compared with scission and cross-linking yields for nonaromatic polymers but conform with the radiation resistance observed in low-molecular-weight aromatic compounds, (*28*) a phenomenon that has been attributed to energy transfer to the benzene ring with subsequent nondegradative dissipation of the energy. A $G(x)$ value of 0.035 ± 0.005 and a $G(s)$ value of 0.01 ± 0.003 would appear to be the most consistent values obtained by previous authors (*10,14,20*) for polystyrene irradiated at 25°C in vacuum. The value of 0.038 for $G(x)$ found from the electron exposure curve is in excellent agreement with these reported values. The $G(s)$ value of 0.020 is somewhat higher than expected for irradiation at 25°C. However, due to the relative insensitivity of PS, high currents were required to generate the electron exposure response curve. Under such conditions a temperature increase in the exposed region would be expected.

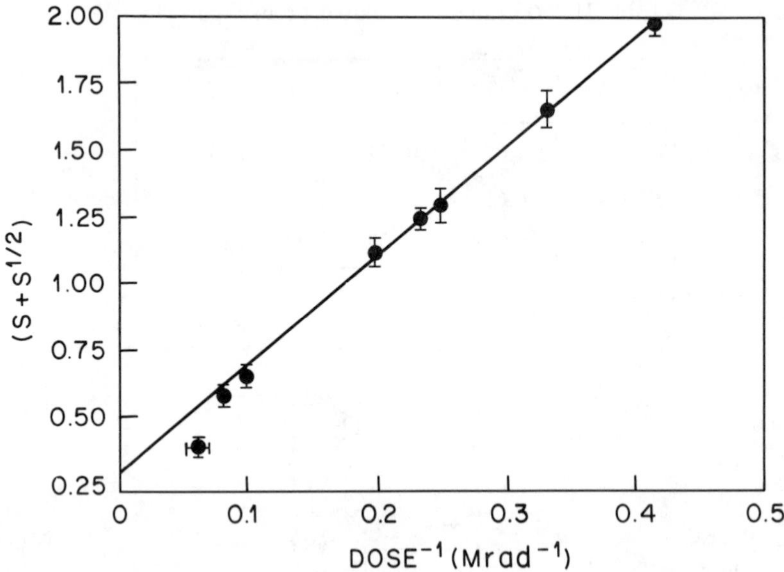

Figure 6. Charlesby-Pinner plot of GMC using data taken from Figure 5.

It is known that polystyrene irradiated at 45°C, for example, has $G(x) = 0.04$ and $G(s) = 0.02$ (*20,29,30*) which are similar values to those found in this study. Increasing the temperature during irradiation, favored the disproportionation reaction of chain scission radicals relative to recombination reactions and thereby increased main chain scission yields (*20*) (Figure 7).

Deviations from linearity in the Charlesby-Pinner plots occur when polymers with non-random distributions of molecular weight were used (*10,32*). Nearly monodisperse polystyrene was also investigated since narrow molecular weight distribution polymers are known to give enhanced contrast (i.e., higher rate of gel formation) and improved pattern resolution. Charlesby-Pinner curves derived from exposure response curves for nearly monodisperse polystyrene showed a 10% steeper slope ($G(x) = 0.043$) for the low dose region of the Charlesby-Pinner plot and will curve upwards towards the $S + S^{1/2}$ axis at high doses, see Figure 8. Inokuti (*32*) has predicted this behaviour from theoretical considerations and it was experimentally observed for polyethylene radiolysis (*10,33*). Therefore the Charlesby-Pinner curves that were generated from the e-beam exposure response curves show the quantitative and qualitative behaviour that is theoretically predicted (*10,32,34*) and observed by conventional techniques (e.g., Soxhlet extraction) (*33*).

Introduction of a chlorine substituent in the meta position of the aromatic ring in polystyrene resulted in a 15-fold increase in the $G(x)$ value and an increase in $G(s)$ from 0.02 to 0.16 (see Table II). Hiraoka and Welsh (*21*) using ESCA data have attributed the higher cross-linking yields found for poly(p-chlorostyrene) to efficient halogen removal followed by liberation of HCl and formation of an α-carbon radical. However analysis of free radical and

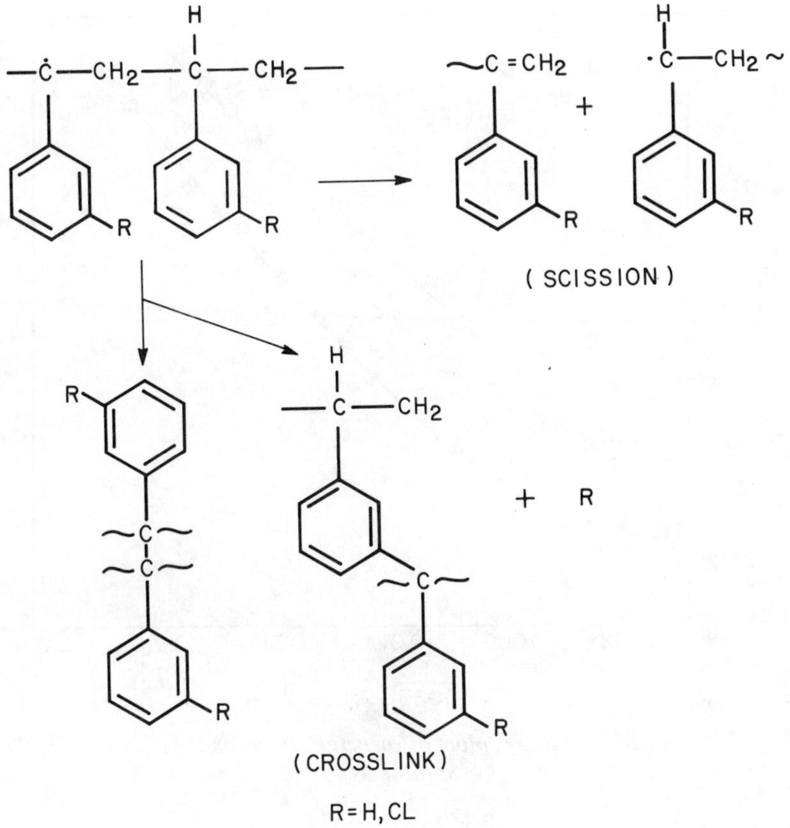

Figure 7. Radiation degradation mechanism of PS and chlorinated PS.

volatile product yields (*10,12,22*) suggested that the chlorine substituent was retained and the enhanced $G(x)$ values arose from a more reactive as well as a more abundant α-carbon radical. The increase in $G(s)$ from 0.02 to 0.16 found for PCLS implies that the α-carbon radical is more reactive and/or more abundant, since the α-carbon radical is a reactive intermediate in the scission reaction for both mechanisms, as shown in Figure 7. Volatile product studies and spectroscopic analysis are in progress to help elucidate the mechanism for our polymers.

Copolymerization of 3-chlorostyrene with glycidyl methacrylate to form GMC, resulted in $G(x)$ increasing from 0.61 to 1.02 and $G(s)$ from 0.16 to 0.42. Polyalkylmethacrylates are well-known to undergo main chain scission upon irradiation; (*10,23-6*) e.g., polymethylmethacrylate has a $G(x) = 0$ and $G(s) = 1.4$(*27*). The increase in $G(s)$ can therefore be attributed entirely to the presence of the methacrylate moiety. The enhanced $G(x)$ value of GMC arises from the epoxide ring opening upon exposure and initiating cross-linking.

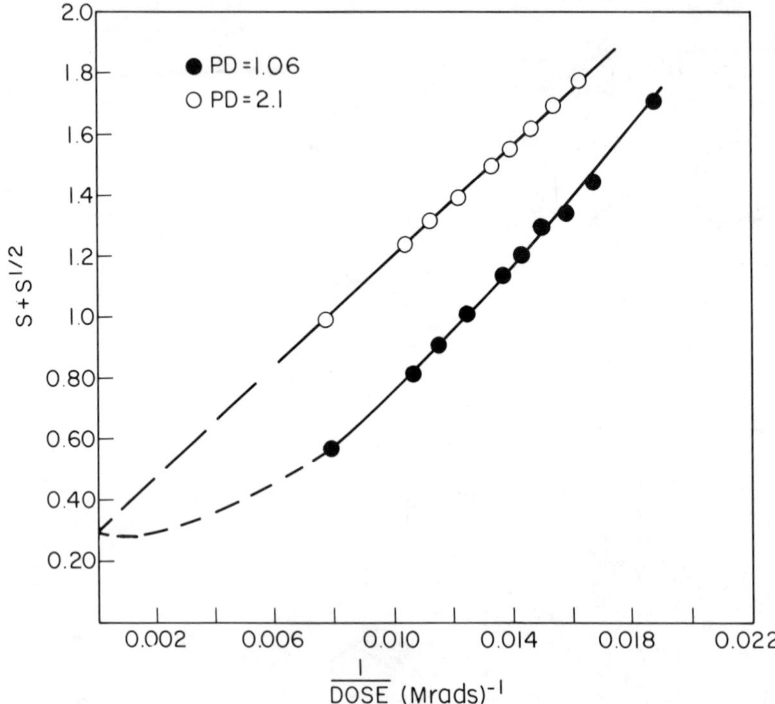

Figure 8. Charlesby-Pinner plot of nearly mono and polydisperse PS of similar M_w.

Conclusions

The electron beam exposure response curve can be used for determining G-values for negative acting resists. Measurements of the film thickness remaining as a function of incident dose provides sol-gel data, from which radiation chemical yields for scission and cross-linking $(G(s), G(x))$ are obtained using the Charlesby-Pinner relationship (*4*). Comparison of this method with the conventional Soxhlet extraction technique shows good agreement (within 10%) between the methods.

Acknowledgments

The authors wish to thank Carl Lochstampfor for EBES-I exposures and M. J. Bowden for many helpful discussions.

Literature Cited

1. Wilson, G. C. "Introduction to Microlithography"; Thompson, L. F.; Bowden, M. J.; Wilson, C. G., Eds.; ACS SYMPOSIUM SERIES No. 219, ACS.
2. Wolf, T.; Taylor, G. N. *Polymer Eng. Sci.* 1980, *20(16)*, 1087.
3. Greenwich, J. S. *J. Electrochem. Soc.* 1975, *122*, 970.
4. Charlesby, A. "Atomic Radiation and Polymers"; Pergamon Press: New York, 1960.

5. Thompson, L. F.; Doerries, E. M. *J. Electrochem. Soc.* 1979, *126*, 1699.
6. Novembre, A. E.; Bowden, M. J. *Polym. Eng. Sci.* 1983, *23(17)*, 977.
7. Bowden, M. J. "Energy Absorption in Polymeric Electron Beam Resists"; private communication.
8. Feit, E. D.; Stillwagon, L. *Polymer Eng. Sci.* 1980, *20*, 1058.
9. Heidenreich, R. D.; Thommpson, L. F.; Feit, E. D.; Melliar, C. M. *J. Appl. Physcis* 1973, *44*, 4039.
10. Dole, M., Ed. "The Radiation Chemistry of Macromolecules"; Academic Press: 1972; (a) Vol. I, p. 224; (b) Vol. II, p. 57,97.
11. Wall, L. A.; Brown, D. W. *J. Phys. Chem.* 1957, *61*, 129.
12. Burlant, W.; Neerman, J.; Serment, V. *J. Polym. Sci.* 1962, *58*, 491.
13. Graessley, W. *J. Phys. Chem.* 1964, *68*, 2258.
14. Parkinson, W. W.; Bopp, C. D.; Binder, D.; White, J. *J. Phys. Chem.* 1965, *69*, 828.
15. Alberino, L. M.; Graessley, W. W. *J. Phys. Chem.* 1968, *72*, 4229.
16. Kells, D. I. C.; Koike, M.; Guillet, J. E. *J. Polym. Sci.* Part, 1968, *A-1* *6*, 595.
17. Heusinger, H.; Rosenberg, A. "Symposium on Large Radiation Sources for Industrial Process"; Beck, E. R. A., Ed.; IAEA, Venna, 1969, p. 151-164.
18. O'Donnell, J. H.; Rahman, N. P.; Smith, C. A.; Winzor, D. J. *J. Polym. Sci., Polym. Chem. Ed.* 1979, *17*, 4081.
19. O'Donnell, J. H.; Rahman, N. P.; Smith, C. A.; Winzor, D. J. *Macromolecules* 1979, *12*, 113.
20. Bowmer, T. N.; O'Donnell, J. H.; Winzor, D. J. *J. Polym. Sci., Polym. Chem. Ed.* 1981, *19*, 1167.
21. Hiraoka, H.; Welsh, L. W., Jr. *Proceedings of the Symposium on Electron and Ion Beam Science and Technology*, Montreal, May 1982, Vol. 83-2, p. 171.
22. Antuf'ev, V. V.; Dokukina, A. F.; Votinov, M. P.; Suntsov, Ee. V.; Boldyrev, A. G. *Vyosokomol. Soyed.* 1965, *7*, 380.
23. Graham, R. K. *J. Polym. Sci.* 1959, *38*, 209.
24. Alexander, P.; Charlesby, A.; Ross, M. *Proc. Roy. Soc.,* (London), 1956, *A223*, 392.
25. Todd, A. *J. Polym. Sci.* 1960, *42*, 223.
26. Kircher, J. F.; Slieners, F. A.; Markle, R. A.; Gager, W. R.; Leininger, R. I. *J. Phys. Chem.* 1965, *69*, 189.
27. Bowmer, T. N.; Reichmanis, E.; Wilkins, C. W., Jr.; Hellman, M. Y. *J. Polym. Sci., Polym. Chem. Ed.* 1982, *20*, 2661.
28. Manion, J. P.; Burton, M. *J. Phys. Chem.* 1952, *56*, 560.
29. Bowmer, T. N., Ph.D. Thesis, Uinversity of Queensland, Queensland Australia 1980.
30. Shimuzu, Y.; Mitsui, H. *J. Polym. Sci., Polym. Chem. Ed.* 1979, *17*, 2307.
31. Hellman, M. Y.; Bowmer, T. N.; Taylor, G. N. *Macromolecules* 1983, *16*, 34.
32. Inokuti, M. *J. Chem. Phys.* 1963, *38*, 2999.
33. Kitamaru, R.; Mandelkern, L.; Fatou, J. *J. Polym. Sci.* 1964, *B2*, 511.
34. Saito, O. *J. Phys. Soc. Japan* 1958, *13*, 198.

Appendix

E(keV), the energy absorbed per incident electron during electron irradiation, is given by:

$$E = -V_a \int_o^f \Lambda\,(f)df$$

where

V_a = accelerating voltage in keV

f = normalized penetration of an electron = Z/R_G

where

Z = film thickness $\quad R_G$ = Grun range of electron

$R_G = \dfrac{0.046\,V_a^{\,1.7}}{\rho}$, $\quad \rho$ = density of polymer

$\Lambda(f)$ = depth dose function if $5 < N < 12$, where N is the atomic number of atoms in polymer

then

$$\Lambda(f) = \int_o^f 0.74 + 4.7f - 8.9f^2 + 3.5f^3$$

For polystyrene of $Z = 0.50\ \mu m$, $\rho = 1.05\ g/cm^3$, and exposed at 20 keV,

$$R_G = 0.046/1.05 \times (20)^{1.75} = 8.29\ \mu m$$

$$f = 0.50/8.29 = 0.0604$$

$$\Lambda(f) = \int_0^{0.0604} 0.74 + 4.7f - 8.9f^2 + 3.5\,f^3$$

$$\Lambda(f) = 0.0526$$

$$\therefore \quad E = (20)\ (0.0526) = 1.05\ \text{keV}.$$

Convert the incident dose (coulombs/cm^2) to absorbed dose (Mrads) as follows: The sheet density (ρ_S) of a polystyrene film = $\rho \times Z = 5.25 \times 10^{-5}\ g/cm^2$, the energy E' absorbed in a volume element comprised of one cm^2 and Z cm thick is then given by

$$E' = \frac{E}{\rho_S} = \frac{E}{5.25 \times 10^{-5}} \text{ eV/g/cm}^2$$

$$= \frac{E}{5.25 \times 10^{-5}} \times 1.60 \times 10^{-12} \text{ rg/g/cm}^2$$

$$= 3.05 \times 10^{-8}\ E \times 10^{-8} \text{ Mrads/cm}^2$$

$$= 3.05 \times 10^{-16}\ E \text{ Mrads/cm}^2 .$$

the number of incident electrons per unit absorbed dose is given by

$$N = \frac{1}{3.05 \times 10^{-16}\ E} = \frac{3.28 \times 10^{15}}{E} \text{ electrons/cm}^2\text{/Mrads}$$

Expressing the incident dose in coulombs/cm²/Mrad, the conversion factor C becomes

$$C = \frac{3.28 \times 10^{15}\ e}{E} \text{ coulombs/cm}^2\text{/Mrad}$$

e = charge of electron

$$C = \frac{5.25 \times 10^{-4}}{E}$$

at 2×10^4 eV, $E = 1.05 \times 10^3$

$$\therefore C = 4.99 \times 10^{-7} \text{ coulombs/cm}^2\text{/Mrad}$$

Table I-A gives the value of C for the three polymers investigated. Table II-A lists the equivalent dose in coulomb/cm² as a function of accelerating voltage for a constant absorbed dose in Mrads.

Table A-I.

Polymer	$\rho_{g/mL}$	$C \times 10^7$ Coulombs/cm²/Mrad
GMC	1.31	4.84
PCLS	1.25	4.89
PS	1.05	4.99

Table A-II.

Absorbed Dose (Mrad)	Incident Dose (Coulombs/cm^2) × 10^7			
	5 kV	10 kV	15 kV	20 kV
0.5	0.677	1.11	1.78	2.42
1.0	1.35	2.23	3.55	4.84
3.0	4.06	6.71	10.7	14.5
5.0	6.77	11.1	17.8	24.2
10.0	13.5	22.3	35.5	48.4
15.0	20.3	33.5	53.3	72.6
20.0	27.1	44.7	71.1	96.8

RECEIVED August 6, 1984

12

Anomalous Topochemical Photoreaction of Olefin Crystals

M. HASEGAWA, K. SAIGO, and T. MORI

Department of Synthetic Chemistry
Faculty of Engineering
The University of Tokyo
Hongo, Bunkyo-ku, Tokyo 113 Japan

A great number of olefinic compounds are known to photodimerize in the crystalline state (*1,2*). Formation of α-truxillic and β-truxinic acids from two types of cinnamic acid crystals was interpreted by Bernstein and Quimby in 1943 to be a crystal lattice controlled reaction (*3*). In 1964 their hypothesis on cinnamic acid crystals was visualized by Schmidt and co-workers, who correlated the crystal structure of several olefin derivatives with photoreactivity and configuration of the products (*4*). In these olefinic crystals the potentially reactive double bonds are oriented in parallel to each other and are separated by approximately 4 Å, favorable for [2+2] cycloaddition with minimal atomic and molecular motion. In general, the environment of olefinic double bonds in these crystals conforms to one of three principal types: (a) the α-type crystal, in which the double bonds of neighboring molecules make contact at a distance of ~3.7 Å across a center of symmetry to give a centrosymmetric dimer ($\bar{1}$-dimer); (b) the β-type crystal, characterized by a lattice having one axial length of 4.0 ± 0.1 Å between translationally related molecules to give a dimer of mirror symmetry (m-dimer); and (c) the γ-type crystal, which is photochemically stable because no double bonds of neighboring molecules are within 4 Å. On the basis of mechanistic and crystallographic results it has been established that in a typical topochemical photodimerization, transformation into the product crystal is performed under a thermally diffusionless process giving the space group quite similar to that of the starting crystal (*5,6*).

At present, it is common knowledge that not only the photoreactivity, but also the stereochemistry, of the photoproduct is predictable from crystallographic information of starting olefin substrates. This ability of olefinic crystals to dimerize has been widely applied to the topochemical photocycloaddition polymerization of conjugated diolefinic compounds, so called "four-center type photopolymerizations" (*7,8*). All the photopolymerizable diolefin crystals are related to the center of symmetry mode (centrosymmetric α-type crystal) and thus give polymers having cyclobutanes with a 1,3-*trans* configuration in the main chain on irradiation.

Recent studies on the photodimerization of olefinic crystals several examples, which deviate from accepted topochemical principles, have been

0097-6156/84/0266-0255$06.00/0

reported. The present report deals with such anomalous behaviors of olefin crystals.

Photodimerization behavior of 4-formyl-, 3,4-dichloro-, and several other cinnamic acid derivatives is greatly influenced by other molecules outside of the crystal (*9,10*). For example, 4-formylcinnamic acid 1 crystallizes in two modifications, photoreactive and photostable forms. The photoreactive crystals of 1 (mp 249°C), on photoirradiation at room temperature in the presence of even a trace of moisture, dimerize to crystalline dimer 2 containing one molecule of water. The continuous change of the x-ray diffraction pattern during the photodimerization indicates a typical crystal-to-crystal transformation process. On the other hand, the same crystal 1 photodimerizes into amorphous dimer 2 in the absence of water. The same cyclobutane derivative is produced in very high yield in both reactions. However, highly crystalline dimer 2 is obtained only by the photodimerization of 1 in the presence of water and is not regenerated by any attempted recrystallization procedures from various aqueous solutions of 2.

Discussion

The characteristic strong peaks (1680 and 1722 cm^{-1}) are observed in the ir spectrum of the crystal 2, whereas only a strong peak is seen at 1680 cm^{-1} in the amorphous dimer 2. The spectral difference may be attributed to hydrogen

CHO CHO
• H_2O 2 (crystalline)
2OHC–⟨○⟩–CH=CHCOOH
1
hν presence of water
hν absence of water
HO–C(=O) C(=O)–OH
dimer 2 (amorphous)

bond formation of the carbonyl group with water in the crystal 2. The crystal structure of the photoreactive form of 1 projected onto the (100) plane is shown in Figure 1 (*11*). A parallel plane-to-plane stack, which is a common feature of the packing in photodimerizable crystals, is seen in the crystal. However the shortest distance between reactive double bonds (4.825 Å) is extraordinarily large for the photoreactive crystals of cinnamic derivatives. The crystal structure of $2 \cdot H_2O$ has not yet been determined.

In addition to the influence on the dimer morphology, the presence of water molecules strikingly affect apparent photoreaction rate and temperature dependence of the rate (*12*). Since the topochemical reaction deteriorates pronouncedly at reaction temperatures close to the melting point of the starting crystal, maximal reaction rate is necessarily observed at a specific temperature for individual crystals, for example, at ca. 20°C for a α-form crystal of cinnamic acid (mp 132°C) (*13*). In an aqueous dispersant the apparent maximal rate of photodimerization of 1 is observed about 15°C while the temperature for maximal rate in a non-aqueous dispersant is about 35°C. The

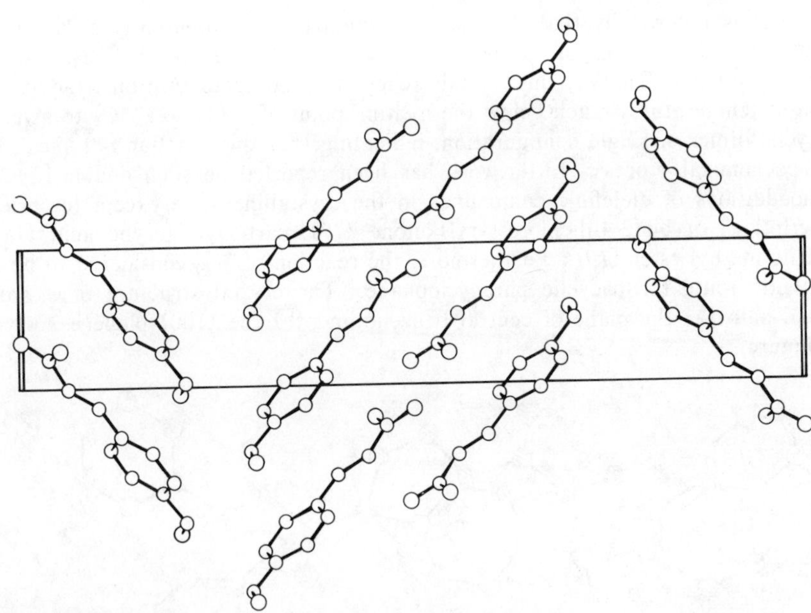

Figure 1. Crystal structure of 1 (β-form) (b−c plane) Crystal data monoclinic, $P21/n$, $a = 6.261\,(2)$, $b = 4.825\,(1)$, $c = 27.614\,(9)$ Å; $\beta = 91.54\,(2)^{\circ}$.

presence of a water molecule appears to lubricate the crystal-to-crystal transformation in the reaction between two distant olefin bonds in the crystal of 1 while retaining the crystalline state of reacting crystal.

Contrary to anomalously distant reacting double bonds, the mixed crystal of 1:1 donor-acceptor type cinnamic acids has been reported to be photostable, in which double bonds of adjacent donor and acceptor components in the stack are within photoreactive distance of each other (3.80 Å) (*14*).

We have found a rare reaction where two topochemical processes occur competitively in a single crystal, that is, a competitive photocyclo-dimerization and -polymerization in the crystal of 1,4-dicinnamoylbenzene 3 (*15*). On photoirradiation with a mercury lamp (100W) at 20°C for 9 h, the crystals of 3 (2.00g), dispersed in heptane (400 ml), are transformed into amorphous substances consisting of a tricyclic dimer, 21,22,23,24-tetraphenyl-1,4,11,14-tetraoxo-2(3),12(13)-diethano[4,4]paracyclophane, 4 (isolated yield 57%) (*16*), a mixture of oligomers (ca. 30%) and unreacted 3 (7%).

$$\text{C}_6\text{H}_5\text{CH=CHC(=O)C}_6\text{H}_4\text{C(=O)CH=CHC}_6\text{H}_5 \;(\mathbf{3}) \xrightarrow{h\nu} \mathbf{4} + \text{oligomers}$$

It is noteworthy that the major reaction path is apparently a "double" photocyclodimerization of 3 which proceeds with remarkable ease and stereoselectivity. That is, the crystals react at a moderate rate on irradiation even at temperatures much below the melting point of 3 (211-212°C) to give a tricyclic dimer of single configuration, indicating that this reaction is typical of a topochemical process. Little work has been reported on such double [2+2] cycloadditions of diolefinic compounds in the crystalline state except for brief descriptions of bis(3,4-dichlorostyryl)ketone (*17*), carbonyldiacrylic acid (*18*) and its methyl ester (*19*). Furthermore, the reaction of 3 is considered to be a synthetic route for tricyclic paracyclophane. The crystal structure of 3 projected onto (a) the plane of central benzene and (b) the (100) plane is shown in Figure 2.

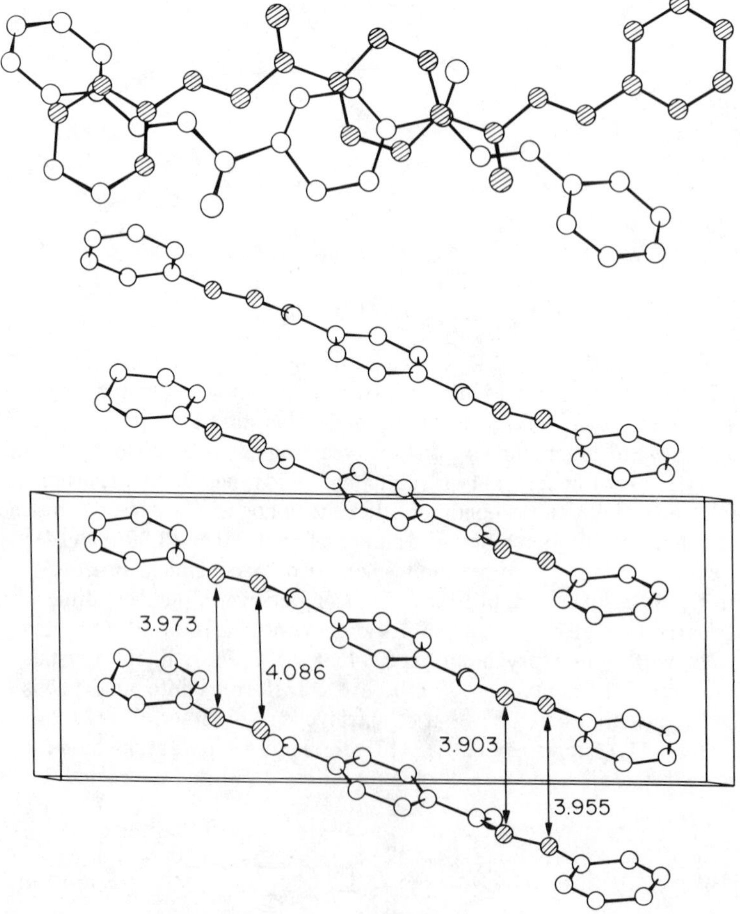

Figure 2. Crystal structures of 3 projected onto (a) the plane of central benzene and (b) the (100) plane *Crystal data* triclinic, $P\bar{1}$, $a = 5.789(1)$, $b = 7.923(1)$, $c = 19.307(6)$ Å, $\alpha = 89.12(2)$, $\beta = 82.12(2)$, $\gamma = 88.67(1)°$.

The reacting molecular pair is arranged skewed to each other, and the distances between the intermolecular photoadductive carbons are 3.973 and 4.086 Å for one cyclobutane ring, and 3.903 and 3.955 Å for the other. The two observed topochemical pathways, to afford 4 or to afford the oligomers, are reasonably interpreted based on the molecular arrangement in the starting crystal 3. In general, the symmetries due to the parallel orientation result from photoadductive double bonds. Therefore, the dimerization of 3, which is regarded as *pseudo* twofold axis symmetry, is an unusual photochemical reaction since a nonparallel arrangement without any symmetry in the reacting carbon-carbon double bonds is observed. Photodimerization between double bonds having such an unfavorable nonparallel arrangement is rare and has been reported only recently for crystals of a few coumarin derivatives (*20*).

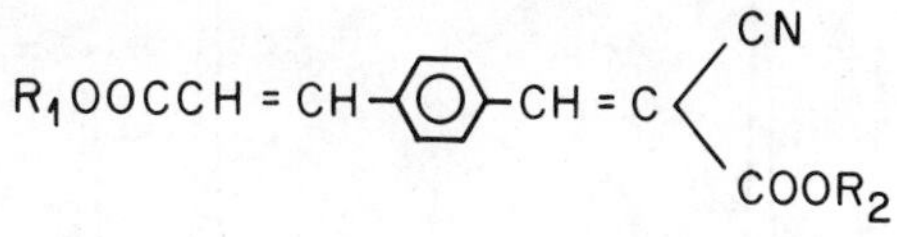

$$R_1OOCCH=CH-C_6H_4-CH=C\begin{matrix}CN\\COOR_2\end{matrix}$$

5

a, R_1 = (R,S)-sec-butyl ; R_2 = ethyl

b, R_1 = 3-pentyl ; R_2 = methyl

Unsuccessful preparation of single crystals of 4 for x-ray crystallographic analysis led us to study the coupling constants of the cyclobutane protons in 3 using a modified LAOCOON III Program (*21*). The derived ^{1}H-NMR spectrum based on the calculation is shown in Figure 3 together with the experimental spectrum. Being almost identical, these two spectra display high values for $J_{AA'}$, J_{AB}, and $J_{BB'}$ indicating that the configuration of the cyclobutane ring in 4 is 1α, 2β, 3α, 4β, namely δ-type, which is in accord with the configuration predicted from the molecular arrangement in the crystal of 3. The formation of a δ-type cyclobutane is extremely rare in the crystalline state photoreaction of olefin derivatives, and only one example has been reported for 1,1′-trimethylenebisthymine (*22,23*). However the photodimerization of 3 is the first report in which δ-type cyclobutane derivatives are produced on a preparative scale.

The photoproduct derived from 3, the new tricyclic [4,4]cyclophane 4, has δ-type cyclobutane rings and has no alternating axis of symmetry, showing it chiral, and the oligomers are considered to have a zig-zag shaped rigid chain structure of alternative (1S*, 2S*, 3R*, 4R*) and (1R*, 2R*, 3S*, 4S*).

From the viewpoint of polymer synthesis it is of great interest to prepare a high polymer possessing the repeating zig-zag structure. Addadi and Lahav have succeeded in preparing an optically active polymer *via* four-center type photopolymerization of unsymmetrically substituted conjugated diolefinic molecule 5, which is related by translation in crystal (*24*). In contrast with a quantitative photopolymerization of the diolefins having a (*pseudo*) centrosymmetry of crystal, in the crystal 3 the unavoidable formation of cyclic dimer 4 will disturb the topochemical chain growth seriously. The reaction schemes of topochemical photoreaction of 3 are illustrated in Equations 1-13.

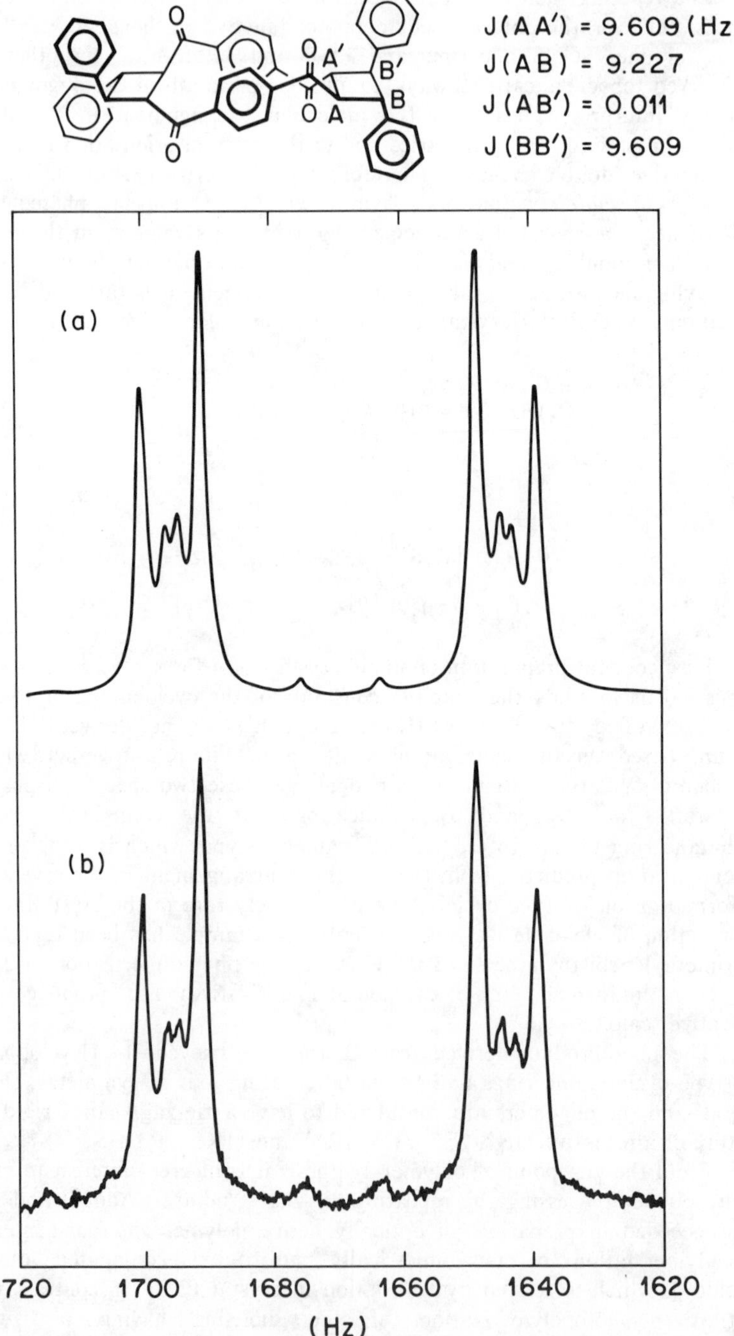

Figure 3. Calculated (a) and experimental (b) ^{1}H-NMR spectra of cyclobutane protons of 4 at 400 MHz.

$$A \xrightarrow{h} A^* \quad (1)$$

$$B \xrightarrow{h\nu'} B^* \quad (2)$$

$$D_n \xrightarrow{h\nu'} D_n^* \quad (3)$$

$$A^* + A \xrightarrow{k_1} B \quad (4)$$

$$A^* + B \xrightarrow{k_2} D_1 \quad (5)$$

$$A^* + D_n \xrightarrow{k_3} D_{n+1} \quad (6)$$

$$B^* \xrightarrow{k_4} C \quad (7)$$

$$B^* + A \xrightarrow{k_5} D_1 \quad (8)$$

$$B^* + B \xrightarrow{k_6} D_2 \quad (9)$$

$$B^* + D_n \xrightarrow{k_7} D_{n+2} \quad (10)$$

$$D_n^* + A \xrightarrow{k_8} D_{n+1} \quad (11)$$

$$D_n^* + B \xrightarrow{k_9} D_{n+2} \quad (12)$$

$$D_n^* + D_m \xrightarrow{k_{10}} D_{m+n+2} \quad (13)$$

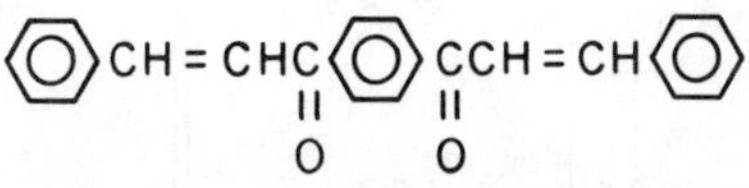

A

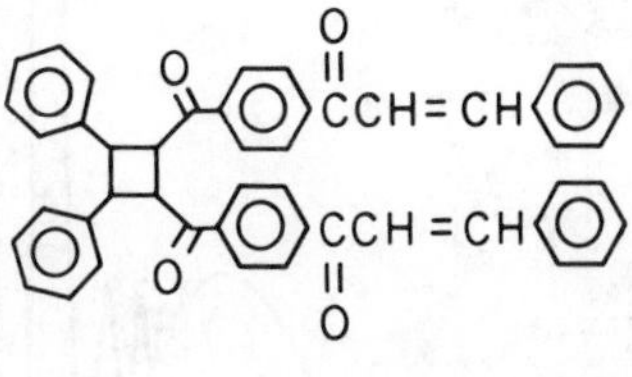

B

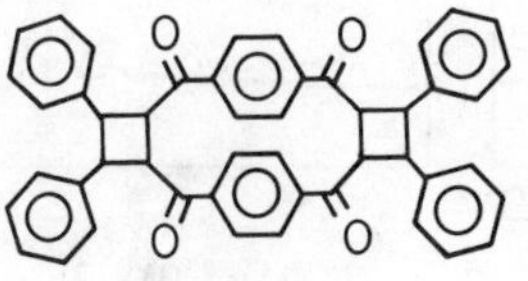

C

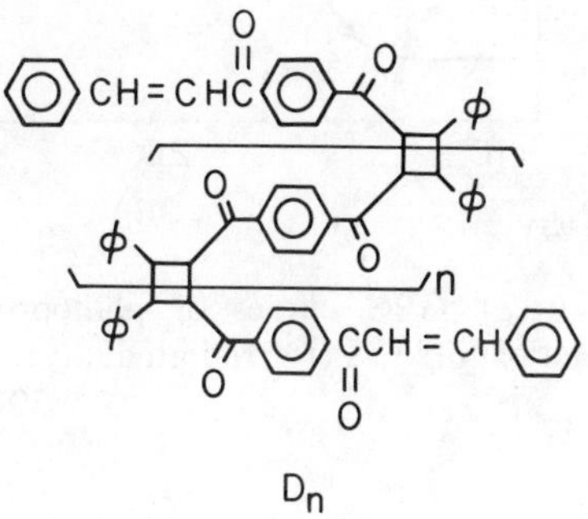

D_n

The electronic transition of B and D is shifted to a higher energy level than that of A. In the above reaction schemes, A^*, B^*, and D^* represent the species A, B, and D respectively in the excited state. Equations 4 and 7 represent a dimerization reaction and a ring-closure reaction to afford 4, and Equations 5, 6, 8-13 represent growth reactions. On photoirradiation at the long wavelength edge of the monomer A $(h\nu)$, only reactions (4), (5), and (6) proceed.

A typical example of GPC curve of the final photoproduct after irradiation on 3 at room temperature for 9 h is shown in Figure 4(a). In Figure 4(a) almost all of the oligomers are distributed from the trimer to the

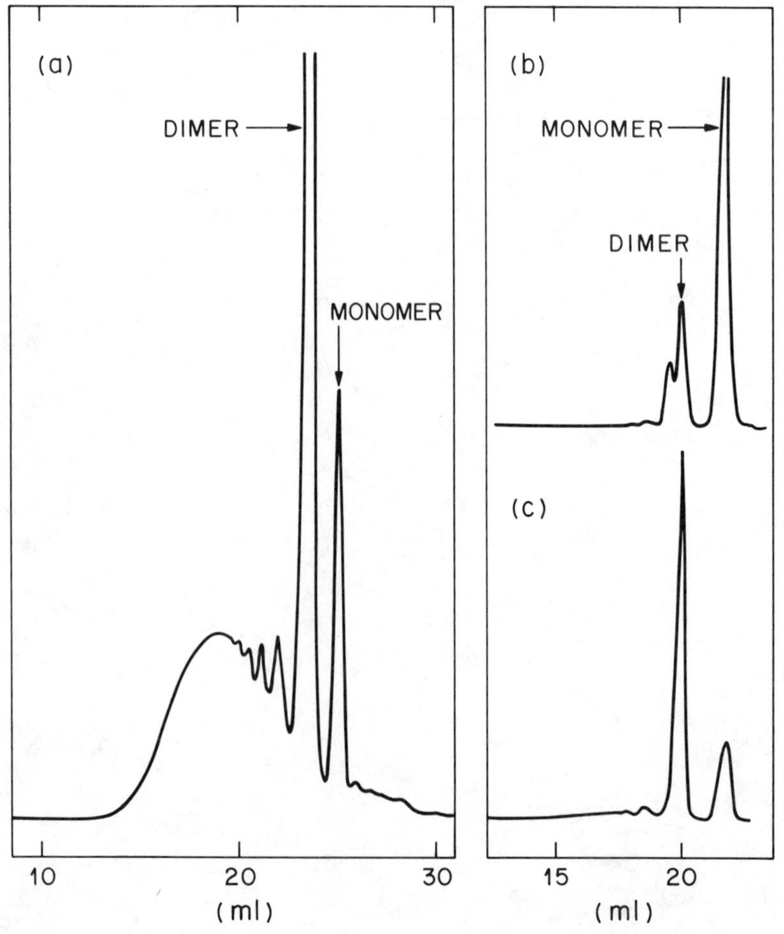

Figure 4. GPC curves of photoproducts obtained from the crystalline state reaction of 3; (a) irradiated at room temperature for 9 h, (b) irradiated at 50°C for 8 h.

molecular weight of 20,000 calculated using polystyrene standards. The degree of polymerization increases predominantly at the later stage of photoirradiation, showing that the reaction is essentially a stepwise mechanism. At the early stage of irradiation, two peaks are observed in the GPC curve at around the retention time for a dimer of which the major peak corresponds to the tricyclic dimer 4, as shown in Figure 4(b). These two peaks are gradually transformed into the single sharp peak of 4 in Figure 4(c). The minor peak us undoubtly attributed to a species *B* which is produced by Equation 4. The photoproduct ratio of 3, 4, and the oligomers varies substantially with reaction temperature, suggesting that each elementary process is influenced to a different individual degree by thermal vibration in the crystal lattice. At −20°C the oligomers decrease to below 5% while unreacted 3 increases to as high as 50%. Such a remarkable depression of reactivity is presumably explained in terms of the restricted local movement of molecule in the crystal. It is rather surprising that in an extreme case the dimer yield attains more than 90%, while the residual monomer yield less than a few percent.

Acknowledgments

We are pleased to acknowledge the valuable contributions of our coworkers, Dr. Hachiro Nakanishi, Mr. Masao Nohara and Mr. Hirofumi Uno.

Literature Cited

1. Mustafa, A. *Chem. Rev.* 1952, *51*, 1.
2. Dilling, W. L. *Chem. Rev.* 1983, *83*, 1.
3. Bernstein, H. I.; Quimby, W. C. *J. Am. Chem. Soc.* 1943, *65*, 1845.
4. Cohen, M. D.; Schmidt, G. M. J. *J. Chem. Soc.* 1964, 1969; for a review see Schmidt, G. M. J. *Pure Appl. Chem.* 1971, *27*, 647.
5. Nakanishi, H.; Hasegawa, M.; Sasada, Y. *J. Polym. Sci., Polym. Phys. Ed.* 1977, *15*, 173.
6. Nakanishi, H.; Jones, W.; Thomas, J. M.; Hursthouse, M. B.; Motevalli, M. *J. Phys. Chem.* 1981, *85*, 3636.
7. Hasegawa, M. *Adv. Polym. Sci.* 1982, *42*, 1.
8. Hasegawa, M. *Chem. Rev.* 1983, *83*, 507.
9. Nakanishi, F.; Nakanishi, H.; Tsuchiya, M.; Hasegawa, M. *Bull. Chem. Soc. Jpn.* 1976, *49*, 3096.
10. Nakanishi, F.; Hirakawa, M.; Nakanishi, H. *Isr. J. Chem.* 1979, *18*, 295.
11. Nakanishi, H.; Saigo, K.; Mori, T.; Hasegawa, M. to be published.
12. Hasegawa, M.; Katsuki, H.; Iida, Y. *Chem. Lett.* 1981, 1799.
13. Hasegawa, M.; Shiba, S. *J. Phys. Chem.* 1982, *86*, 1490.
14. Desiraju, G. R.; Sarma, J. A. R. P. *J. Chem Soc., Chem. Commun.* 1983, 45.
15. Hasegawa, M.; Nohara, M.; Saigo, K.; Mori, T.; Nakanishi, H. *Tetrahedron Lett.*, 1984, *25*, 561.
16. 4: mp 348-352°C. ^{1}H-NMR (DMSO-d_6) δ 7.74 (pseudo-d, 4H, $J = 8$ Hz), 7.38 (m, 20H), 7.27 (m, 4H), 4.24 (pseudo-d, 4H, $J = 9$ Hz), and 4.11 (pseudo-d, 4H, $J = 9$ Hz) ppm. IR (KBr) 1675, 915, 750, and 700 cm^1. UV (CH_2Cl_2) λ_{max} 267 (ϵ 25,500) and 232 (ϵ 19,200) nm. Mass spectrum, m/e 676 (M^+).

17. Green, B. S.; Schmidt, G. M. J. *Tetrahedron Lett.* 1970, p. 4249.
18. Midorikawa, H. *Bull. Chem. Soc. Jpn.* 1953, *26*, 302.
19. Stobbe, H.; Färber, E. *Ber. Dtsch. Chem. Ges.* 1925, *58*, 1548.
20. Ramasubbu, N.; Row, T. N. G.; Venkatesan, K.; Ramamurthy, V.; Rao, C. N. R. *J. Chem., Soc., Chem. Commun.* 1982, 178.
21. Modified LAOCOON III program by Dr. T. Hirano.
22. Frank, J. K.; Paul, I. C. *J. Am. Chem. Soc.* 1973, *95*, 2323.
23. Leonard, N. J.; McCredie, R. S.; Logue, M. W.; Cundall, R. C. *J. Am. Chem. Soc.* 1973, *95*, 2320.
24. Addadi, L.; van Mil, J.; Lahav, M. J. *J. Am. Chem. Soc.* 1982, *104*, 3422 and references therein.

RECEIVED August 6, 1984

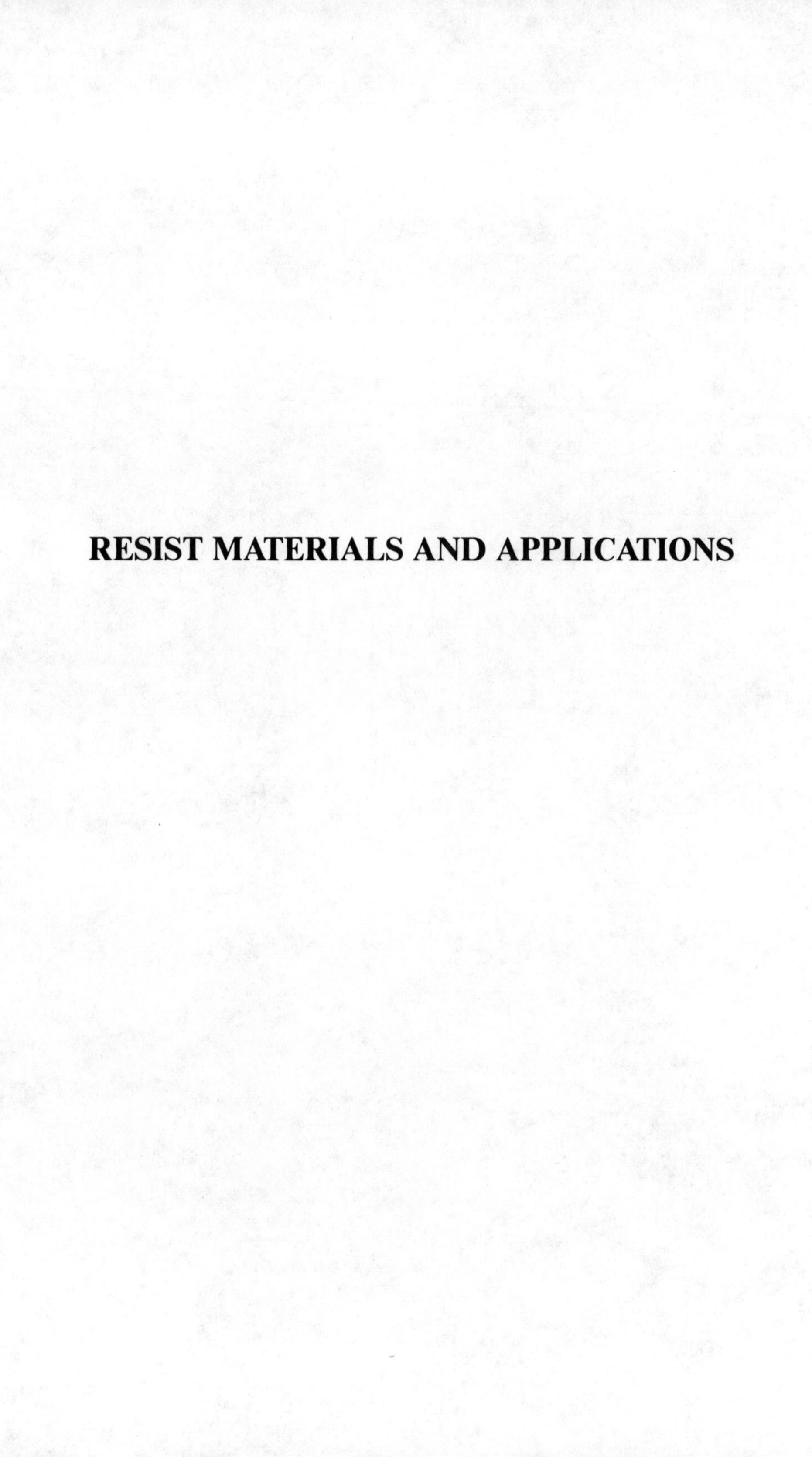

RESIST MATERIALS AND APPLICATIONS

Resist Materials and Applications

The design and development of resist materials for use in microcircuit fabrication can be considered to be in its third evolutionary stage. The imaging materials technology used to launch the semiconductor industry evolved from the printing industry. In the 1960's devices were fabricated with negative-tone resist materials developed for that industry which function on the basis of cross-linking or network formation. Chief among these were the bis-arylazide/rubber resists and polyvinyl cinnamate-based systems. The 70's saw the introduction of diazonaphthoquinone-sensitized novolak systems which offered higher resolution and greater process flexibility. Emergence of the first materials exclusively designed by and for the microelectronics industry appeared in the early 70's. Noteworthy among these are the e-beam resists COP and PBS that were developed to support mask making by Bell Laboratories. During the 80's there has been a dramatic increase in research activities directed toward the development of materials tailored specifically for use in microcircuit fabrication. Several trends in resist design have emerged from such studies, and the third part of this book contains accounts of recent advancements in a number of these. Throughout the work reported in Part III, particular emphasis has been placed on the fundamental understanding of the relation between lithographic behavior and chemical properties of the imaging materials.

As the industry continues its advance toward higher circuit densities, minimum linewidths in the patterned resist films used to fabricate such devices have become smaller, but the resist film thicknesses required have not undergone a proportionate reduction. This is because the minimum acceptable film thickness is determined by the need for covering structures placed on the substrate in previous operations, and by the need to minimize defect-producing pinholes in the film. The resulting demand for greater aspect ratios has been a driving force behind the design of multilayer resist (MLR) systems. Chapters 14 and 15 describe the synthesis and characterization of two new MLR materials where the imaging layers are novel radiation-sensitive organosilicon polymers and the final high-aspect images are formed by anistropic image transfer of the pattern into a thick planarizing underlayer.

Much recent activity has been directed toward designing photoresists capable of efficient imaging with mid- or deep-UV light in order that the enhanced optical resolution attendant with shorter wavelength exposures can be utilized. Chapter 19 describes photochemical studies of ketone polymer films for such a potential lithographic application. This work effectively demonstrates how knowledge gained from basic mechanistic studies of a system (reviewed in Chapter 6 of Part II) can be used to design imaging materials and to understand their behavior.

The epoxy-based negative e-beam resist systems such as COP are widely used in the manufacturing of photomasks. These materials now seem to be giving way to the halogenated or halomethyl-substituted polystyrenes which offer both higher contrast and higher etch resistance. Chapter 18 reports on the relationship between polymer properties and lithographic performance for such a system. This Chapter and Chapter 5 of Part II form a set of complementary studies on the application of polychlorostyrene as an e-beam resist and the mechanism of its radiation chemistry.

In general, negative tone systems that function on the basis of network formation suffer from resolution limitations associated with swelling phenomena that occur during development. This problem can be minimized by careful choice of the developer (and rinse) solvents but it cannot be eliminated. New chemistry for negative tone imaging has recently evolved that escapes this limitation. These systems are based on either a radiation-induced change in the polarity of a polymer side chain, or the use of azide sensitization with phenolic polymer matrix resins. New examples of advances in both these areas are provided in Chapters 13 and 21, respectively. These new design concepts have served to extend the potential of negative-tone imaging systems for both optical and e-beam applications.

Chapters 16 and 20 contain descriptions of new positive e-beam resists that are based on copolymers of maleic anhydride and substituted benzyl methacrylates, respectively. Chemical modifications of novolak resins, and how these modifications alter the lithographic properties of resists prepared from such polymers, have been reviewed in Chapter 17.

As with the other parts of this book, this section is not intended to be a comprehensive assessment of the current state of resist materials development. It does provide evidence of the continuing and accelerating evolution of this field.

13

The Photo-Fries Rearrangement and Its Use in Polymeric Imaging Systems

T. G. TESSIER and J. M. J. FRECHET

Department of Chemistry, University of Ottawa
Ontario, KIN-9B4, Canada

C. G. WILLSON and H. ITO

IBM Research Laboratory, Dept. K42-282
San Jose, CA, 95193

The photo-Fries rearrangement which was first reported (*1,2*) in the early 1960's has some potential in the design of photoresist materials since it involves the transformation of molecules such as phenolic esters into free phenols, thereby providing a route to selective image development by differential dissolution. Initial interest in the photo-Fries rearrangement of aromatic polyesters was mainly due to the fact the reaction is accompanied by the formation of *o*-hydroxy aromatic compounds which possess great photostability. This interesting property of the rearranged products can be used to design polymers containing photostabilizing *ortho*-hydroxy aromatic groups (*3-6*); the photostabilizing action of such compounds has been recently reviewed (*7-8*).

Scheme 1 shows the reaction which occurs when an aromatic polyester such as [I] is subjected to UV irradiation (*5*). The polymer first undergoes main-chain cleavage with subsequent rearrangement to polymer [II] which is photostable and can be used as a thin coating to protect efficiently other substrates which are normally photodegradable.

A considerable amount of attention has also been paid to the photo-Fries rearrangement of polymer pendant groups. For example, the rearrangement of poly(phenyl acrylate) (*10,11*) in solution or in the solid-state, is usually incomplete and results in the formation of both the *ortho* and the *para*-hydroxyphenone rearranged products in amounts which vary with the conditions of the photolysis. A concurrent side-reaction, which we term the Fries degradation, also results in the liberation of small amounts of phenol (Scheme 2). Similar results have been obtained with poly(phenyl methacrylate) and other substituted aryl acrylates (*4,9,12*).

The enhanced stability of the photo-Fries rearrangement products was again confirmed in a thorough study by Guillet and co-workers (*13*) who attributed it to the high extinction coefficient of both the *ortho* and the *para* photoproducts and to their ability to dissipate the absorbed energy by non-photochemical pathways.

0097-6156/84/0266-0269$07.25/0

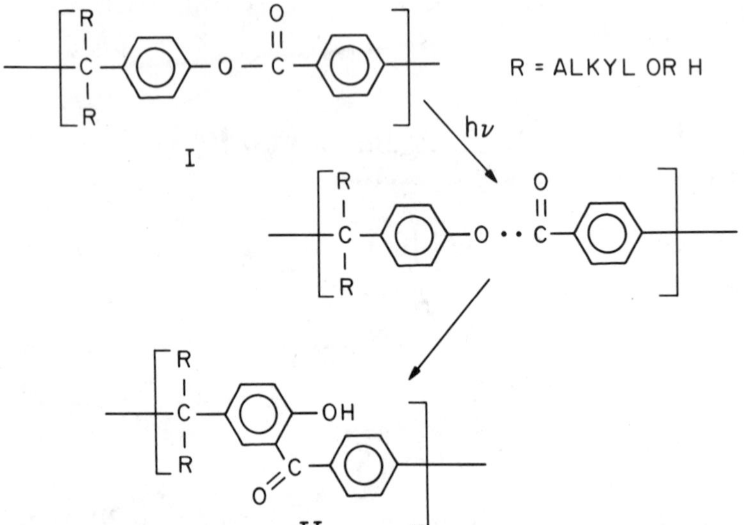

R = ALKYL OR H

SCHEME 1.

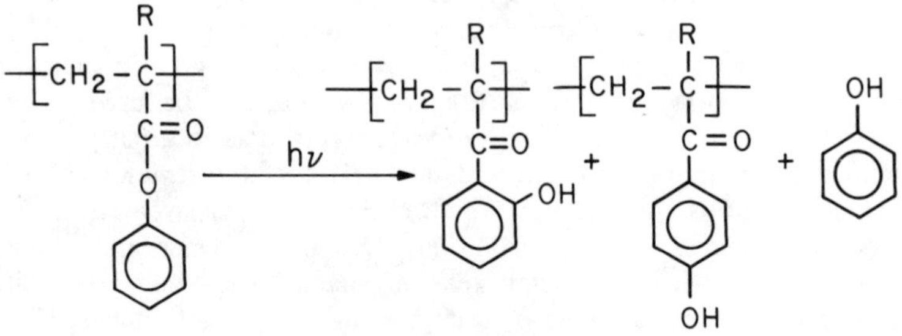

SCHEME 2.

Although the photo-Fries rearrangement and the concurrent degradation of a number of polymers has been reported in the literature (*3-13*), essentially no previous attempts at using this reaction in lithographic processes have been reported. The lithographic potential of the photo-Fries reaction rests on the ability to selectively dissolve either the exposed or the unexposed areas of a polymer film. Typically, it is expected that in the case of polymers containing phenolic ester groups, the photoproducts, being substituted phenols, should dissolve in aqueous base while the unchanged starting polymer should remain undissolved. Similar results should also be within reach using the photo-Fries reaction on polymers containing aromatic amide groups. In addition, the photodegradation component of the photo-Fries reaction which often results in a decrease in the molecular weight of the irradiationed polymer, would also be expected to contribute to the solubility difference between exposed and unexposed areas of an aromatic polyester or polyamide film. Although it is anticipated that, in some cases, the photostabilizing effect of the rearrangement products might have a deleterious effect on the imaging characteristics of the polymers, it is nevertheless likely that imaging of thin films should remain possible. The polymers used in this study included poly(*p*-acetoxystyrene) [III], poly(*p*-formyloxystyrene) [IV], poly(*p*-acetamidostyrene) [V], poly(phenyl methacrylate) [VI], and poly(methacryl anilide) [VII].

Preparation of the Polymers

The *p*-acetoxystyrene monomer, precursor of polymer III, is prepared from *p*-hydroxyacetophenone using the procedure of Corson et al. (*14*) which involves acetylation of the phenolic group followed by catalytic hydrogenation of the ketone and dehydration of the resulting benzylic alcohol as shown in Scheme 3.

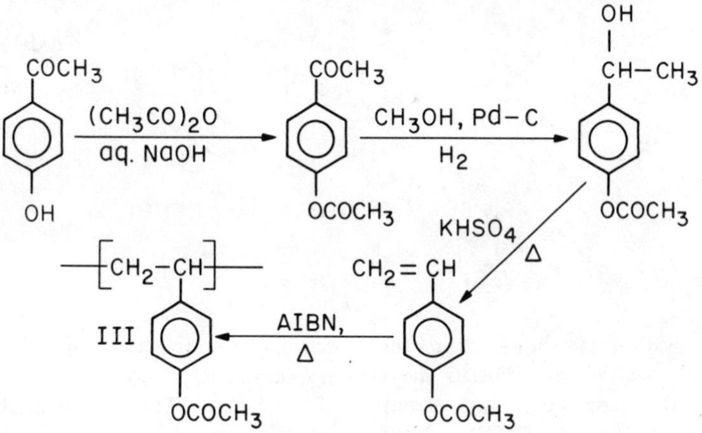

$COCH_3$ / OH → $(CH_3CO)_2O$, aq. NaOH → $COCH_3$ / $OCOCH_3$ → CH_3OH, Pd–C, H_2 → OH, CH–CH_3 / $OCOCH_3$ → $KHSO_4$, Δ → CH_2=CH / $OCOCH_3$ → AIBN, Δ → –[CH_2–CH]– / $OCOCH_3$ III

SCHEME 3.

This reaction sequence is satisfactory although the overall yield is approximately 50%. A different route for the preparation of poly(*p*-acetoxystyrene) involves the direct acetylation of poly(*p*-hydroxystyrene) with acetic anhydride. The main problem with this approach is the lack of commercial availability of high purity poly(*p*-hydroxystyrene).

Similarly, poly(*p*-formyloxystyrene) (IV) can be prepared by formylation of poly(*p*-hydroxystyrene) using formic acid-acetic anhydride mixture as a formylating agent (Scheme 4). The formylation reaction is best

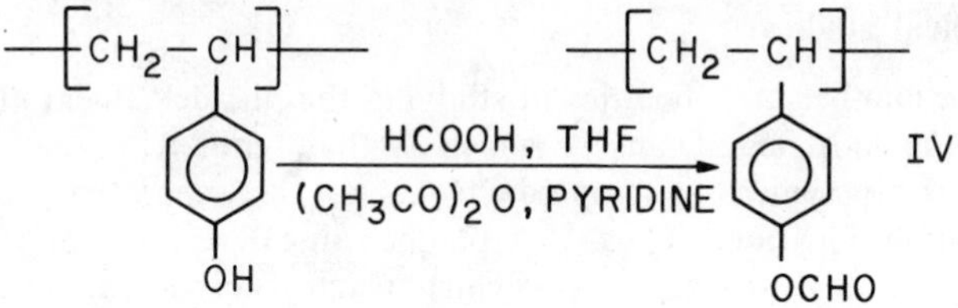

–[CH_2–CH]– / OH → HCOOH, THF, $(CH_3CO)_2O$, PYRIDINE → –[CH_2–CH]– / OCHO IV

SCHEME 4.

carried out in the presence of a small amount of pyridine and is almost complete. Model experiments with phenol have shown that up to 98% formylation is obtained while no acetylated material is obtained. Alternately, the *p*-formylstyrene monomer can be prepared by formylation of *p*-hydroxystyrene using the same reagent mixture and polymer IV is obtained by free-radical polymerization (Scheme 5). The infrared spectrum of poly(*p*-

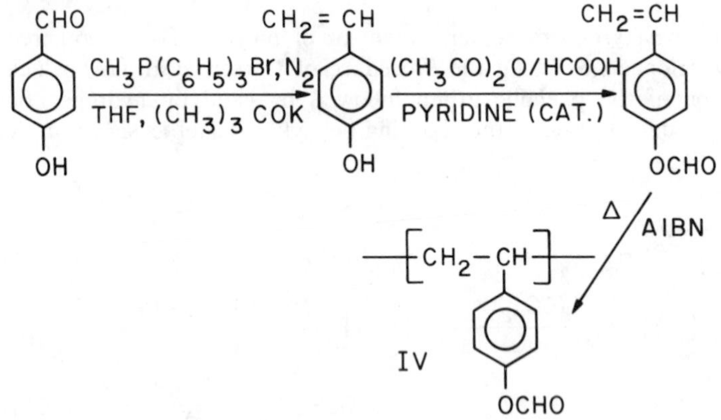

CHO / OH → $CH_3P(C_6H_5)_3Br, N_2$ / THF, $(CH_3)_3$ COK → $CH_2=CH$ / OH → $(CH_3CO)_2O/HCOOH$ / PYRIDINE (CAT.) → $CH_2=CH$ / OCHO → Δ / AIBN → $-[CH_2-CH]-$ / OCHO IV

SCHEME 5.

formyloxystyrene) shows characteristic twin carbonyl absorptions at 1738-1761 cm^{-1} which are due to the presence of both possible s-cis and s-trans conformations of the formate group in the polymer. The partial double bond character of the ester C-O bond due to electron delocalization is responsible for this phenomenon (*17*) which has also been observed with numerous carboxylic acids and amides (*18*). In contrast, the NMR spectrum of the polymer shows only a single formyl signal at room temperature due to the rapid interconversion between the conformers.

Poly(*p*-acetamidostyrene) (V) is prepared from *p*-nitrobenzyl bromide as shown in Scheme 6. Homopolymer V has very little solubility in common organic solvents and thus it is difficult to use; attempts at increasing the solubility of V by incorporation of up to 40% styrene units in copolymers such as Va do not result in any significant improvement in solubility. Poly(phenyl methacrylate) (VI) and poly(methacryl anilide) (VII) are prepared from the corresponding monomers according to literature procedures (*19-20*).

Photochemical Studies

There are a number of difficulties in studying the photochemical modification of polymers, the most significant of which is that, unlike low molecular weight materials, the polymeric photoproducts cannot be separated from unreacted moieties for purification. Thus, if a photochemical reaction only reaches 50% conversion, the final product is a polymer which incorporates equal amounts of modified and unmodified units. In addition, side-reactions give rise to small

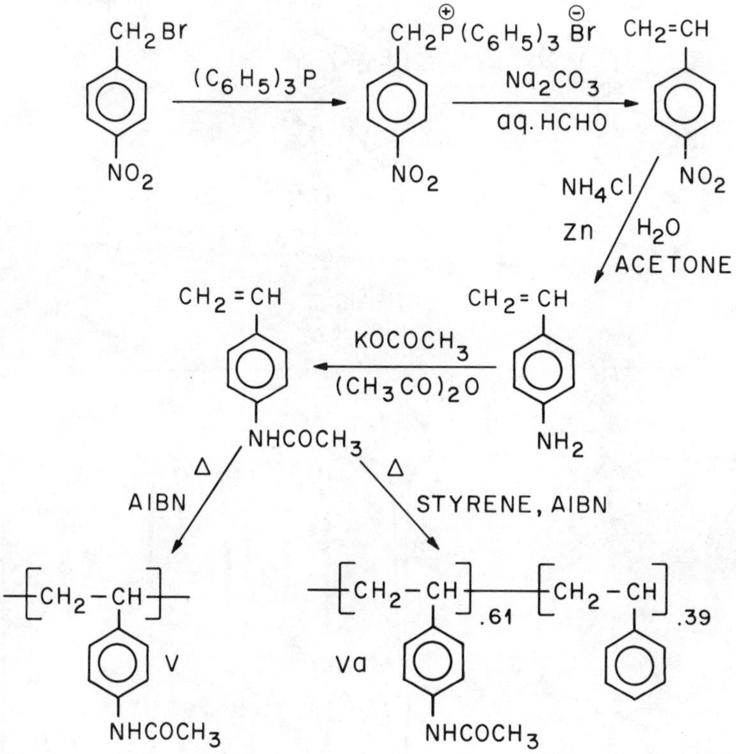

SCHEME 6

amounts of uncontrollably modified units which also remain incorporated in the final polymer. The net result is a greatly decreased ability to characterize fully the products and monitor accurately the reaction. As our study of the photo-Fries reaction was aimed at an evaluation of its potential use in polymeric imaging systems full chemical characterization of the modified polymers was only required in some cases once the exposure was completed. However, UV spectrometry as well as quantitative FT-IR spectrometry were used extensively to monitor the formation of photoproducts at various times during the solution or solid-state photolysis experiments.

Reactive Pendant Groups on Styrene Backbones. The photolysis of *p*-acetoxystyrene (III) in acetonitrile solution gives results which are similar to those obtained in a comparable study in the solid-state using 1 μm thick films; in both cases the expected rearrangement is taking place as shown on Scheme 7.

IR monitoring of the reaction (Figure 1) shows a continuous decrease in the intensity of the starting acetoxy peak at 1765 cm^{-1} with increasing

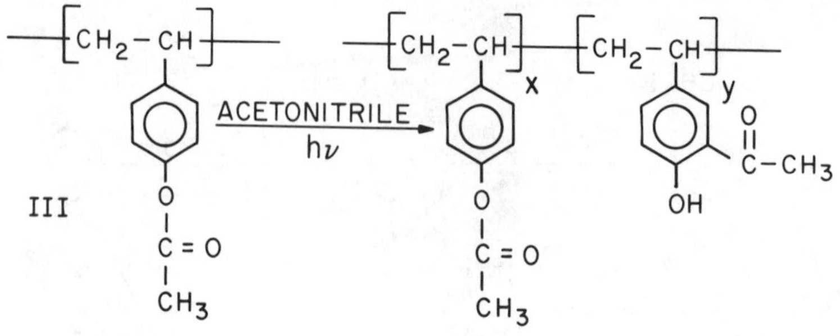

III

ACETONITRILE

$h\nu$

SCHEME 7.

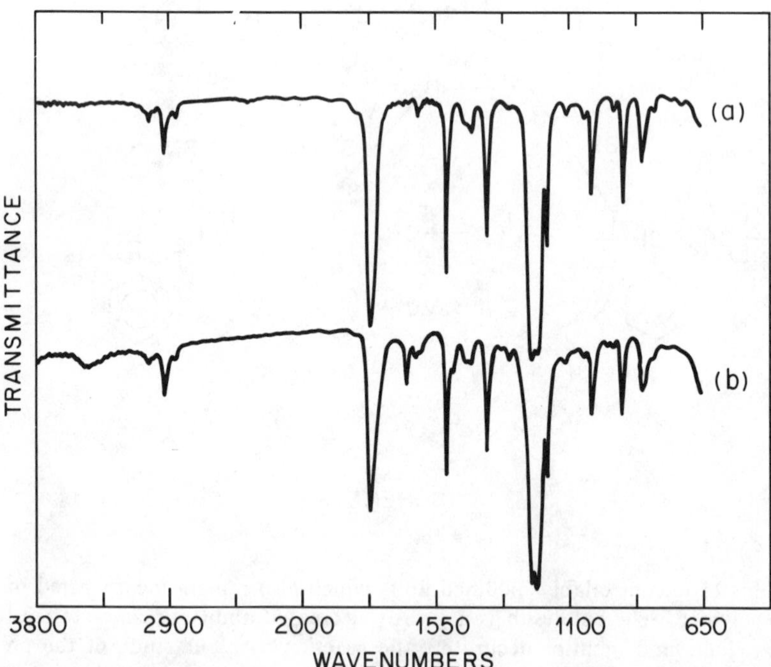

Figure 1. Infrared spectrum of poly(p-acetoxystyrene) (a) unexposed (b) 2 J/cm²

exposure, while a new carbonyl band corresponding to the *o*-rearranged ketone appears at 1643 cm^{-1}. As the *para* position is blocked by the polymer backbone the rearrangement occurs exlcusively in the *ortho* position and the appearance of a weak hydrogen bonded hydroxyl band at 3483 cm^{-1} suggests that little or no degradation to hydroxystyrene units has occurred.

As the reaction proceeds, the presence of increasing amounts of photoproducts results in a drastic decrease in the reaction rate (*21*). NMR analysis of the polymer after prolonged exposure shows that a maximum conversion of approximately 50% is obtainable. However, such prolonged irradiation also gives rise to secondary polymer chain reactions such as radical chain couplings with concurrent increases in the molecular weight of the polymer (e.g. from 11,500 to 45,000). As was mentioned earlier, the emergence of the weak hydrogen bonded hydroxyl at 3483 cm^{-1} confirms that recombination of the acetyl radical is the predominant reaction. This may be attributed to the strong cage effect of the solvent which favors recombination to the starting or rearranged material and prevents escape of the acetyl radicals as is the case for low molecular weight aryl esters (*22*). In the case of the thin films of III, the low mobility of the acetyl radicals in the solid polymer matrix also favors recombination of the radical pair to the starting or rearranged product. The photostabilizing effect of the *o*-rearranged product is further evidenced by the change in UV absorption of a thin film of poly(*p*-acetoxystyrene) upon exposure (Figure 2).

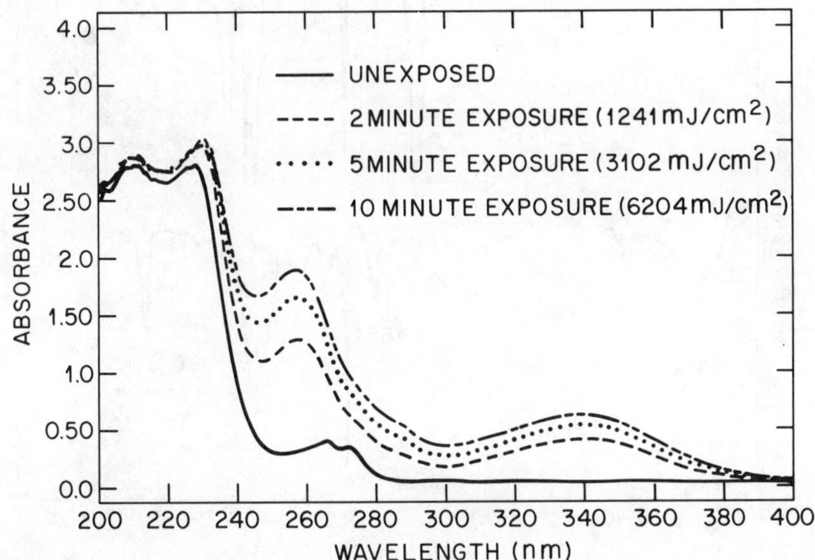

Figure 2. Changes in UV absorption upon irradiation of a 1.1 μm film of poly(p-acetoxystyrene).

The sharp increase in adsorbance which is observed near 256 and 340 nm is characteristic of the *o*-rearranged product. Similar results are also obtained when the irradiation experiments are carried out under nitrogen atmosphere, indicating that the photo-oxidation reported by Hiraoka and Pacansky (*23*) for poly(*p*-hydroxystyrene) is not a factor in this instance.

Although the acetate ester of poly(*p*-hydroxystyrene) is more readily prepared than the corresponding formate (IV), the occurrence of a photostabilizing rearrangement during photolysis of III makes it ultimately ill-suited for use as an imaging system. A survey of the literature on the photo-

Fries rearrangement of aromatic formate esters suggests that phenols are the major products (*24*) obtained in the reaction. As poly(*p*-hydroxystyrene) is remarkably clear in the deep UV, it is likely that poly(*p*-formyloxystyrene) will not suffer from the same problem of photostabilization upon exposure as was the case with poly(*p*-acetoxystyrene). This expectation was confirmed by our study of the photo-Fries reaction of *p*-cresyl formate: no *ortho* rearranged product was isolated after reaction while *p*-cresol and a small amount of starting material were obtained.

Once again photolysis experiments carried out with thin films or solutions of poly(*p*-formyloxystyrene) (IV) in dioxane or acetonitrile give essentially the same results. IR monitoring of the exposure (Figure 3) shows a

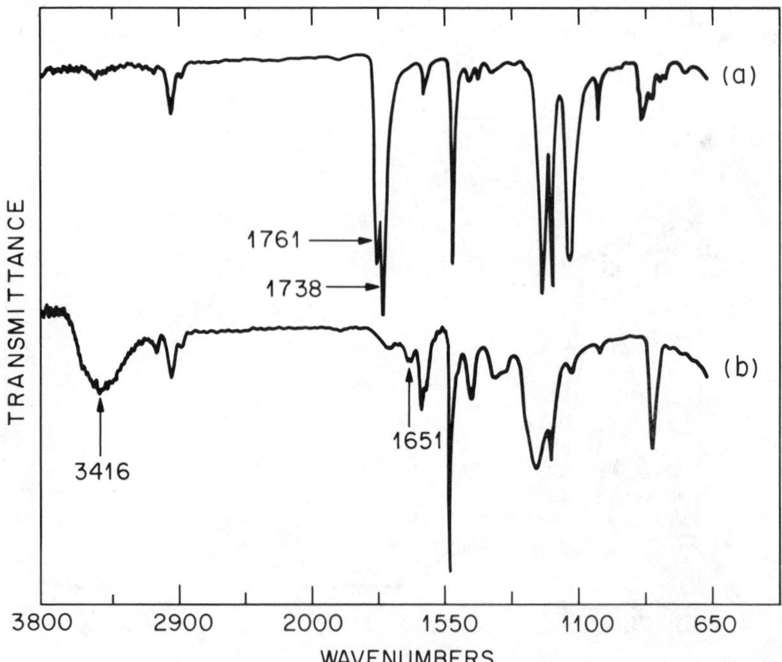

Figure 3. Infrared spectrum of poly(p-formyloxystyrene) (a) unexposed (b) 100 mJ/cm²

rapid decrease in the intensity of the formate carbonyl bands at 1761 and 1738 cm^{-1} as the decarbonylation reaction of IV proceeds to near completion. A strong hydroxyl band which appears at 3416 cm^{-1} confirms the formation of poly(*p*-hydroxystyrene). A very small carbonyl band at 1651 cm^{-1} is observed at very low exposure dose suggesting that some hydrogen-bonded rearranged *o*-hydroxyaldehyde is also formed in the early stages of the reaction.

As expected, the UV spectrum of IV (Figure 4) also changes with exposure, however the photoproduct is still fairly transparent near 250 nm and the decabonylation reaction can proceed to near completion.

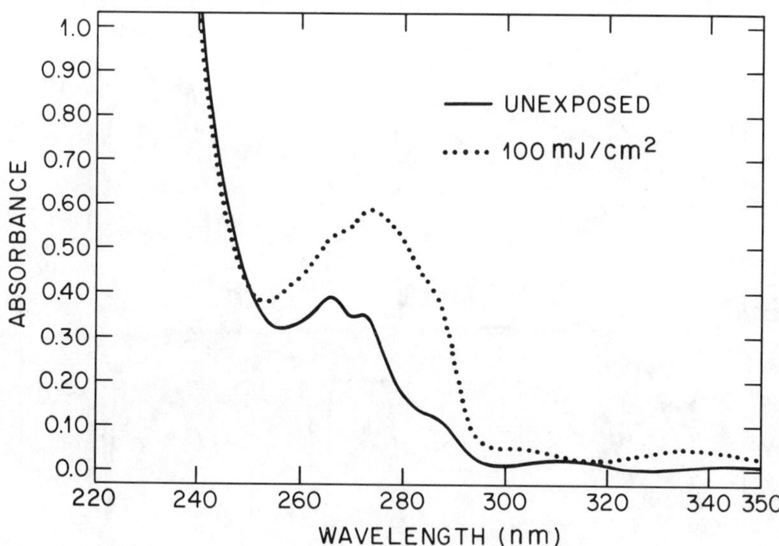

Figure 4. Change in UV absorption upon irradiation of a 1.1 μm film of poly(p-formyloxystyrene).

^{13}C and ^{1}H NMR spectra of the polymer isolated after irradiation show the formation of essentially pure *p*-hydroxystyrene polymer but provide no evidence supporting the formation of any *ortho* rearranged photoproduct. As is the case for phenyl formate (*17*), a detailed analysis of the IR data shows that the s-cis conformer of the formate units in the polymer is less stable than the other conformer and the carbonyl band at 1761 cm^{-1} disappears faster than the band at 1738 cm^{-1}. The large difference in the nature of the photoproducts which are obtained upon exposure of the acetate and formate esters of poly(*p*-hydroxystyrene) is likely only a reflection of the very different stabilities of the acetyl and formyl radical which are formed in the first stage of the Fries reaction. Even in a strongly caging solvent, the very unstable formyl radical is decomposed rapidly thereby preventing recombination of the radical pair; the same decomposition with loss of carbon monoxide is observed in the solid-state.

A thorough study of the photo-Fries rearrangement of poly(*p*-acetamido styrene) was made difficult by the lack of solubility of the polymer in solvents other than DMSO or DMF. Nevertheless, photoexposure of thin films of V cast from DMF solution confirmed the lack of reactivity of the polymer as only minimal conversions could be achieved even after long periods of exposure.

Reactive Pendant Groups on Acrylic Backbones. Our study of the photolysis of poly(phenyl methacrylate) in dioxane confirms the results of previous workers (*4,9*). The expected photorearrangement is observed and can again be monitored conveniently by IR spectroscopy (Figure 5). The intensity of the ester carbonyl band at 1747 cm^{-1} decreases while a small hydroxyl and a new carbonyl band appear at 3384 and 1626 cm^{-1} respectively. This new carbonyl band corresponds to the hydrogen-bonded *ortho* rearranged ketone. Although

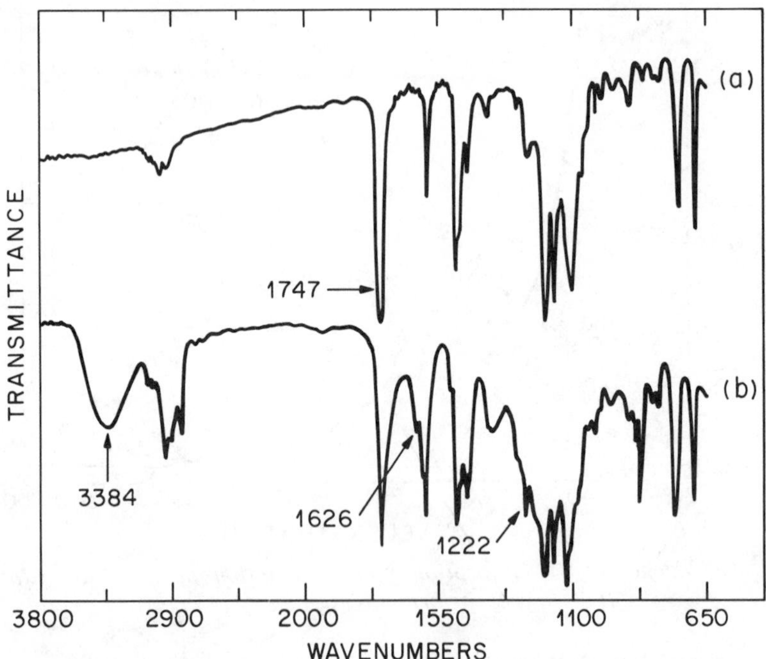

Figure 5. Infrared spectrum of poly(phenyl methacrylate) (a) unexposed (b) 4 J/cm²

the *para* position is unoccupied and available for rearrangement, no *para*-substituted photoproduct is observed.

While a maximum conversion of only approximately 25% of the polymer's repeating units can be obtained in solution, prolonged exposure times also result in a significant decrease in the molecular weight of polymer VI as some degradation also occurs with the liberation of phenol. The mechanism for this degradation is likely to resemble that proposed by Hiraoka and others (*25,26*) for the radiolysis of PMMA. The ^{13}C NMR spectrum of the polymeric photoproduct shows clearly that the polymer contains a majority of unreacted starting phenyl methacrylate units with also some *o*-hydroxy ketone units.

In the solid-state, the rearrangement can proceed a little further with conversion of up to 38% of the starting repeating units. In this case however, a small amount of the *p*-hydroxy ketone product is also formed as evidenced by the appearance of a small carbonyl band at 1663 cm^{-1} in the IR spectrum of an exposed film of VI. As was the case in the solution studies, the formation of a photostabilizing product is clearly seen in the UV spectrum (Figure 6) of the exposed polymer with new absorptions of increasing intensities appearing at 261 and 330 nm during exposure. The overall reaction which occurs upon irradiation of VI is shown in Scheme 8.

Finally, in the case of poly(methacryl anilide) (VII), the rearrangement also occurs, but at an extremely slow rate with a decrease in amide carbonyl

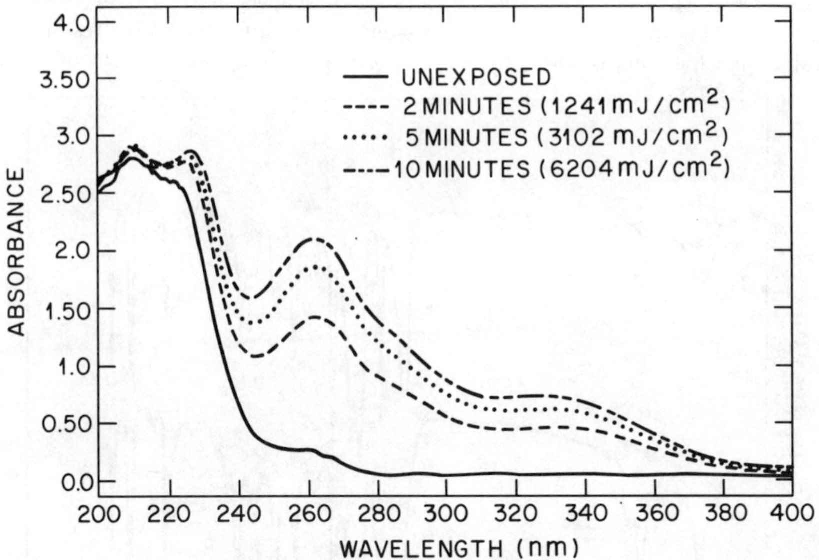

Figure 6. Change in UV absorption upon irradiation of a 1.1 μm film of poly(phenyl methacrylate).

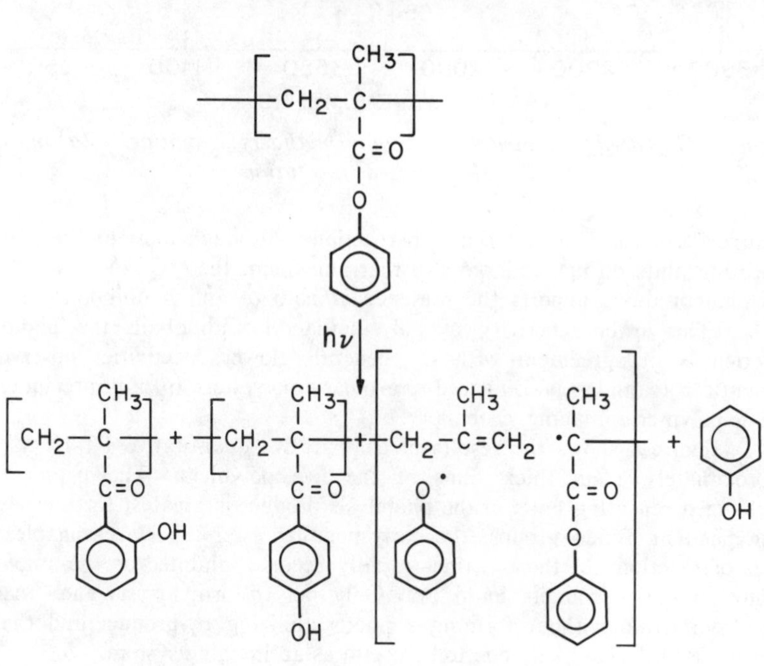

SCHEME 8.

absorption at 1679 cm^{-1} and the appearance of a new small carbonyl band at 1654 cm^{-1} (Figure 7). As expected, prolonged exposure times also result in a considerable lowering of the polymer's molecular weight due to the concurrent

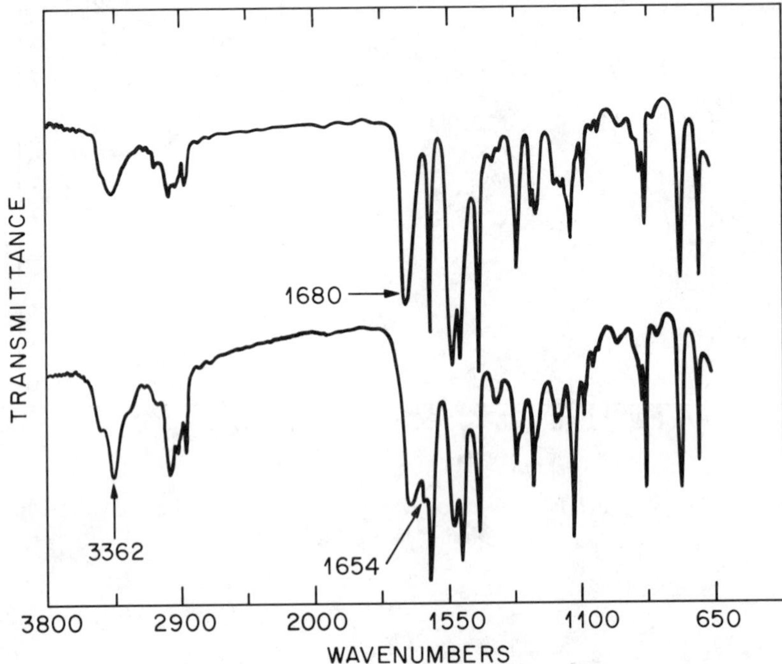

Figure 7. Infrared spectrum of poly(methacryl anilide) (a) unexposed (b) exposed in solution

occurrence of the Fries degradation reaction. Although most of the polymer's repeating units do not undergo any rearrangement, the ^{13}C-NMR spectrum of the photoproduct supports the presence of both *o*- and *p*-aminophenyl ketone units. The lower reactivity of poly(methacryl anilide) in the photo-Fries reaction is in agreement with the generally lower reactivities observed for aromatic polyamides as compared to aromatic polyesters (*6*) and precludes their use as polymeric imaging systems.

Figure 8 shows the results of comparative quantitative FT-IR study on approximately 1 μm thick films of the five polymers. The percentage of unchanged repeating units in the photolysis product is greatest for the polymers with pendant amido groups; the corresponding esters show reasonable initial rates of reaction but the reactions rapidly become inhibited as the amounts of photostabilizing products build up. Only poly(*p*-formyloxystyrene) reacts to near completion without forming a photostabilizing by-product and thus this polymer is the most likely one to be useful as an imaging system.

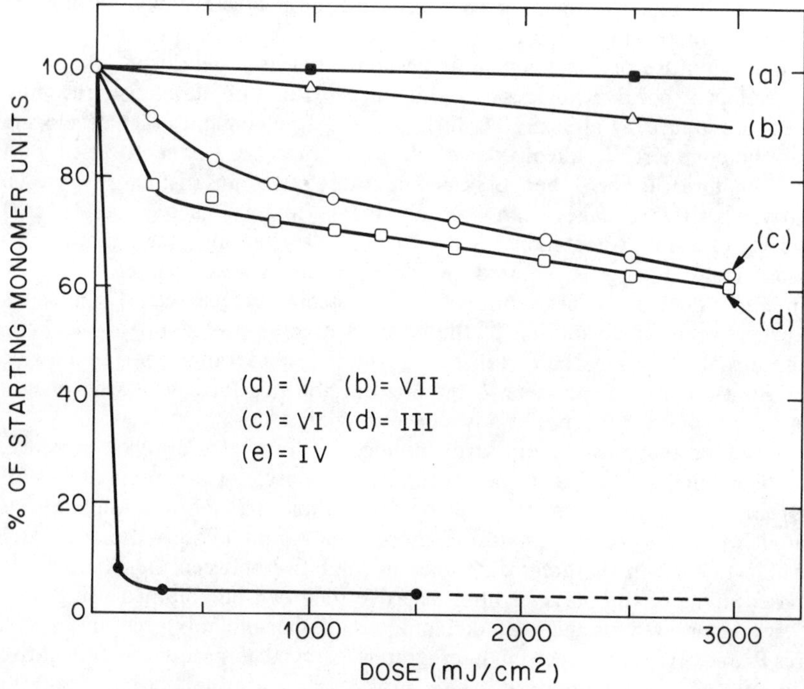

Figure 8. Decrease in % of starting monomer units upon irradiation of polymers III-VII.

Imaging Experiments

The results shown in Figure 8 can be used to predict the imaging potential and characteristics of polymers III-VII. The higher photoconversion of poly(*p*-formyloxystyrene) relative to the other polymers tested makes it the most attractive for use as a photoresist as the great difference in solubility and polarity between the starting polymer and the poly(*p*-hydroxystyrene) which is obtained by photo-Fries reaction should facilitate image development. In the case of the two other polymers with pendant ester groups, III and VI, much greater doses are required to reach conversions of only 20-40%. As a result, the low conversion to phenolic moieties is expected to hamper image development by differential dissolution; slightly better imaging properties may be obtained with VI as the photo-Fries degradation which accompanies the rearrangement results in a decrease in the molecular weight of the polymer through main chain cleavage. For the same reason, it might be possible to obtain images from poly(methacryl anilide) VII despite its very low sensitivity.

In all cases, the imaging experiments were carried out using unfiltered UV light from a high pressure mercury-xenon lamp. Preliminary experiments consisted of contact printing of 1 μm thick films of the various polymers through a quartz mask, while the poly(*p*-formyloxystyrene) was also tested in projection printing using a Perkin Elmer 500 aligner.

As predicted, the imaging characteristics of poly(*p*-acetoxystyrene) are poor, best results being obtained for a very high exposure dose (3.1 J/cm^2) using a 1:1 mixture of 3-heptanone and isopropanol as a developer. Although fully developed positive images are obtained with this developer the film thickness is reduced by almost 50% during image development; other developers such as the commercial aqueous basic solutions which are frequently used with positive photoresists or other organic mixtures give unsatisfactory results. Similarly, positive images can be obtained from films of poly(phenyl methacrylate) using 3-heptanone as a developer. In this instance again a very high dose (ca. 3 J/cm^2) is required and development is accompanied by a 30% loss in film thickness. The nature of the developing solvent clearly indicates that the preferential dissolution of the exposed areas is the result of the Fries degradation of the main chain rather than the Fries rearrangement of the side chain. Both the amide polymers V and VII did not give fully developed images even after prolonged exposure at doses exceeding 10 J/cm^2.

As expected, best results are obtained with poly(*p*-formyloxystyrene); 1 μm thick films of IV can be imaged with doses of approximately 75-80 mJ/cm^2 and positive image development without loss of film thickness is accomplished using a 10:1:1 mixture of isopropanol-ammonium hydroxide-water as developer. Due to the great difference in solubility between the exposed and unexposed areas of the polymer film, negative tone can be obtained using non-polar developer systems such as dichloromethane-hexane mixtures or anisole. Figures 9 and 10 show electron micrographs of typical positive and negative images which are obtained by projection printing using poly(*p*-formyloxystyrene).

Poly(*p*-formyloxystyrene) was also subjected to quantitative sensitivity analysis using a calibrated multidensity resolution mask. A plot of the normalized thickness remaining as a function of dose is provided in Figure 11. The resist exhibits a sensitivity of approximately 70 mJ/cm^2 in the deep UV and has a contrast (γ) comparable to that of the classical diazonaphtoquinone-novolac positive resists that are commonly employed in semiconductor manufacturing.

These results confirm the potential of poly(*p*-formyloxystyrene) as a useful resist material combining such interesting properties as ability to be imaged in both positive or negative tone, ease of preparation, and activity in the deep UV, with a moderate sensitivity and good contrast.

Experimental

The quantitative FT-IR studies were carried out using a Nicolet MX-1 spectrometer; all NMR spectra were obtained using Varian EM360, CFT-80 or XL200 spectrometers while UV spectra were recorded using Varian or Hewlett Packard spectrometers. Molecular weight determinations were performed using a Wescan 231 membrane osmometer using toluene or 1,2-dichloroethane as solvents, while GPC measurements were obtained on a Waters 150 chromatograph equipped with 6 microstyragel columns and using THF as the mobile phase. Photolysis experiments were performed in solution under N_2 atmosphere using a 500 ml Ace Glass reactor with an unfiltered 100 or 500

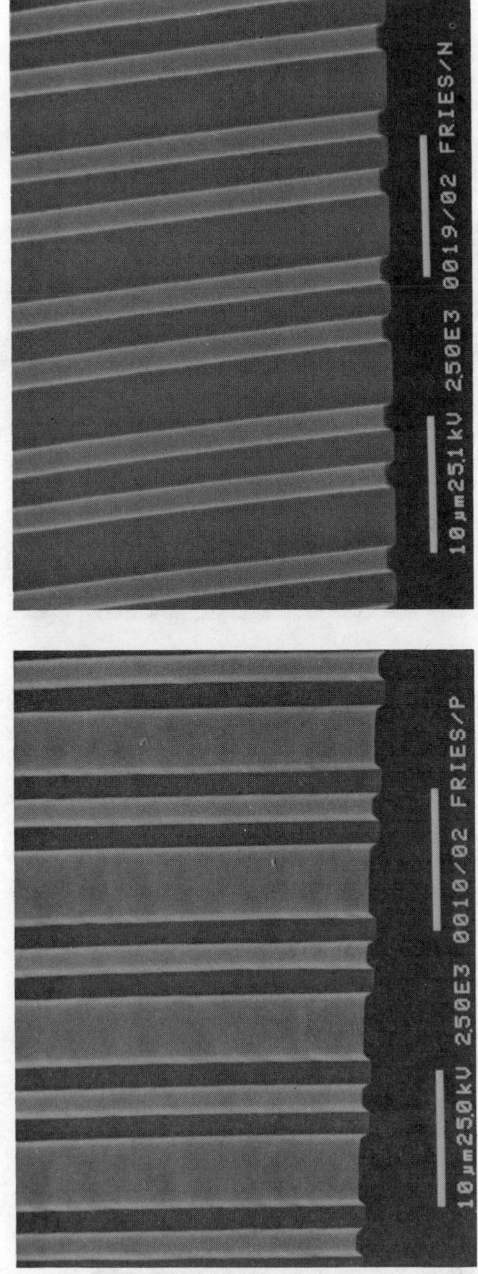

Figure 9. 1.75 µm images in positive (left) and negative (right) tone projection printed in poly(p-formyloxystyrene).

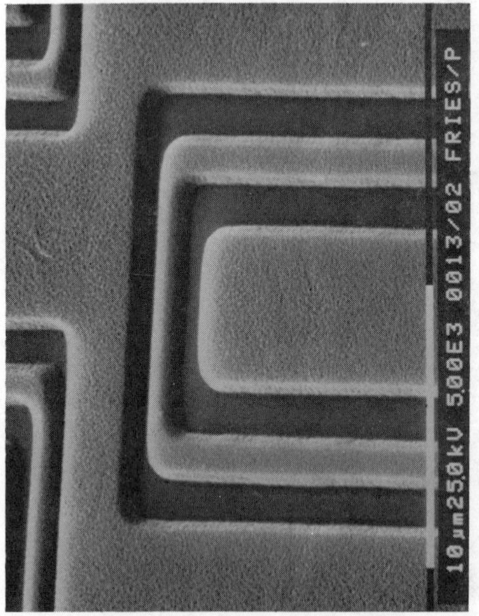

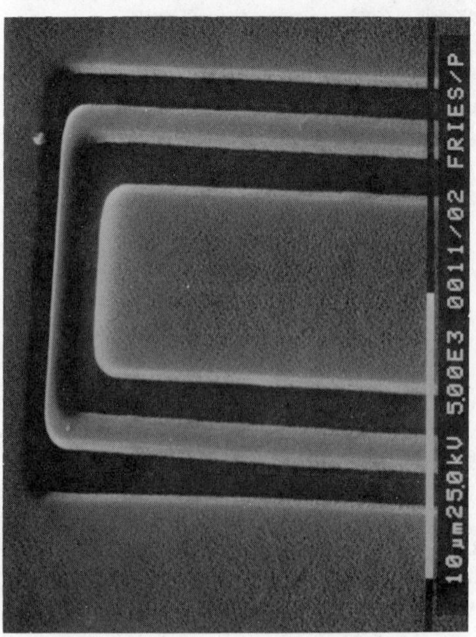

Figure 10. 1.5 μm (left) and 1.25 μm (right) positive tone images projection in printed poly(p-formyloxystyrene).

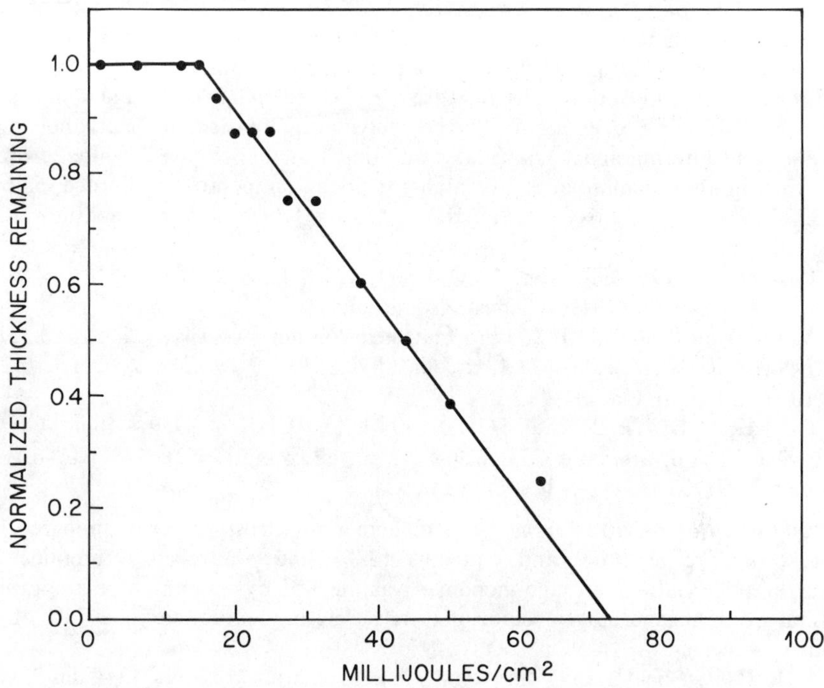

Figure 11. Characteristic curve for poly(p-formyloxystyrene).

watt Hanovia low pressure mercury source. Aliquots of the solutions were withdrawn at regular time intervals for IR or UV monitoring of the progress of the reaction. In addition solid-state studies were also performed in air on approximately 1 μm thick films of the various polymers cast from appropriate solvents onto quartz, sodium chloride, or silicon wafers. Irradiation was carried out using a high pressure mercury-xenon source in an OAI deep UV source or Perkin-Elmer 500 aligner. Additional experimental details are available elsewhere (*31*).

Preparation of p-Acetoxystyrene: This compound was prepared by a modification of the procedure of Corson et al. (*14*), using *p*-hydroxyacetophenone as a starting material. Fractional distillation of the crude material gave a 46% yield of *p*-acetoxystyrene with b.p. 69-72°C/0.5 mm Hg (lit. 73-75°C/0.6 mm Hg[14]).

FT-IR (NaCl, neat): 1765 cm^{-1} (vs, C=O ester); 1631 cm^{-1} (m, C=C vinylic); 1212, 1195 & 1014 cm^{-1} (s, C-O ester).

^{1}H-NMR ($CDCl_3$): 2.13 (3H, s, -CH_3), 6.96 (2H, d, aromatic, J=8.5 Hz), 7.34 (2H, d, aromatic, J=8.5 Hz), AMX pattern for vinyl protons with $_A$ = 5.15, $_M$ = 5.59, $_X$ = 6.51 with J_{AM} = 1.2 Hz, J_{AX} = 11.0 Hz and J_{MX} = 17.0 Hz.

M. S.: m/e 162 · (M^+), 120 (base peak, C_8H_7OH), 43 (CH_3CO^+), 27 (CH_2=CH^+).

Preparation of Poly(p-Acetoxystyrene) (III): A 9.95 g aliquot of *p*-acetoxystyrene and 0.0997 g AIBN were added to 60 ml of toluene in a 100 ml 2-necked round-bottomed flask equipped with a water-cooled condenser, nitrogen inlet and outlet. The mixture was heated to 70-75°C under nitrogen for 24 hours. Precipitation of the polymeric solution in methanol and subsequent filtering and washing with additional methanol gave a white powder. Drying in the vacuum oven overnight at room temperature, afforded 5.46 g (55%) of poly(*p*-acetoxystyrene) **(III)**. ($M_n = 1.17 \times 10^4$, $M_z = 2.06 \times 10^4$, polydispersity = 1.76 (gpc), $M_n = 1.85 \times 10^4$).
FT-IR (KBr): 1761 cm^{-1} (vs, C=O ester); 1218, 1206 & 1017 cm^{-1} (vs, C-O ester), 846 cm^{-1} (s, C-H *p*-disbust. aromatic).
UV Spectrum (1 μm film): 234 nm (cut-off), 265 nm (A=0.41), 271 (A=0.35).
^{1}H-NMR ($CDCl_3$): 1.03-1.97 (3H, bm, allylic H's), 2.19 (3H, s, $-CH_3$), 6.23-7.13 (4H, bm, aromatic).
^{13}C-NMR ($CDCl_3$): 21.2 (q, $-CH_3$), 40.01 (d, -CH), 40.1-49.4 (bm, $-CH_2$), 121.9 (122.6, d, aromatic C3), 129.3 (127.0, d, aromatic C2), 142.3 (146.7, s, aromatic C1), 148.7 (149.1, s, aromatic C4).

Preparation of p-Nitrostyrene: The monomer *p*-Nitrostyrene was prepared by the procedure of Broos and Anteunis (*27*) using *p*-nitrobenzyl bromide as starting material. The crude monomer was purified by column chromatography using a 9:1 hexane/ethyl acetate mixture as eluting solvent to yield 78% of *p*-nitrostyrene.
FT-IR (NaCl, neat): 1597 cm^{-1} (vs, C=C aromatic); 1514 & 1344 cm^{-1} (vs, NO_2); 990 & 926 cm^{-1} (m, monosubst. alkene).
^{1}H-NMR ($CDCl_3$): 7.45 (2H, d, aromatic, J=9.0 Hz), 8.09 (2H, d, aromatic, J=9.0 Hz), AMX pattern for vinyl protons with $_A = 5.41$, $_M = 5.84$, $_X = 6.73$ ppm with $J_{AM} = 1.0$ Hz, $J_{AX} = 10.6$ Hz, and $J_{MX} = 17.0$ Hz.

Preparation of p-Acetamidostyrene: *p*-Acetamidostyrene was prepared by a modification of the procedure reported by Boyer and Alul (*28*) using the *p*-nitrostyrene prepared above. Recrystallization of the crude product in ethyl acetate, with drying in a vacuum oven for 24 hours afforded 13.08 g (60%) of *p*-acetamidostyrene (m.p. 132.5°C, lit. 134°C (*28*).
FT-IR (KBr pellet): 3287 cm^{-1} (m, NH); 1661 cm^{-1} (vs, C=O amide); 1625 cm^{-1} (s, C=C vinylic); 1599 cm^{-1} (vs, C=C aromatic); 1512, 1541 cm^{-1} (vs, N-H & C-N of 2° amide); 906 cm^{-1} (s, monosubst. alkene); 843 cm^{-1} (s, *p*-disubst. aromatic).
^{1}H-NMR ($CDCl^3$): 2.13 (3H, s, $-CH_3$), 7.27 (2H, d, J=8.2 Hz), 7.44 (2H, J=8.2 Hz, aromatic), 7.87 (1H, bs, N-H), AMX pattern for vinyl protons with $_A = 5.15$, $_M = 5.59$, and $_X = 6.64$ ppm with $J_{AM} = 1.5$ Hz, $J_{AX} = 10.6$ Hz, and $J_{MX} = 17.2$ Hz.
M. S.: m/e 161 (M^+), 119 (base peak, $C_8H_7NH_2$), 43 ($C_2H_3O^+$), 27 ($C_2H_3^+$).

Preparation of Poly(p-Acetamidostyrene) (V): Into a 3-necked 100 ml round-bottomed flask equipped with condenser and nitrogen inlet and outlet, was dissolved 3.59 g of *p*-acetamidostyrene in 75 ml toluene and 0.0393 g of AIBN was added, and the reaction stirred for 24 hours at 72°C under nitrogen. A

heavy white suspension formed during the polymerization was dissolved by addition of DMF to the reaction mixture. Precipitation in methanol gave a yellow precipitate. The polymer was washed with methanol, and dried in the vacuum oven overnight at 65°C to give 2.54 g (71%) of poly-(*p*-acetamidostyrene) (**V**) containing 8.41% N (6.00 meq N/g-P). **V** was insoluble in methylene chloride, acetonitrile, dioxane, ethanol and soluble in only DMF and DMSO.
FT-IR (KBr pellet): 3299 cm^{-1} (m, N-H); 1667 cm^{-1} (vs, C=O amide); 1602 cm^{-1} (vs, C=C aromatic); 1536 & 1514 cm^{-1} (vs, N-H & C-N); 832 cm^{-1} (m, *p*-disubst. aromatic).
^{13}C-NMR (DMSO-d6): 23.92 (q, $-CH_3$), 119.14 (118.6, d, aromatic C3), 127.45 (126.0, d, aromatic C2), 136.83 (137.1, s, aromatic C4), 139.61 (142.7, s, aromatic C1), 168.02 (s, C=O amide).

Preparation of Poly(p-Acetamidostyrene-co-Styrene) (Va): A mixture of 1.30 g of styrene, 1.96 g of *p*-acetamidostyrene, 15 ml of ethanol, and 0.0208 g AIBN was loaded into a 50 ml three-necked round-bottomed flask equipped with condenser, and nitrogen inlet and outlet. The mixture was stirred under nitrogen at 76°C for 48 hrs, diluted with toluene and precipitated in ether. Washing with ether and drying the product in the vacuum oven overnight at room temperature gave 2.09 g (64%) of poly(*p*-acetamidostyrene-co-styrene) (**Va**) containing 6.10% N (3.77 meq N/g-P) for a molar ratio of 0.61:0.39. Copolymer **Va** was insoluble in methylene chloride, dioxane, tetrahydrofuran and acetonitrile, slightly soluble in ehtanol and toluene, and very soluble in DMF.
FT-IR (KBr pellet): 3302 cm^{-1} (m, N-H); 1668 cm^{-1} (vs, C=O amide); 1602 cm^{-1} (vs, C=C aromatic); 1536 and 1514 cm^{-1} (vs, N-H & C-N); 832 cm^{-1} (m, *p*-disubst. aromatic); 761 & 701 cm^{-1} (m, monosubst. aromatic).

Preparation of p-Hydroxystyrene: An aliquot of 17.85 g methyltriphenylphosphonium bromide in 10 ml THF was combined with 12.01 g potassium t-butoxide at room temperature under N_2. Next 6.10 g of *p*-hydroxygenzaldehyde was added while maintaining the reaction temperature below 25°C. After one hour of stirring the reaction mixture was poured into 50 ml of ice water and extracted with ethyl acetate. The aqueous phase was acidified to pH 4 and re-extracted with ethyl acetate. Evaporation of the solvent gave 18.60 g of a liquid residue. A silica gel column using a 6:4 hexane/ethyl acetate mixture as eluting solvent gave 5.02 g (83.6%) of pure *p*-hydroxystyrene having a m. p. of 71-73°C (lit. 69-70°C (*30*).
FT-IR (KBr pellet): 3383 cm^{-1} (s, O-H); 1629 cm^{-1} (m, C=C vinylic); 1261 & 1255 cm^{-1} (s, C-O alcohol). **^{1}H-NMR** ($CDCl_3$): 5.10 (1H, s, -OH), 6.86 (2H, d, aromatic, J=8.0 Hz), 7.35 (2H, d, aromatic, J=8.0 Hz), AMX pattern for vinyl protons with $_A = 5.16$, $_M = 5.62$, $_X = 6.66$ ppm, with $J_{AM} = 0.75$ Hz, $J_{AX} = 11.0$ Hz and $J_{MX} = 18.0$ Hz.
^{13}C-NMR ($CDCl_3$): 110.79 (6, $-CH_2$), 116.30 (116.0, d, aromatic C3), 128.37 (127.9, d, aromatic C2), 130.71 (130.7, s, aromatic C1), 137.74 (d, -CH), 158.37 (154.5, t, aromatic C4).
M. S.: m/e 120 (M^+, base peak), 27 ($CH_2 = CH^+$).

Preparation of p-Formyloxystyrene: An aliquot of 4.30 g of formic acid was combined with 9.54 g acetic anhydride (*29*). After cooling the mixture to room temperature 5.20 g of *p*-hydroxystyrene and 0.15 g pyridine were added. The mixture was stirred in the dark for 3 days. A TLC obtained of the crude reaction mixture using an 8:2 hexane/ethyl acetate mixture indicated the reaction had gone to completion. Removal of the acids and anhydrides at low temperature under vacuum left a yellow residue which upon fractional distillation gave 6.39 g (66.5%) of *p*-formyloxystyrene (b. p. 83 °C at 3 mm Hg).
FT-IR (NaCl, neat): 1763 & 1739 cm^{-1} (vs, C=O ester); 1631 cm^{-1} (m, C=C vinylic); 1298, 1194, 1170 & 1105 cm^{-1} (vs, C-O ester).
^{1}H-NMR ($CDCl_3$): 6.98 (2H, d, aromatic, J=8.6 Hz), 7.35 (2H, d, aromatic, J=8.6 Hz), 8.17 (1H, s, -OCHO), AMX pattern for vinyl protons with $_A = 5.17$, $_M = 5.63$, $_X = 6.55$ ppm with $J_{AM} = 1.0$ Hz, $J_{AX} = 10.6$ Hz & $J_{MX} = 17.0$ Hz.
M. S.: m/e 148 (M^+), 120 (base peak, C_8H_7OH), 29 (HCO^+), 27 ($CH_2{=}CH^+$).

Preparation of Poly(p-formyloxystyrene) (IV): A sample of 6.11 g *p*-formyloxystyrene and 0.0652 g AIBN were combined in a 25 ml round-bottomed flask equipped with an air cooled condenser and N_2 bubbler, and heated at 80 °C overnight. The resulting polymeric mass was dissolved in THF and precipitated in petroleum ether. Repeated washing, followed by drying of the polymer overnight in a vacuum oven at room temperature yielded 4.85 g (79%) of poly(*p*-formyloxystyrene) ($M_n = 1.09 \times 10^4$, $M_z = 1.65 \times 10^4$, polydispersity = 1.51 (gpc)).
FT-IR (KBr pellet): 1761 & 1738 cm^{-1} (vs, C=O ester); 1205, 1169 & 1107 cm^{-1} (vs, C-O ester).
UV Spectrum (1 μm film): 241 nm (cut-off), 266 nm (A = 0.39), 272 nm (A = 0.36), 286 nm (A = 0.12).
^{1}H-NMR ($CDCl_3$): 0.91-2.24 (3H, bm, allylic H's), 6.23-7.24 (4H, bm, aromatic), 8.22 (1H, s, -OCHO).
^{13}C-NMR ($CDCl_3$): 40.0-49.5 (bm, $-CH_2$), 40.21 (d, -CH), 121.82 (122.6, d, aromatic C3), 128.64 (127.0, d, aromatic C2), 142.92 (146.7, s, aromatic C1), 148.01 (149.1, s, aromatic C4), 159.44 (s, -COO-).

Solution Photolysis of Poly(p-acetoxystyrene): A 2.576 g sample of **III** was dissolved in 350 ml of acetonitrile and placed in a 500 ml Hanovia reactor equipped with a water-cooled quartz finger and 500 watt mercury arc lamp, condenser, nitrogen inlet and bubbler. The solution was irradiated for a total of 39.5 hours with hourly aliquot samplings. The aliquots were concentrated on the flash evaporator. The residues were dissolved in CH_2Cl_2, and films cast on NaCl disks. The wafers were placed in a vacuum oven at room temperature for 30 minutes to remove trace solvent. FT-IR spectra of these aliquots were used to monitor the progress of the reaction. After 39.5 hours, the IR data indicated the photorearrangement had gone to about 50% completion. The acetonitrile was removed on the flash evaporator, and the yellow syrupy residue was dissolved in CH_2Cl_2. Precipitation in methanol afforded a light yellow powder.

The precipitate was filtered, washed with additional methanol and dried overnight in the vacuum oven at 40°C to give 1.82 g of polymer (71% recovery). ^{1}H-NMR spectral analysis of the product confirmed that it contained both *p*-acetoxystyrene and 2-hydroxy-5-vinylacetophenone repeating units in a molar ratio of 0.52 : 0.48. The polymer had $M_n = 4.50 \times 10^4$, $M_z = 3.62 \times 10^5$, polydispersity = 8.06 (gpc). The spectral data which is listed below corresponds to the product after 39.5 hrs. of irradiation.
FT-IR (NaCl, films): 3483 cm^{-1} (m, -OH); 1763 cm^{-1} (s, C=O ester); 1642 cm^{-1} (s, C=O ketone); 1615 cm^{-1} (m, C=C aromatic); 1254 cm^{-1} (s, C-O phenol); 1217 & 1201 cm^{-1} (vs, C-O ester); 912 & 850 cm^{-1} (m, C-H *p*-disubst. aromatic); 874 & 832 cm^{-1} (m, C-H 1,2,4-trisubst. aromatic).
^{1}H-NMR ($CDCl_3$): 0.97-1.77 (3H, bm, allylic H's), 2.20 (1.57H, s, $-CH_3$ ester), 2.31 (1.43H, s, $-CH_3$ ketone), 6.17-7.03 (3.75H, bm, aromatic).
^{13}C-NMR ($CDCl_3$): 21.60 (q, $-CH_3$ ester), 26.53 ($-CH_3$ ketone), 40.43 (d, -CH), 40.0-48.50 (m, $-CH_2$, 118.34 (115.8, d, aromatic C5), 122.06 (122.6, d, aromatic C3), 122.17 (124.9, s, aromatic C3), 129,32 (127.0, d, aromatic C2), 129.75 (129.6, d, aromatic C2), 135.44 (133.7, d, aromatic C6), 142.23 (146.7, s, aromatic C1), 148.75 (149.1, s, aromatic C4), 160.51 (s, C-OH), 169.30 (s. C=O ester), 204.62 (s, C=O ketone).

Solid-State Photolysis of Poly(p-Acetoxystyrene) (III): A 35% solution of **III** in diglyme was prepared and 1.10 μm thick films on NaCl disks and quartz wafers were obtained by spin-coating at 2000 rpm for 30 seconds. Trace solvent was removed from the films by prebaking at 100°C for 30 minutes. The films of **III** were exposed in air using an unfiltered 500 watt OAI deep UV source, with the output measured using a 254 nm probe. The quartz wafers were exposed to doses ranging from 0 to 18.6 J/cm^2 with UV spectra recorded at regular intervals. A parallel FT-IR study was also carried out for doses between 0 and 2.94 J/cm^2. Spectral data confirmed the formation of a *p*-acetoxystyrene 2-hydroxy-5-vinylacetophenone copolymer. Quantitative FT-IR interpretation was used to evaluate the relative proportions of the starting and photorearranged polymer units as a function of exposure dose.
FT-IR (NaCl, 1.26 μm film, dose = 2.94 J/cm^2): 3483 cm^{-1} (w, -OH); 1764 cm^{-1} (s, C=O ester); 1643 cm^{-1} (m, C=O ketone); 1615 cm^{-1} (w, C=C aromatic); 1254 cm^{-1} (w, C-O phenol); 1217 & 1201 cm^{-1} (vs, C-O ester); 912 & 846 cm^{-1} (m, C-H *p*-disubst. aromatic); 836 cm^{-1} (m, C-H 1,2,4-trisubst. aromatic)
UV Spectrum (1.10 μm, dose = 3.10 J/cm^2): 238 nm (cut-off), 256 nm (A = 1.63), 271 nm (A = 0.79), 287 nm (A = 0.95), 340 nm (0.55).

Solution Photolysis of Poly(p-Formyloxystyrene): A 2.0 g sample of poly(*p*-formyloxystyrene) was dissolved in 300 mls of spectral grade acetonitrile, and the solution irradiated in the Hanovia apparatus. Small aliquots of the irradiating polymer solution were withdrawn and analyzed at regular intervals as described above.

Solid-State Photolysis of Poly(p-Formyloxystyrene): A 10% solution of poly(*p*-formyloxystyrene) (**IV**) (M_n 18,000) in diglyme was prepared under

yellow light. The polymer solution was spin-coated onto NaCl disks and quartz wafers so as to obtain 1 μm thick films. The films were prebaked for 30 minutes at 100°C on a hot plate. The films were exposed to unfiltered deep UV doses of 100, 300 and 1500 mJ/cm^2 as measured by the 254 nm probe of the exposure monitor. The progress of the photolysis was monitored by both FT-IR and UV spectrophotometry. The spectral data obtained indicated almost complete converison to poly(*p*-hydroxystyrene) after the 100 mJ/cm^{-1} dose.
FT-IR (NaCl, film): 3414 cm^{-1} (s, O-H); 1727 cm^{-1} (w, C=O ester); 1652 cm^{-1} (w, C=O aldehyde); 1219 & 1170 cm^{-1} (s, C-O phenolic).
UV Spectrum (1 μm film): 241 nm (cut-off), 266 nm (A = 0.52), 276 nm (A = 0.59), 286 nm (A = 0.41).

Lithographic Testing of Poly(p-Acetoxystyrene): A 37% solution of **III** in diglyme gave 0.97 μm thick polymer films on silicon wafers for a spin-speed of 3500 rpm for 30 seconds. After a 30 minute prebake at 100°C, the films of **III** were exposed to a deep UV dose of 3.10 J/cm^2 through a quartz mask using a 500 watt OAI deep UV source, and then postbaked at 100°C for 15 minutes. Pattern development was attempted in various mixtures of 3-heptanone, 2-methoxyethylacetate and N-butyl acetate with little success due to rapid dissolution of the unexposed areas of the film. Dip development for 90 seconds in a 1:1 3-heptanone/isopropanol mixture, followed by an isopropanol wash gave fully developed images, while reducing the unexposed film thickness to 0.48 μm. MF-312 and AZ-400K aqueous alkali developers were also tried but after 3 minutes in these developers extensive film swelling and cracking occurred without any apparent pattern development.

Lithographic Testing of Poly(p-Formyloxystyrene): A 15% solution of **IV** in diglyme was used to coat silicon wafers with a 1.01 μm thick film. After a 30 minute prebake at 100°C, the films were exposed through a contact quartz mask to varying doses of unfiltered deep UV light, measured with the 254 nm probe of the exposure monitor. For a dose of 100 mJ/cm^2, a clean positive image was obtained upon development in a 10:1:1 isopropanol/ammonium hydroxide/water solution for 90 seconds. Negative image development of irradiated films of **IV** was achieved using methylene chloride or anisole-based developers, diluted with hexane. Similar results were obtained using wafers exposed in a Perkin Elmer 500 projection aligner. The characteristic curve for positive tone imaging shown in Figure 11 was obtained by exposing a 1.0 μm film of the resist on silicon wafer to exposure through a multidensity step wedge quartz mask. The exposed resist was developed to clear the highest dose pads. The resist thickness remaining in the lower dose pads was measured using a Talystep profilometer.

Acknowledgments

Partial support of this research by the Natural Sciences and Engineering Research Council of Canada in the form of an operating grant and equipment grant (to JMJF) is gratefully acknowledged. Additional support from the Government of Ontario under the BILD program is also acknowledged.

Finally, the authors wish to thank Mr. N. Clecak for sharing his expertise in the imaging experiments and Mr. W. Rolls for samples of poly(*p*-formyloxystyrene).

Literature Cited

1. Anderson, J. C.; Reese, C. B. *Proc. Chem. Soc.* 1960, 217.
2. Kobsa, H. *J. Org. Chem.* 1962, *27*, 2293. Stenberg, V. I. "Organic Photochemistry"; Chapman, O. L., Ed.; Marcel Dekker: New York, 1967; Vol. 1, pp. 127-154.
3. Maerov, S. B. *J. Polym. Sic., Part A* 1965, *3*, 487.
4. Bellus, D.; Slama, P.; Hrdkovic, P.; Manasek, Z.; Durisinova, L. *J. Polym. Sci., Part C* 1969, *22*, 629.
5. Cohen, S. M.; Young, R. H.; Markhart, A. H. *J. Polym. Sci., Part A-1* 1971, *9*, 3263.
6. Carlsson, D. J.; Gan, L. H.; Wiles, D. M. *J. Polym. Sci. Polym. Chem. Ed.* 1978, *16*, 2353.
7. Allen, N. S. *Polymer Photochemistry* 1983, *3*, 167.
8. Ranby, B.; Rabek, J. F. "Photodegradation, Photo-oxidation and Photostabilization of Polymers"; Willey Press: New York, 1975.
9. Okawara, M.; Tani, S.; Imoto, E. *Kogyo Kagaku Zasshi* 1965, *68*, 223.
10. Li, S-K. L.; Guillet, J. E. *Macromolecules* 1977, *10*, 840.
11. Gilazhov, E. G.; Ivanova, N. P.; Zubko, N. V.; Arbuzova, I. A. *Akad. Nauk. Kaz, SSR, Ser. Khim.* 1978, *28*, 81.
12. Gilazhov, E. G.; Ivanova, N. P.; Zubko, N. V.; Arbuzova, I. A. *Plast. Massy* 1979, *6*, 26.
13. Guillet, J. E.; Merle-Aubry, L.; Holden, D. A.; Merle, Y. *Macromolecules* 1980, *13*, 1138.
14. Corson, B. B.; Heintzelman, W. J.; Schwartzman, L. H.; Tiefenthal, H. E.; Lokken, R. J.; Nickels, J. E.; Atwood, G. R.; Pavlik, F. J. *J. Org. Chem.* 1958, *23*, 544.
15. Fréchet, J. M. J.; Eichler, E.; Ito, H.; Willson, C. G. *Polymer* 1983, *24*, 995.
16. Stevens, W.; Van, A. *Es. Rec. Trav. Chim.* 1965, *84*, 1247.
17. Oki, M.; Nakanishi, H. *Bull. Chem. Soc. Jpn.* 1970, *43*, 2558.
18. Oki, M.; Hirota, M. *Bull. Chem. Soc. Jpn.* 1960, *33*, 119; *Bull. Chem. Soc. Jpn.* 1961, *34*, 374.
19. Sumrell, G.; Campbell, P. G.; Ham, G. E.; Schramm, C. H. *J. Am. Chem. Soc.* 1959, *81*, 4310.
20. Patai, S.; Bentov, M.; Reichman, M. E. *J. Am. Chem. Soc.* 1952, *74*, 845.
21. Bellus, D.; Hrdlovic, P. *Chem. Rev.* 1967, *67*, 599.
22. Kalmus, C. E.; Hercules, D. M. *J. Am. Chem. Soc.* 1974, *96*, 449.
23. Hiraoka, H.; Pacansky, J. *J. Electrochem. Soc.* 1981, *128*, 2645.
24. Horspool, W. M.; Pauson, P. L. *J. Chem. Soc.* 1965, 5162.
25. Hiraoka, H. *IBM J. Res. Develop.* 1977, *21*, 121.
26. Willson, C. G. In "Introduction to Microlithography"; Thompson, L. F.; Willson, C. G.; Bowden, M. J., Eds.; ACS SYMPOSIUM SERIES No. 219, Washington, D.C., 1983, p. 88.

27. Broos, R.; Anteunis, M. *Synth. Comm.* 1976, *6*, 53.
28. Boyer, J. H.; Alul, H. *J. Am. Chem. Soc.* 1946, *68*, 2136.
29. Sofuku, S.; Muramatsu, I.; Hagitani, A. *Bull. Chem. Soc. Jpn.* 1967, *40*, 2942.
30. Havinga, E.; De Jongh, R. O.; Dorst, W. *Rec. Trav. Chim. Pays-Bas* 1956, *75*, 378.
31. Tessier, T. G. M.Sc. Thesis, University of Ottawa, Canada, 1984.

RECEIVED October 8, 1984

14

Soluble Polysilane Derivatives: Interesting New Radiation-Sensitive Polymers

R. D. MILLER, D. HOFER, D. R. McKEAN and C. G. WILLSON

IBM Research Laboratory, San Jose, CA 95193

R. WEST and P. T. TREFONAS III

Department of Chemistry, University of Wisconsin
Madison, WI 53706

Organopolysilanes (i.e., high molecular weight polymers which contain only silicon in the backbone) are old materials which have renewed interest because of improvements in synthetic and characterization techniques. The first reported aromatic organosilane polymers were described by Kipping in 1924 (*1*). Twenty-five years later, Burhard reported the preparation of permethylated polysilane (*2*). These materials were, however, highly crystalline, insoluble white solids which evoked little scientific interest until recently when it was discovered that silane polymers could be used as thermal precursors to β-silicon carbide fibers (*3-5*). In this regard, Yajima and co-workers reported that poly(dimethyl)silane could be converted by the two-step process shown below to β-silicon carbide, a structural material of considerable industrial importance.

$$Me_2SiCl_2 \xrightarrow{Na} (Me_2Si)_n \xrightarrow[Ar]{320^\circ C} \left[\begin{matrix} H & & H & \\ | & & | & \\ Si & -CH_2- & Si & -CH_2 \\ | & & | & \\ CH_3 & & CH_3 & \end{matrix} \right]_n$$

$$\xrightarrow[air]{200^\circ C} \text{surface oxidation} \xrightarrow{1300^\circ C} \beta\text{–SiC} \qquad (1)$$

Although the chemistry of the initial thermal transformation is obviously quite complex, it was determined that considerable carbon insertion into the Si-Si bonds occurs, resulting in an intermediate carbosilane which can be drawn into fibers. At this point, brief oxidation results in the formation of a surface oxide which imparts dimensional stability, and subsequent heating to 1300°C produces silicon carbide fibers (*3,5*).

The observation that more soluble, less crystalline materials containing larger alkyl substituents (*6*) or aromatic groups (*7,8*) could be synthesized suggested additional applications for polysilane derivatives. In this regard, West and co-workers have reported that a soluble copolymer *1* produced by the

0097–6156/84/0266–0293$06.00/0

co-condensation of dimethyl and methyl phenyl dichlorosilane (*7*) could be used as an impregnating agent for ceramic materials (*9,10*). In addition, they found that the same copolymer could be doped to a semiconducting level by treatment with oxidizing agents such as arsenic pentafluoride (*7*). We were intrigued by the observation that *1*, unlike the related carbon polymer polystyrene, absorbs *strongly* from 200-350 nm. Furthermore, this material was described as photosensitive, and irradiation produced insoluble material which indicated that cross-linking had occurred (*7*).

The unique spectral properties of *1*, coupled with the reported photosensitivity, suggested that soluble polysilane derivatives might comprise a new class of radiation sensitive polymers with lithographic potential. In particular, it seemed that materials of this type, which have a relatively high Si/C ratio and can be imaged in an efficient manner, could be useful in dry etch image transfer processes utilizing oxygen reactive ion etching (O_2-RIE) (*11*). In addition, the strong adsorbance from 200 to 350 nm suggested multilayer lithographic applications which are currently popular as a technique for generating high resolution images where the dimensions of chip topography approach the imaged feature sizes (*12*).

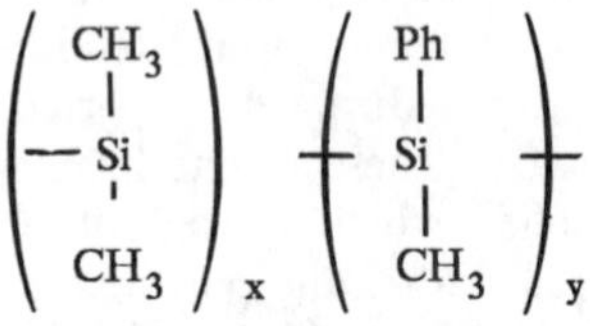

$$\left(-\underset{CH_3}{\overset{CH_3}{\underset{|}{\overset{|}{Si}}}} \right)_x \left(-\underset{CH_3}{\overset{Ph}{\underset{|}{\overset{|}{Si}}}}- \right)_y$$

1

Synthesis and Characterization

We have recently synthesized and characterized a variety of soluble, substituted polysilane homopolymers by the condensation of appropriately substituted methylsilyl dichlorides with sodium dispersion as shown below and in Table I (*13*).

$$RMeSiCl_2 \xrightarrow[\text{Toluene} \; \Delta]{2Na(5\% \; xs)} \left[-\underset{R}{\overset{Me}{\underset{|}{\overset{|}{Si}}}}- \right]_n + 2NaCl \qquad (2)$$

2 R = Ph	*6* R = n-Butyl
3 R = p-Tolyl	*7* R = n-Hexyl
4 R = β-Phenethyl	*8* R = n-Dodecyl
5 R = n-Propyl	*9* R = Cyclohexyl

Table I. Yields and Molecular Weights (Relative to Polystyrene) for Both Fractions of the Bimodal Distribution from Gel Permeation Chromatography in THF of Organosilane Polymers.

Polymer[c]	M_n $\times 10^{-3}$	M_w $\times 10^{-3}$	M_z $\times 10^{-3}$	M_w/M_n	R[a]	% Yield
(*2*)	107 3.3	193 5.6	313 9.9	1.81 1.69	0.72	55
(*3*)	66 4.7	213 5.9	421 7.0	3.23 1.26	0.56	25
(*4*)	134 3.3	286 4.4	489 6.2	2.13 1.36	2.30	35
(*5*)	297 7.4	644 13.3	1160 22.1	2.17 1.79	0.27	32
(*6*)	50 4.4	110 5.9	218 7.6	2.19 1.36	1.50	34
(*7*)	281 14.6	524 20.5	811 27.2	1.86 1.40	2.40	11
(*8*)	172	483	881	2.81	b	8
(*9*)	300 3.2	804 4.5	1419 5.9	2.67 1.40	8.7	10

a. R is the ratio of the high MW fraction of the bimodal distribution to the low MW fraction from the GPC elution profile.

b. This polymer displayed a broad, monomodal distribution.

c. See Equation 2 for specific structure.

The polymers were produced in yields which ranged from 8-55%. It was observed in most cases that the highest molecular weights were obtained when the sodium dispersion was added to the monomer in toluene (4:1 toluene/monomer) in what we term "inverse addition." Normal addition (i.e., addition of the monomer to the sodium) usually resulted in slightly improved yields of lower molecular weight material. In either case, it is advantageous to avoid a large excess of sodium in the reaction mixture, since this usually results in the formation of small amounts of toluene insoluble gels which are difficult to filter, thus complicating the polymer purification. Most of the polymers isolated using the "inverse addition" technique show distinct bimodal molecular weight distributions (see Table I). In this regard, *8* is anomalous, as the material isolated in the usual fashion has a broad "monomodal distribution." All of the polymers are soluble in toluene, although *9* was considerably more soluble in a 3:1 mixture of toluene-ethylcyclohexane. The polymers containing an aromatic substituent (*2-4*) as well as the cyclohexyl derivative *9* are hard, brittle, high melting solids when completely solvent free. The other materials are generally soft, sticky or elastomeric. The spectral and analytical data for these polymers are as expected and are described in detail in Reference 13. The hard brittle materials *2-4* and *9* have glass transition temperatures in excess of 75°C as measured by differential scanning calorimetry (DSC) or by thermomechanical analysis (TMA). Thermal gravimetric analysis (TGA) shows that the polysilanes, in general, are quite stable in a nitrogen atmosphere and suffer little weight loss at temperatures below 325°C.

Examination of Table I reveals another interesting feature of the anionic polymerization process. As the steric bulk of the substituents in the monomer increases, the isolated yields of high polymer drop precipitously. The remaining material is, in such cases, isolated as a mixture of low molecular weight materials including cyclic oligomers. We felt that the poor yields of high polymer might be due to a slow propagation reaction between the sterically hindered growing anionic chain and the tertiary silyl electrophile, a feature that would be exacerbated by the presence of a nonpolar solvent such as toluene. To test this hypothesis, the polymerization medium was modified by the incorporation of varying quantities of diethyleneglycol dimethyl ether (diglyme) in hope of facilitating the propagation step. The results of this procedure for the preparation of *9* are shown in Table II. The addition of diglyme (~25%) results in an approximately three-fold increase in the yield of high polymer. At the same time, this modification allows the control of molecular weight, since the increased solvent polarity results in an early precipitation of polymer. The successful solvent modification for the preparation of *9* suggests possible utility for the generation of polymers from very sterically hindered monomers.

Spectral Properties

Examination of the absorption spectra of the new polysilane materials reveals a number of interesting features (*14*). As shown in Table III, simple alkyl substituted polymers show absorption maxima around 300-310 nm. Aryl substitution directly on the silicon backbone, however, results in a strong bathochromic shift to 335-345 nm. It is noteworthy that *4*, which has a pendant aromatic side group that is buffered from the backbone by a saturated spacer atom, absorbs in the same region as the peralkyl derivatives. This red shift for the silane polymers with aromatic substituents directly bonded to the backbone is reminiscent of a similar observation for phenyl substituted and terminate silicon catenates relative to the corresponding permethyl derivatives

Table II. Solvent Effect on Yields of Poly(Cyclohexyl Methyl) Silane.

$$\text{MeSiCl}_2(\text{C}_6\text{H}_{11}) + \text{Na} \xrightarrow{\Delta} \left(-\text{Si(Me)(C}_6\text{H}_{11})-\right)_x$$

Entry	Na/Si (molar ratio)	Solvent	% Yield	$M_w \times 10^{-3}$	$M_n \times 10^{-3}$	M_w/M_n
1	2.05	b	14	840[a]	300	2.67
				4.5	3.2	1.40
2	2.0	b	11	1.101[a]	486	2.27
				6.1	4.6	1.33
3	2.0	c	30	23.1	11.1	2.08
4	2.0	d	31	16.5	6.04	2.73

a. Bimodal Distribution

b. Toluene

c. 25% Diglyme
75% Toluene

d. 75% Diglyme
25% Toluene

(*15*). In the model compounds, this red shift has been ascribed to a combination of σ-π mixing of the HOMO of the silicon backbone with the π-orbitals of the aromatic substituent coupled with a decrease in the LUMO energy due to π^*-(σ^*, *d*) interactions (*15,16*). Further examination of the data in Table III shows that the absorption maximum of the cyclohexylmethyl derivative, *9*, is also somewhat red-shifted relative to the other alkyl polymers suggesting that the steric bulk of the substituents and/or conformational effects may also influence the polysilane absorption spectrum.

Earlier work has shown that materials with catenated Si-Si bonds and only alkyl substituents absorbed strongly from 200-300 nm (*16,17*). This transition has been described as either a σ-σ^* or a σ-$3d\pi_{\text{Si-Si}}$ transition which rapidly shifts to longer wavelengths with increased catenation until it

Table III. Ultraviolet Absorption Properties of High Molecular Weight Poly(organosilanes).

Polymer	λ_{max} [a]	M_2 [b]
$[PhMeSi]_n$ (*2*)	341	193,000
[(p-toly) (Me) Si]$_n$ (**3**)	337	75,000
[(β-phenethyl) (Me) Si]$_n$ (*4*)	303	286,000
[(n-Pr) (Me) Si]$_n$ (*5*)	306	644,000
[(n-Bu) (Me) Si]$_n$ (*6*)	304	110,000
[(n-Hex) (Me) Si]$_n$ (*7*)	306	524,000
[(n-dodecyl) (Me) Si]$_n$ (*8*)	309	483,000
[(Cyc) (Me) Si]$_n$ (*9*)	326	804,000

a. UV of *2-8* was measured in spectrograde THF.
UV of *9* was measured in cyclohexane.

b. M_w determined by GPC and are relative to polystyrene.

approaches a limiting value in the region of 20-24 monomer units (*18*). We have extended the range of this study using photodegraded samples of *8* (*vide infra*) of known molecular weight and the results are shown in Figure 1. Since the λ_{max} of polysilane derivatives approaches a limiting value at a relatively low degree of polymerization, the spectral characteristics of high polymers of different molecular weights can be directly compared with relatively little error. Pitt and co-workers have discussed the theoretical aspects of this spectral leveling phenomenon (*19*).

Figure 2 displays a plot of the molar absorptivities per Si-Si bond *versus* the chain length for two typical polysilanes, *2* and *8*. In a manner similar to that observed for the absorption maxima, the molar absorptivities of both polysilanes increase rapidly with molecular weight, but ultimately approach a limiting value. It is interesting that not only does *2* absorb at longer wavelengths than *8*, but it is also more strongly absorbing. The behavior of the molar absorptivities with molecular weight is somewhat unexpected based on a related study involving linear permethyl silanes $Me(Me_2Si)_nMe$, $n \leqslant 24$ which indicated that although the total absorptivity increased with n, the absorptivity/Si-Si actually decreased (*17,18*).

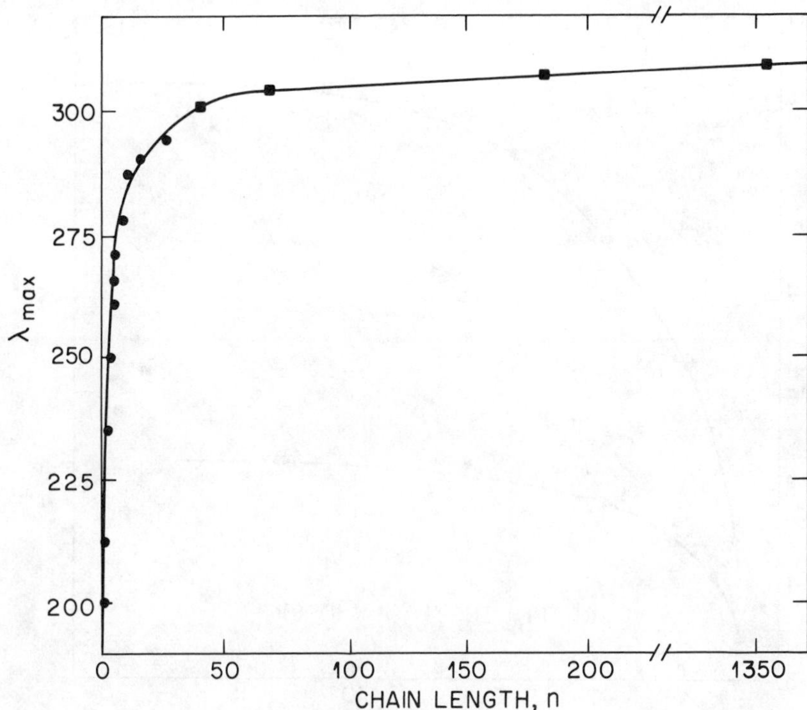

Figure 1. Plot of UV absorption maxima versus chain length(n) for poly(alkylsilane). The (●) represent $Me(Me_2Si)_n$ and the (■) represent $[(\underline{n}\text{-dodecyl})(Me)Si]_n$.

Photochemistry

The dependence of both the λ_{max} and the molar absorptivity on the degree of polymerization has interesting consequences for any radiation induced process which significantly lowers the molecular weight. This is dramatically demonstrated in Figure 3 which shows the effect of continued photolysis on the absorption spectrum of a solid film of *2*. The continuous shift in the absorption maximum to shorter wavelengths and the attending decrease in the peak intensity strongly suggests that the polymer is undergoing a significant reduction in molecular weight. This fact was confirmed by GPC examination (micro-styragel column, polystyrene standards) of THF solutions of the photodegraded polymer films and is particularly interesting in light of the report that the copolymer *1* undergoes extensive cross-linking upon exposure, particularly in the solid-state (*7*).

In an effort to quantify the effect of photolysis, polysilanes *2* and *8* were selected as typical models. For this experiment, samples of *2* were fractionated by repeated precipitation from toluene using isopropanol. No additional

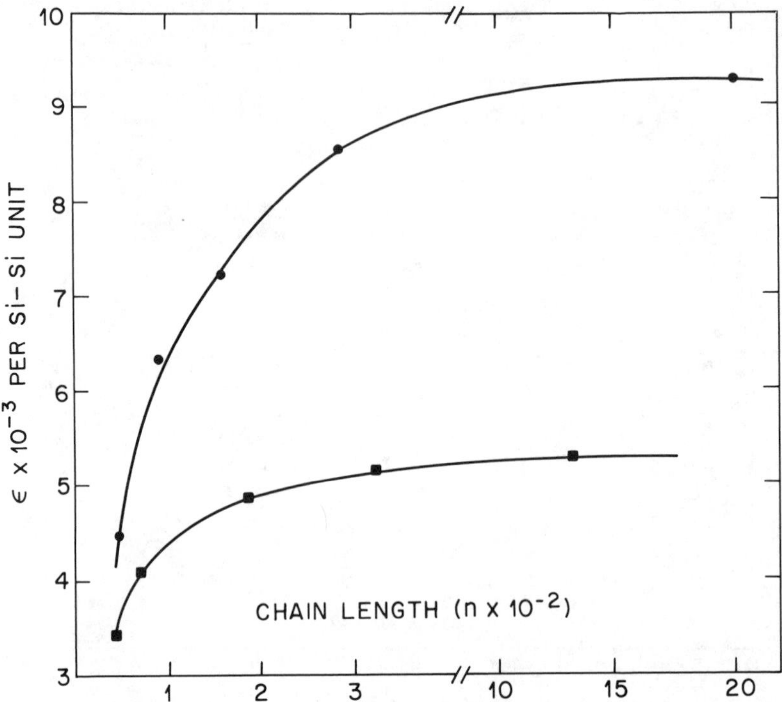

Figure 2. Plot of absorptivity per Si-Si bond at M_{max} versus chain length n. The circles represent $[PhMeSi]_n$ and the squares represent $[(\underline{n}\text{-dodecyl})(Me)Si]_n$.

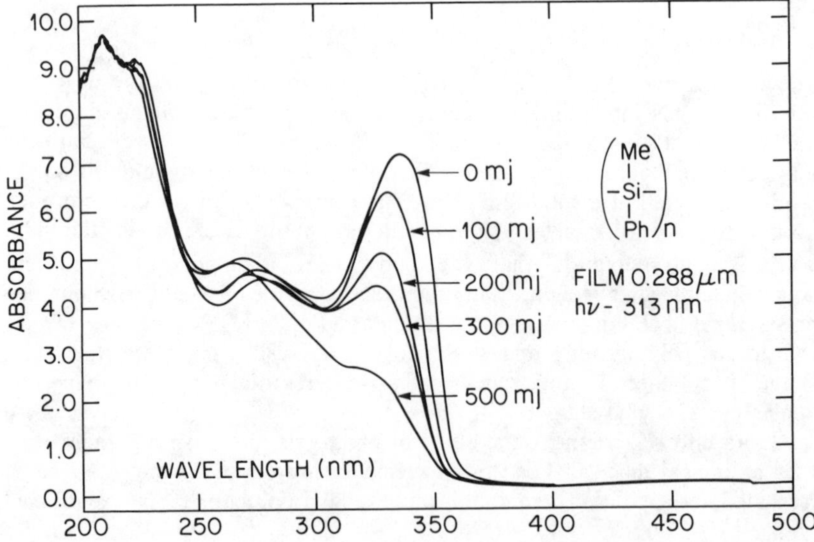

Figure 3. Photobleaching of a pristene film of poly(methylphenyl) silane (313 nm).

fractionation of *8* was necessary, since it was generated as a monomodal distribution in the synthesis (see Table I). Solutions of *2* and *8* (0.005% by weight) were prepared in spectrograde THF. At the same time, *2* was also investigated in the solid-state by the irradiation of films (0.43 μm) spun on 3" silicon wafers. In the latter case, after each irradiation the polymer was dissolved from the wafer and analyzed by GPC analysis. All samples were irradiated at 313 nm and the changes in molecular weight were recorded as a function of dose. The inverse molecular weights were then plotted against the absorbed dose to produce linear plots (see Figure 4 for a representative example). Measurement of the slopes of these plots allowed the calculation of the quantum efficiencies for both scission ($\phi(s)$) and cross-linking ($\phi(x)$) by the application of Equations 3 and 4 (*20,21*) where N_A is Avogadro's number.

$$\frac{1}{M_n} = \frac{1}{M_n^0} + \phi(s) - \phi(x)\,\frac{D}{N_A} \tag{3}$$

$$\frac{1}{M_w} = \frac{1}{M_w^0} + \frac{\phi(s) - 4\phi(x)}{2}\,\frac{D}{N_A} \tag{4}$$

M_n^0 = Initial number average molecular weight
M_w^0 = Initial weight average molecular weight
D = Dose
N_A = Avogadro's number

The results of these measurements are shown below. As expected, polymer *8* undergoes only scission to lower molecular weight fragments upon exposure. On the other hand, *2*, which has pendant unsaturation, shows a significant cross-linking component both in solution and in the solid-state. In

Summary of Photodegradation Experiments on Polysilane Samples at 313 nm.

Photodegradation Experiment	$\phi(s)$ - $\phi(x)$	$\phi(s)$	$\phi(x)$	$\frac{\phi(s)}{\phi(x)}$
0.005% soln of *8*[a]	0.20[b]	0.20[b]	0.00[b]	∞
0.005% soln of *2*[a]	0.85[c]	0.97[d]	0.12[g]	7.9[h]
0.43 μ film of *2*	0.013[e]	0.017[f]	0.0036[e]	4/7[i]

a. Spectrograde THF was used as the solvent.
b. ±0.01
c. ±0.02
d. ±0.15
e. ±0.001
f. ±0.003
g. ±0.03
h. ±1.0
i. ±0.5

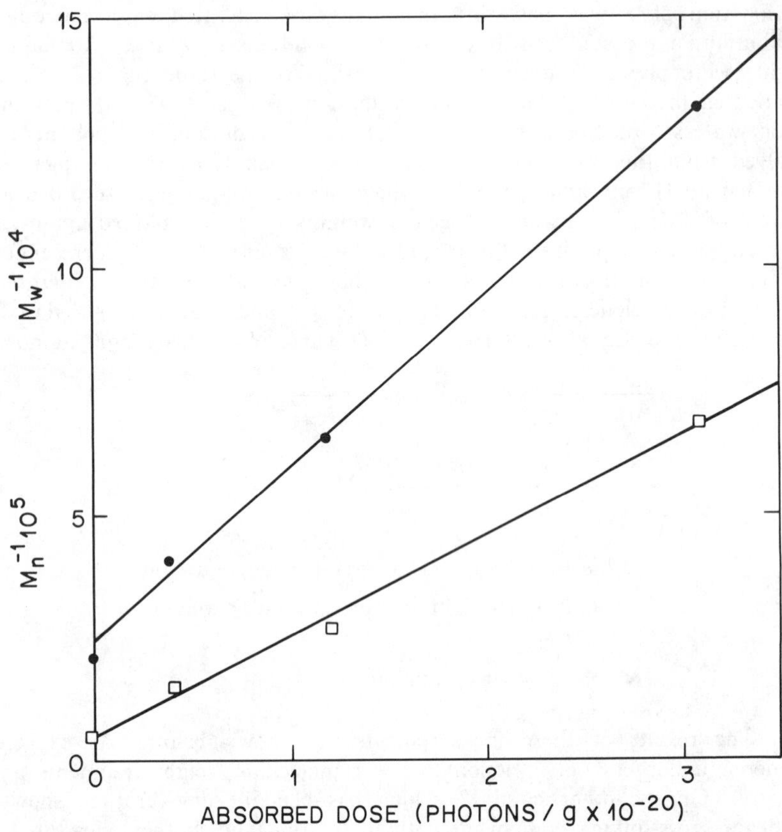

Figure 4. Plot of (●) M_n^{-1} and (□) M_w^{-1} versus absorbed radiation dose at 313 nm for a 0.006% solution of [(n-dodecyl)(Me)Si]$_n$ in THF.

solution, *2* appears to be approximately four to five times as sensitive as *8*. Obviously, the presence of aromatic substituents not only alters the absorption spectrum, but also effects the photolability of the polysilane. The photoefficiency for both scissioning and cross-linking drops precipitiously in the solid-state presumably due to chain repair that is facilitated by solid-state cage effects which greatly impede polymer motion (*22*). Similarly, the ratio $\phi(s)/\phi(x)$ decreases markedly in the solid-state due to the restricted mobility and the lack of hydrogen donor sources. The detection of a significant cross-linking component for *2* both in solution and in the solid-state is consistent with the previous qualitative observations on the copolymer *1*.

The mechanism of the photochemical degradation of catenated silicon derivatives has received considerable attention (*23*). Substituted cyclic derivatives photochemically extrude a silylene fragment which can be intercepted by appropriate trapping reagents (e.g., trialkylsilanes or 2,3-dimethyl butadiene). This extrusion results in the formation of the corresponding ring contracted cyclopolysilane. The process continues upon additional irradiation until a cyclotetrasilane results which then undergoes radical ring opening. A similar extrusion process has also been reported for acylic polysilane derivatives (*23*). In the latter case, the isolation of silyl hydrides with fewer silicon atoms than the starting materials is taken as evidence for the presence of linear radical intermediates.

A related process could be evoked to explain the facile photodegradation of the polysilane high polymers (see Figure 5). According to this scheme, light absorption by the polymer would result in the formation of a disubstituted silylene fragment leaving behind the two radical ends of the scissioned polymer. These radicals could then abstract hydrogen from an *in situ* source, undergo further photodegradation, thermally extrude additional silylene fragments or recombine.

The key intermediates in the process are the silylenes and the radical fragments, both of which should be amenable to trapping experiments. With this in mind, two representative, polysilane derivatives *6* and *9*, were irradiated (254 nm) in the presence of excess triethylsilane, and the results of this experiment are shown in Figure 6. Since the polysilane *9* was not soluble in the trapping reagent, it was dissolved in cyclohexane which contained a 100 fold molar excess of triethylsilane. In each case, the major isolable volatile products (*24*) were the expected insertion adducts of triethylsilane and the corresponding silylene intermediate (*10,13*). The isolation of the disilanes *11* and *14* suggests a possible radical abstraction process involving polymer chain ends. The disilanes, which are weakly absorbing at 254 nm, are expected to accumulate in

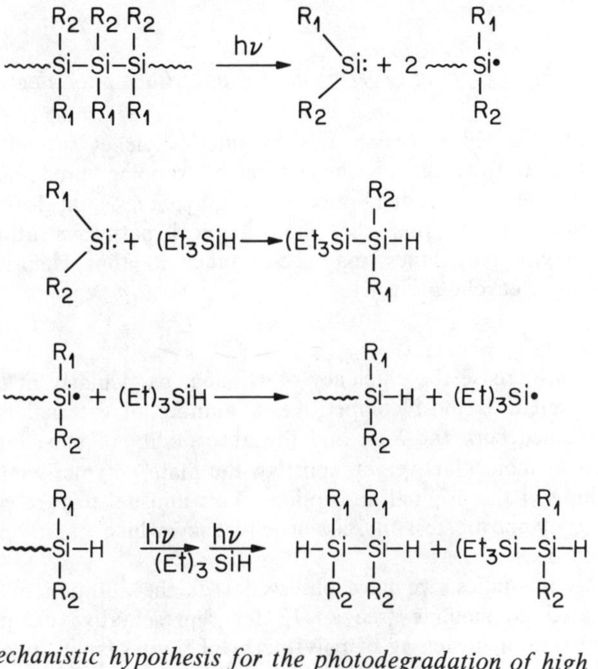

$$\sim\!\mathrm{Si}(R_2)(R_1)\!-\!\mathrm{Si}(R_2)(R_1)\!-\!\mathrm{Si}(R_2)(R_1)\!\sim \xrightarrow{h\nu} R_1R_2\mathrm{Si}\!: + 2 \sim\!\mathrm{Si}^{\bullet}(R_1)(R_2)$$

$$R_1R_2\mathrm{Si}\!: + (\mathrm{Et_3SiH} \longrightarrow (\mathrm{Et_3Si}\!-\!\mathrm{Si}(R_2)(R_1)\!-\!\mathrm{H}$$

$$\sim\!\mathrm{Si}^{\bullet}(R_1)(R_2) + (\mathrm{Et})_3\mathrm{SiH} \longrightarrow \sim\!\mathrm{Si}(R_1)(R_2)\!-\!\mathrm{H} + (\mathrm{Et})_3\mathrm{Si}^{\bullet}$$

$$\sim\!\mathrm{Si}(R_1)(R_2)\!-\!\mathrm{H} \xrightarrow{h\nu} \xrightarrow[(\mathrm{Et})_3\,\mathrm{SiH}]{h\nu} \mathrm{H}\!-\!\mathrm{Si}(R_1)(R_2)\!-\!\mathrm{Si}(R_1)(R_2)\!-\!\mathrm{H} + (\mathrm{Et_3Si}\!-\!\mathrm{Si}(R_1)(R_2)\!-\!\mathrm{H}$$

Figure 5. Mechanistic hypothesis for the photodegradation of high molecular weight polysilane derivatives.

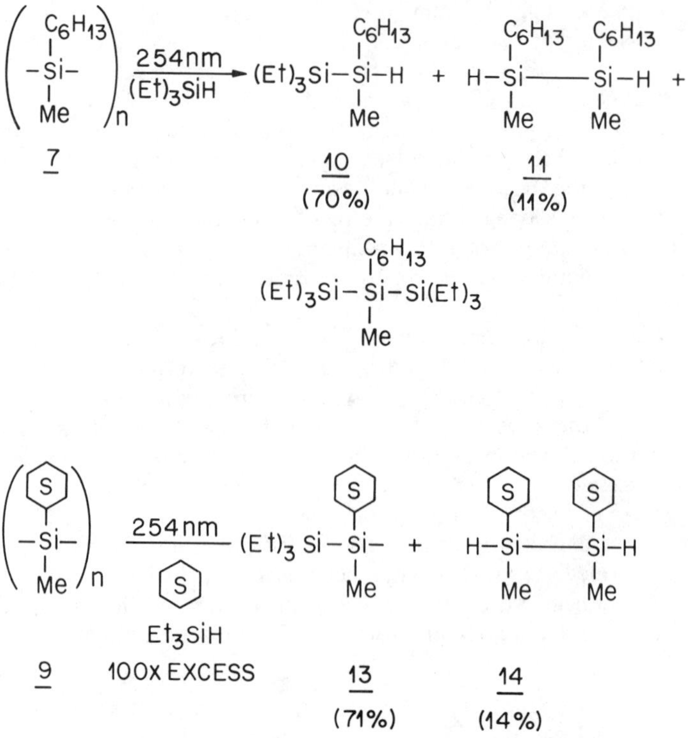

$(-Si(C_6H_{13})(Me)-)_n$ **7** $\xrightarrow[(Et)_3SiH]{254\,nm}$ $(Et)_3Si-Si(C_6H_{13})(Me)-H$ **10** (70%) + $H-Si(C_6H_{13})(Me)-Si(C_6H_{13})(Me)-H$ **11** (11%) + $(Et)_3Si-Si(C_6H_{13})(Me)-Si(Et)_3$

$(-Si(S)(Me)-)_n$ **9** $\xrightarrow[\text{S},\ Et_3SiH\ 100\text{X EXCESS}]{254\,nm}$ $(Et)_3Si-Si(S)(Me)-$ **13** (71%) + $H-Si(S)(Me)-Si(S)(Me)-H$ **14** (14%)

Figure 6. Trapping experiments for polysilane photodegradation.

solution after repeated extrusion of substituted silylenes from the photolabile fragments (*25*). Isolation of the expected trapping products lends some credence to the silylene-radical mechanistic hypothesis, at least for solution processes, and suggests that the photochemical pathways utilized by high molecular weight polysilanes may be similar to that reported for lower molecular weight acyclic derivatives.

Sensitization

In an effort to increase the efficiency of scission, particularly in the solid-state where chain repair is most competitive, a number of external additives were auditioned. Since both the λ_{max} and the absorptivity of polysilane derivatives are functions of molecular weight, splitting the main polymer chain results in a rapid bleaching of the original absorption. This unusual feature can be used as a qualitative, diagnostic test of scission efficiency which greatly facilitates the testing of additives.

In recent studies, we have observed that the addition of a number of polyhalogenated compounds (see *15-17* for representative examples) greatly increases the rate of bleaching of poly(methylphenyl) silane *2* upon irradiation. This effect is dramatically demonstrated by comparison of Figures 3 and 7. For the example shown in Figure 7, <95% of the incident light (313 nm) is absorbed by the polymer.

CCl_3 / CCl_3 — 15

CCl_3 / N / N / CCl_3 / N / CCl_3 — 16

CCl_3 / N / N / N / CCl_3 / OMe / OMe — 17

Since our recent mechanistic studies on the photochemical degradation of silane high polymers suggested the formation of silylenes and radical fragments, it seemed possible that the role of the additives was simply one of chain transfer, thus preventing the rapid recombination of the polymeric radical fragments. For a number of reasons, we now feel that the mechanism is more complex than simple chain transfer of chlorine atoms from the additives to the polymer radical chains. For example, the incorporation of *17* (which absorbs strongly in the visible) and irradiation at 404 nm such that only the sensitizer absorbs the light, still results in a rapid bleaching of the polymer absorption spectrum. Since energy transfer from the sensitizer to the polymer seems unlikely on energetic grounds, (*26*) some other mode of activating (scissioning) the polymer must be operative in this case. Another inconsistency in the simple chain transfer hypothesis is the observation that the incorporation of polyhaloalkanes into films of saturated alkyl polymers such as *9* not only failed

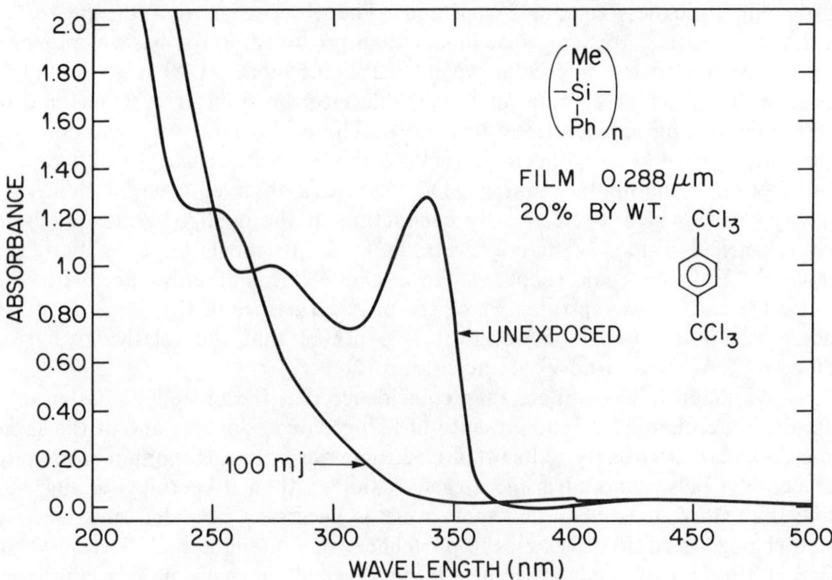

Figure 7. Photobleaching of a film of poly(methylphenyl) silane containing 20 wt.% 1,4-bis-trichloromethylbenzene (313 nm).

to accelerate the photoscission but actually seemed to stabilize the polymer somewhat. The same effect was observed for the phenethyl derivative *4* where the aromatic substituent is buffered from the main chain by pendant methylene groups. Since we have already demonstrated through trapping experiments that silylene fragments are extruded and that the production of radical chain ends upon irradiation of *9*, seems likely it is unlikely that chlorine atoms are abstracted efficiently by the radicals generated from *2* but not by those produced from *4* and *9*.

To resolve this dilema, we propose that the polymer is interacting with the additive in the excited state, (*27*) perhaps *via* electron transfer, and that this interaction leads to the irreversible degradation of the polymer. The direct interaction of photoexcited monomeric polysilanes with halogen derivatives resulting in the cleavage of Si-Si bonds had been reported (*28*). In a similar fashion, we must conclude either that this interaction does not occur with the alkyl silane polymers or that it does not result in rapid polymer degradation.

The suggestion that electron transfer may play a role in the increased photolability of *2* in the presence of halogenated additives is intriguing, as perhalogenated materials are known to be good electron acceptors and catenated silicon chains are easily oxidized (*29*). Since little was known about the oxidation potentials of high molecular weight polysilanes in the solid-state, *2, 4* and *9* were coated (~ 400Å) on platinum electrodes and examined by cyclic voltammetry in acetonitrile. All three samples show a highly irreversible oxidation wave. The oxidation wave for *2* is reproducibly ~0.5V lower (+0.95V *versus* SCE, 50 mV/sec) than for the alkyl substituted polymers *4* and *9*. The oxidation values were reproducible from one sample to the next as long as the film thickness remained constant. The observation that the measured oxidation potentials are somewhat higher than predicted on the basis of previous solution studies on low molecular weight polysilane models (*28*) is attributed to the dependence on sweep rate for irreversible electron transfer reactions and to the nonconducting nature of the films (*30*). The peak shapes and potentials for thin films often differ significantly from solution values because the diffusion of electrolyte into the film is limiting. This feature is observed even for thin films of polypyrrole which is electrically conducting in the oxidized state and is in good electrical contact with the electrode (*31*). In the latter case, both the potentials and the peak shape of the cyclic voltammograms vary with the electrolyte and the sweep rate. Since the physical nature of the films from *2, 4* and *9* was quite similar, however, it is expected that the relative oxidation potentials (i.e., *2* < *4* and *9*) are accurate (*32*).

Although it seems more than coincidence that the photodegradation of *2* is strongly accelerated by the presence of halogenated additives and at the same time *2* exhibits a strongly red-shifted electronic absorption maximum relative to the peralkyl polymers and is more easily oxidized than its peralkyl analogs by more than 0.5V, it is clear that more work is necessary to understand the role of polyhalogenated additives on the photochemistry of silane high polymers. In spite of the lack of understanding of the detailed mechanism, it seems from these preliminary experiments that in some cases appropriate additives can be used not only to increase the efficiency of fragmentation in the solid-state, but also to modify the region of spectral sensitivity.

Lithography

The photochemical studies suggested that polysilane derivatives might have lithographic potential. Since the current trend in lithography is away from classical wet development and toward dry plasma processing for reasons of improved adhesion and resolution as well as for environmental reasons (*33*), we were pleased to note that while the polysilanes are easily etched in fluorocarbon plasmas, (CF_4-O_2) they are, in contrast, quite stable in an oxygen discharge due to the rapid formation of a passivating surface layer of silicon dioxide.

For this reason, we have exercised the silane polymers in two multilayer configurations: (1) as thin oxygen reactive ion etch (O_2-RIE) resistant barriers in trilevel structures and (2) as thin imagable layers for O_2-RIE image transfer into a thick planarizing polymer layer. In the first case, the polysilanes represent easily processable candidates for the replacement of materials such as spin-on-glass and CVD deposited silicon nitride. In this configuration, we have demonstrated utility by the production of 1.25 μm aluminum lines by the use of a standard metal lift-off process employing polysilanes as the O_2-RIE barrier (see Figure 8).

It is, however, the second application, i.e., their use as a thin imagable layer for O_2-RIE image transfer, that is most exciting. In this configuration, a thin film (0.20-0.25) of a photosensitive polysilane cast on a thick planarizing polymer layer can be imaged at 313 nm in a *positive* tone using a commercial 1:1 projection printer at doses as low as 50-100 mJ/cm^2. Transfer of these high resolution images into the underlying planarizing polymer using anisotropic O_2-RIE has produced submicron images with vertical wall profiles (Figure 9 and 10). The bleaching of the polysilane absorption at the irradiation frequency is a key feature in this application, since it allows light to ultimately penetrate to the bottom of the film even though this layer is opaque at the exposure wavelength prior to irradiation. This feature permits the small features to be cleanly developed to the planarizing polymer surface prior to the image transfer step.

Figure 8. Electron micrographs of a trilevel aluminum lift off process employing a typical polysilane as the O_2-RIE barrier. Key: left, electron-beam imaged; and right, optically imaged Mann step and repeat.

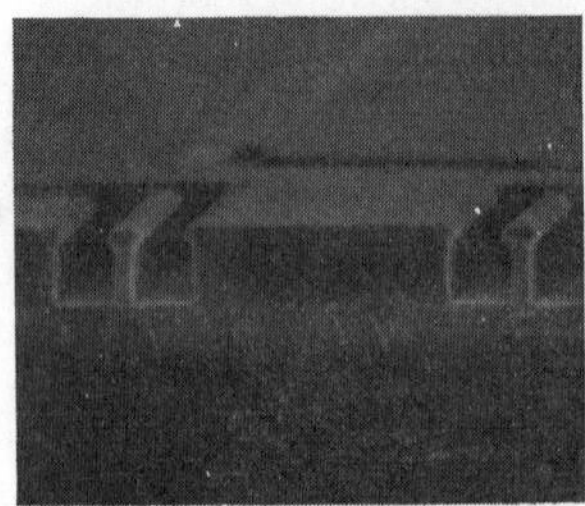

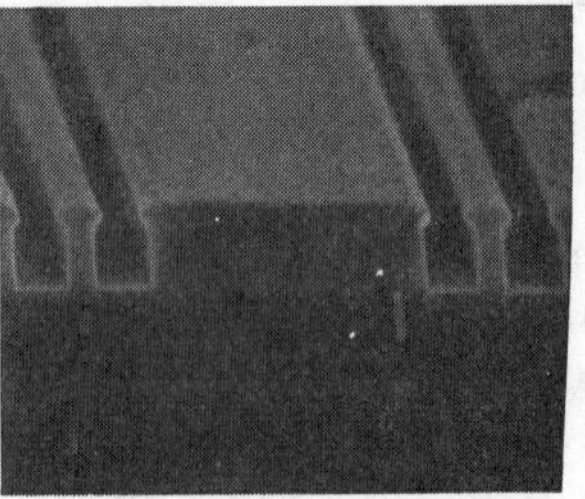

Figure 9. Polysilane bilayer resist, top layer 0.25 μm experimental polysilane, bottom layer 1.0 μm of hardbaked photoresist, projection printed at 313 nm with O_2-RIE image transfer.

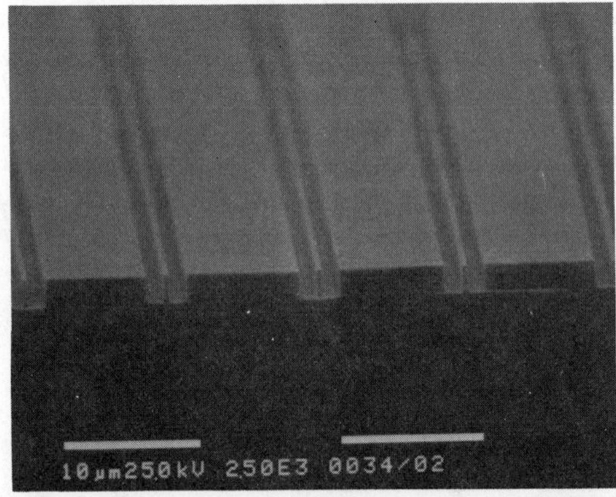

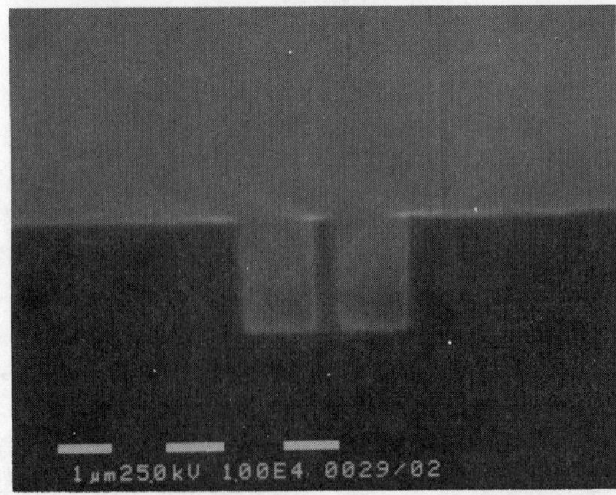

Figure 10. Polysilane bilayer resist, top layer 0.25 μm of an experimental polysilane, bottom layer 1.0 μm of hardbaked photoresist, projection printed at 313 nm with O_2-RIE transfer. The oxidized polysilane layer has been removed with a buffered HF wash followed by a toluene rinse.

It is important to note that the polysilanes represent the *first* class of silicon backbone polymers to provide a O_2-RIE resistant barrier for processing which can be imaged in a *positive* tone using conventional light sources. The use of positive resists, with their inherently higher resolution and transfer of the developed image using anisotropic plasma processes is consistent with the needs of the electronic industry for higher resolution, higher aspect ratio images.

Literature Cited

1. Kipping, F. S. *J. Chem. Soc.* 1924, *125*, 2291.
2. Burkhard, C. A. *J. Am. Chem. Soc.* 1949, *71*, 963.
3. Yajima, S.; Omori, M.; Hayashi, J.; Okamura, K.; Matsuzawa, T.; Liau, L. *Chem. Lett.* 1976, 435.
4. Wession, J. P.; Williams, T. C. *J. Polym. Sci., Polym. Chem. Ed.* 1981, *19*, 65, and references therein.
5. Yajima, S.; Hasegawa, Y.; Hayashi, J.; Iimora, M. *J. Mater. Sci.* 1980, *15*, 720.
6. Wesson, J. P.; Williams, T. C. *J. Polym. Sci., Polym. Chem. Ed.* 1980, *18*, 959.
7. West, R.; David, L. D.; Djurovich, P. I.; Stearley, K. L.; Srinivasan, K. S. V.; Yu, H. *J. Am. Chem. Soc.* 1981, *103*, 7352.
8. Trujillo, R. E. *J. Organomet. Chem.* 1980, *198*, C27.
9. Mazdyasni, K. S.; West, R.; David, L. D. *J. Am. Cer. Soc.* 1978, *61*, 504.
10. West, R.; David, L. D.; Djurovich, P. I.; Yu, H.; Sinclair, X. *J. Am. Chem. Soc.*, in press.
11. Shaw, J.; Hatzakis, M.; Paraszczak, J.; Liutkus, J.; Babich, E. *Polym. Eng. and Sci.* 1983, *23*, 8.
12. Lin, B. J. "Introduction to Microlithography"; Thompson, L. F.; Willson, C. G.; Bowden, M. J., Eds.; ACS SYMPOSIUM SERIES No. 219, American Chemical Society: Washington, D.C., 1983; Chap. 6.
13. Trefonas, P., III; Djurovich, P. I.; Zhang, X-H; West, R.; Miller, R. D.; Hofer, D. *J. Polym. Sci., Polym. Lett. Ed.* 1983, *21*, 819.
14. Trefonas, P., III; West, R.; Miller, R. D.; Hofer, D. *J. Polym. Sci., Polym. Lett. Ed.* 1983, *21*, 823.
15. Pitt, C. F.; Carey, R. N.; Toren, Jr., E. C. *J. Am. Chem. Soc.* 1972, *94*, 3806.
16. Pitt, C. G.; Bursey, M. M.; Rogerson, P. F. *J. Am. Chem. Soc.* 1970, *92*, 519.
17. Kumada, M.; Tomao, K. *Adv. Organometal. Chem.* 1968, *6*, 80.
18. Boberski, W. G.; Allred, A. L. *J. Organometal. Chem.* 1974, *71*, C27.
19. Pitt, C. G.; Jones, L. I.; Ramsey, B. G. *J. Am. Chem. Soc.* 1967, *89*, 5471.
20. Kilb, R. W. *J. Phys. Chem.* 1959, *63*, 1838.
21. Schanabel, W.; Kuvi, J. "Aspects of Degradation and Stabilization of Polymers"; Jellinek, H. H. G., Ed.; Elsevier: New York, 1978; Chap. 4.
22. Mita, I. In "Aspects of Degradation and Stabilization of Polymers"; Jellinek, H. H. G., Ed.; Elsevier: New York, 1978; Chap. 6, p. 255.

23. Ishikawa, M.; Kumada, M. *Adv. Organometal. Chem.* 1981, *19*, 51.
24. Spectral data consistent with all new structures was obtained.
25. One alternative route to the observed disilane derivatives could be *via* the silylenes by hydrogen abstraction and dimerization. This process, although we consider it to be less likely, might be one pathway anticipated if triplet silylenes are generated in the reaction media.
26. Turro, N. J. "Modern Molecular Photochemistry"; Benjamin/Cumming Publishing Co., Inc.: Menlo Park, California, 1978; Chap. 7.
27. Turro, N. J. "Modern Molecular Photochemistry"; Benjamin/Cumming Publishing Co., Inc.: Menlo Park, California, 1978; Chap. 7.
28. Ishikawa, M.; Kumada, M. *J. Organomet. Chem.* 1972, *42*, 325.
29. Boberski, W. G.; Allred, A. L. *J. Organometal. Chem.* 1975, *88*, 65.
30. Fukui, M.; Kitani, A.; Degrand, C.; Miller, L. L. *J. Am. Chem. Soc.* 1982, *104*, 28.
31. Diaz, A. F.; Kanazawa, K. "Extended Linear Chain Compounds"; Miller, J. S., Ed.; Plenum Publishing Corp.: 1983; Vol. 3, Chap. 8.
32. A detailed study of the electrochemistry of polysilane high polymer films will be reported in the future.
33. Mucha, J. A.; Hess, D. W. "Introduction to Microlithography"; Thompson, L. F.; Willson, C. G.; Bowden, M. J., Eds., ACS SYMPOSIUM SERIES No. 219, American Chemical Society: Washington, D.C., 1983; Chap. 5.

RECEIVED August 6, 1984

15

Preparation and Resolution Characteristics of a Novel Silicone-Based Negative Resist

A. TANAKA, M. MORITA, S. IMAMURA, T. TAMAMURA and O. KOGURE

Polymer Section, Ibaraki Electrical Communication Laboratory, Nippon Telegraph and Telephone Public Corporation, Tokai, Ibaraki, 319-11 Japan

Electron beam resists to be used in direct wafer writing for submicron devices need significant improvement in sensitivity, resolution and dry etching durability. Multilayer resist (MLR) systems are now regarded as the most important technology to perform practical submicron lithography for VLSI fabrication (*1-3*). Many advantages in MLR compared with one layer resists (1LR) are listed here:

1) The bottom thick layer has the ability of planarizing the topographic surface of the substrate and reducing the backscattering effect from the substrate in electron beam lithography.

2) The topmost imaging layer can be as thin as possible to obtain the high resolution pattern.

3) A high aspect ratio pattern can be obtained because the bottom layer is etched by a reactive ion etching O_2 plasma (O_2 RIE) where the etching takes place anisotropically.

Vital disadvantages in three layer resist (3LR) as MLR systems are a substantial increase in process steps and the need for the evaporation equipment that is used for forming the middle isolation layer. 3LR system consisting of a thin topmost imaging layer, a middle isolation layer and planarizing layer needs more than 11 steps to form a resist pattern (*1*). To minimize the processing steps, several materials for two layer resist (2LR) system have been proposed (*4-8*). A 2LR system can eliminate the middle isolation layer. It takes only 6 steps to accomplish the resist process and it is just a 2 step increase compared with 1LR system.

We have developed a new silicone based negative resist (SNR) by introducing the chloromethyl group into polydiphenylsiloxane. SNR has a high T_g (170°C), good electron beam sensitivity and excellent durability to O_2 RIE (*9*). In this paper we describe the SNR preparation and characteristics, and demonstrate a high resolution of 2LR system using this resist.

0097–6156/84/0266–0311$06.00/0

Molecular Design

The topmost imaging resist in 2LR system needs not only high sensitivity and high resolution but excellent dry etching durability to O_2 RIE. In addition to this, it must have the ability to coat a uniform, pin-hole free film on a substrate. It is preferable that the resist is solid at room temperature to be handled easily. In order to satisfy these characteristics, we designed a molecular structure of the topmost imaging resist as shown in Figure 1.

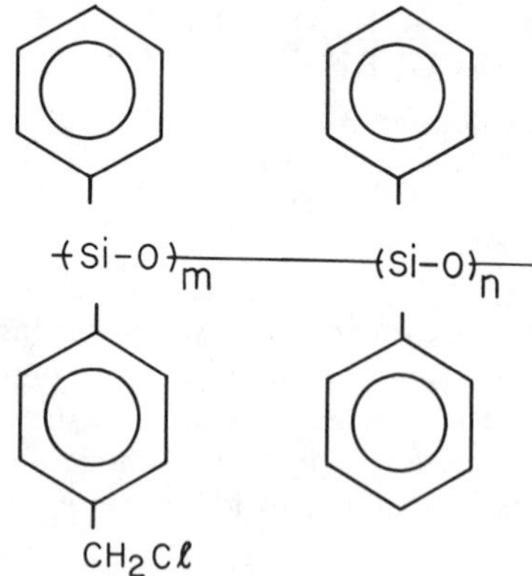

Figure 1. Structure of silicone based negative resist (SNR).

The main chain of SNR consists of the polydiphenylsiloxane (PDS) structure, because PDS is one of the few silicone polymer to show a high T_g (150°C). The high T_g of the polymer also eliminates difficulty in the handling of spin-coated wafers as compared to wet films of conventional silicone resins. SNR patterns are expected to show high resolution because the swelling of resist during development is suppressed by a high T_g. The siloxane structure is expected to have excellent dry etching durability to O_2 RIE. It was reported that a silicon containing polymer exhibited minimal film thickness loss during O_2 RIE when it contained more than 10 wt% of silicon (*7*). A polymer having a siloxane main chain is converted to SiO_2 by O_2 plasma and the etching rate becomes almost zero because of the high silicon content. As shown in Figure 1, SNR contains 12 wt% of silicon.

As in the case of polystyrene, various cross-linkable units may be incorporated in PDS structure. We choose the chloromethyl group, which is known to give polystyrene the largest sensitization among various halogen groups (*10*). The chloromethyl group is also known to have no post-irradiation polymerization effect because of the step-wise reaction (*11*). The unsaturated groups have more sensitivity than the chloromethyl group but tend to result in post-irradiation polymerization because of a chain reaction.

Experimental

SNR was prepared by direct chloromethylation of PDS. Chloromethylation of PDS was successfully performed in chloromethylmethylether solution using stannic chloride as a catalyst. The PDS obtained from Petrarch Systems was terminated with OH groups at the end of siloxane chain. A typical chloromethylation method is described as follows. PDS (30 g) was dissolved in chloromethylmethylether (500 ml), then stannic chloride (25 ml) was added to the solution. The solution was stirred for 17 hours at 0°C, and polymers in the solution were precipitated and washed with methanol. White polymer powders were purified by reprecipitation in methanol. An average yield of polymer was about 60% after two reprecipitations.

Polymer molecular parameters were determined relative to polystyrene standards by gel permeation chromatography (GPC). The chloromethyl groups were determined by IR spectroscopy and X-ray fluorescence analysis.

AZ-1350 photoresist was used as a thick bottom layer polymer. AZ resist, thicker than 1.0 μm was spin-coated on silicon wafer (oxide coated) or substrate with topographic features. The resist was hard-baked for 1 hour at 200°C. SNR film was then spin-coated on a hard-baked AZ resist layer from 5 wt% solution in methylisobutylketone.

An electron beam exposure machine ELS-5000 (Elionix) operated at 20 kV was used for lithographic evaluation. Sensitivity curves to X-ray and deep UV (254 nm) were determined with a X-ray exposure machine SR-1 (Mo target), developed in our laboratory (*12*) and a Spectro Irradiator, CRM-FA (JASCO), respectively. After exposure, SNR was developed with (3/1) methylethylketone-isopropanol solvent mixture and rinsed with isopropanol.

The SNR patterns were transferred into the bottom layer (AZ resist) by O_2 RIE using an RIE machine (DEM-451 ANELVA). RIE conditions were chosen to etch the bottom layer only in a direction perpendicular from the surface of the bottom layer to the substrate. The conditions are 50 SCCM of gas flow, 10 mTorr of gas pressure, 0.1 W/cm^2 of RF power and 0.5 kV of voltage between electrodes.

Molecular Parameters

SNR samples with a variety of controlled weight average molecular weight (M_w) were prepared by chloromethylation of PDS. The new absorption at 1260 cm^{-1} in an IR spectrum shows that the chloromethyl group was successfully introduced into PDS. Table I summarizes M_w and molecular weight distribution of SNR samples. SNR samples have higher M_w and broader molecular weight distributions when compared to the PDS starting polymer. As the reaction time is increased, M_w of SNR is increased. This result indicates that chloromethylation is not the only reaction and at the same time a side reaction such as cross-linking between polymer chain is inevitable. This side reaction was not observed in chloromethylation of polystyrene under mild conditions. The terminating OH group of PDS has an important contribution in their polymerization. The result of our investigation of the chloromethylation and polymerization mechanisms of PDS will be reported at a later date.

Exposure Characteristics

Several electron beam sensitivity curves are shown in Figure 2. From these sensitivity curves, we estimate the dose for starting point of gel formation (D_0), dose required for 50% gel formation ($D_{0.5}$) and contrast (γ-value) of SNR.

Table I. Molecular Parameter of SNR and PDS

Item	Weight average molecular weight (M_w)	Molecular weight distribution (M_w/M_n)
SNR 1	1.2×10^4	1.8
SNR 2	1.4×10^4	2.1
SNR 3	3.8×10^4	3.1
SNR 4	7.6×10^4	4.7
SNR 5	1.1×10^5	8.5
PDS	1.4×10^3	1.8

SNR: silicone based negative resist

PDS: polydiphenylsiloxane

Contrast is derived from the slope of the linear portion of the sensitivity curve, (Figure 2) and can be used to predict the resolution capability of a resist with higher contrast values resulting in higher resolution. Figure 3 shows the dependence of sensitivity ($D_{0.5}$) on molecular weight (M_w) and Figure 4 shows the dependence of contrast (γ) on molecular weight. As with many negative electron resists, an increase in M_w results in an increase in sensitivity, but gradually decreases the resist contrast.

SNR with M_w of 3.8×10^4 has a sensitivity of 5 $\mu C/cm^2$ at $D_{0.5}$ and contrast of 2.0, which clearly indicates the utility of SNR in practical electron beam lithography. As $D_{0.5}$ of chloromethylated polystyrene (CMS) with a molecular weight of 6.0×10^4 is 4.5 $\mu C/cm^2$, the cross-linking reactivity of SNR is even higher than CMS. The contrast of SNR is somewhat lower than that of CMS in 1LR, because the molecular weight distribution of SNR is much broader than that of CMS. But resist contrast of the top imaging layer in 2LR system is not as important as that in 1LR system because of its low thickness. As mentioned later, 2LR with SNR/AZ resist exhibits higher resolution than CMS in 1LR does. Resist contrast would be improved if a narrower molecular weight distribution of SNR was obtained by fractional precipitation.

It is confirmed that the SNR shows high sensitivity to X-ray (Mo target) and deep UV (254 nm) as well as electron beam. Table II summarizes D_0, $D_{0.5}$ and contrast of SNR with M_w of 3.8×10^4 using three typical exposure machines. SNR has about 3 times larger absorption than CMS for X-ray (Mo target), resulting in high sensitivity. In deep UV (254 nm), the strong absorption of phenyl group gives SNR a high sensitivity.

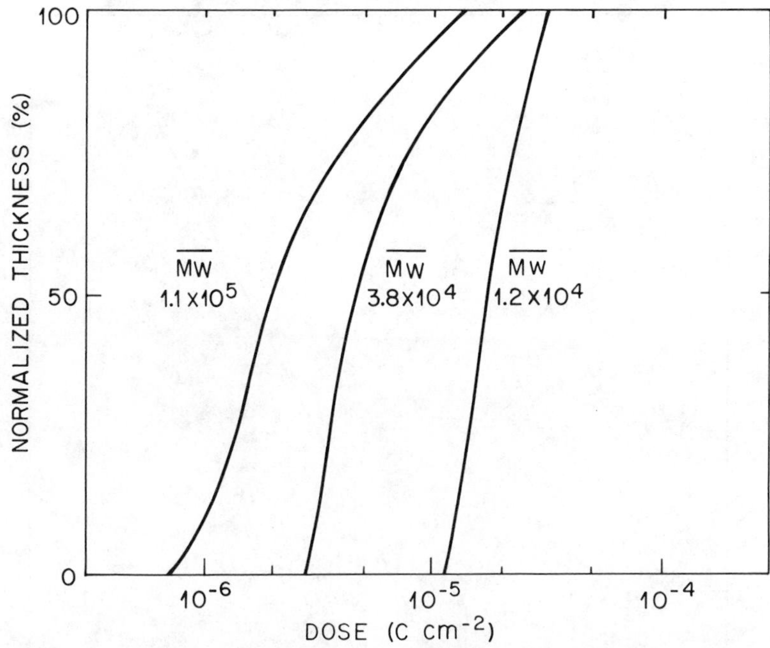

Figure 2. SNR sensitivity curves to electron beam exposure using an accelerated voltage of 20 kV.

Post-irradiation Polymerization

To evaluate the post-irradiation polymerization effect, the thickness of the developed pattern, after leaving a sample in an electron beam exposure machine for a given period, was compared with resist thickness of a pattern developed immediately. The results are shown in Figure 5, and compared with those of methacrylated PDS. Methacrylated PDS was prepared by substituting the methacryloyl group with chlorine on the chloromethyl group of SNR. This reaction was carried out with potassium methacrylate, SNR and ethyl trimethyl ammonium iodide catalyst. The post-irradiation polymerization in SNR was negligible, while after 3 hours in the vacuum, methacrylated PDS thickness increased by 1.3 times the initial thickness. As discussed later, the cross-linking reaction in SNR proceeds mainly by a step-wise reaction rather than by a chain reaction, because of its high T_g and the absence of unsaturated groups. The chain reaction enhances the sensitivity but decreases the resolution and causes post-irradiation polymerization. It is recognized that the product of D_0 and M_w, which is almost constant for a particular polymer regardless of its molecular weight and gives the best measure of polymer reactivity (*13*). The lower the $M_w \cdot D_0$ product the higher the sensitivity to an electron beam. The $M_w \cdot D_0$ product of methacrylated PDS is 0.010 C/cm^2, whereas that of SNR is 0.073 C/cm^2. Although methacrylated PDS is more sensitive than SNR, it shows post-irradiation polymerization effect which is detrimental for electron beam resists.

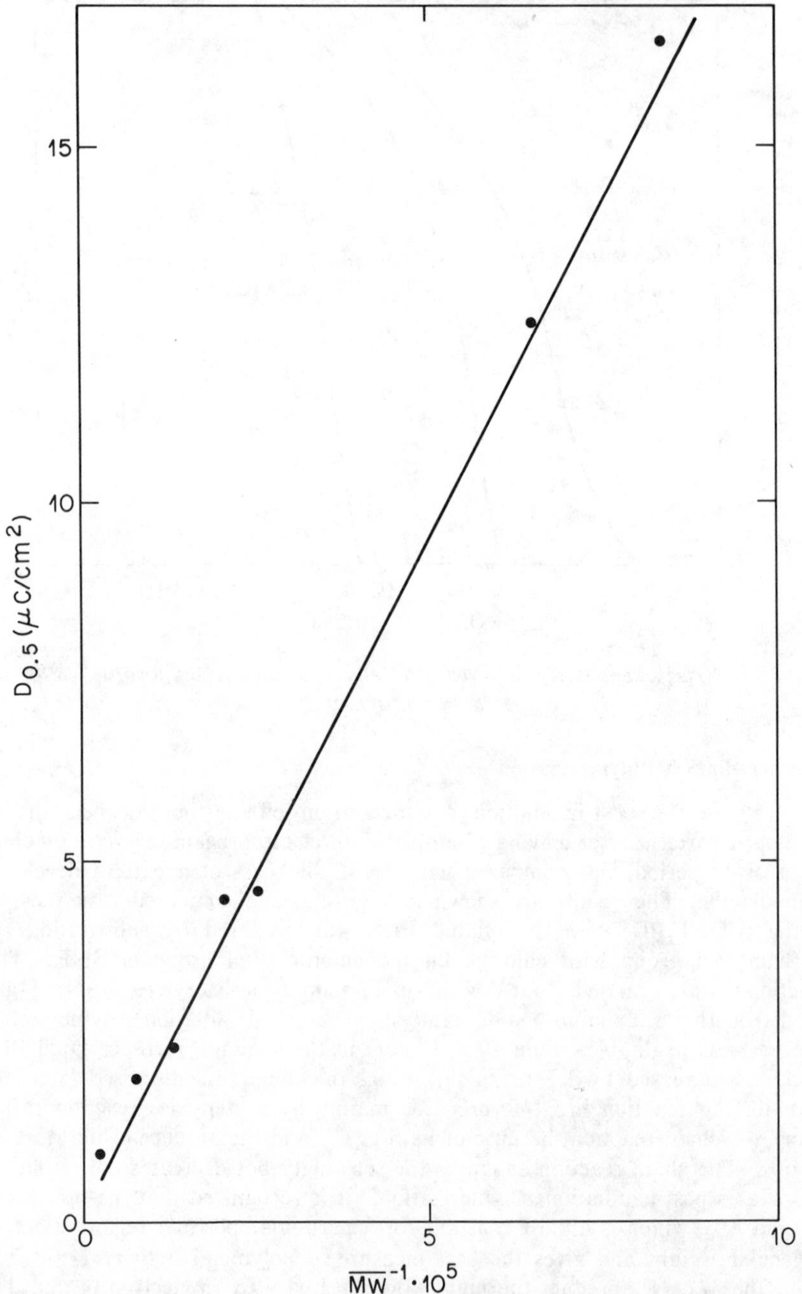

Figure 3. Molecular weight dependence of sensitivity to electron beam exposure.

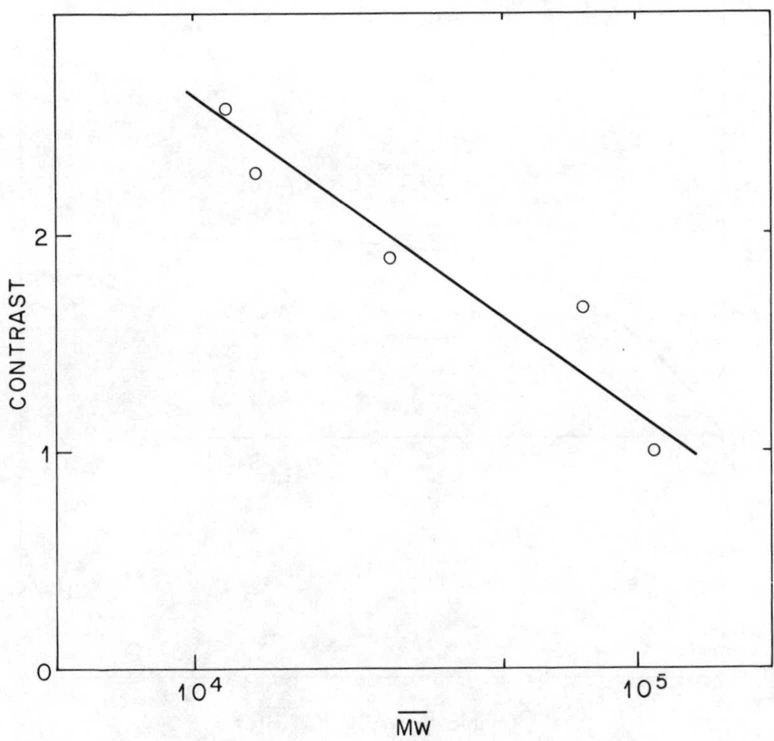

Figure 4. Molecular weight dependence of contrast to electron beam exposure.

Table II. Exposure Characteristics of SNR with M_w of 3.8×10^4

Item	Dose D_0	Dose $D_{0.5}$	Contrast (γ-value)
Electron beam	2.8 $\mu C/cm^2$	5.0 $\mu C/cm^2$	2.0
X-ray (Mo)	32 mJ/cm^2	55 mJ/cm^2	1.9
Deep UV (254 nm)	4 mJ/cm^2	10 mJ/cm^2	1.3

Dry Etching Durability

SNR has excellent dry etching durability to O_2 RIE. The etching rate of SNR is less than 3 nm/min for the O_2 RIE condition used, whereas that of AZ resist is more than 80 nm/min. SNR selectivity ratio for AZ resist is more than 20. This result allows the fabrication of thick AZ resist patterns with a mask of

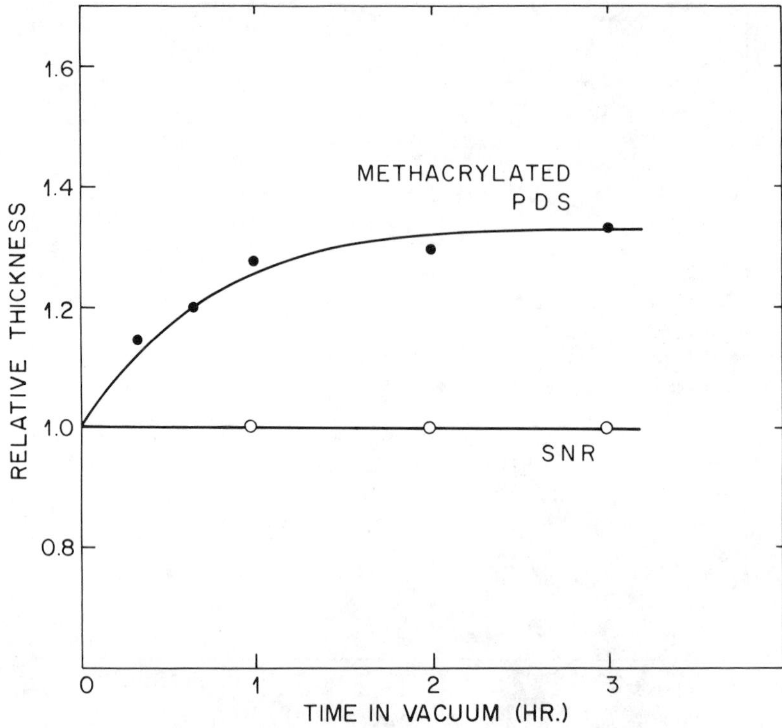

Figure 5. Post-irradiation polymerization effect in SNR and methacrylated PDS.

very thin, high resolution SNR patterns. A 0.2 μm thickness of SNR is enough to fabricate submicron patterns with 2 μm thickness of AZ resist.

To clarify the mechanism of O_2 RIE of SNR, we examined the conversion of SNR to SiO_2 by O_2 RIE. Depth profiles of chemical elements in several SNRs etched by O_2 RIE with different times were analyzed by Auger Electron Spectroscopy (AES). The depth profile of an non-etched sample (Figure 6(a)) indicated the same elemental composition from the surface to bottom of SNR layer whereas, the surface of etched samples show an extreme decrease of carbon content and increase of oxygen content, indicating conversion of SNR to SiO_2 even after one minute of etching (Figure 6(b)). There were no differences in silicon content between non-etched and etched samples as observed by AES. It should be noted that the oxidized regions were at most 10 nm from the surface, and is independent with etching time from one to ten minutes. This indicates that the surface of SNR is converted to SiO_2 by O_2 RIE and SNR is hardly etched because of the barrier layer of SiO_2. The SiO_2 formed on the SNR is slightly etched by the sputtering effect and then the SNR just under the SiO_2 is oxidized to SiO_2 so that the thickness of surface SiO_2 is kept constant. Since organic polymers, not containing silicon, are not able to form the barrier layer (SiO_2) during O_2 RIE, they are etched very rapidly compared with SNR.

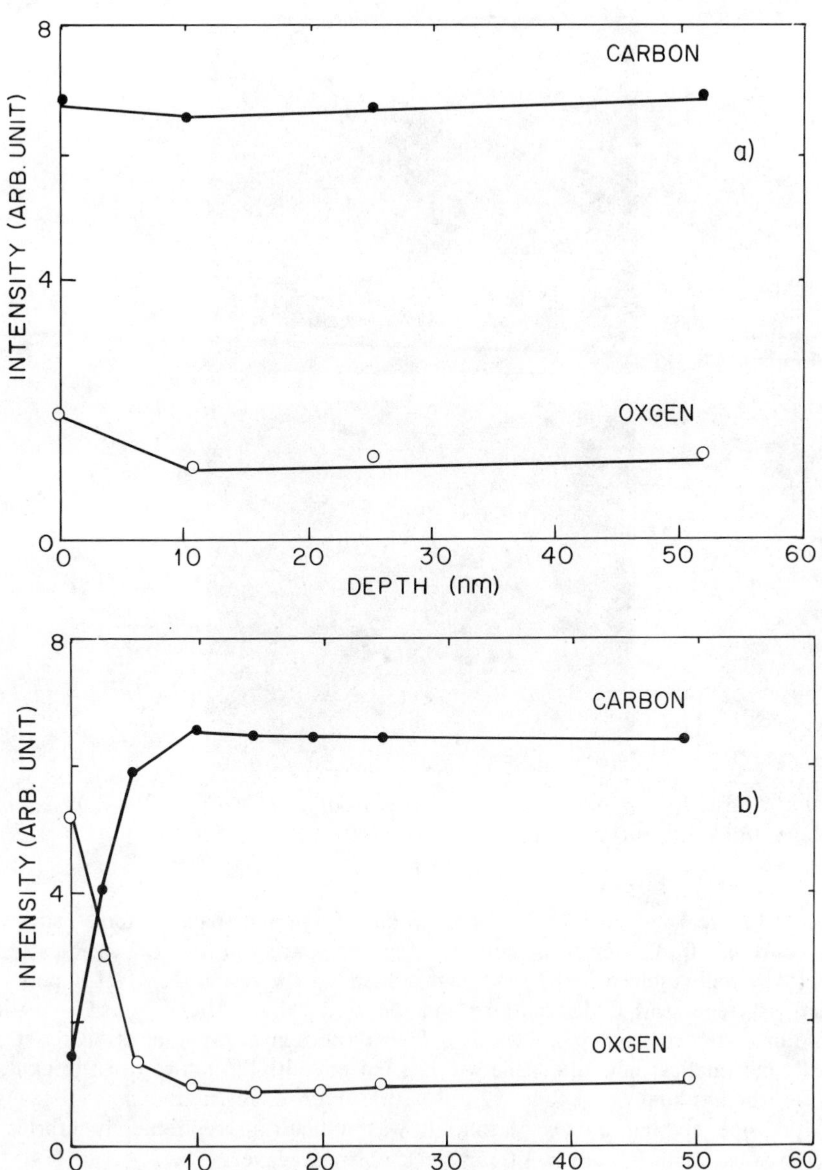

Figure 6. Depth profile of SNR layer by Auger Electron Spectroscopy (AES); (a) no-etching, (b) 1 minute O_2 RIE.

Resolution of Two Layer Resist with SNR/AZ Resist

The resolution of 2LR with SNR/AZ resist where SNR had M_w of 3.8×10^4 was evaluated as a well-resolved line and space width on the Si wafer. A 0.2 μm SNR layer was coated on a hard-baked AZ resist with 1.0 μm thickness and exposed with an electron beam, then developed. The obtained pattern of SNR is transferred to the AZ resist layer by O_2 RIE.

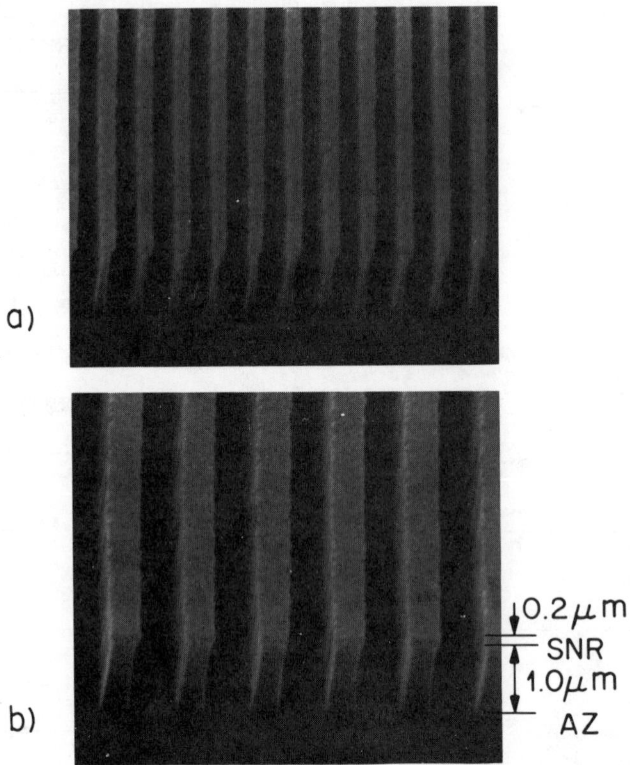

Figure 7. Submicron line and space patterns of 2LR with SNR/AZ resist of 1.0 μm thickness; (a) line and space width 0.2 μm, (b) line and space width 0.4 μm.

Figure 7 shows the SEM photographs of line and space patterns using an exposure of 10 $\mu C/cm^2$. The smallest line and space width of the 2LR pattern that was well-resolved is 0.2 μm. Figure 8 shows the resolution of 1LR poly-α-methylstyrene (αM-CMS) pattern and the 2LR using SNR/AZ resist. In the case of αM-CMS that is known as a high resolution negative electron resist in 1LR, the smallest line and space width is 0.4 μm with 0.6 μm of resist thickness (*14*), whereas that of 2LR is 0.2 μm with 1.0 μm of resist thickness. Because of the side etching by O_2 plasma, it was difficult to reproducibly fabricate patterns less than 0.2 μm in 1.0 μm thick AZ resist layer.

As shown in Figure 8, a high resolution pattern is obtained by reducing the resist thickness in 1LR system, but in VLSI fabrication process resist thickness more than 1.0 μm is needed to planarize the topographic surface, so that 2LR system is more useful for submicron fabrication than 1LR system.

Figure 9 demonstrates submicron lithography on a substrate with topographic features. A 2.0 μm thick AZ resist almost completely planarizes the surface structure where 0.6 μm wide and 0.8 μm high SiO_2 patterns were already fabricated. A 0.2 μm thick SNR is coated on AZ resist. High resolution lithography can be accomplished with this process because of high

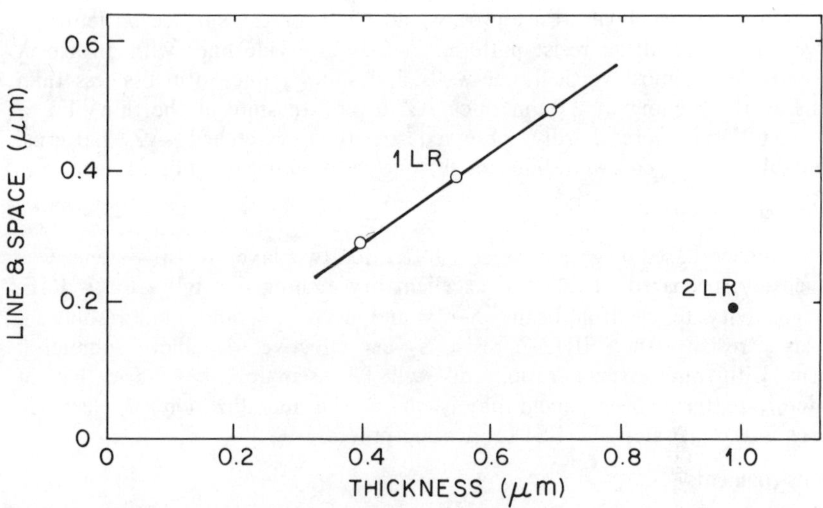

Figure 8. Resolution comparison between 1LR with α M-CMS and 2LR with SNR/AZ resist.

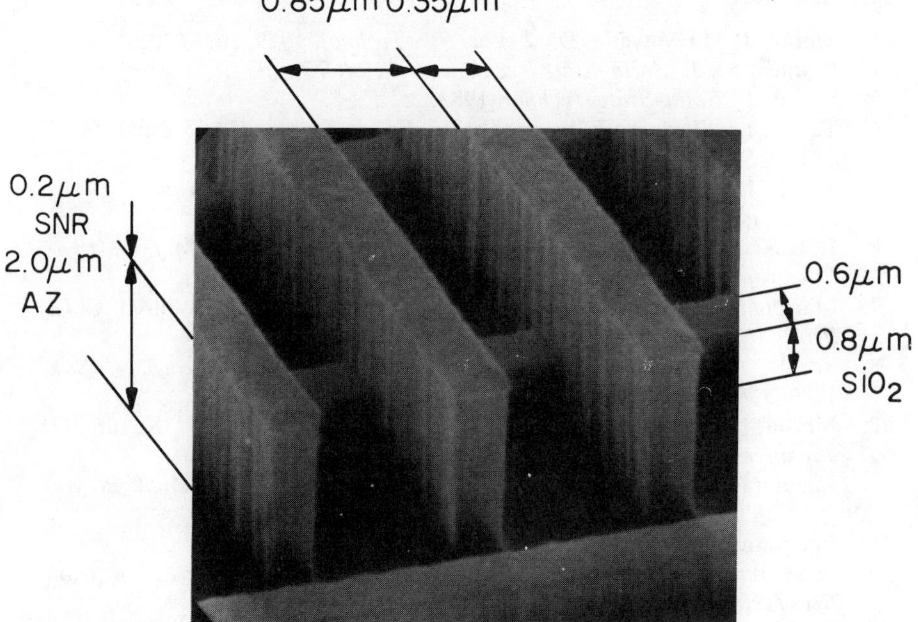

Figure 9. Submicron pattern on a substrate with topographic features.

performance of SNR resist characteristics and the usage of the thin resist on a thick organic layer, which reduces the backscattering effect in electron beam exposure.

With the two layer resist process, no effect of the surface structure is observed in the resulting resist pattern. A 0.35 μm wide lines with an aspect ratio of 6 have almost vertical side-walls and show a linewidth loss less than 0.1 μm by the etching of 2.0 μm thick AZ layer. In spite of the thin film of SNR, pin-holes were hardly observed in the dry-etched AZ patterns, presumably due to good wettability of silicone resin on organic film surface.

Conclusion

A new silicone-based negative resist (SNR) for two layer resist systems was designed and prepared. It showed excellent dry etching durability to O_2 RIE, high sensitivity to electron beam, X-ray and deep UV, and high resolution. Two layer resist with SNR/AZ resist is very effective to achieve submicron patterns with high aspect ratio, and will be used for the fabrication of submicron patterns over topography such as the metallization of electrode patterns in the last step of VLSI fabrication process.

Acknowledgments

The authors are greatly indebted to Mr. K. Matsuyama for his continuous encouragement. The authors are also indebted to Mr. K. Murase for his valuable discussion.

Literature Cited

1. Moran, J. M.; Maydan, D. *J. Vac. Sci. Technol.* 1979, *16*, 1620.
2. Bowden, M. J. *Solid-State Technol.* 1981, *24*, 73.
3. Lin, B. J. *Solid-State Technol.* 1983, *26*, 105.
4. Tai, K. L.; Sinclair, W. R.; Vadimsky, R. G.; Moran, J. M.; Rand, M. J. *J. Vac. Sci. Technol.* 1979, *16*, 1977.
5. Tai, K. L.; Ong, E.; Vadimsky, R. G. *Proc. Symp. Inorganic Resist Systems*, 1982, p. 9.
6. Hatzakis, M.; Paraszczak, J.; Shaw, J. *Proc. Int'l Conf. Microlithography*, 1981, p. 386.
7. MacDonald, S. A.; Steinmann, F.; Ito, H.; Lee, W-Y.; Willson, C. G. *Proc. ACS Div. Polym. Mater. Sci. Eng.*, 1983, p. 104.
8. Suzuki, M.; Saigo, K.; Gokan, H.; Ohnishi, Y. *J. Electrochem. Soc.* 1983, *130*, 1962.
9. Morita, M.; Tanaka, A.; Imamura, S.; Tamamura, T.; Kogure, O. *Japanese J. Appl. Phys.* 1982, *27*, 937.
10. Imamura, S.; Tamamura, T.; Harada, K.; Sugawara, S. *J. Appl. Polym. Sci.* 1982, *27*, 937.
11. Imamura, S. *J. Electrochem. Soc.* 1979, *126*, 1628.
12. Hayasaka, T.; Ishihara, S.; Kinoshita, H. *Proc. Symp. Elec. & Ion Beam Sci. Technol.*, 1982, p. 347.
13. Charlesby, A. "Atomic Radiation and Polymers"; Oxford, 1960; p. 452.
14. Sukegawa, K.; Tamamura, T.; Sugawara, S. *Proc. Symp. Elec. & Ion Beam Sci. Technol.*, 1983, p. 193.

RECEIVED August 6, 1984

16

Positive-Working Electron-Beam Resists Based on Maleic Anhydride Copolymers

K. U. POHL and F. RODRIGUEZ

School of Chemical Engineering, Olin Hall

Y. M. N. NAMASTE and S. K. OBENDORF

Department of Design and Environmental Analysis
Martha Van Rensselaer Hall
Cornell University, Ithaca, NY 14853

Many studies of structure versus radiation sensitivity have led to certain generalizations regarding possible candidates for positive-working electron-beam resists. One such generalization is that polymers bearing hydrogens on adjacent chain carbons are not suitable since they are likely to cross-link (*1*). However, the generalization is based mainly on observations of vinyl and vinylidene structures. Maleic anhydride and its derivatives provide polymer structures in which adjacent chain carbons have only one hydrogen each. It has been found that this structure does not lead to cross-linking, and, in fact, undergoes chain scission, at least in the polymers reported in the present study.

Maleic Anhydride

Maleic anhydride is a five membered ring anhydride (structure I) containing an

```
O      O      O
 \\  /   \  //
   C       C        I
   |       |
   CH  ==  CH
```

olefinic double bond. It was commonly believed until the early 1960s that this monomer would not homopolymerize. Since then it has been shown that maleic anhydride can be polymerized with both gamma and UV radiation, free-radical initiators, pyridine-type bases, electrochemically, and under shock-waves (*2*). However, the yields are generally poor and the molecular weights are low.

Copolymerization, on the other hand, is very easy with maleic anhydride. It copolymerizes by a free-radical reaction with a wide variety of monomers and many of the copolymers are perfectly alternating. This tendency of MA to form alternating copolymers derives from the participation of a donor-acceptor complex formed by the two reacting monomers. The term is used to describe

0097-6156/84/0266-0323$06.00/0

an intermolecular complex between maleic anhydride, the acceptor, and the other monomer, the donor. The interaction of the donor and the acceptor can be summarized by the equilibrium equation:

$$D + A \underset{}{\overset{K}{\rightleftharpoons}} DA \tag{1}$$

The magnitude of the equilibrium constant, K, is a measure of the strength of interaction between the two molecules. The equilibrium constant has been determined for a variety of complexes (*3*). The rate of formation of this complex is usually much higher than the rate of polymerization. The polymerization proceeds by adding donor-acceptor units to the growing chain.

It is generally accepted that a good positive resist should have a high G(scission) value, the $G(s)$ value being a measure of the number of main chain scissions per 100 eV of absorbed energy. There are some empirical rules for polymer structures according to which one can predict degradation versus cross-linking of a given polymer. Polymers of the general structure II (where

$$-CH_2-\overset{\displaystyle R_1}{\underset{\displaystyle R_2}{\overset{|}{\underset{|}{C}}}}- \qquad \text{II}$$

R_1 and R_2 are substituents other than hydrogen) are expected to degrade predominantly upon electron beam radiation. Copolymers containing MA do not conform with this structure since MA contains one hydrogen on each olefinic carbon.

Characterizing Sensitivity

A variety of techniques have been used in the present work to establish the relative sensitivity of positive electron-beam resists made from copolymers of maleic anhydride (Table I). The term sensitivity is used rather loosely at times. In the most practical sense, sensitivity is a comparative measure of the speed with which an exposure can be made. Thus, the exposure conditions, film thickness, developing solvent and temperature may be involved. Most often, the contrast curve is invoked as a more-or-less objective measure of sensitivity. The dose needed to allow removal of exposed film without removing more than about 70% of the unexposed film can be a measure of sensitivity. The initial film thickness and the developing conditions still must be specified so that this measure is not, strictly speaking, an intrinsic property of the polymeric material.

Another measure of sensitivity can be obtained from the rate of dissolution of exposed versus unexposed film at various doses. The amount of energy needed to obtain some arbitrary ratio of rates or the exponent of the dissolution rate ratio versus dose at high doses may be used. In the present study, both a contrast curve and a solubility rate ratio versus dose data are reported for the copolymer of maleic anhydride with alphamethylstyrene. However, these tests are burdensome when many materials are to be screened

Table I. Methods and Hardware used in Characterizing the Sensitivity and Radiation Properties of the Polymers

Exposure tools:	Gamma radiation from ^{60}Co source Flood exposure to 50 keV electrons Pattern exposure to 20 keV electrons
Measures of change:	Viscosity measurements Intrinsic viscosity Viscosity-average molecular weight Molecular weight distribution, GPC Solubility Rates Thickness changes on development

as resist candidates. The change in molecular weight on exposure to measured doses of electron radiation can be used to calculate $G(s)$, the number of chain scissions per 100 eV of absorbed energy. Plotting $1/M_n$ versus incident dose yields a straight line with a slope which is proportional to $G(s)$ (*4*). A "depth-dose function" is used to convert incident dose to absorbed dose (*5*). In the present work a 3-inch diameter silicon wafer (oxide-coated) is spin-coated with a polymer solution and baked. This step requires adjustment of viscosity and concentration to give a consistent final film thickness of about a micron. Also, not every solvent yields a homogeneous film without cracks or holes. The solvent must also be removable at some reasonable baking temperature. A wafer coated in this manner bears about 5 mg of polymer. Flood exposure of the wafer surface to the calibrated, 50 keV, defocused beam of electrons in a conventional transmission electron microscope (RCA Model EMV-3) results in an adequate amount of exposed polymer for a number of molecular weight measurements. The preferred technique is gel permeation chromatography, GPC, using a high performance liquid chromatograph, HPLC, since this gives a complete molecular weight distribution from which M_n and M_w can be derived. Once again, the polymer being tested has to be soluble in the chosen eluting solvent, in our case, tetrahydrofuran, THF. Moreover, the polymer must not exhibit any tendency to associate with the column packing lest it provide false, low values of molecular weight (long elution times) or, worse yet, not elute at all. Several copolymers of maleic anhydride do show these unfortunate tendencies.

Another exposure tool is available in gamma radiation. While the correlation is not always perfect, there is a high degree of similarity in the response of polymers to the radiation from ^{60}Co and from 50 keV electrons. Because of the penetrating nature of gamma rays, the exposure is not restricted to thin films or small amounts of polymer. Also, the absorbed dose is not complicated by the depth-dose function which must be used when electron-beam radiation absorption is used. Since films do not have to be spun, it is not necessary to work out a spinning solvent.

Gamma radiation can be used with macroscopic amounts of polymer. This is particularly welcome when polymers are not compatible with the GPC technique. Larger samples can be characterized by viscosity changes, usually measured in dilute solutions. All that is needed is a suitable solvent. If the Mark-Houwink parameters are known, it is possible to calculate viscosity-average molecular weight, M_v, from dilute solution viscosities. However, even the raw viscosity-concentration data in terms of the reduced viscosity may be enough to indicate the sensitivity of a given polymer in qualitative terms. The reduced viscosity at concentrations c is η_{sp}/c where η_{sp} = (solution viscosity − solvent viscosity)/solvent viscosity.

Copolymer with Alpha-Methylstyrene

There are two generalizations often made concerning poly(alpha-methylstyrene), (PAMS). The first is that it degrades by random chain scission when exposed to high energy radiation and is in contrast to the cross-linking which predominates when polystyrene is so exposed. The second is that, because of the delocalization of energy by phenyl rings, PAMS is inherently more stable than any non-aromatic, carbon-chain counterpart. The radiation-chemical yield cited (*6*) most often is $G(s) = 0.25$ scissions/100 eV. In the present work, values of 0.23 and 0.25 were obtained by gamma radiation and electron beam radiation (50 keV), respectively.

A copolymer of MA and AMS was made by free-radical polymerization, see Table II. Seymour and Garner (*7*) have shown that the alternating copolymer is invariably obtained below polymerization temperatures of 80°C although random copolymers are obtained above 100°C.

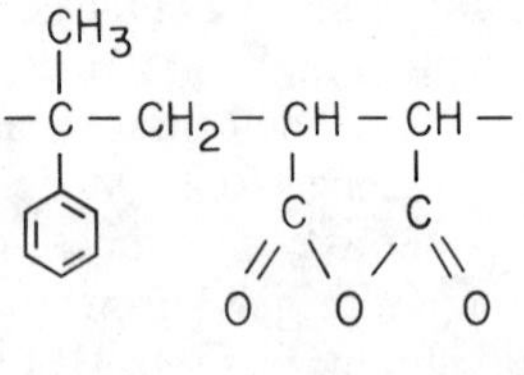

```
      CH3
      |
    — C — CH2 — CH — CH —     P(AMS-MA)
      |          |    |
     Ph          C    C
               // \  / \\
              O    O    O
```

Table II. The Effect of Maleic Anhydride (MA) and the Methyl Half-Ester of Maleic Acid (MM) on Alphamethylstyrene (AMS) Polymer Sensitivity

Yield	PAMS	P(AMS-MA)	P(AMS-MM)
$G(s)$, scissions per 100 eV	0.25	0.62	1.59
Relative yield	1.0	2.5	6.4

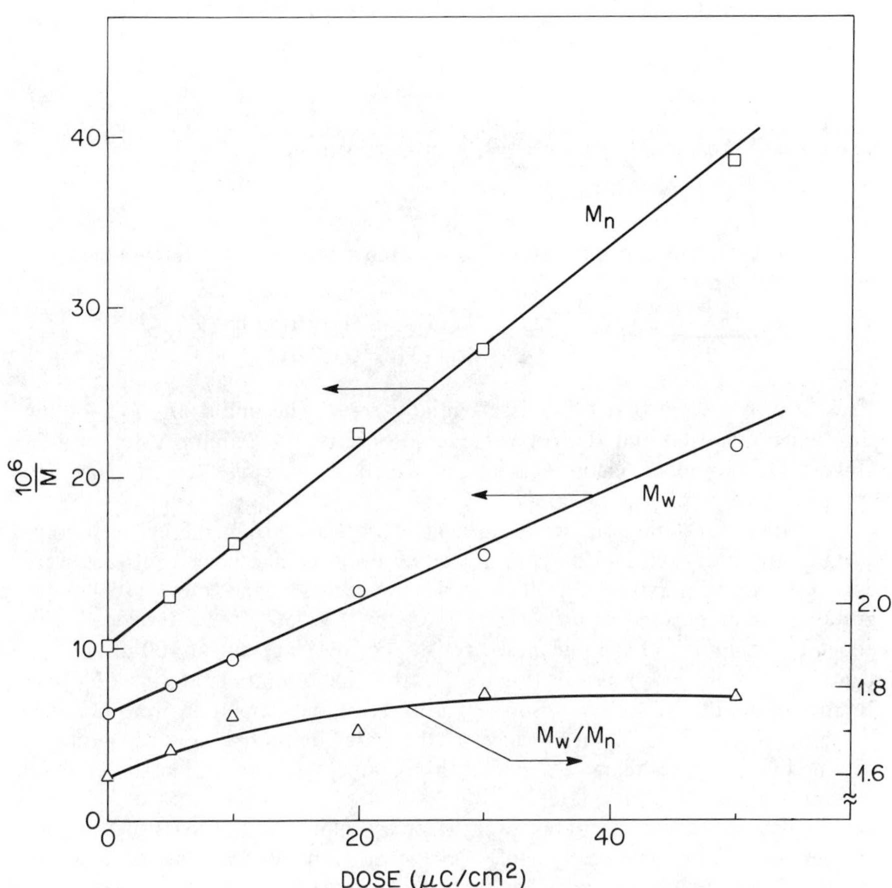

Figure 1. Sensitivity curve for a perfectly alternating copolymer of maleic anhydride with α-methylstyrene. A 50 keV beam was used.

The copolymer definitely is more sensitive to electron beam radiation than the homopolymer of AMS (Figure 1). There is, of course, a certain hazard in comparing polymers made by different initiating systems. For example, the thermal stability of poly(methylmethacrylate) is known to vary with the initiator used in its polymerization (*8*). However, in the present work, the circumstances dictate such a comparison because AMS does not homopolymerize by free-radical initiation and the alternating copolymer is most conveniently prepared this way. There can be little doubt that the predominant effect of radiation is chain scission with a $G(s) = 0.62$. This result indicates that MA enhances the scissioning tendency of the copolymer despite the fact that the MA unit does not belong to the category of polymers of the general structure II discussed above. Whether or not there is a small amount of cross-linking is difficult to pin down. The slope of the $(M_n)^{-1}$ curve (Figure 1) is 1.9 times that of the corresponding $(M_w)^{-1}$ curve. According to the equations of Saito and others (*9*) the ratio of the slopes can be used to ascertain the relative amounts of scissioning and cross-linking. The slope of the $(M_n)^{-1}$ curve, $\sigma(n)$, is proportional to

$$G(s) - G(x) \tag{2}$$

The slope of the $(M_w)^{-1}$ curve, $\sigma(w)$, is proportional to

$$G(s)/2 - 2G(x) \tag{3}$$

at least when the initial distribution is the "most probable." It follows that the ratio of the slopes is

$$\frac{\sigma(n)}{\sigma(w)} = 1.9 = \frac{G(s) - G(x)}{G(s)/2 - 2G(x)} = \frac{G(s)/G(x) - 1}{G(s)/G(x) - 4} \times 2 \tag{4}$$

Thus, it would seem that $G(x)$ is essentially zero. The initial M_w/M_n for the copolymer was 1.60 and the ratio increased only to a maximum value of 1.78 (Figure 1). Random chain scission is expected to cause the M_w/M_n to approach a value of two.

Films for lithographic evaluation were cast from methyl cellosolve acetate, and prebaked at 120°C in a vacuum oven for one hour. Patterns were developed using mixtures of ethyl cellosolve acetate and methyl cellosolve acetate. Areas exposed at a dose of 80 $\mu C/cm^2$ (20 keV) were developed with about 10% thinning of the unexposed resist. For development at 100 $\mu C/cm^2$, a contrast of 2.1 was observed (Figure 2), and the resolution at this dose was limited to about one micron. Superior sensitivity and resolution were obtained using a MA-AMS copolymer formed on the wafer by prebaking a copolymer of the methyl half ester of maleic acid with alphamethylstyrene. This latter resist system is discussed a little later.

The conventional wisdom predicts cross-linking on the basis of radiation studies of ordinary vinyl polymers. On the other hand, there seems to be no previous mention of the radiation sensitivity of maleic anhydride polymers in the literature.

There is some reason to expect that conversion of the anhydride to a half-ester might reduce the sensitivity of the copolymers. Hiraoka (*10*) determined the relative sensitivities of PMMA, PMA (polymethacrylic acid) and PMA AN (polymethacrylic anhydride) by measuring the gaseous products (CO, CO_2, and H_2) given off when these polymers were exposed to electron beam radiation of 2.5 keV at 297°K. He found that the G values (number of chemical events produced per 100 eV of absorbed radiation) for the removal of side groups are 2.0, 7.4 and 16 for PMMA, PMA and PMA AN, respectively. Anderson (*11*) found a similar relative order of sensitivity. For copolymers of methylmethacryate with 25% dimethylitaconate, 25% monomethyl itaconate or 25% itaconic acid (or anhydride) the $G(s)$ values were 1, 2, 3, respectively. For the copolymer of alpha-methylstyrene and monomethyl maleate, on the other hand, we find an increase in sensitivity by a factor of 2.5 over the corresponding anhydride as described below.

The copolymer of AMS with the methyl-ester of MA, MM, was prepared by refluxing a previously-obtained AMS-MA copolymer in methanol

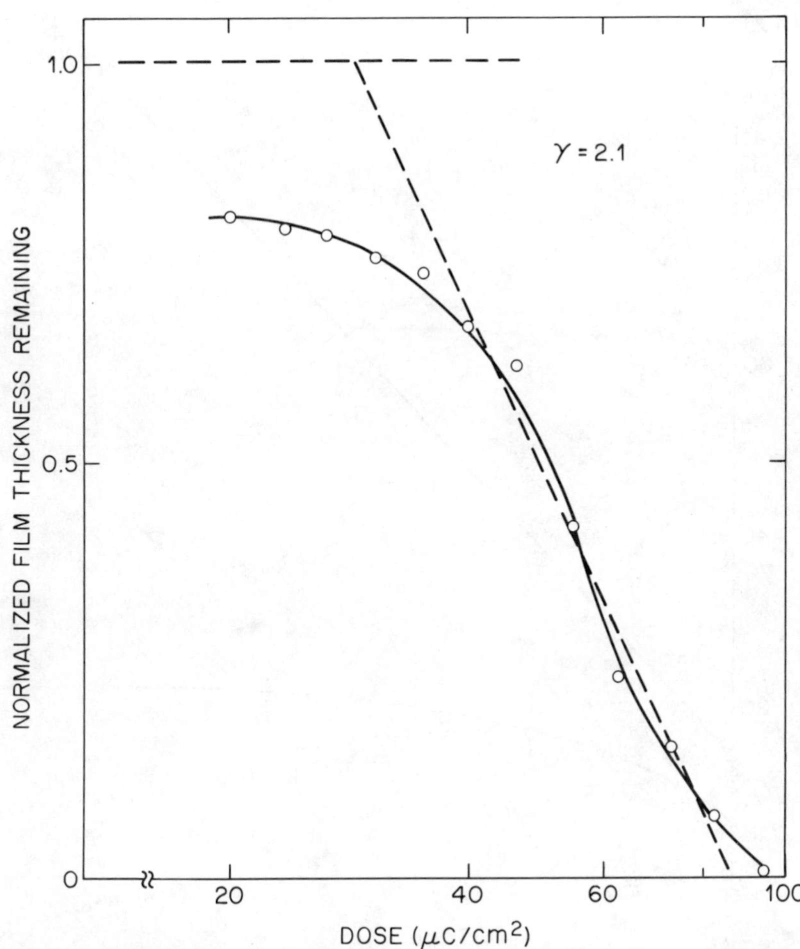

Figure 2. Contrast curve for the copolymer of Figure 1. Films were cast from methyl cellosolve acetate, prebaked at 120° C (in vacuum), and developed using a mixture of ethyl and methyl cellosolve acetates. Incident doses of 20 keV electrons were imposed on a film about one micrometer thick.

for two days until all polymer had dissolved. The direct copolymerization of AMS with MM was not attempted since there are reports that the polymerization results only in poor yields of low-molecular weight copolymers (*12-14*). The rationalization is that MM is a poorer electron acceptor than the anhydride. The IR spectrum of the AMS-MM copolymer shows absorption peaks at 1740 cm^{-1} and 1710 cm^{-1}. These are characteristics of ester and carboxylic acid groups. Titration shows 92% of calculated equivalents of acid in the polymer. The titration was conducted using standardized sodium methoxide in methanol with thymol blue as an indicator.

The copolymer was spun on silicon wafers from methylcellosolve acetate followed by baking at 125°C in a vacuum oven for 1 hour. The $G(s)$ value is

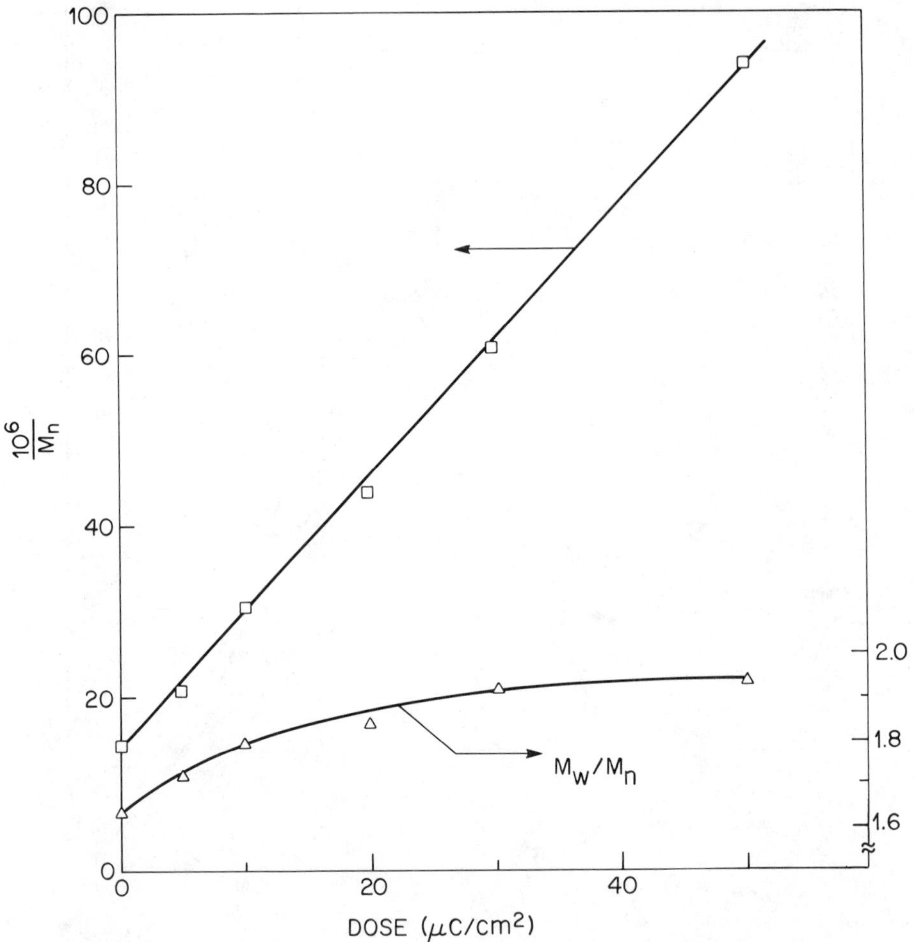

Figure 3. Sensitivity curve for the perfectly alternating copolymer of the methyl half-ester of maleic acid with α-methylstyrene.

1.59 (Figure 3, Table II) and the polydispersity again rises towards 2.0. This result shows that main chain scissioning is easier when the anhydride ring is already opened providing for a more facile decarboxylation of one of the chain fragments. This explanation is speculation until it can be confirmed by some other means such as mass spectroscopy of irradiated polymer. Decarboxylation is well-established as taking place when poly(methylmethacrylate) is irradiated since quantitative amounts of CO_2 are produced (*15*). It also appears that the ring in maleic anhydride is more stable than those of itaconic anhydride and methacrylic anhydride.

For lithographic evaluation, films of P(MM-AMS) were cast from solutions in 2-ethoxyethanol and prebaked one hour at 150°C. Under these prebake conditions, the ester was converted largely back to the anhydride as confirmed by infrared analysis. However, the properties of the films, and thus, the lithographic performance of this resist system were quite distinct from that of the P(MA-AMS) discussed previously. Developing with pure cyclohexanone, areas exposed at a dose of 45 $\mu C/cm^2$ (20 keV) were developed with a 10% thinning of the unexposed resist and a contrast of 2.0 was obtained for development at 35 $\mu C/cm^2$. A resolution of 0.5 μm was observed at an exposure dose of 50 $\mu C/cm^2$ and vertical-walled profiles were produced at a dose of 90 $\mu C/cm^2$. These results are very similar to those reported by Hatzakis for PMMA (*16*).

Copolymer with Methyl Vinyl Ether

The material used is a commerically available polymer obtained from GAF Corp., GANTREZ AN-169. According to the manufacturer, it is an

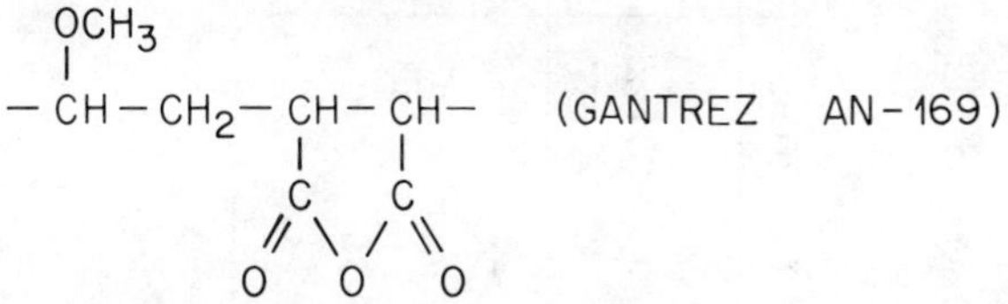

```
  OCH3
  |
— CH — CH2— CH— CH—      (GANTREZ  AN-169)
            |    |
            C    C
           //\  / \\
          O   O    O
```

alternating copolymer which is entirely reasonable since this particular pair of monomers almost form a classical donor-acceptor pair (*17*). Neither monomer polymerizes by free radical initiation to any great extent but the alternating copolymer is easily made with relatively high molecular weight. Only preliminary tests were made, but they indicate that maleic anhydride contributes a tendance toward chain scissioning rather then cross-linking.

A cobalt source was used to provide gamma radiation for a powdered polymer sample under a nitrogen atmosphere. A total dose of 20 Mrad was applied. Solutions with 1.0 dl/g were made of the initial and exposed polymer in THF at 30°C. In the usual fashion, the reduced viscosity, η_{sp}/c was calculated for each. From an original value of 1.81, the η_{sp}/c fell to 0.14 after irradiation.

According to most sources, the homopolymer of the vinyl alkyl ethers react to ionizing radiation predominantly by cross-linking (*18*). It is safe to conclude that maleic anhydride has contributed a predominance of chain scissioning to the copolymer and that, if cross-linking does take place, it is in very small proportion.

Copolymer with Isopropenyl Acetate

Like the vinyl ethers, isopropenyl acetate does not readily form homopolymers by free radical initiation. It does participate in the donor-acceptor mode of polymerization with maleic anhydride and the copolymer was made by free radical polymerization (see Table III). When attempts were made to measure molecular weight by GPC using THF as a solvent on STYRAGEL columns, the polymer did not elute. Therefore, viscosity in acetone at 30°C was used as a measure of molecular size changes on irradiation. The changes on exposure

Table III. Polymerization Conditions and Properties of Maleic Anhydride Copolymers

Reactants					Polyz. Conditions		Polymer	
Maleic Anhydride gms.	Comonomer type	Comonomer wt gms.	Solvent vol. (ml.)	Init. wt,mg.	°C	Time hr	Yield %	MN × 10^4
18	AMS	18	55[b]	3[e]	56	24	30	9.8
7.5	IB	15	70[c]	5.5[e]	70	2	26	5.0
8.5	MMA	20	50[b]	5[e]	56	40	14.5	15.0
5	IPA	5	100[d]	4.5[a]	50	48	11	—

a = 2,2[1] — azo bis (2,4 dimethyl-valenonitrite)
b = acetone
c = benzene
d = toluene
e = Bz_2O_2

```
  CH3
  |
 -C- CH2 -CH - CH-    P(IPA-MA)
  |        |    |
  O        C    C
  |      // \  / \\
 C=O    O    O    O
  |
  CH3
```

to gamma radiation are not big. Doses of 25 and 50 Mrad decreased the η_{sp}/c from an original value of 0.245 to 0.167 and 0.128 dl/g, respectively. If it is assumed that the initial molecular weight is on the order of 40,000, (substituting the values for poly(vinylacetate) in the Mark-Houwink equation), the total effect of radiation in this experiment might amount to a $G(s)$ of about 1.2 times that of PMMA. Some previous unpublished work in our laboratory in which isopropenyl acetate was copolymerized with methyl methacrylate had also led to the conclusion that poly(isopropenylacetate) would be only slightly more sensitive than PMMA. It is not possible to check for cross-linking when only viscosity is measured on irradiation. However, there was nothing in the dissolution behavior of the exposed copolymer to indicate the formation of any gel fraction.

Copolymers with Olefins

The preparation of alternating copolymers of maleic anhydride with olefins has been reported extensively (*19*). In the present work, two copolymers have been examined. The ethylene copolymer is manufactured by the Monsanto Company as a dispersant and thickener, mainly for aqueous applications. The particular example chosen was Monsanto's EMA 31. Measurement of molecular weight by GPC was not possible since it was not soluble in THF. Therefore, dilute solution viscosity was used. The original polymer had a reduced viscosity, η_{sp}/c, of 0.975 dl/g at a concentration of 1.02 dl/g in dimethyl formamide at 30°C and is within the range reported by the manufacturer (*20*). On exposure of the powdered polymer under nitrogen to 20 Mrad in the cobalt gamma cell, the η_{sp}/c dropped to 0.464 dl/g. While no quantitative measure of $G(s)$ can be estimated, it certainly appears that the change in molecular weight is less than that for the methyl vinyl ether copolymer. For that latter copolymer, the same radiation dose changed the η_{sp}/c from 1.81 to 0.14 albeit in a different solvent system.

It is of some interest to note that the structure of the copolymer is that of a head-to-head homopolymer of acrylic anhydride: It is, of course, well established that the predominant reaction of the acrylates to irradiation is cross-linking. Such is apparently not the case for the head-to-head anhydride.

```
 -CH2-CH-CH-CH2-              -CH2-CH-
      |  |                         |
    //C\ /C\\                     C=O
   O    O   O                      |
                                   O
                                   |
     COPOLYMER                     H
                            POLY(ACRYLIC ACID)
```

A second olefin copolymer with a more promising structure is that with isobutylene (2-methylpropene-1). Polyisobutylene itself is a chain-scissioning polymer which has been studied often. It is not much used as a positive resist despite its $G(s)$ of 1.5 to 5 (*21*) because its T_g is so low, about -60°C.

Maleic anhydride and isobutylene, IB, polymerize to give an alternating copolymer p(IB-MA) following the procedure suggested by Bacskai (*22*) (see Table III). The copolymer had an intrinsic viscosity of 0.90 dl/g (dimethyl formamide, 30°C). Judging from the viscosity measurement the GPC molecular weight is 5×10^4. This value appears to be too low by about a factor of two. When we use Frank's (*23*) value for the Mark-Houwink relation for the propylene copolymer with maleic anhydride we obtain a $M_v = 1.1 \times 10^5$ based on intrinsic viscosity. Furthermore, the monomethylester of p(IB-MA) was made by us and gave a (GPC) M_n of 1.2×10^5. It would appear that the anhydride is retarded in the GPC columns by some form of complexation which results in an apparent low molecular weight while the monomethylester does not seem to be retarded. It was not possible to find a suitable solvent for spinning films of p(IB-MA) on wafers so electron beam radiation was not tried. However, the polymer powder under nitrogen was exposed to gamma radiation. A dose of 12.8 Mrad resulted in a $\eta_{sp}/c = 0.46$. Thus the change in molecular size was about the same as that for the ethylene copolymer but with about half the dose.

Copolymer with Methyl Methacrylate

In contrast to the other copolymers in the present work, the copolymers with methyl methacrylate are expected to be random rather than perfectly alternating. Since both monomers contain an electron-poor bond, both would be classified as "acceptors."

The reactivity ratios in the copolymerization of MA and MMA vary according to the solvent used (*24*). The composition of the copolymer described here (see Table III) as calculated from reactivity ratios (as given for methylethylketone as solvent: $r_{mma} = 3.85$; $r_{ma} = 0.02$) is 90.9% MMA and 9.1% MA. Wafers were spin-coated with a 12% solution of polymer in methylcellosolve acetate and baked at 122°C for one hour in a vacuum oven.

The $G(s)$ value is 2.5 times that of PMMA under the same conditions (Figure 4). Our own value of $G(s)$ for PMMA matches the 0.8 value reported by Hatzakis for electron-beam exposure (*25*). Values of 1.3 or even higher are often quoted for gamma-radiated PMMA (*26*). It is significant that MA increases the sensitivity of MMA and that the plot of $(M_n)^{-1}$ versus dose is quite linear. The gradually increasing value of M_w/M_n also is consistent with random chain scission. Thermal stability is known to be changed by copolymerization, copolymers of MMA and maleic anhydride undergo chain scission at a substantially higher rate than PMMA at 200°, 220° and 240°C (*27*). Although maleic anhydride units accelerate chain scission they are said to inhibit volatilization (chain unzipping).

The higher sensitivity indicated by the increased $G(s)$ value can also be tested by measuring the solubility rate ratio, S/S_o where

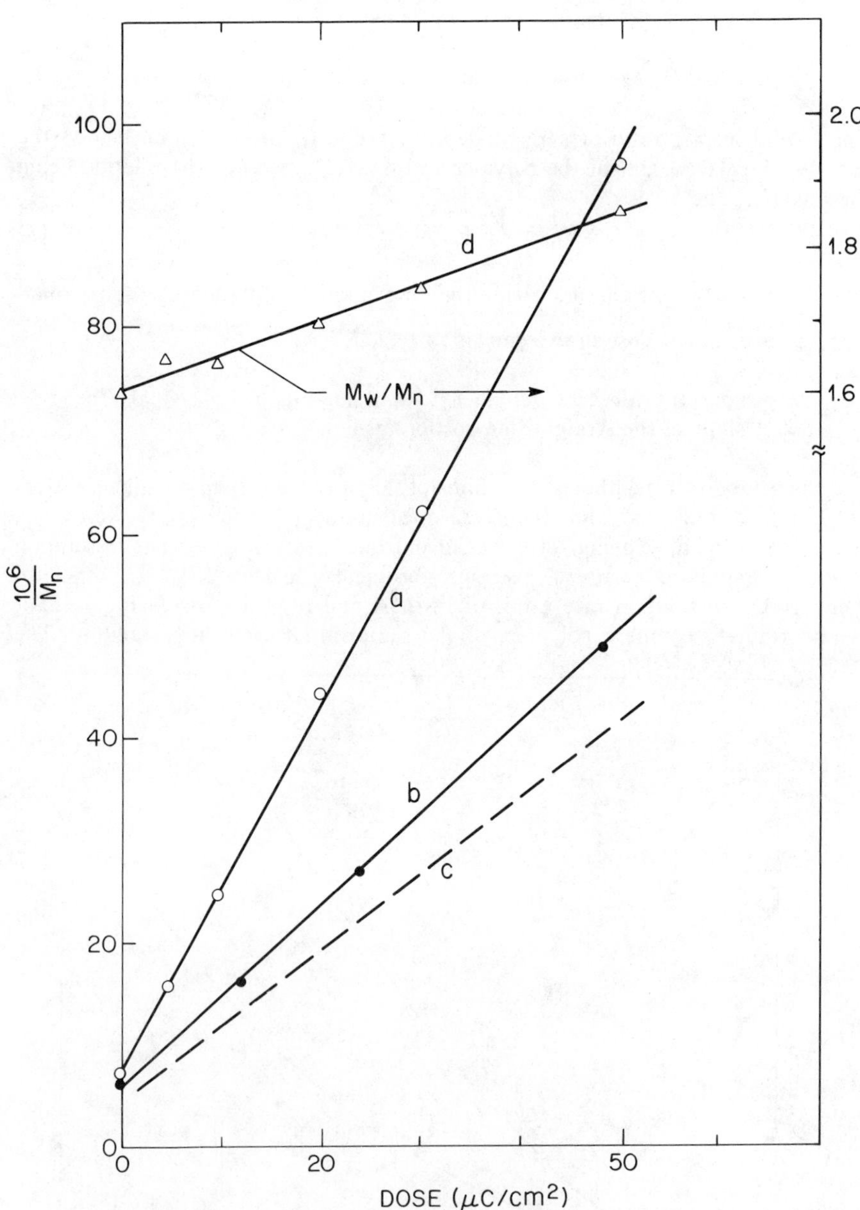

Figure 4. Sensitivity curves for (a) P(MMA-10% MA), (b) P(MMA-10% IPA), and (c) PMMA. Also shown in is (d) the M_w/M_n behavior for P(MMA-10% MA).

S = dissolution rate of exposed polymer

S_o = dissolution rate of unexposed polymer

The solubility rate ratio of a resist depends not only on $G(s)$, but also on the initial molecular weight of the polymer and several constants, the relation being the following (*25*):

$$S/S_o = (1 + FMn_oG(s)D)^a \tag{5}$$

where F is a constant characterizing the absorption of radiation by the polymer

D = incident dose in $\mu C/cm^2$

a = constant for each polymer. (In a log-log plot a equals the slope of the straight line portion at high doses.)

For the solubility rate studies, thin films of polymer were spun on silicon wafers as described previously for the $G(s)$ measurement. The wafers were then immersed in the developing solvent at a controlled temperature. The dissolution rate was monitored with a laser-interferometer device (*28*). In order to compare the dissolution rate ratios of PMMA and p(MMA-MA) (Figure 5) it is important to compare polymers of the same initial molecular weights. The

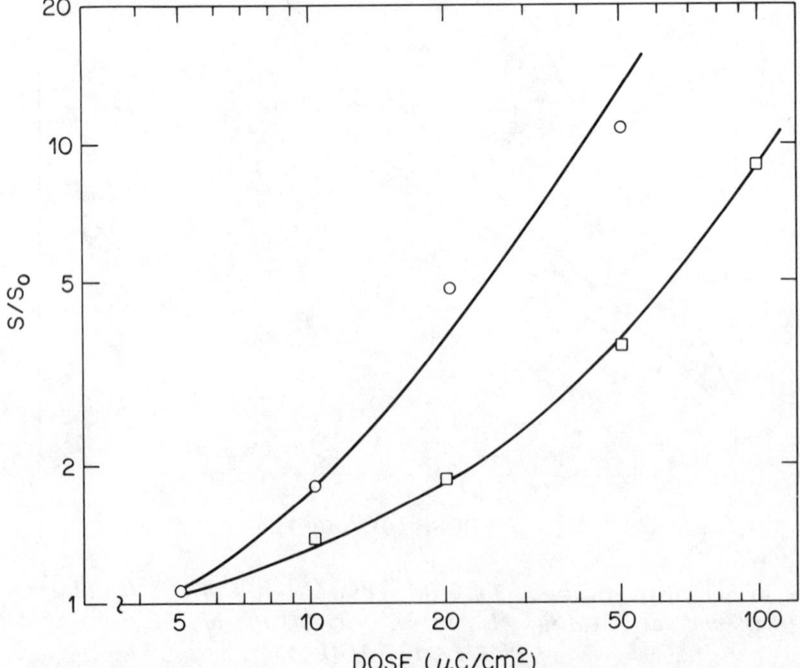

Figure 5. Dissolution rate ratio, S/S_o, for PMMA in methylethylketone (squares), and for P(MMA-10% MA) in methylisobutylketone (circles), both at 27°C. The 50 keV electron beam was used with films about one micrometer thick.

solubility data for PMMA was available for Mn_o = 320,000. It was scaled down to Mn_o = 150,000, the initial molecular weight of the copolymer. Figure 5 shows higher values for the solubility rate ratios at comparable dose for the polymer containing 10% MA but not a steeper slope in the linear part of the log-log plot. In other words the slope "*a*" in the equation above is about the same for PMMA and the copolymer. The value of "*a*" is very important. It can be related to the traditional contrast parameter γ. To produce a resist more sensitive than PMMA, both a higher $G(s)$ value and a steeper slope "*a*" are desirable (*29*).

Conclusions

From this exploratory study, we can state confidently that maleic anhydride enhances the radiation sensitivity of its copolymers, both the random and the perfectly alternating ones. Moreover, the methyl half ester seems even more sensitive than the anhydride. However, this latter conclusion is based on very limited evidence. Sensitivity measured by $G(s)$, or even by a contrast curve is still a far cry from being a demonstration of lithographic utility. A great deal of effort is needed to optimize solvent systems for spinning and especially for development of patterns. Another potential asset of the structure under study here is its reactivity towards amine-isocyanate, and hydroxyl-bearing compounds. Reaction with these compounds *after* development of patterns could be used to convert the copolymer film into a layer which is thicker, and which might have high thermal stability or resistance to plasma etching.

Acknowledgments

The authors thank The International Business Machine Corporation for their financial support.

Literature Cited

1. Dole, M. "The Radiation Chemistry of Macromolecules"; Academic Press: New York, 1973; Vol. II, p. 98.
2. Trivedi, B. C.; Culbertson, B. M. "Maleic Anhydride"; Plenum Press: New York, 1982, p. 239.
3. Hih, D. J. T. *et al.*, *Prog. Polym. Sci.*, 1982, *8*, 215.
4. Thompson, L. F.; Willson, C. G.; Bowden, M. J. "Introduction to Microlithography"; ACS SYMPOSIUM SERIES No. 219, American Chemical Society: Washington, D.C., 1983, p. 56.
5. Everhart, T. E.; Hoff, P. H. *J. Appl. Phys.* 1971, *42*, 5837.
6. Chapiro A. "Radiation Chemistry of Polymeric Systems"; Interscience Wiley: New York, 1962; p. 538; Schnabel, W. In "Aspects of Degradation and Stabilization of Polymers"; Jellinek, H. H. G., Ed.; Elsevier, Amsterdam, Oxford, New York, 1978, p. 159.
7. Seymour, R. B.; Garner, D. P. *Polymer* 1976, *17*, 21.
8. McNeil I. C. In "Developments in Polymer Degradation — 1"; Grassie, N., Ed.; Applied Science: London, 1977; p. 47.
9. Saito, O. *J. Phys. Soc. Jpn.* 1968, *13*, 198.
10. Hiraoka, H. *IBM J. Res Dev.* 1977, *21*, 121.

11. (a) Anderson, C. C.; Krasicky, P. D.; Rodriguez, F.; Namasté, Y. M. N.; Obendorf, S. K. In "Polymers in Electronics"; Davidson, T., Ed.; ACS SYMPOSIUM SERIES No. 242, American Chemical Society: Washington, D.C., 1984, p. 119.

(b) Namasté, Y. M. N.; Obendorf, S. K.; Anderson, C. C.; Rodriguez, F. paper presented at Santa Clara Meeting of SPIE in March, 1984.

12. Kurokawa, M.; Minoura, Y. *J. Polym. Sci. - Polym. Chem. Ed.* 1979, *17*, 473.
13. Cabaness, W. R.; Chang, C H. *ACS Polymer Chem. Div. Preprints* 1977, *18*, 545.
14. Urushido, K. *et al.*, *J. Polym. Sci. - Polym. Chem. Ed.* 1981, *19*, 245.
15. Chapiro, A. "Radiation Chemistry of Polymeric Systems"; Interscience Wiley: New York, 1962, p. 538; Schnabel, W. In "Aspects of Degradation and Stabilization of Polymers"; Jellinek, H. H. G., Ed.; Elsevier, Amsterdam, Oxford, New York, 1978, p. 159.
16. Hatzakis, M. *J. Vac. Sci. Tech.* 1975, *12*, 1276.
17. Trivedi, B. C.; Culbertson, B. M. "Maleic Anhydride"; Plenum Press: New York, 1982, p. 239.
18. Bovey, F. A. "The Effects of Ionizing Radiation on Materials and Synthetic High Polymers"; Interscience Wiley: New York, 1958, p. 256.
19. Trivedi, B. C.; Culbertson, B. M. "Maleic Anhydride"; Plenum Press: New York, 1982, p. 239.
20. "EMA," Tech. Bull. No. 1C/FP-7, Monsanto Company, St. Louis, Mo., 1980.
21. Schnabel, W. "Aspects of Degradation and Stabilization of Polymers"; Jellinek, H. H. G., Ed.; Elsevier, Amsterdam, Oxford, New York, 1978, p. 159.
22. Bacskai, R. *et al.*, *J. Polym. Sci.* 1972, *A-1*, 1297.
23. Frank, H. P. *Makromol. Chem.* 1968, *114*, 113.
24. Case, C.; Loucheux, C. *J. Macromol. Sci. Chem* 1978, *A12*, 1501.
25. Hatzakis, M.; Ting, C.; Viswanathan, N. "Fundamental Aspects of E-Beam Exposure of Polymer Resists"; Electron, Ion and Photon Beam Science and Technology, Sixth International Conference, San Francisco, May 1974.
26. Schnabel, W.; Sotobayashi, H. *Prog. Polym. Sci.* 1983, *9*, 315.
27. Grassie, N.; Davidson, A. J. *Polymer Degradation and Stability* 1980, *3*, 25.
28. Krasicky, P. D.; Rodriguez, F. paper to be presented at 18th ACS Middle Atlantic Regional Meeting, Newark, May 1984.

RECEIVED August 6, 1984

17

Functionally Substituted Novolak Resins: Lithographic Applications, Radiation Chemistry, and Photooxidation

H. HIRAOKA

IBM Research Laboratory, San Jose, CA 95193

Since the days of Bakeland many varieties and applications of phenol-formaldehyde resins have been found (*1*). In this paper we are concerned only with recent applications of functionally substituted phenol-formaldehyde-Novolak resins as applied to lithographic uses for micro-image fabrication.

Apart from multi-level layer resist systems, conventional positive-tone resists can be classified into two categories: one-component and two-component systems. Classical examples of the former systems are poly(methyl methacrylate), and poly(butene-1-sulfone) (*2,3*). Typical examples of the latter system are AZ-type photoresists, which are mixtures of cresol-formaldehyde-Novolak resins and a photoactive compound acting as a dissolution inhibitor (*4*). The most popular photoactive compound, **1**, is a substituted diazo-naphthoquinone shown below together with a cresol-formaldehyde Novolak resin, **2** (*5*). There are many varieties of photoactive compounds that generate

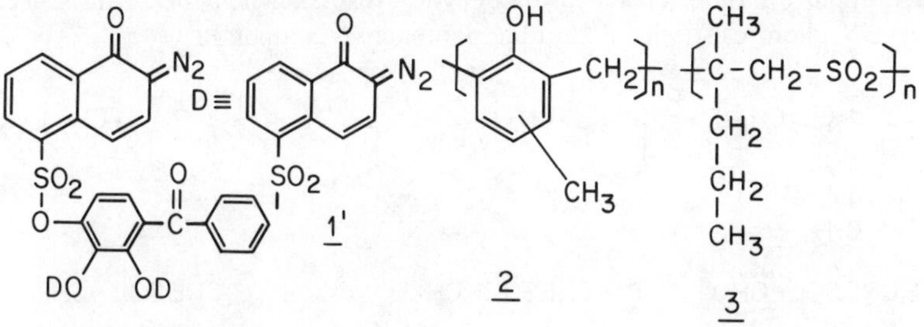

positive tone patterns. More recently a polymeric dissolution inhibitor, **3**, has been reported for electron beam exposure (*5*); a similar mixture has been used for deep UV exposure (*6*). The photochemistry of the diazo-naphthoquinones is described in the following reactions (*7,8*). The photoactive compounds are no

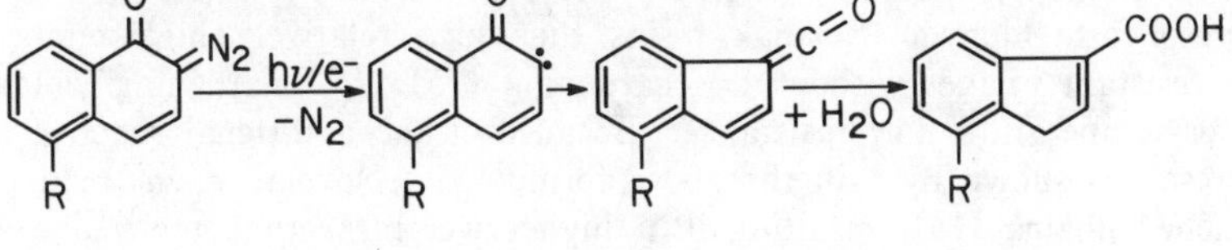

0097–6156/84/0266–0339$06.75/0

longer dissolution inhibitors after irradiation, but act as dissolution promoters in an alkaline developer because of the acquired acid functionality. This reaction also takes place under electron beam exposure, providing positive-tone polymer images (*9-11*).

The UV-insensitive Novolak resins are expected to be inert upon exposure. Despite this expectation, the selection of proper Novolak resins is very important for such overall resist performance as lithographic sensitivity, resolution, thermal image stability, adhesion, and so on. The reasons why Novolak resins are so important as resist materials and why specific kinds of Novolak resins are being investigated are the main subject of this paper.

Cresol-Formaldehyde Novolak Resins

A mixture of three isomeric cresols is used in a commercially available cresol-formaldehyde Novolak resin. This mixed Novolak resin, Varcum resin (*12*), provides adequate properties as a host resin for near-UV- and mid-UV-photoresist applications. Gipstein and his co-workers prepared pure cresol-formaldehyde Novolak resin from each isomeric cresol and compared their spectroscopic and resist characteristics (*13*). Their data on the UV-absorption spectra of each cresol-formaldehyde Novolak resin together with the commercially available Varcum resin are as follows: the absorbances of 0.2 μm thick Novolak films at 250 nm are 0.165(Varcum), 0.096(o-cresol), 0.092(m-cresol), and 0.055(p-cresol). The so-called "window" in the UV absorption at around 250 nm is a maximum with the *p*-cresol-formaldehyde Novolak resin, while the other isomeric cresol and formaldehyde Novolak resins yielded similar UV absorptions at this wavelength. The smallest UV absorption at 254 nm is an advantage for the *p*-cresol-formaldehyde Novolak when the resin is used for a deep UV photoresist with a suitable photoactive compound (*14*).

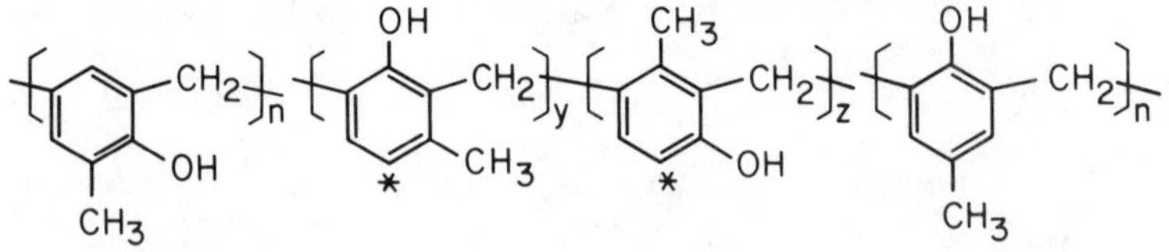

o-CRESOL-CH_2O m-CRESOL-CH_2O p-CRESOL-CH_2O

MAJOR STRUCTURES OF CRESOL-CH_2O NOVOLAK RESINS

Among the three isomeric cresols, however, *m*-cresol provides the Novolak with the highest molecular weight and the highest melting point, as shown in Table I. This is because *m*-cresol provides three reactive sites, whereas *o*- and *p*-cresol provide only two reactive sites. Photoresists made of *m*-cresol-formaldehyde Novolak resins thus give relatively high temperature flow resistant images without UV hardening (*15*). UV hardening yields even higher temperature flow resistance. Because of the additional reactive site in *m*-cresol, as shown by * in the above formula, its Novolak resins tend to give branched chains (*16*), resulting in a higher weight-average molecular weight and a higher molecular dispersity, as shown in Table I.

Table I. Comparison of Molecular Weights and Absorbances of Cresol-Formaldehyde Novolak Resins[a]

Resin	M_w [b]	M_n [c]	Absorbance[d] at 250 nm
Varcum	11,000	1,000	0.165
o-Cresol	540	490	0.096
m-Cresol	1,050	450	0.092
p-Cresol	650	460	0.055

a CH_2O/Cresol molar ratio 0.83; catalyst: oxalic acid; 140° C/22hr.

b LALLS measurements.

c VPO measurements.

d Absorbance at 250 nm of 0.2-μm thick films.

Bulky Alkyl/Aryl Substituted Phenol-Formaldehyde Novolak Resins. Low molecular weight cresol-formaldehyde Novolak resins tend to have high solubility rates in alkaline developers. To increase developer resistance, Novolak resins containing a hydrophobic chain incorporated on a portion of the phenol group were synthesized, as shown below (*17*). As the number of alkyl carbons per phenol group increased from 1.0 to 1.7, developer resistance of the resin increased considerably. However, incorporation of too many hydrophobic groups resulted in non-uniform wetting on silicon dioxide substrates. The *p-t*-butyl phenol, *o-t*-butyl, *p*-phenyl, *p*-nonyl, and *m*-pentadecyl phenols were incorporated into the resins, and their performance tested. In resist systems formulated from these resins and with a fixed *t*-butyl ratio and fixed sensitizer concentration, the molecular weight and molecular weight distribution were shown to influence photoresist sensitivity. With a reduced polydispersity, the resist photosensitivity was demonstrated to increase, provided other factors like sensitizer concentration and molecular weight were the same.

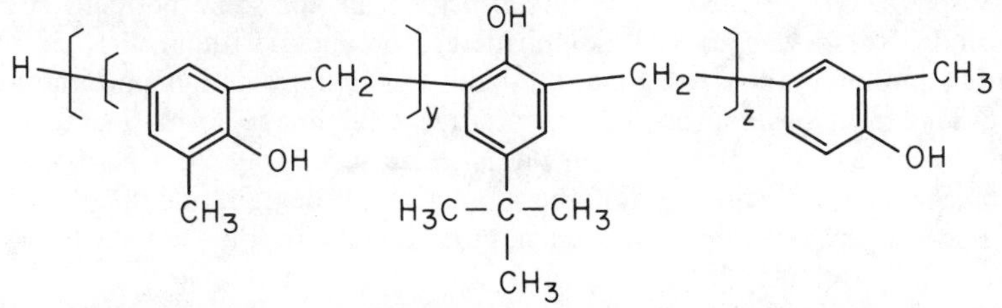

The structural variations of Novolak resins also influence how well they mix or form solid solutions with a dissolution inhibitor when resist films are cast onto substrates. This is a crucial problem for resist formulation. Usually, cresol-formaldehyde Novolak resins mix well with photoactive compounds like a mono-, di-, or tri-esters of 2,3,4-trihydroxybenzophenone with naphthoquinone-diazide sulfonates, provided the concentration of the photoactive compound is kept below about 20 wt%. The solubility of the sensitizer in the Novolak decreases with an increase in the degree of esterification. Solubility also depends on the chemical structure to which the diazo-naphthoquinone moiety is attached, substituents on the Novolak resin, and casting solvent.

When two polymeric systems are mixed together in a solvent and are spin-coated onto a substrate, phase separation sometimes occurs, as described for the application of poly(2-methyl-1-pentene sulfone) as a dissolution inhibitor for a Novolak resin (*4*). There are two ways to improve the compatibility of polymer mixtures in addition to using a proper solvent: modification of one or both components. The miscibility of poly(olefin sulfones) with Novolak resins is reported to be marginal. To improve miscibility, Fahrenholtz and Kwei prepared several alkyl-substituted phenol-formaldehyde Novolak resins (including 2-*n*-propylphenol, 2-*t*-butylphenol, 2-*sec*-butylphenol, and 2-phenylphenol). They discussed the compatibility in terms of increased specific interactions such as formation of hydrogen bonds between unlike polymers and decreased specific interactions by a bulky substituent, and also in terms of "polarity matches" (*18*). In these studies, 2-ethoxyethyl acetate was used as a solvent (*4,18*). Formation of charge transfer complexes between the Novolak resins and the poly(olefin sulfones) is also reported (*6*).

Another interesting aspect is the claim that a bulky alkyl/aryl substituted phenol-formaldehyde Novolak resin itself provides positive tone polymer images after the samples are exposed to electron beams at a dose of about 5×10^{-6} C/cm^2 even when maintained in air for some period, and are processed in wet developer (*19*). The images obtained without the use of a dissolution inhibitor appeared good, although no high resolution patterns were shown. The mechanism proposed for positive tone image formation involved oxygen attachment to radical sites on the Novolak chain to yield a peroxide; the same mechanism is proposed for thermo-oxidative degradation as described later. Because the reactions involve a number of factors, the reproducibility of good quality images may be a problem.

Novolak resins alone usually provide negative tone polymer images when they are exposed to high doses of electron beams, about 1×10^{-3} C/cm^2, followed by image development in an alkaline solution. As discussed later, high dose electron or UV-light exposures of AZ-type photoresists yields negative tone polymer images because of cross-linking (*11*). There are several methods for image reversal, which may or may not involve Novolak resins in the reactions (*20*).

Cross-Linking Novolak Resins. Under normal exposure doses of electron beams, below 5×10^{-5} C/cm^2, or under mid- to near-UV-light exposure, cross-linking of phenol/cresol-formaldehyde Novolak resins is insignificant. However, there are increasing demands for readily cross-linkable Novolak resins for packaging applications such as sealing of semiconductor devices, and for other applications. This can be done by thermally curing AZ-type photoresist images above 160 °C for several hours. Another way is to use epoxyphenol-formaldehyde Novolak resins. Epoxide Novolaks obtained by the reaction of

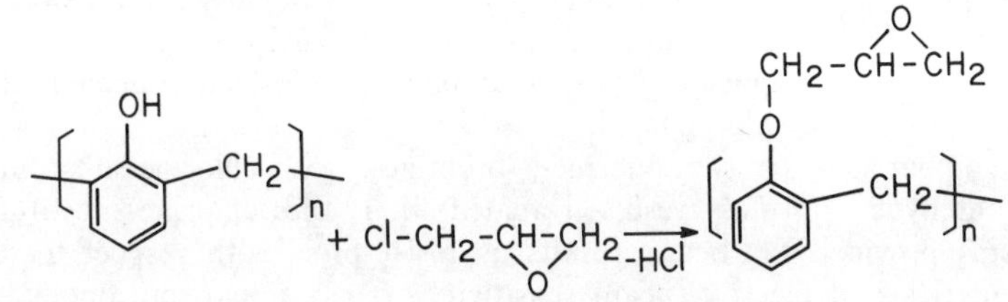

Novolak resins with epichlorohydrin are thermally cross-linked, and the resultant polymer layers show excellent mechanical and dielectric properties (*21*). Another type of negative tone Novolak resin was obtained by direct incorporation of the photoactive chromophore into the polymer chains as shown below (*21*):

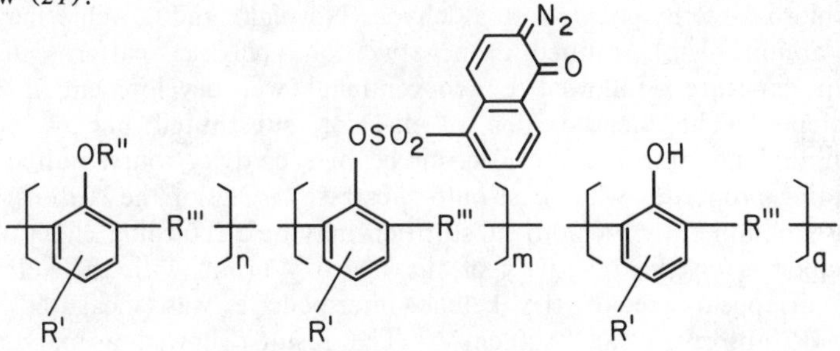

R': -HYDROGEN, HALOGEN, ALKYL, PHENOXY, ARALKYL, $-CH_2-C_6H_4-O-C_6H_5$

R'': -ALKYL, HALOGEN-ARYL, ARYL, ARALKYL $-CH_2-C_6H_4OC_6H_4-CH_2-CH-CH_2$ (epoxide, O)

R''': $-CH_2-$, $-CH_2-C_6H_4-CH_2-CH_2-C_6H_4OC_6H_4OC_6H_4-CH_2-$

m = 70-30 mol%; n = 20-70 mol%; q = 0-10 mol%

Chlorinated Novolak Resins. Mixtures of a cresol formaldehyde Novolak resin and a photoactive compound cross-link at electron doses far smaller than the dose required for the Novolak resin alone (*11*). The reason for this accelerated cross-linking is the reactions between the ketene (an intermediate formed from the photoactive compound upon irradiation) and the Novolak resin. This reaction may be reduced by using a Novolak resin modified for this purpose, or by using certain additives. The rationale for developing a halogen-substituted Novolak resin is the control of the reaction between the intermediate ketene and the Novolak.

Many halogen-substituted cresol/phenol-formaldehyde Novolak resins were prepared and tested for their electron beam sensitivity when mixed together with a photoactive compound. The following halogen-substituted phenols and cresols were used as starting materials: *o*-, *m*-, *p*-chlorophenols; *o*-, *m*-, *p*-bromophenols; *o*-, *m*-, *p*-fluorophenols; *p*-iodophenol; 2-chloro-5-methylphenol; 4-chloro-3-methylphenol; 2-chloro-3-methylphenol; 2-bromo-4-methylphenol; 2,5-dichlorophenol; 3,4-dichlorophenol; 2.3-dichlorophenol; 2,4-dichlorophenol; and 3,5-dichlorophenol.

The comparisons of the fluorine-, bromine-, and chlorine-substituted phenol-formaldehyde Novolak resins revealed that the chlorine-substituted Novolak resins provided far better results as resist films with respect to film forming properties and electron beam sensitivity. The comparison among the chlorine-substituted phenol-formaldehyde Novolak resins showed ortho-substitution to be the best choice. The chlorine-substituted cresol-formaldehyde Novolak provided better films than the chlorine-substituted phenol-formaldehyde Novolaks. The best result was obtained with the 2-chloro-5-methylphenol-formaldehyde Novolak resin. Some of the Novolak resins (for instance, 4-chloro-3-methylphenol-formaldehyde Novolak and a chlorinated product of Varcum resin) resulted in negative tone polymer patterns after electron beam exposures, followed by conventional wet development in an alkaline developer. The incorporation of di-chloro-substituted phenols into Novolak resins did not improve the electron beam sensitivity, but resulted in poor film forming properties when cast onto substrates. One of the reasons for the good results obtained by *o*-chloro-substitution may be a retarding effect of a neighboring chlorine on the reactivity of the hydroxy group with the ketene. The rate of disappearance of the ketene intermediate was measured by following its IR absorption at 2100 cm^{-1}. The results showed a markedly slower disappearance of the ketene in the ortho-chlorinated Novolak in comparison to other types of Novolak resins (*23*).

Experimental

Preparations of Resins. Most of the Novolak resins studied were prepared in sealed tubes, following the known procedures (*24*). Comparisons were made on samples prepared in this way. However, the 2-chloro-5-methylphenol-formaldehyde Novolak resin was made for more comprehensive studies in the following manner. The warm, preformed mixture of 107 g of 2-chloro-5-methylphenol and 40.5 g of trioxane was added dropwise to a mixture of 107 g of the phenol and 4.9 g of concentrated sulfuric acid (96% assay) at 125 to 130°C in a nitrogen atmosphere. The thickening mixture was then heated to 160 to 170°C, and finally the mixture was placed under vacuum at 180°C for 30 min. The reaction is as follows:

$$\text{2-chloro-5-methylphenol} + [CH_2O]_3 \xrightarrow{\text{HEAT, } H^+} \text{-[}C_6H(CH_3)(Cl)(OH)\text{-}CH_2\text{]}_n\text{-}$$

After the mixture cooled, the solid mass was dissolved in 600 ml of acetone, then filtered and precipitated in water. The polymer was again dissolved in dichloromethane and was precipitated by the addition of hexane to produce a fine, white powder. Further purification of the polymer resin has been carried out with liquid column chromatography over silica gel, using chloroform.

Formulation of Resist Solutions. Forty grams of a Novolak resin was mixed with 10 g of the photoactive compound, and dissolved in 100 g of bis-2-methoxy-ethylether. After wafers were spin-coated, the samples were immediately placed on a hot plate at 82°C for 14 min. The formulation procedure of a composite resist of poly(2-methyl-1-pentene sulfone) in the Novolak resin is as follows; the polysulfone was mixed with the resin (13 wt% solid), and then dissolved in 2-methoxyethyl acetate; the films were spin-coated onto silicon wafers, and then baked at 100°C for 20 min prior to electron beam exposure.

Development of Resist Patterns. Development was done in AZ2401 developer diluted with 2 to 5 times its volume of water; AZ2401 is an aqueous solution of KOH with a surfactant. When the resist films were exposed to electron beam doses of 5 $\mu C/cm^2$ at 25 keV, it usually took 1.5 to 2.0 min for complete development of the images using a diazo-naphthoquinone sensitizer with *o*-chloro-cresol-formaldehyde Novolak resin in (1:3) AZ2401/water developer. With poly(2-methyl-1-pentene sulfone) the chlorinated Novolak resin exposed to 1 $\mu C/cm^2$, it took 2.0 min in (1:4) AZ2401 developer for complete image development.

Lithographic Performances. The high-molecular-weight chlorinated Novolak resin has $Mn = 1500$ as determined by vapor pressure osmometry and a $Mw/Mn = 1.2$ determined by gel permeation chromatography (*25*). The chlorinated Novolak resin prepared in sealed tubes had $Mn = 1000$ and $Mw/Mn = 1.4$. These samples prepared in sealed tubes did not have exhibit an induction period for dissolution in a dilute AZ2401 developer. With an extended induction period, resists provide high aspect ratio images because of a large difference in the dissolution rate. Even the resist made from the unfractionated high-molecular-weight resin did not have an induction period, whereas the fractionated resin from the high-molecular-weight sample showed an induction period for dissolution rates in unexposed areas of the resist films. This result is shown in Figure 1. The chlorinated Novolak resist with a high-molecular-weight fraction, after exposure in an El 2 electron beam exposure system at a dose of 5.6 $\mu C/cm^2$, developed in a non-linear fashion. Typical resist images obtained with the chlorinated Novolak resin with a diazo-naphthoquinone sensitizer are shown in Figure 2 exposed with various electron beam doses at 25 keV.

For further enhancement of electron beam sensitivity, the chlorinated Novolak resin was studied using poly(2-methyl-1-pentene sulfone) as a dissolution inhibitor. The chlorinated Novolak resin mixed well with the polysulfone, and there was no phase separation observed when the films were spin-coated. With 13 wt% of the polysulfone, the chlorinated Novolak resist cast from a cellosolve acetate solution yielded fully developed images with $R/R_o = 9.2$ after exposure to 2 $\mu C/cm^2$. It gave fully developed images with $R/R_o = 3.2$ at a dose of 1 $\mu C/cm^2$, as shown in Figure 3. There are some problems with this resist system; some cracking of the developed resist images

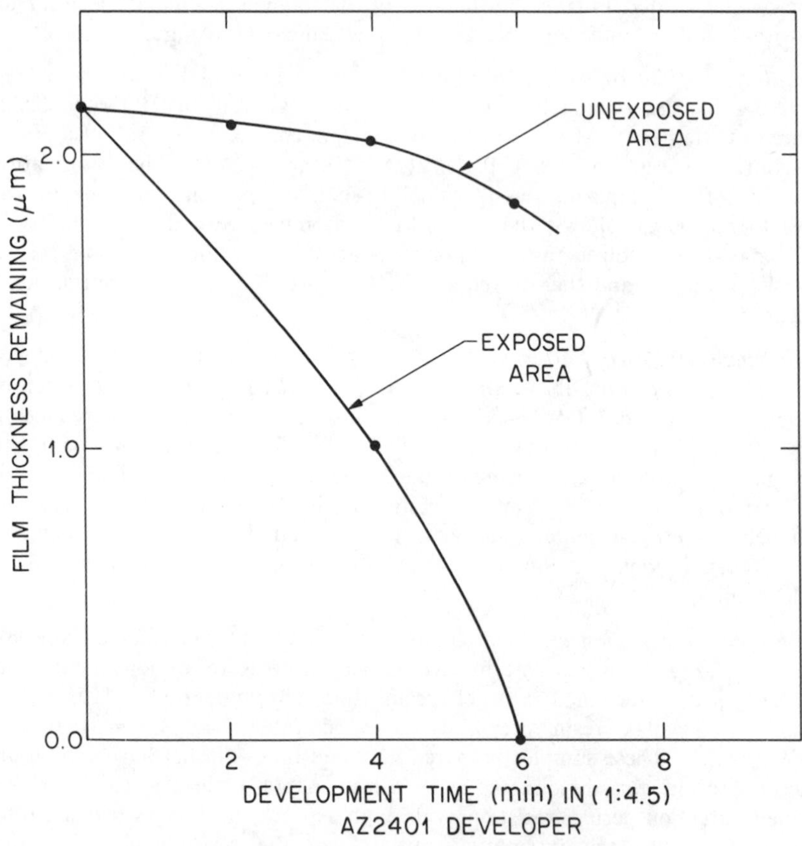

Figure 1. Dissolution rates of a composite resist made of a diazonaphthoquinone sensitizer and o-*chloro*-m-*cresol-formaldehyde Novolak resin after 5 μC/cm² electron beam exposures. Note: this kind of an induction period appeared only in the high-molecular-weight fraction resin.*

appeared during image development in the developer. This problem has not yet been completely solved.

Acetaldehyde, butylaldehyde, and furfuraldehyde are reported to replace formaldehyde for special purpose Novolak resins (*1*). However, the formaldehyde Novolak resins are reported to release formaldehyde as the major gaseous product upon heating at 450°C (*26*). Formaldehyde is a well-known toxic compound and should be avoided if possible. Benzaldehyde has never been reported as a component for Novolak resins, but has been found to replace formaldehyde in Novolak resin preparations to yield fairly good resist materials.

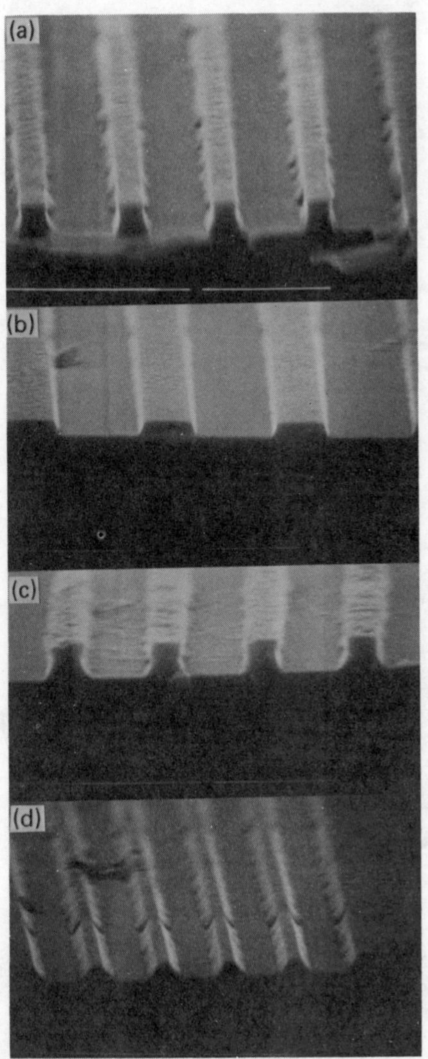

Figure 2. SEM pictures of a composite resist made of a diazo-naphthoquinone sensitizer and o*-chloro-*m*-cresol-formaldehyde Novolak resin after various electron beam doses: (a) 7 μC/cm², (b) 5 μC/cm², (c) 5 μC/cm², (d) 3 μC/cm².*

The preparation method is similar to that for the cresol-formaldehyde Novolak resin with a molar ratio of cresol/benzaldehyde = 1.1 in acidic conditions. We have prepared varieties of substituted *m*-cresol-benzaldehyde Novolak resins, and 1-, and 2-naphthol-4-hydroxybenzaldehyde Novolak resins in the same manner. Almost all of these benzaldehyde Novolak resins give excellent resist films when spin-coated onto silicon or silicon dioxide substrates after being dissolved, together with a photoactive compound, in a solvent like 2-methoxyethyl ether.

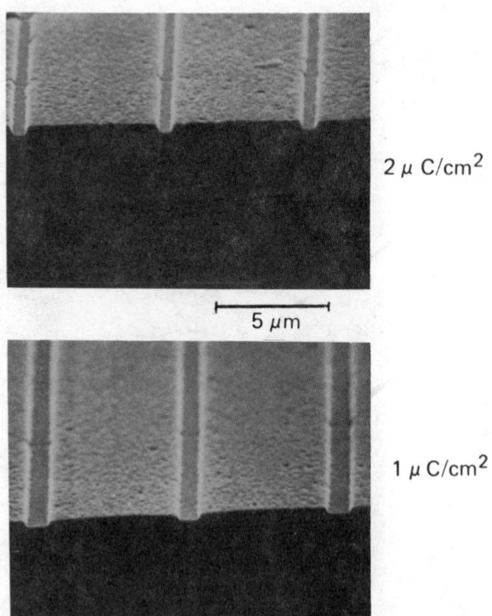

Figure 3. SEM pictures of a composite resist made of poly(2-methyl-1-pentene sulfone) and o-*chloro*-m-*cresol-formaldehyde Novolak resin after electron beam exposures with doses of (a) 2 μC/cm², and (b) 1 μC/cm².*

HEAT, H^+

HEAT, H^+

X: −H, OH, CN, NO_2, Cl, Br, OCH_3.

The glass transition temperatures of these Novolaks are given in Table II. Except for the methoxy group substituted benzaldehyde Novolak resins, their glass transition temperatures are reasonably high, and the elemental analysis shows one-to-one condensation of *m*-cresol and substituted benzaldehyde under acidic polymerization conditions (*27*).

Table II. Glass Transition Temperatures of m-Cresol-Substituted Benzaldehyde Novolak Resins

X	T_g, °C	M_w	M_n
H	112	840	550
2-Br	137	–	–
3-Br	112	–	–
4-Br	84	–	–
2-OH	137	–	–
3-OH	181	–	–
4-OH	160	–	–
2-MeO	120	–	–
3-MeO	72	–	–
4-MeO	52	–	–
3-CN	150	–	–
4-CN	200	–	–
2-Cl	142	–	–
3-Cl	140	–	–
4-NO_2	132	–	–

The UV-absorption spectra of these Novolak resins vary widely depending upon substitution. However, the *m*-cresol-benzaldehyde Novolak resin is characteristic in its transparency within 300-320 nm in comparison with the cresol-formaldehyde resin. The chlorinated Novolak resin made of 2-chloro-5-methylphenol and formaldehyde has a slightly stronger UV absorption in this wavelength range, but weaker absorption in the range of 250 and 300 nm in comparison with a commercially available cresol-formaldehyde Novolak resin, as shown in Figure 4.

Typical resist images obtained after a mid UV-exposure with an UV3 Perkin Elmer Exposure System are shown in Figure 5 using the *m*-cresol-benzaldehyde Novolak and the diester photoactive compound. The experiments were not carried out under optimum conditions, and the exposure dose at 313 nm was 500 mJ/cm^2. Although this resist system was not the fastest one at this wavelength region, it clearly provides usable patterns.

Although the Novolak resin of the *m*-cresol-benzaldehyde failed to show a marked increase in CF_4 plasma etching resistance, the Novolak resins of hydroxy-naphthalene-hydroxybenzaldehyde showed a remarkable increase in the plasma etching resistance. The resist films also yielded excellent patterns when used together with a diazo-naphthoquinone sensitizer; almost non-diluted AZ2401 developer had to be used for image development due to the hydrophobic nature of the naphthalene group.

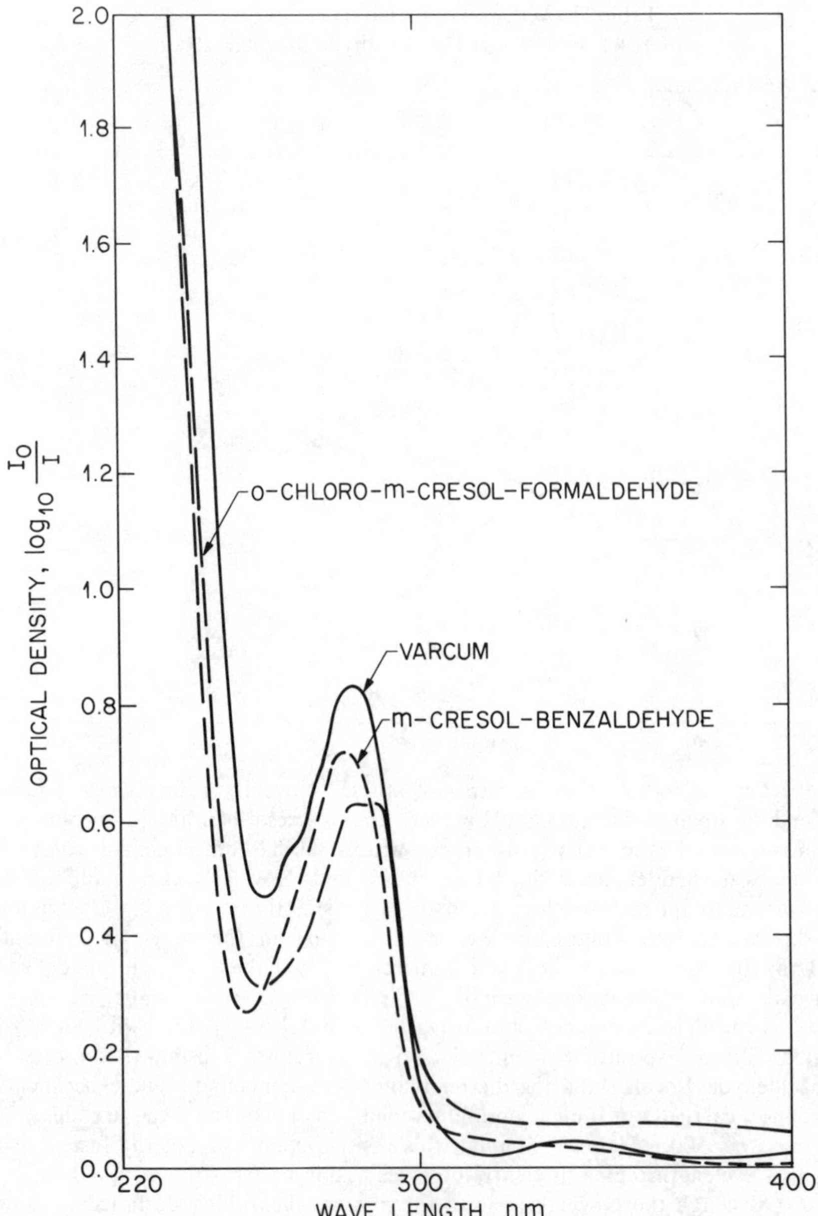

Figure 4. UV-absorption spectra of Varcum resin (solid line), o*-chloro-*m*-cresol-formaldehyde Novolak resin (long dotted line), and* m*-cresol-benzaldehyde Novolak resin (short dotted line); 0.32-µm thick films were coated on quartz plates.*

Figure 5. Resist images made of m-*cresol-benzaldehyde Novolak resin with a photoactive compound (15 wt%); 500 mJ/cm² at 313 nm, and 70 sec in (1:4.5) AZ2401/water developer.*

Results and Discussion

Radiation Chemistry and Thermal Degradation. When high energy radiation (including gamma- and X-rays, electrons, protons and other ions) impinges on polymeric films, ions and radicals are formed with subsequent bond scission and cross-linking. The differences in the bond scission and cross-linking rates determine the resistance to high energy radiation. The rates of degradation are much slower for polymers containing aromatic rings. With an increase in aromaticity, the polymers become more resistant to high energy radiation. Because Novolak resins are thermosetting and contain aromatic rings in the main chain, the polymer is resistant to radiation-initiated and thermal degradation.

In CF_4 plasma etching at 0.5 to 1 Torr pressure, the active species are neutral free radicals like F atoms and CF_3 (*28*). Novolak resins have fairly strong CF_4 plasma etch resistance (*29*), which, because of widespread use of the dry etching process, is one of the required features for resins used as resist materials for IC manufacturing processes.

In CF_4 reactive ion etching of 1×10^{-4} to 1×10^{-1} Torr pressure and -300 to -500 V bias potential, the active species are mostly CF_3^+ ions. Under these RIE conditions, the Novolak resins exhibited strong etch resistance, as shown in Figure 6 (*30*). When formaldehyde was replaced by benzaldehyde, the resins did not show any significant improvement of CF_4 RIE resistance. When benzene rings were replaced by naphthalene, however, substituted hydoxynaphthalene-benzaldehyde Novolak resins showed a marked improvement in CF_4 RIE resistance. A similar improved etching resistance by incorporation of naphthalene rings has been reported for argon ion (300 eV) milling of poly(vinyl naphthalene) (*31*).

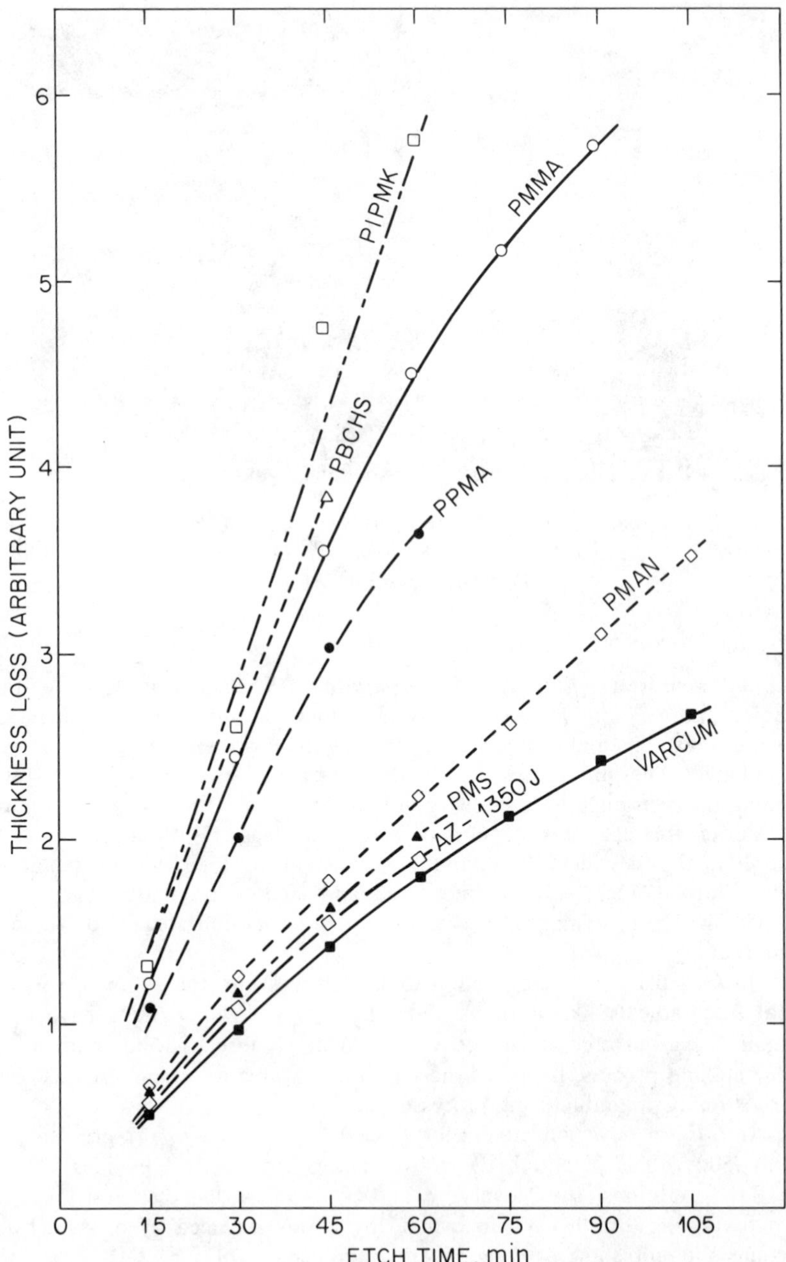

Figure 6. Polymer etching rates in CF_4 reactive ion etching: 0.21 W/cm^2 R.F. power density, 4×10^{-4} Torr, -500 V bias potential. The thickness loss is shown in the unit of 1.6×10^{-6} *m.*

The pyrolysis of Novolak resins is reported to take place in three stages (*1,31*). In the first stage in which temperatures of up to 300°C are used, significant degradation takes place only by releasing water, phenol and formaldehyde which were trapped during curing. In the second stage, with temperatures of up to 600°C, water, CO, CO_2, CH_4, cresol, and xylene are split off in significant amounts, while a strong increase in the amount of the ketone/carboxylic group (IR: 6.05 μm) is observed. In the third stage, with temperatures above 600°C, CO_2, CH_4, water, benzene, toluene, phenol, cresol, and xylene are released, causing a significant shrinkage. In the second and third stages, thermo-oxidative degradation is assumed to occur regardless of whether the pyrolysis occurs in oxidative or inert atmosphere (*32*). Enough oxygen is present in the resins even in vacuum. Lee, however, reported significant release of formaldehyde at 450°C, which was a minor product at 350°C, as shown in Table III (*26*). Our experiments showed only a trace amount of formaldehyde in the pyrolysis of Varcum resin at 450°C. The proposed mechanism of thermo-oxidative degradation is the following:

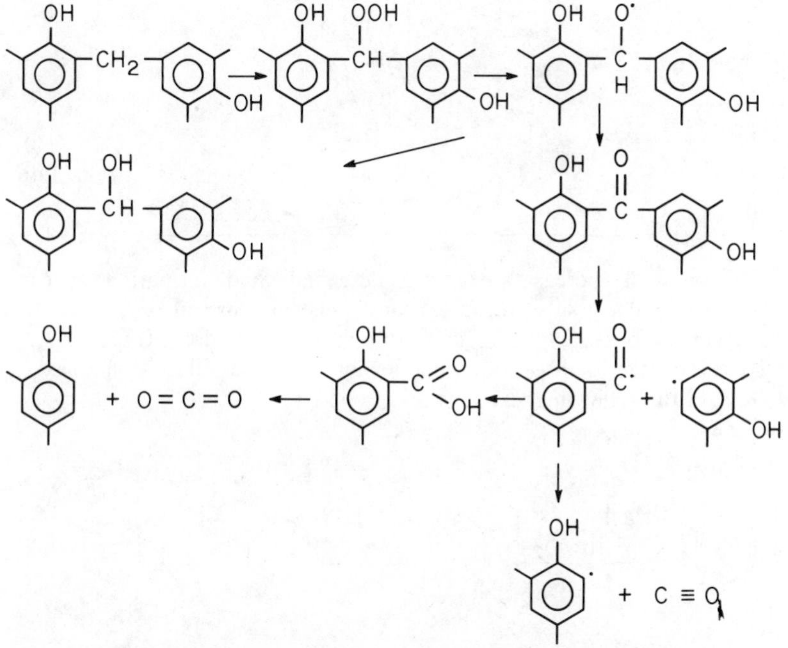

In the report of positive-tone Novolak resins containing no dissolution inhibitor, Fahrenholtz proposed a similar mechanism for image development (*19*).

Photo-Oxidation and UV-Image Hardening. Novolak resins show changes in ESCA data very similar to those of poly(styrenes) when they are exposed in air to UV-light (*33*). With the chlorinated Novolak resin, deep-UV exposures resulted in the elimination of chlorine and the simultaneous formation of carbonyl carbons, as shown in the C_{1S} core level signal at 289 eV (Figure 7). In Figure 8 the C_{1S} core level signal of a cresol-formaldehyde Novolak resin is shown as a function of the UV-exposure time. The increase of the signal at 290

Table III. Volatile Products from Pyrolysis of a Novolak Resin (Ref. *25*)

Compounds	At 350 °C (Mole%)	At 450 °C (Mole%)
Hydrogen	–	0.47
Methane	0.45	0.75
Water	87.40	10.50
Ethylene	0.14	0.51
Carbon Monoxide	2.95	40.20
Formaldehyde	0.38	10.10
Propylene	0.05	1.66
Acetaldehyde	0.19	0.96
Carbon Dioxide	5.84	30.60
Acetone	0.12	–
Propionaldehyde	–	0.30
Acetic Acid	0.64	1.71
Pentadiene	0.16	0.46
Toluene	1.44	1.46
Phenol	–	0.12

and 287 eV with increased exposure dose indicated formation of carbonyl carbons and hydroxy substituted carbons. The increase in oxygen content, of carbonyl or carboxyl types, is shown in the ESCA data (Figure 9), which showed the change of O_{1S} as a function of UV dose. Based on these ESCA and IR data the following reactions are proposed:

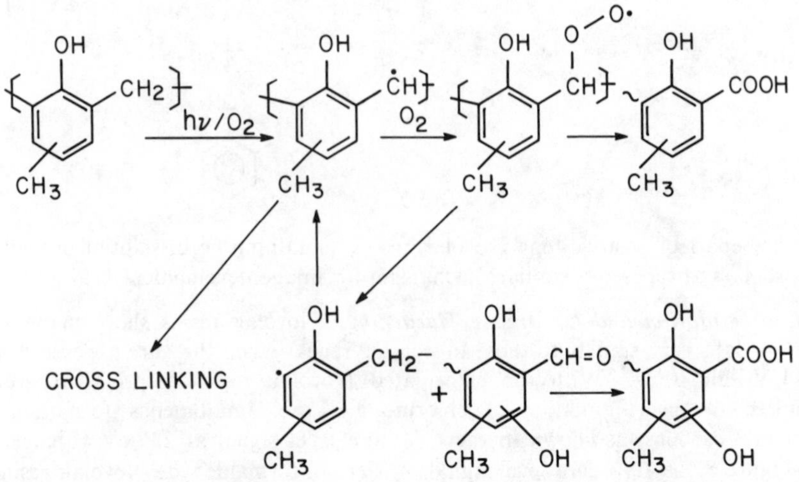

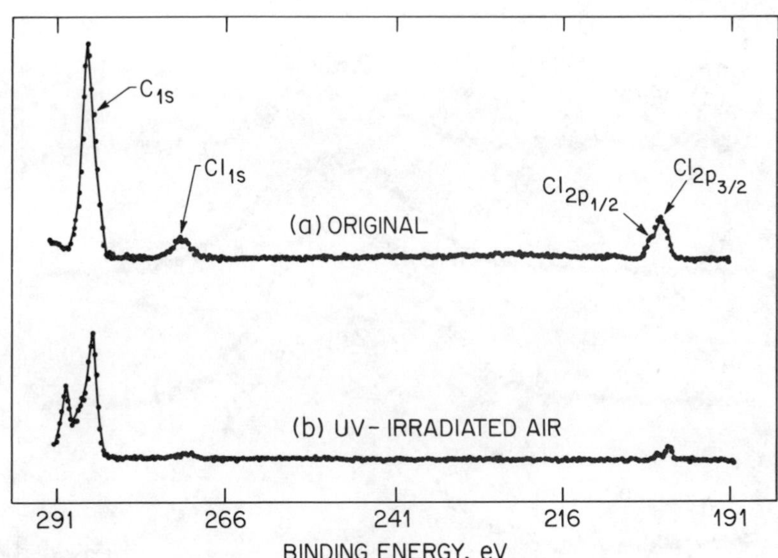

Figure 7. Wide band ESCA spectra of o*-chloro-*m*-cresol-formaldehyde Novolak resin: (a) original films, (b) after 1.5 hr UV exposure.*

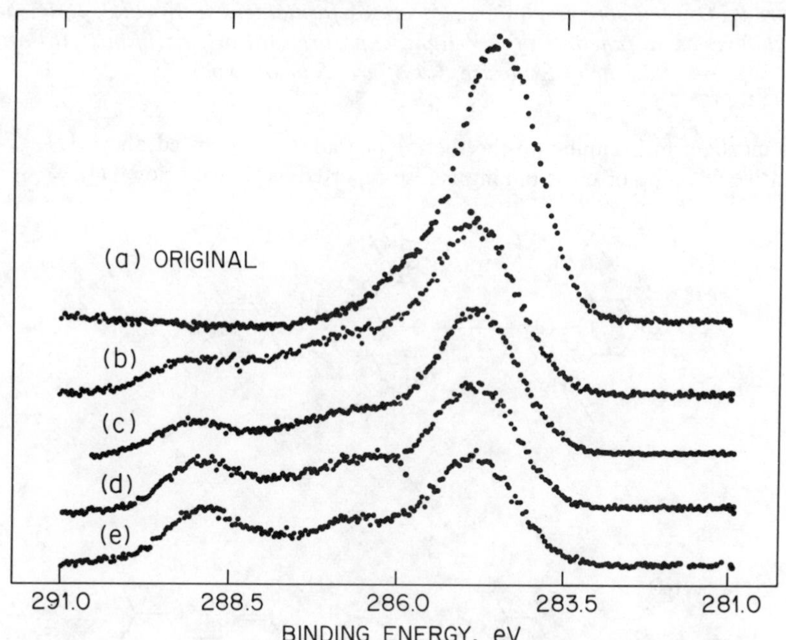

Figure 8. Carbon core level signal of cresol-formaldehyde Novolak resin and its change as a function of UV-light exposure (1 mW/cm^2 at 254 nm): (a) original films, (b) after 0.5 hr of exposure, (c) after 1 hr of exposure, (d) after 1.5 hr of exposure, (e) after 4 hr of exposure.

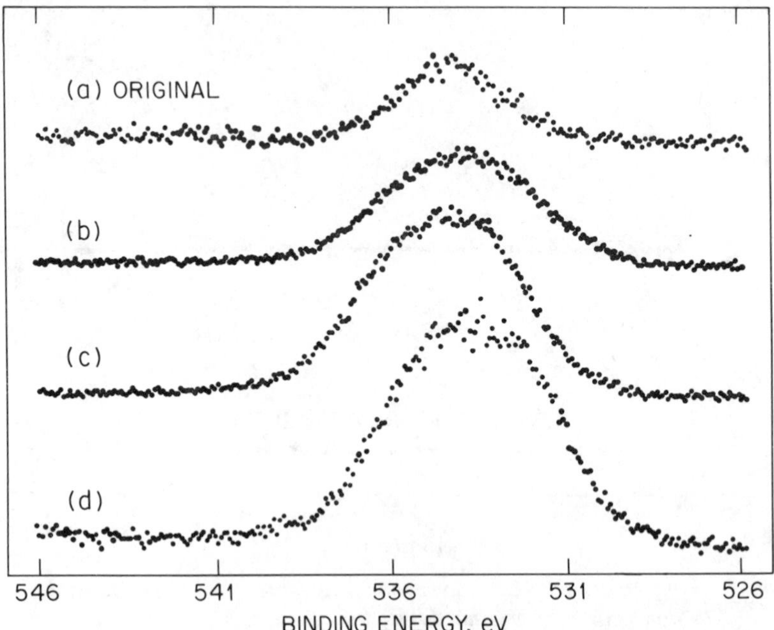

Figure 9. Oxygen core level signal of cresol-formaldehyde Novolak resin, and its change as a function of UV-light exposure: (a) original films, (b) after 0.5 hr of exposure, (c) after 1.5 hr of exposure.

This mechanism is similar to the thermo-degradation discussed above (*1*). The oxidative coupling or cross-linking is also reported as shown below (*1*):

DIPHENOQUINONE

These photo-oxidation and oxidative couplings lead to UV-image hardening (*15,34-5*). An example of UV-hardening of micron-sized images is shown in Figure 10. The UV-hardened resist films still dissolve in certain organic solvents, while thermally cross-linked Novolak resin-based resist images or films are insoluble in organic solvents. These solubility differences reveal different degrees of contribution of the coupling reactions in UV-hardening vs thermosetting. UV-image hardening was originally proposed only for AZ-type photoresists (*15*). However, as demonstrated from the ESCA data, polystyrenes and Novolak resins have similar surface structures after deep-UV exposures. It

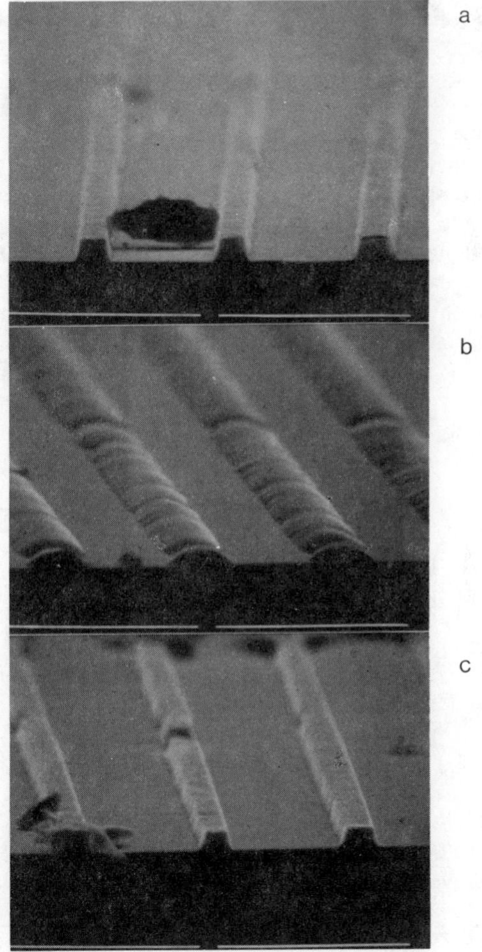

Figure 10. UV-image hardening of AZ2400 patterns: (a) original patterns, (b) 155° C heating for 30 min, (c) 155° C heating for 30 min after 10 min of UV exposure.

was expected, therefore, that UV-hardening would work for substituted polystyrene-based resists. UV-image hardening has now been demonstrated for poly(p-hydroxystyrene)-based resists, (*33*) and also for poly(chloromethyl styrene)-based resists (*36*). The images can be stable to temperatures as high as 350°C for poly(hydroxystyrene)-based photoresist.

UV-image hardening is most efficient when UV light at about 250 nm is used, because the resins have an optical window at this wavelength (Figure 4). The UV light can penetrate resist films a few microns thick, which makes UV-hardening more reliable and wrinkling free in comparison with CF_4 plasma hardening (*37-38*). Plasma hardening depends entirely on surface modifications. Because the UV-hardened surfaces are entirely different in nature, an over-coat layer can be applied on top of the hardened resist layers, making the multi-level resist technique applicable with ease.

Summary

Recent developments in substituted phenol-formaldehyde Novolak resins as applied to microlithography processes have been reviewed, with some emphasis on our own work. Ortho-chloro-substituted *m*-cresol-formaldehyde Novolak resin has been studied for improved electron beam sensitivities with a diazo-naphthoquinone-type dissolution inhibitor and with poly(2-methyl-1-pentene-sulfone). Benzaldehyde-based Novolak resins have been studied for improved lithographic and plasma etching resistance purposes. With increased use of Novolak resins, an understanding of their thermal degradation, photo- and radiation chemistry, and photo-oxidation becomes important in the areas of reactive ion etching, ion milling, packaging, and image hardening. A useful application like UV-image hardening was described in terms of photo-oxidation and oxidative coupling. Its extension to poly(hydroxystyrene)- and poly(chloromethyl styrene)-based resists was also discussed.

Literature Cited

1. Knop, A.; Scheib, W. "Chemistry and Applications of Phenolic Resins"; Springer-Verlag: New York, 1979.
2. Haller, I.; Hatzakis, M.; Srinivasan, R. *IBM J. Res. Develop.*, 1968, *12*, 251.
3. Bowden, M. J.; Thompson, L. F. *J. Appl. Poly. Sci.* 1973, *17*, 3211.
4. Bowden, M. J.; Thompson, L. F.; Fahrenholz, S. R.; Doerries, E. M. *J. Electrochem. Soc.* 1981, *128*, 1304.
5. DeForest, W. S. "Photoresist Materials and Processes"; McGraw-Hill: New York, 1975.
6. Hiraoka, H.; Welsh, L. W. *Amer. Chem. Soc. Div. Org. Coatings and Appl. Poly. Sci.*, Papers, 1983, *48*, 48.
7. Süs, O. *Ann.* 1944, *556*, 65.
8. Pacansky, J.; Lyerla, J. *IBM J. Res. Develop.* 1979, *23*, 42.
9. Hatzakis, M.; Shaw, J. *Electrochem. Soc.*; Extended Abstract Vol. 78-1, 1978, p. 927.
10. Pacansky, J.; Coufal, H. *J. Am. Chem. Soc.* 1980, *102*.
11. Hiraoka, H.; Gatierrez, A. R. *J. Electrochem. Soc.* 1979, *126*, 860.
12. Varcum Corporation trade mark.

13. Gipstein, E.; Ouano, A. C.; Tompkins, T. *J. Electrochem. Soc.* 1982, *129*, 201.
14. Willson, C. G.; Clecak, N. J.; Grant, B. D.; Twieg, R. J. *Electrochem. Soc.*; Extended Abstract Vol. 80-1, May 1980, No. 275.
15. Hiraoka, H.; Pacansky, J. *J. Electrochem. Soc.* 1981, *128*, 2645; *J. Vac. Sci. Technol.* 1981, *19*, 1132.
16. Winkler, E. L.; Parker, J. A. "Molecular Configurations and Pyrolysis of Phenolic Novolaks"; in "Rev. Macromol. Chem."; Butler, G. B.; O'Driscoll, T. F.; Shen, M., Eds.; Marcel Dekker, Inc., New York, 1971, *6*, p. 245.
17. Fahrenholtz, S. R.; Goldrick, M. R.; Hellman, M. Y.; Long, D. T.; Pitetti, R. C. *Am. Chem. Soc., Div. Org. Coat. Plast. Chem.*, Paper, 1975, *35(2)*, 306.
18. Fahrenholtz, S. R.; Kwei, T. K. *Macromolecules* 1981, *14*, 1076.
19. Fahrenholtz, S. R. *J. Vac. Sci. Technol.* 1981, *19*, 1111.
20. Willson, C. G. "Organic Resist Materials"; in "Introduction to Microlithography"; Thompson, L. F.; Willson, C. G.; Bowden, M. J., Eds.; ACS SYMPOSIUM SERIES No. 219, American Chemical Society: Washington, D.C., 1983.
21. Jpn Kokai Tokyo Koho, JP 57, 100, 128; assigned to matsushita Elect. Works, Ltd.
22. Larrischev, V. D. et al., USSR Author's Certificate No. 226, Sept. 5, 1968, 402.
23. Hiraoka, H.; Harada, A. unpublished data.
24. Sorenson, W. R.; Campbell, T. W. "Preparative Methods of Polymer Chemistry"; John Wiley and Sons: New York, 1961.
25. Molecular weights were measured at Springborn Testing Institute, Inc., and also by Lyerla, J. R.; Mathias, D. of our laboratory.
26. Lee, L-H. "Mechanism of Thermal Degradation of Phenolic Condensation Polymers"; in *Proceedings of Battelle Sym.*, on "Thermal Stability of Polymers"; Dec. 5,6, 1963.
27. Elemental analyses were carried out at Childers Laboratories, and the glass transition temperatures of polymers were measured by Carothers, J. A. of our laboratory.
28. Coburn, J. W. "Plasma Etching and Reactive Ion Etching"; Am. Vac. Soc. Monograph Series, 1982.
29. Pederson, L. A. *J. Electrochem. Soc.* 1982, *129*, 205.
30. Hiraoka, J.; Welsh, L. W., Jr. *Radiat. Phys. Chem.* 1981, *18*, 907.
31. Ohnishi, Y.; Tanigaki, K.; Furuta, A. *Am. Chem. Soc. Div. Org. Coat. Appl. Poly. Sci. Proceedings*, 1983, *48*, 184.
32. Conley, R. T. "Thermal Stability of Polymers"; Marcel Dekker Inc.: New York, 1970.
33. Hiraoka, H.; Welsh, L. W. Jr. *Proc. Tenth Intl. Conf. Electron and Ion Beam Sci. Technol., Electrochem. Soc. Proc.*, 1983, *83-2*, p. 171.
34. Birch A. D.; Matthews, J. C. "Microelectronics Manuf. Test."; 1983, Vol. 6, (4) p. 20; Ury, M. G.; Matthews, J. C.; Wood, C. H. *SPIE Proceedings, Optical Microlithography*, Stover, H. L., Ed.; 1983; Vol. 344, p. 241.

35. Allen, R.; Foster, M.; Yen, Y-T. *J. Electrochem. Soc.* 1982, *129*, 1379.
36. Choong, H. S.; Kahn, F. T. *J. Vac. Sci. Technol.* 1983, *B1*, presented at "1983 Intl. Sym. Elect. Ion."; Photon Beams: Los Angeles, CA.
37. Ma, W. *SPIE Proceedings, Submicron Lithography*, Blais, P. D., Ed.; 1983; Vol. 333, p. 20.
38. Moran, J. M.; Taylor, G. N. *J. Vac. Sci. Technol.* 1981, *19*, 1127.

RECEIVED August 6, 1984

18

Synthesis, Characterization and Lithographic Evaluation of Chlorinated Polymethylstyrene

R. TARASCON, M. HARTNEY and M. J. BOWDEN[1]

AT&T Bell Laboratories, Murray Hill, NJ 07974

The requirements of resist materials for submicron lithography include high sensitivity, high resolution and dry-etching resistance. Negative resists have been designed with high sensitivity and dry-etching durability, a combination of properties that is difficult to achieve with positive resists that undergo chain scission. Although positive resists generally exhibit higher resolution, submicron resolution has nevertheless been demonstrated in several negative resists. Polystyrene, for example, is a weakly sensitive negative resist which has excellent plasma-etch resistance because of its aromatic content and exhibits resolution of less than 1 μm. The sensitivity of polystyrene can be markedly enhanced by substitution on the ring with certain substituents, particularly halogen or halomethyl groups. Such studies have produced a new generation of negative electron resists that exhibit high sensitivity while maintaining the other desirable characteristics of polystyrene, viz., excellent dry-etching resistance, negligible post-cure reaction and good resolution.

Resists based on chloromethyl substitution have been extensively studied in the past few years (*1-7*). Halogen and halomethyl groups have been introduced by a variety of methods. Choong and Kahn (*1*) synthesized polychloromethylstyrene (PCMS) by free radical polymerization of chloromethylstyrene and reported a sensitivity ($D_g^{0.5}$) of 0.4 $\mu C/cm^2$ at 20 kV with a contrast of 1.5 for materials with molecular weight of about 400,000 (*1*). The molecular weight distribution of these polymers, all of which contained one chloromethyl group per repeat unit, was about 2. Fractionation of the polymer resulted in improved contrast as a result of narrowing the distribution.

In an alternative approach, Imamura et al. (*2*) chloromethylated polystyrene that had been synthesized by anionic polymerization. This procedure has the advantage of starting with a very narrow molecular weight distribution polymer and obviates the need to fractionate. Imamura et al. found that the sensitivity of the resist, which they called CMS, increased rapidly in the initial stage of chloromethylation but for degrees of chloromethylation (chlorination ratio (CR)) above ~40%, no further increase was observed. In addition, the molecular weight distribution of the chloromethylated product broadened significantly above a CR of 40% with a resultant loss of resolution. They reported a sensitivity ($D_g^{0.5}$) of 0.71 $\mu C/cm^2$ at 20 kV for materials with molecular weight of about 300,000 and a CR of 40%. Recently,

[1] Current address: Bell Communications Research, Inc., Murray Hill, NJ 07974

0097-6156/84/0266-0361$08.25/0

Harita et al. (*3*) obtained a very sensitive deep-UV resist by chlorination of anionically polymerized polymethylstyrene. Although details of their chlorination procedure were not reported, preliminary studies showed that chlorination took place not only on the methyl group attached to the aromatic ring but also on the main chain.

These different synthetic approaches result in structurally different materials and it is important to quantify these effects in terms of lithographic performance so that an optimum synthetic procedure might be identified. In particular, it is important to quantify lithographic performance as a function of the position of chlorine substitution on the polymer molecule.

Chlorination can be effected by a variety of methods most of which involve free-radical mechanisms (*8*). In the present work, a chlorinating agent was sought that would preferentially attack the methyl group of polymethylstyrene so that a comparison with polychloromethylstyrene could be made. Both sulfuryl chloride (SO_2Cl_2) and t-butyl hypochlorite (t-BuOCl) were suitable reagents for this purpose.

The work reported in this paper concerns the synthesis of a series of chlorinated polymethylstyrenes, prepared by chlorination of both anionically and free radically synthesized polymethylstyrene, and their subsequent characterization by ^{1}H NMR, ^{13}C NMR, IR and elemental analysis. The preliminary lithographic evaluation of the chlorinated materials is also discussed.

Experimental

Polymer Preparation. Poly-para-methylstyrene (P-p-MS) was prepared by anionic polymerization in benzene at 50°C initiated by n-butyllithium (*9*) or in THF at 25°C initiated by sodium naphthalene (*10*). Polymerizations in benzene allowed preparation of more monodisperse materials than those prepared in THF since the propagation rate is slower relative to the initiation rate in the nonpolar solvent (*11*). Two different molecular weight materials were chlorinated (P-p-MS1 and P-p-MS2).

Polymethylstyrene (PMS) and PCMS (both prepared from mixtures of meta and para isomers) were prepared by free radical polymerization of the respective monomers in toluene initiated by benzoyl peroxide at 85°C (*12*). Chlorination of P-p-MS was carried out in carbon tetrachloride (CCl_4) at 60°C in presence of AIBN using the desired amount of SO_2Cl_2 (*13*). Chlorination of PMS was carried out in CCl_4 at 50°C under irradiation from a 60W incandescent light with the requisite amount of t-BuOCl (*14*).

Polymer Characterization.

- Molecular weight and polydispersity were determined by gel permeation chromatography. The values reported are in terms of polystyrene equivalent molecular weights.
- Elemental analyses for the determination of chlorine, carbon and hydrogen were performed on each sample.
- ^{1}H NMR spectra were obtained on a Varian NMR spectrometer T-60A from samples in $CDCl_3$ solutions.

- ^{13}C NMR spectra were recorded on a JEOL Fourier Transform NMR spectrometer FX-90Q from samples in solutions in $CDCl_3$.
- IR spectra were obtained on a Model 10MX Nicolet Fourier Transform infrared spectrometer. IR films were spin-coated from polymer solutions in chlorobenzene on either KBr discs or silicon wafers polished on both sides. The samples were baked in vacuum at 90°C for at least 1 hour to ensure solvent removal. Film thicknesses were approximately 1 μm, sufficient to remove interference fringe effects from the spectra.
- Glass transition temperature (T_g) measurements were carried out on a Perkin Elmer DSC2 differential scanning calorimeter. A heating rate of 10°C/min was used with the T_g being taken as the midpoint of the temperature interval over which the discontinuity took place (*15*).

Lithographic Evaluation. Films were spin-coated onto silicon substrates from 10% solutions in chlorobenzene and prebaked at temperatures between 90°C and 100°C for 1 hour to ensure solvent removal. The thickness of each film was about 5000 Å. Electron beam exposures were performed on the AT&T Bell Laboratories electron beam exposure system (EBES-I) operating at 20 kV with a beam adress and spot size both equal to 0.25 μm. A minimal cure time was required since there is no post-exposure reaction (*4,16*).

The exposed samples were dip-developed in appropriate mixtures of methylethylketone and ethanol, rinsed in pure ethanol and finally postbaked above their respective T_g's for 1 hr. Film thicknesses were measured optically using a Nanometrics Nanospec/AFT microarea thickness gauge. Characteristic exposure response curves were plotted as normalized film thickness remaining vs. log dose (expressed in $\mu C/cm^2$).

Results and Discussion

Chlorination. Chlorination conditions were varied to give a series of polymers with a range of chlorination for each base polymer. The degree of chlorination varied between 0.5 (15% w/w Cl) and 2.5 (44% w/w Cl) per monomer unit. Samples with approximately one chlorine per repeat unit (23.3% w/w Cl) were prepared in order to compare their lithographic performance with that of polychloromethylstyrene and chloromethylated polystyrene.

Figure 1a shows the proton NMR spectrum of PMS. A similar spectrum was also obtained from the P-p-MS's. The broad peak from 1.35 to 1.60 ppm arises from the protons in the polymer main chain although it is not possible to distinguish between the methylene and methine protons from these spectra. The intense peak at 2.16 ppm is attributed to the protons of the methyl group substituted on the benzene ring. The aromatic protons appear as a broad multiplet with peaks centered at 6.42 and 6.9 ppm. The small sharp peak appearing at 7.25 ppm is due to proton impurities in $CDCl_3$.

During chlorination by SO_2Cl_2 or t-BuOCl, a new peak, which is assigned to -CH_2Cl, appears at 4.40 ppm. The intensity of this peak increases with reaction time (see Figures 1b-c-d) at the expense of the methyl peak. A weak and broad band appears slightly upfield from 4.40 ppm, which is attributed to a shift in the peaks associated with the main chain protons caused

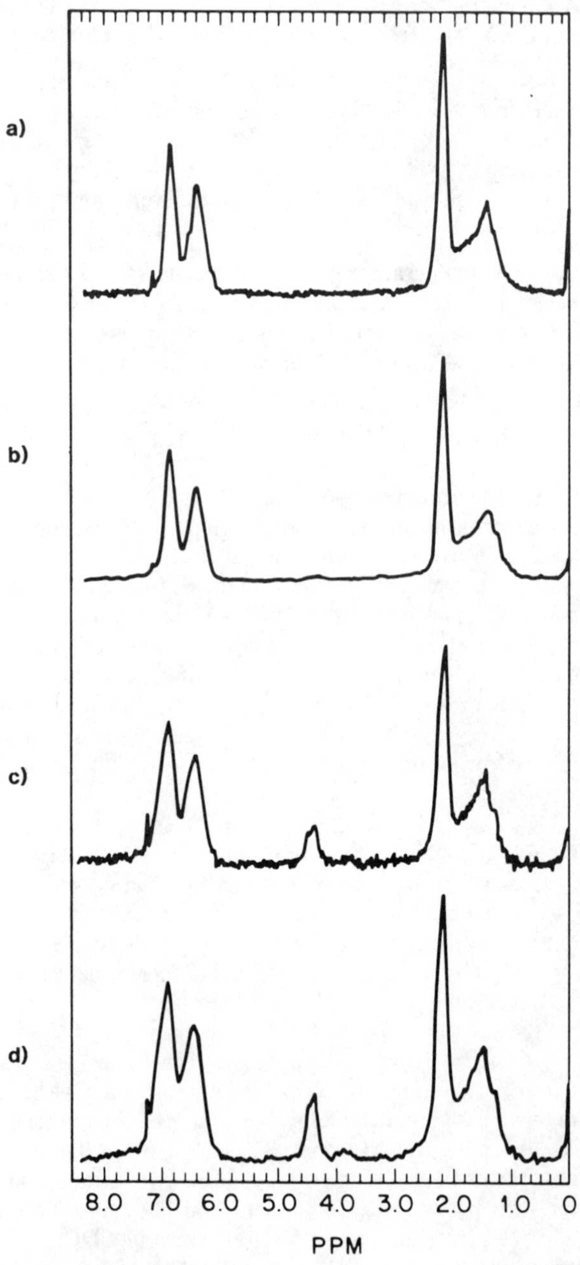

Figure 1. ^{1}H NMR spectra showing the effect of chlorination with 5 ml of t-BuOCl on the peak positions in PMS. a) before chlorination, b) after 50 min., c) after 155 min. and d) after 335 min.

by chlorination. The latter shift is particularly evident in polymers prepared with molar ratios of chlorinating agent to polymer greater than 6.

The $-CH_2Cl$ peak intensity is plotted in Figure 2 as a function of time in the case of chlorination by t-BuOCl, for two different initial concentrations of t-BuOCl. The initial rate of chlorination is extremely rapid and approaches a limiting value after about 3 hours. Clearly the degree of substitution on the methyl group is proportional to the initial t-BuOCl concentration within the concentration limits shown in Figure 2. The limiting degree of monochlorosubstitution ($-CH_2Cl$) decreases at higher t-BuOCl concentrations (molar ratio of t-BuOCl to PMS greater than 3) as shown in Figure 3. However, the total chlorine content (as determined by elemental analysis and expressed in terms of number of chlorines per monomer unit) increases with increasing molar ratio of t-BuOCl to PMS (see Figure 4), indicating that the decrease in intensity of the $-CH_2Cl$ group (Figure 3) is caused by additional chlorination of the monosubstituted polymer.

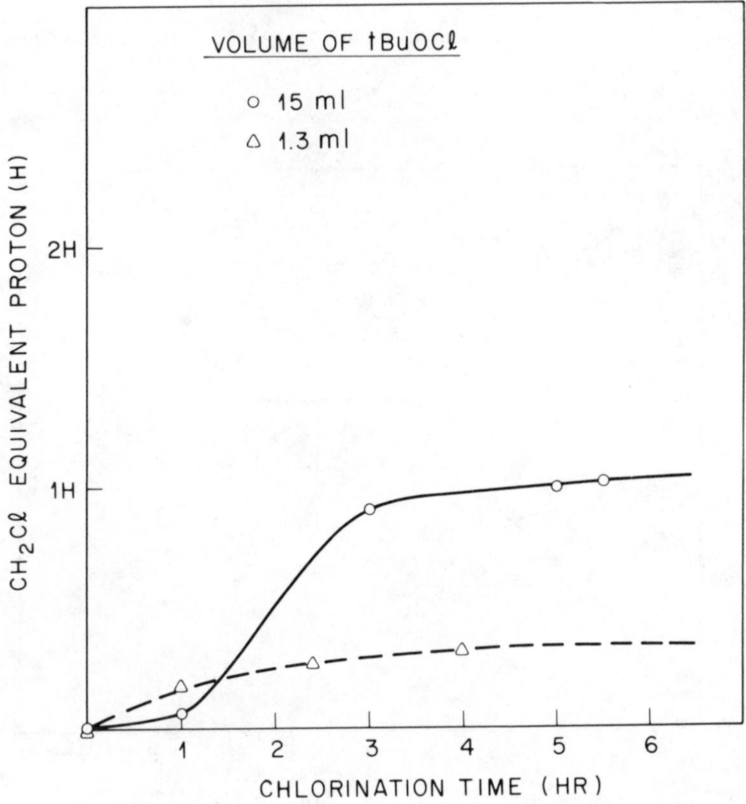

Figure 2. Plot of CH_2Cl equivalent proton vs. chlorination time for two t-BuOCl concentrations.

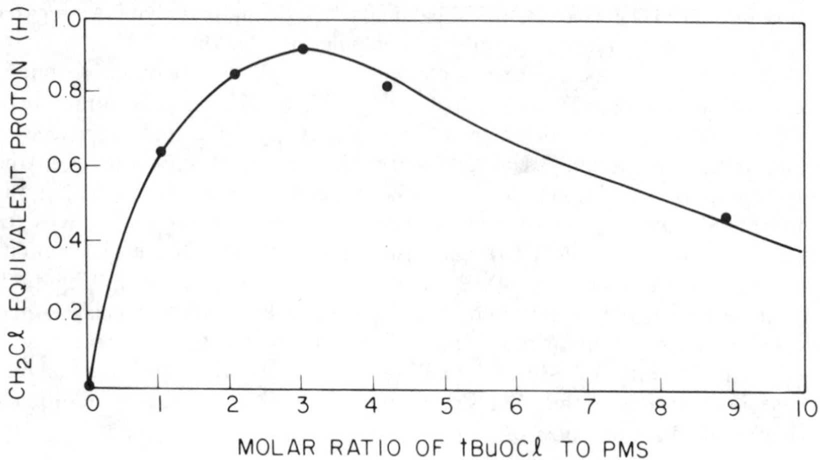

Figure 3. Plot of CH_2Cl equivalent proton vs. molar ratio of t-BuOCl to PMS.

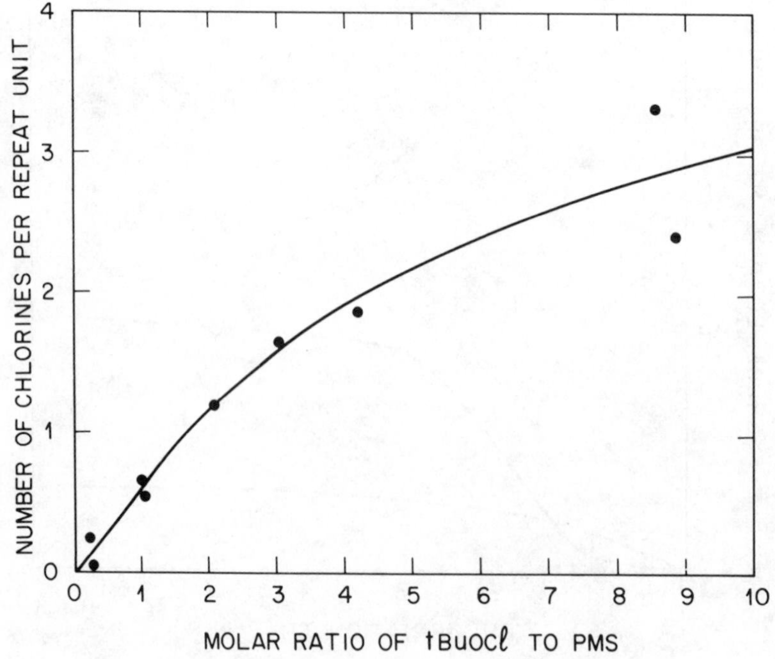

Figure 4. Plot of chlorine content of PMS vs. molar ratio of t-BuOCl to PMS.

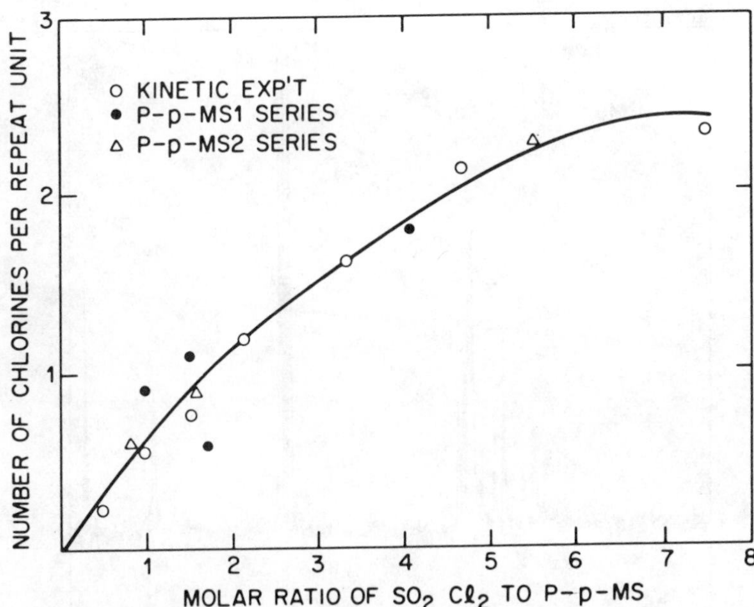

Figure 5. Plot of chlorine content of P-p-MS vs. molar ratio of SO_2Cl_2 to P-p-MS.

In the case of chlorination by SO_2Cl_2, the reaction occurs even more rapidly. Samples taken after 5 minutes of reaction have almost the same chlorine content as the final product taken after 2 hours of reaction. A plot of total chlorine content as a function of the molar ratio of SO_2Cl_2 to P-p-MS (Figure 5) also shows the total chlorine content per monomer unit to increase linearly with concentration of SO_2Cl_2. At high concentration the intensity of the $-CH_2Cl$ peak again decreases as observed in the case of chlorination by t-BuOCl implying degrees of chlorination greater than monosubstitution.

Further evidence for di- and trisubstitution as well as substitution at the α and β-carbon positions is seen from ^{13}C NMR and IR analyses. Figures 6a and 6b show the ^{13}C NMR spectra of PMS and PCMS respectively. Peak assignments for PMS (and P-p-MS (Figure 8a)) are relatively straightforward (*17,18*). Aliphatic carbons appear in the region of the spectrum from 20-80 ppm while peaks due to the aromatic carbons appear between 120 and 150 ppm. It may be noted that the spectrum of P-p-MS (Figure 8a) is identical to PMS except that the peak at 21.5 ppm appears as a singlet rather than a doublet. This peak is due to the methyl carbon attached to the benzene ring, the doublet reflecting the mixture of para and meta isomers in PMS. The sharp signal at 40.4 ppm is attributed to the α-(methine) carbon while the broad signal centered at about 44.3 ppm is attributed to the β-(methylene) carbon. The sharp peaks centered at 77.0 ppm are from the $CDCl_3$ solvent. Comparison of the two spectra in Figures 6a and 6b shows clearly the shift of

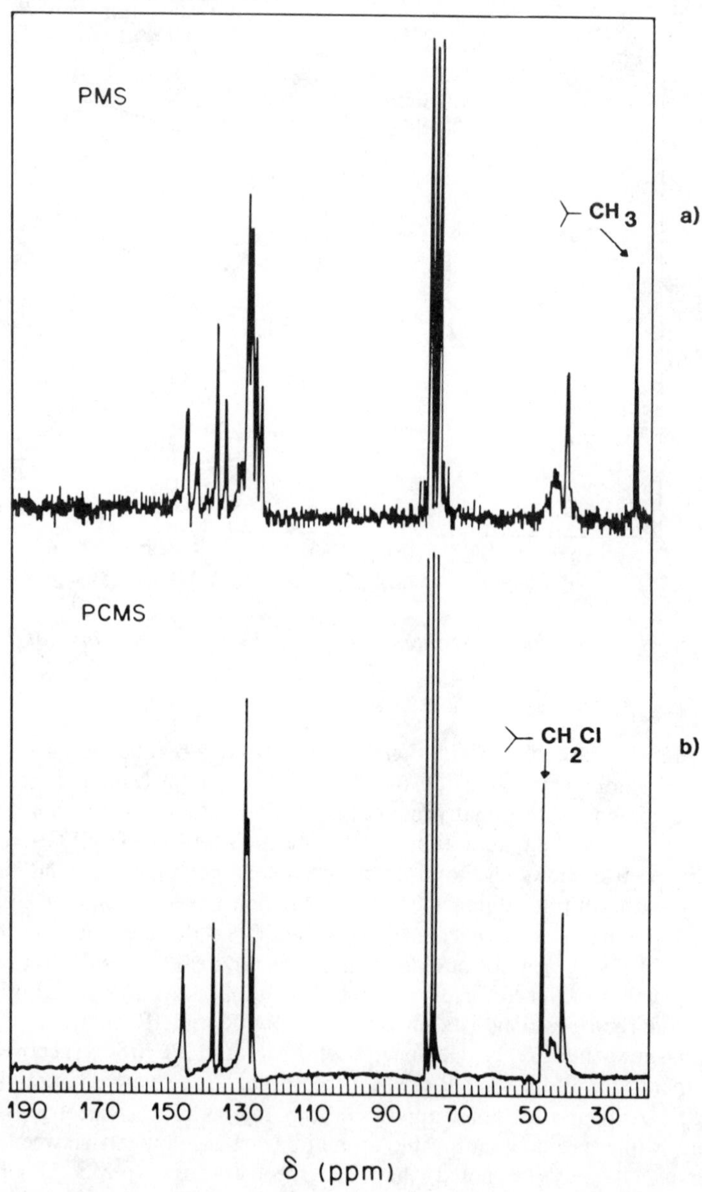

Figure 6. ^{13}*C NMR spectra a. PMS b. PCMS.*

the methyl carbon from 21.5 to 46.1 ppm on monochlorination. Figures 7(a-d) and 8(a-d) show the ^{13}C NMR spectra of a series of chlorinated polymethylstyrenes and poly-para-methylstyrenes respectively, with degrees of chlorination ranging from 0.6 to 2.4 chlorine per monomer unit. Upon chlorination, the sharp peak at 46.1 ppm due to -CH_2Cl, increases with increasing chlorine content at the expense of the methyl signal at 21.1 ppm. A

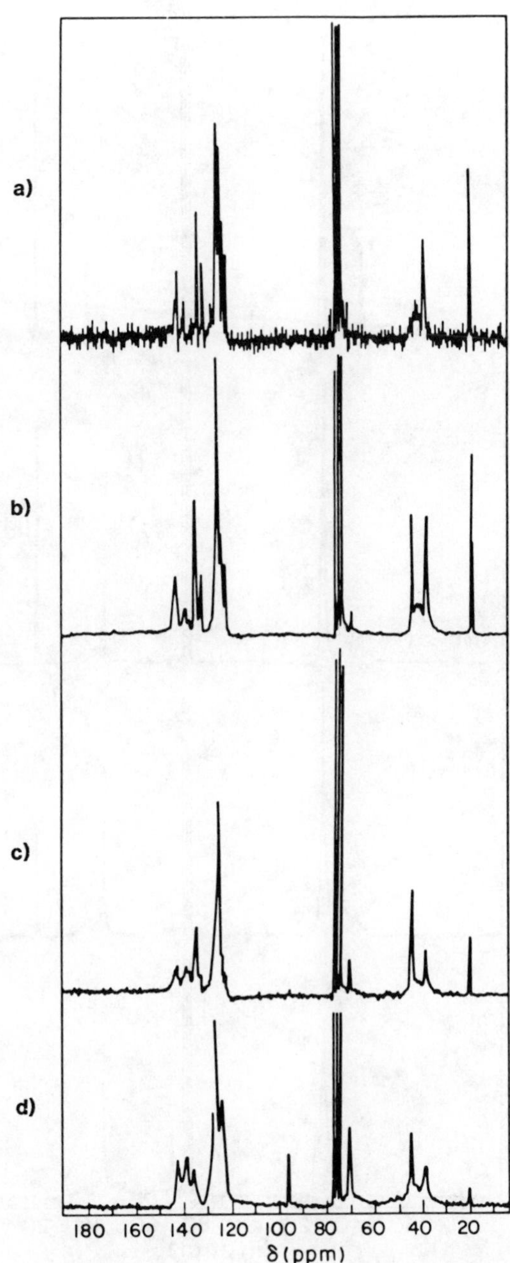

Figure 7. ^{13}C *NMR spectra of a series of chlorinated PMS samples. a) 0 Cl/m.u., b) 0.56 Cl/m.u, c) 1.2 Cl/m.u and d) 2.42 Cl/m.u.*

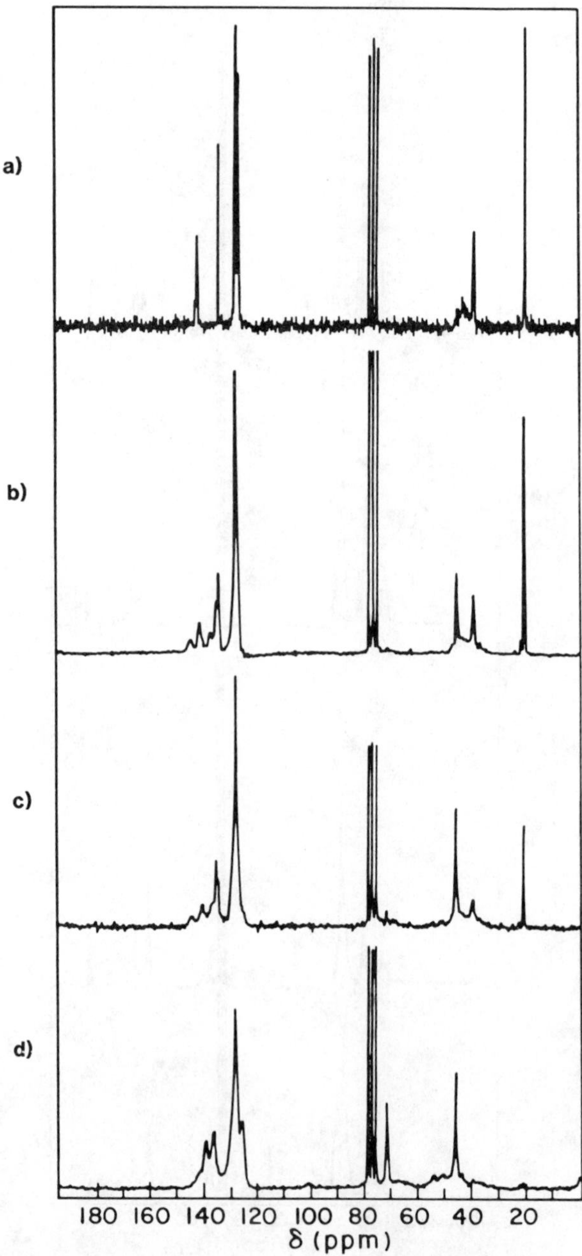

Figure 8. ^{13}C *NMR spectra of a series of chlorinated P-p-MS. a) 0 Cl/m.u, b) 0.6 Cl/m.u, c) 1.1 Cl/m.u and d) 2.3 Cl/m.u.*

comparison of the spectra in Figures 7b-d and 8b-d indicates that, for chlorine contents greater than ~0.6 chlorine per monomer unit, the distribution of chlorines on the different chlorination sites depends to some extent on the chlorinating agent used. In the case of the chlorination with t-BuOCl, a sharp peak attributable to -CH(Cl)$_2$ appears at 71.8 ppm and is even evident in the polymer containing the lowest chlorine content (0.56 Cl/monomer unit). With SO_2Cl_2, this peak does not appear until chlorine concentrations greater than 1.1 Cl per monomer unit have been reached (see Figure 8c). In Figure 7c, the peak intensity at 71.8 ppm has increased further and a new peak at 96 ppm attributed to -CCl_3 has also appeared. Thus, at the highest chlorine content investigated, peaks due to mono, di- and tri-chlorination on the pendant methyl group are present, (see Figure 7d). There is no evidence of trisubstitution with SO_2Cl_2 even at degrees of chlorination as high as 2.3 (Figure 8d). These assignments were verified by comparison with spectra taken on a series of model compounds of mono-, di- and tri-α-chlorinated toluenes.

Although the ^{13}C NMR provides quantitative information on the chlorination at the methyl site, it is less definitive for the backbone carbons. The multiplet centered at 44.3 ppm attributed to the methylene carbon in the main chain decreases at high chlorine contents as does the α-carbon peak. At the same time a close examination of the ^{13}C NMR reveals the appearance of broad weak signals at about 54 and 68 ppm. These peaks may be due to chlorine substitution of the protons on the polymer main chain but this assignment is not definitive since the shift in peak position relative to the unchlorinated carbon is not as great as it would be expected for chlorine substitution.

The structural assignments made on the basis of the ^{13}C NMR analysis were confirmed by IR studies. The IR spectra of the chlorinated PMS and P-p-MS are shown in Figures 9a-d and 10a-d respectively. The assignments of the principal bands of the spectra are given in Table I.

Although the absorption spectra of PMS and P-p-MS are similar, there are two major difference's. First, the C-C stretching band in the 1500 cm^{-1} region appears as a sharp peak in the case of P-p-MS and as a doublet for PMS. This dissimilarity in absorption is due to the different isomeric structures of the polymers. For the same reason in P-p-MS only one out of plane ring bending band due to the para isomer absorption appears at 790 cm^{-1}. In the case of PMS one noticed the presence of a second absorption band at 700 cm^{-1} due to the meta isomer. These bands are interesting because one may use them for the quantitative determination of the relative concentrations of isomers in mixtures.

Chlorination of PMS and P-p-MS results in the appearance of a new absorption band at 1266 cm^{-1} that increases in intensity with increasing chlorine content to about 1.2 chlorine per monomer unit. This band is attributed to CH_2 wagging of -CH_2Cl (*18*) and its growth reflects the substitution at the methyl group. The intensity ratios of the different bands in the polymer containing 1.2 chlorines per repeat unit (Figure 10c) are quite different from PCMS (Figure 11), suggesting more than simple monochlorosubstitution on the pendant methyl group is occurring. For

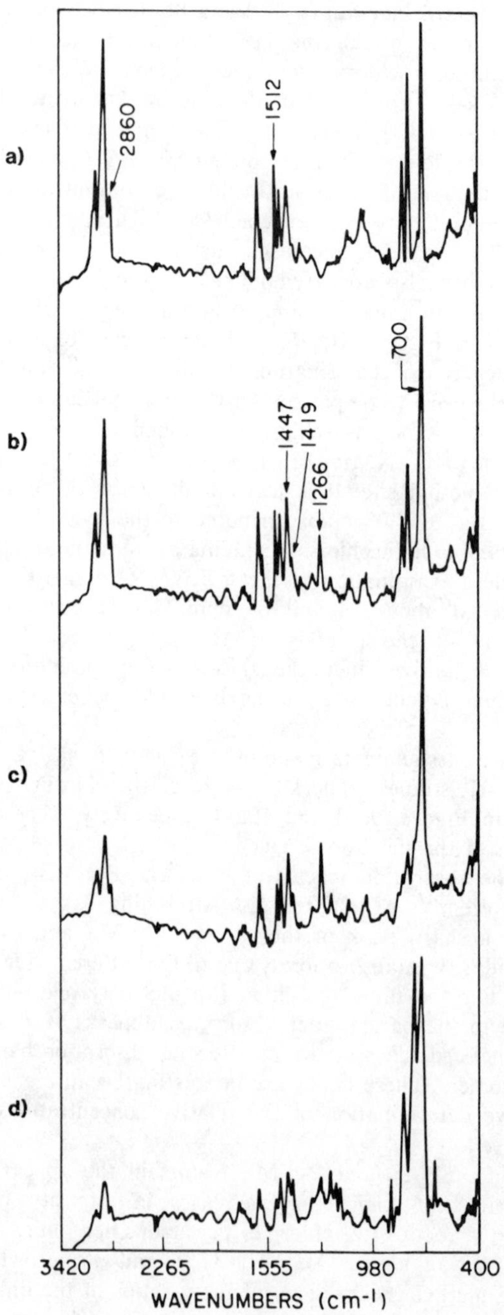

Figure 9. IR spectra of the series of chlorinated PMS. a) 0 Cl/m.u, b) 0.56 Cl/m.u, c) 1.20 Cl/m.u and d) 2.42 Cl/m.u.

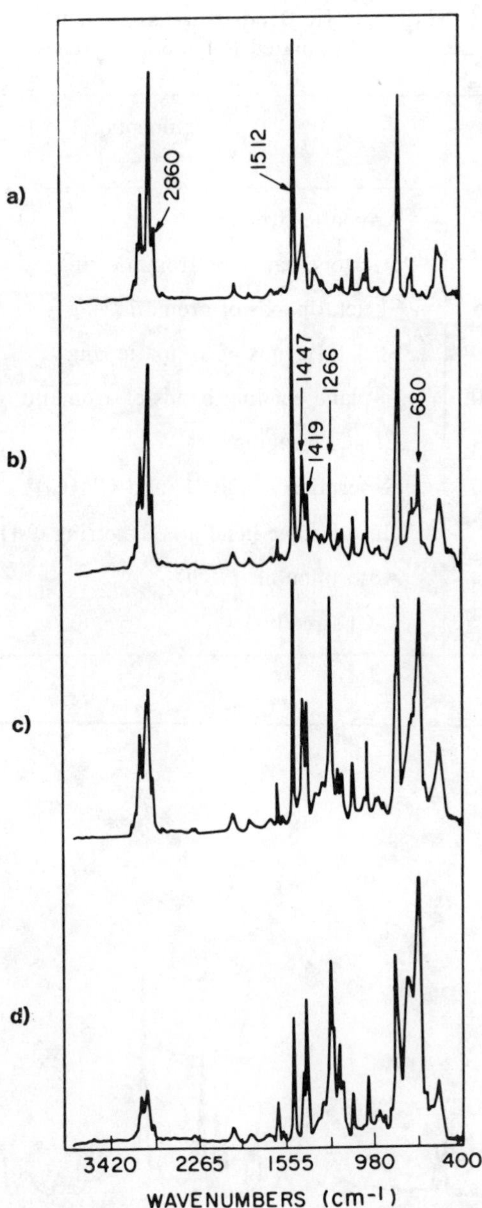

Figure 10. IR spectra of the series of chlorinated P-p-MS. a) 0 Cl/m.u, b) 0.6 Cl/m.u, c) 1.1 Cl/m.u, and d) 2.3 Cl/m.u.

Table I. Principal IR Bands and Assignments for the Series of Chlorinated Polymethylstyrenes

Frequencies (cm^{-1})	Assignments
3100-3000	Aromatic stretching
2000-1650	Overtone bands of aromatic ring
1600-1585	Skeletal bands of aromatic ring
1500-1400	Skeletal bands of aromatic ring
1300-1000	In plane bending bands of aromatic ring
1266	CH_2Cl wagging
1150-1180	Absorption of $C(Cl)_3$ and $CH(Cl)_2$
900-700	Out-of-plane bending of the ring C-H bonds
751	Absorption of $-C(Cl)_2-$
710-680	C-Cl stretching

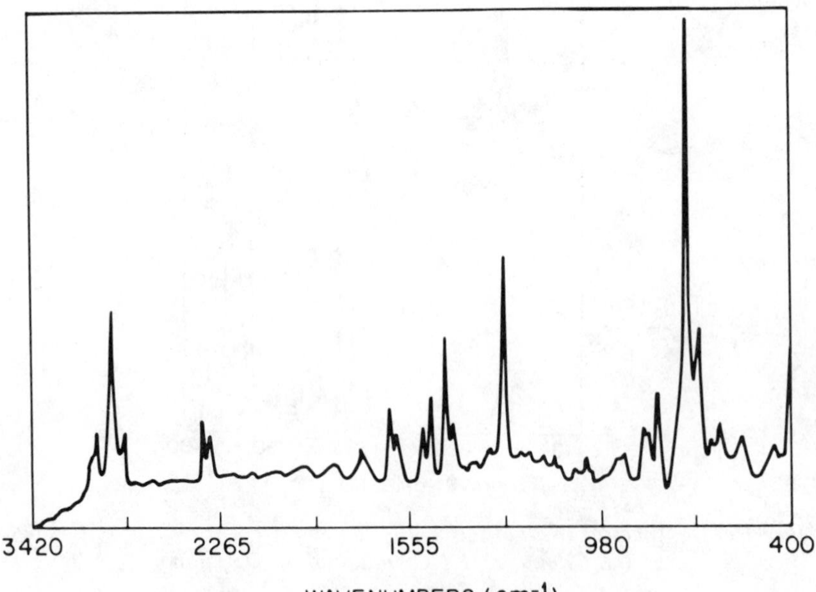

Figure 11. IR spectra of PCMS.

example, the two peaks in the C-H deformation and wagging region that appear at 1447 and 1419 cm^{-1} for both PMS and P-p-MS are due to aromatic ring vibration modes and their intensities vary with the electronic nature of the substituents, and thus become more intense as chlorination proceeds (*19*). The fact that the intensity ratios of these two peaks are different in the chlorinated polymers compared with PCMS implies greater degrees of chlorine substitution in the former. An intense band appears at ~700-710 cm^{-1} in chlorinated PMS and at 680 cm^{-1} in P-p-MS. This band is assigned to the C-Cl stretching vibration and its growth is indicative of an increase in chlorine content. At high chlorine contents, absorption bands appear at ~1180 cm^{-1} which are attributed to $-CH(Cl)_2$ and/or $-C(Cl)_3$ group. In the case of chlorination with t-BuOCl an absorption band appears at 751 cm^{-1} for chlorine contents greater than 1.8 chlorine per monomer unit. This band is attributed to dichlorination at the β-carbon. The peak at 2860 cm^{-1} corresponds to the C-H stretch of the α-carbon (*20*) and decreases quantitatively with increasing degree of chlorination. All these observations on the intensity of the different absorption bands are relative to the aromatic C-C stretch band at 1512 cm^{-1} which is unaffected by chlorination.

Table II lists the composition of each chlorinated species, determined by the methods described above. Values obtained from elemental analysis and

Table II. Composition of Chlorinated Products.

Sample	Total chlorine[d]	Relative chlorine at each position					
		CH_3	CH_2Cl	$CHCl_2$	CCl_3	α-Cl	β-Cl
1[a]	0.04	0.96	0.04	0	0	0	0
2[a]	0.56	0.58	0.32	0.10	0	0.07	0.07
3[a]	1.20	0.25	0.40	0.30	0.05	0.24	0.21
4[a]	1.88	0.10	0.38	0.30	0.22	0.55	0.43
5[a]	2.42	0.05	0.25	0.55	0.15	0.50	0.97
1[b]	0.58	0.70	0.30	0	0	0.15	0.13
2[b]	0.92	0.56	0.44	0	0	0.21	0.27
3[b]	1.15	0.29	0.71	0	0	0.23	0.21
4[b]	1.79	0.13	0.57	0.30	0	0.55	0.17
5[b]	2.28	0	0.50	0.30	0	0.60	0.18
1[c]	0.58	0.56	0.44	0	0	0.08	0.06
2[c]	0.88	0.56	0.44	0	0	0.21	0.25
3[c]	2.24	0	0.50	0.50	0	0.52	0.22

a = PMS samples
c = P-p-MS2 samples
b = P-p-MS1 samples
d = Per monomer unit

FTIR are accurate to about 5 and 10% respectively. Values obtained from ^{13}C NMR are accurate to 10% at low values of chlorination and probably only 20% at higher degrees of chlorination. The average of the relative reactivities of the methyl, methylene and methine carbons to chlorination are 56%, 21% and 23% respectively in the case of chlorination with t-BuOCl, and 58%, 23% and 19% in the case of chlorination with SO_2Cl_2. The reactivities of the different chlorination sites are therefore similar, although the distribution of chlorines on the methyl site differs, particularly at high chlorine contents.

All evidence therefore suggests that chlorination of PMS with t-BuOCl and of P-p-MS with SO_2Cl_2 occurs not only on the methyl group but also on the main chain of the polymer and that the degree of chlorination is determined by the initial concentration of t-BuOCl or SO_2Cl_2. The distribution of chlorine atoms depends on the chlorinating agent. The structure is clearly very complex and is best represented in generalized form as shown in Figure 12b.

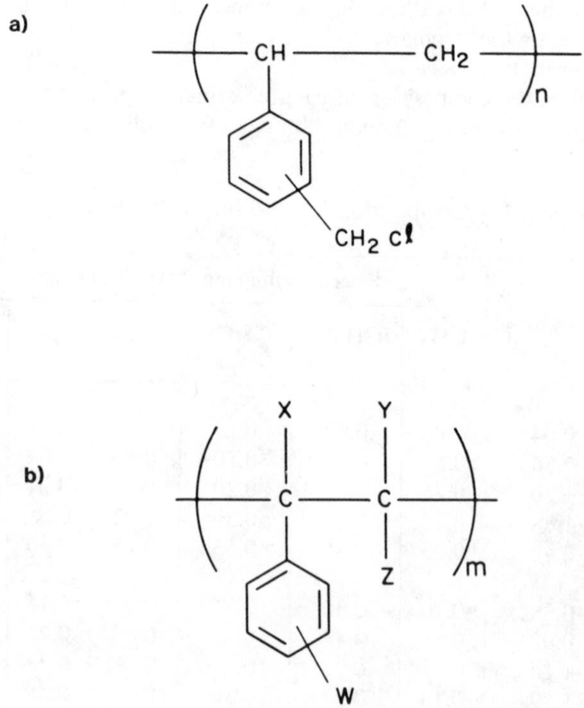

a)

b)

WHERE W = $-CH_2Cl, -CHCl_2, -CCl_3$

X = H, Cl

Y = H, Cl

Z = H, Cl

Figure 12. Generalized copolymer structures.

Glass Transition Temperature. Tables III and IV list the glass transition temperature (T_g) of both chlorinated PMS and P-p-MS respectively. A plot of T_g as a function of chlorine content is shown in Figure 13. The glass transition temperature increases with degree of chlorination as a result of increasing polarity (*21*). This increase in T_g is a linear function of the extent of chlorination.

The glass transition temperature is also affected slightly by the number average molecular weight (*20*). P-p-MS2 which has a molecular weight twice that of P-p-MS1 has a T_g that is a few degrees higher at similar extents of chlorination. The shape of the DSC response changes too, from a sharp

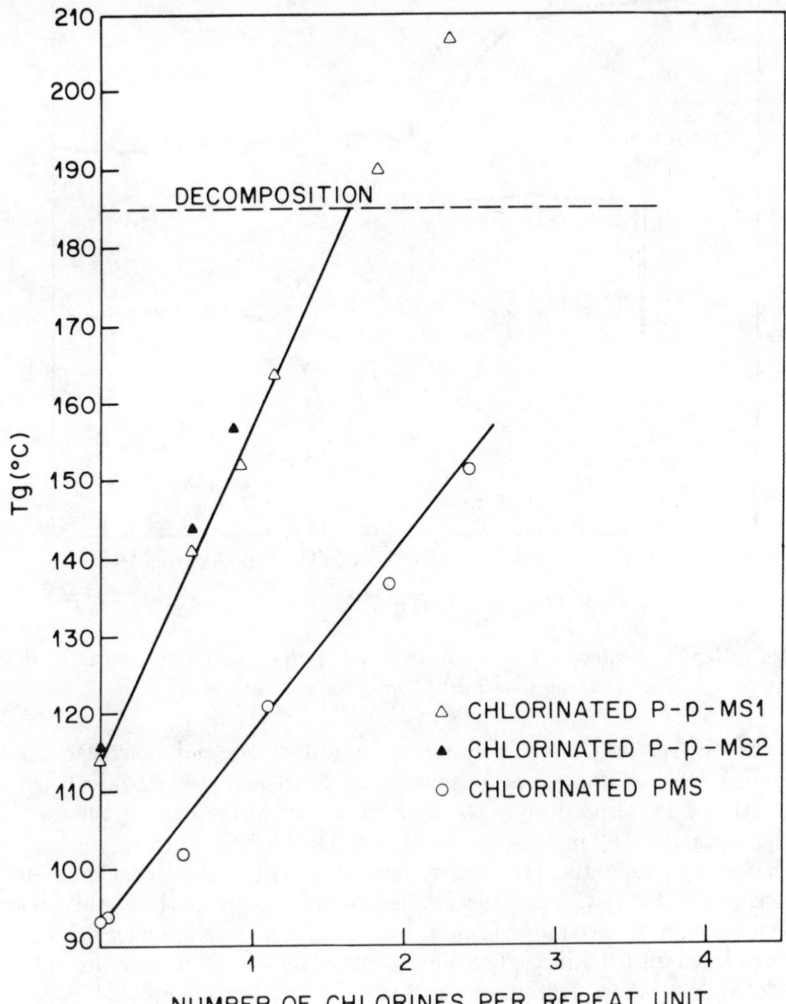

Figure 13. Plot of the glass transition temperature T_g vs. chlorine content of the series of chlorinated PMS and P-p-MS.

discontinuity for the starting material to a broader and weaker transition for the material with the highest degree of chlorination, (see the example of chlorinated PMS in Figure 14). In addition, it is worth noting that the glass transition of the mixture of meta and para isomers is 20°C lower than that of paramethylstyrene homopolymer, and is 20°-40°C lower for the chlorinated polymers at similar degrees of chlorination.

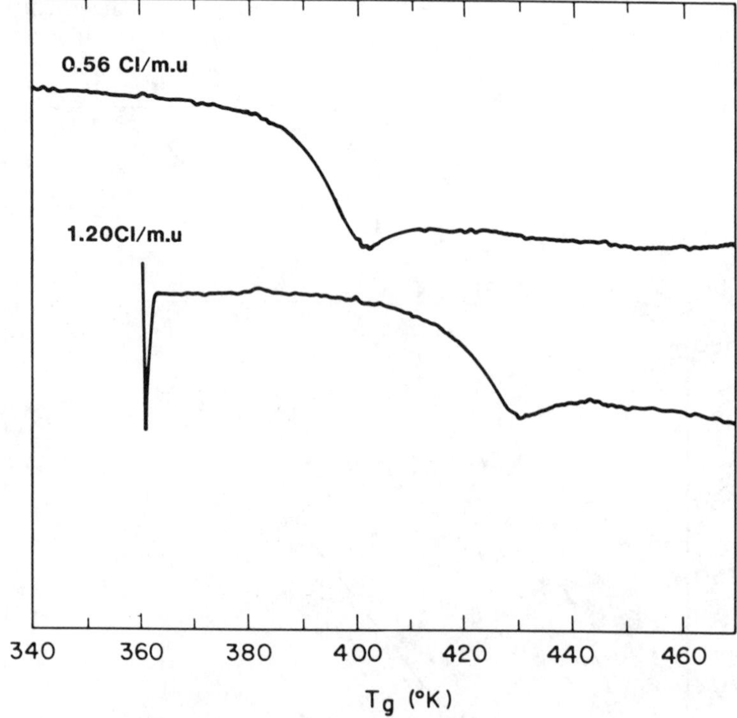

Figure 14. DSC spectra of two chlorinated polymethylstyrenes with 0.56 Cl/m.u and 1.20 Cl/m.u respectively.

Molecular Weight (M_w). The polystyrene equivalent M_w and molecular weight distribution (PD) of the series of chlorinated PMS and P-p-MS are listed in Tables III and IV respectively. The data show that in the case of chlorination with t-BuOCl the M_w increases with increasing chlorine content, whereas the PD changes only slightly, broadening somewhat with increasing chlorination. Chlorination with SO_2Cl_2 causes some degradation particularly of the highest molecular weight material as evidenced by a small decrease in molecular weight and broadened distribution. The final molecular weight distribution of the chlorinated P-p-MS is still narrower than that of the chlorinated polymers prepared from free radically synthesized PMS and the former should therefore be expected to show improved contrast and resolution in resist applications.

Table III. Physical and Lithographic Properties of Chlorinated PMS.

Chlorinated PMS	% Cl (w/w)	Cl/monomer unit	T_g (°C)	$\overline{M_w}$ (×10^{-3})	PD	D_g^i μC/cm²)	$D_g^{0.5}$ (μC/cm²)	γ
PMS	0	0	93.0	220	1.75	9.7	23	1.39
PMS-1	1.3	0.04	94.0	240	1.95	0.76	1.65	1.62
PMS-2	14.5	0.56	103.5	270	2.05	0.39	0.94	1.37
PMS-3	26.5	1.20	121.0	245	1.82	0.53	1.15	1.57
PMS-4	36.5	1.88	137.0	280	2.20	0.58	1.35	1.35
PMS-5	42.5	2.42	151.5	290	2.00	0.88	2.15	1.38

Table IV. Physical and Lithographic Properties of Chlorinated P-p-MS.

Sample	% Cl (w/w)	Cl per monomer unit	T_g (°C)	M_w (10^{-3})	PD	D_g^i [d]	$D_g^{0.5}$ [d]	γ [a]
0[b]	0	0	114	110	1.25	0.26	38	2.9
1[b]	14.8	0.58	141	110	1.31	0.96	1.7	2.6
2[b]	21.7	0.92	152	110	1.41	0.78	1.3	1.8
3[b]	25.8	1.15	164	160	1.37	0.96	1.5	2.4
4[b]	35.3	1.79	190	130	1.32	1.2	1.8	2.3
5[b]	41.0	2.28	207	110	1.59	1.5	2.6	2.1
0[c]	0	0	116	260	1.09	17	28	2.4
1[c]	14.8	0.58	144	230	1.18	0.72	1.0	2.5
2[c]	21.0	0.88	157	180	1.34	0.46	0.94	1.8
3[c]	40.7	2.24	>210	220	1.41	0.66	1.5	1.6

a $\gamma = \left[\log \frac{D_g^{1.0}}{D_g^i}\right]^{-1}$

b = P-p-MS1

c = P-p-MS2

d = μC/cm²

Electron Beam Evaluation. Figure 15 shows exposure curves of a series of PCMS's with M_w's ranging from 50,000 to 300,000 from which sensitivity and contrast data were obtained and are summarized in Table V. As expected, the polymer sensitivity increases with increasing molecular weight (Figure 16a) but contrast decreases (Figure 16b). The results are broadly in agreement with those of Choong and Kahn (*1*).

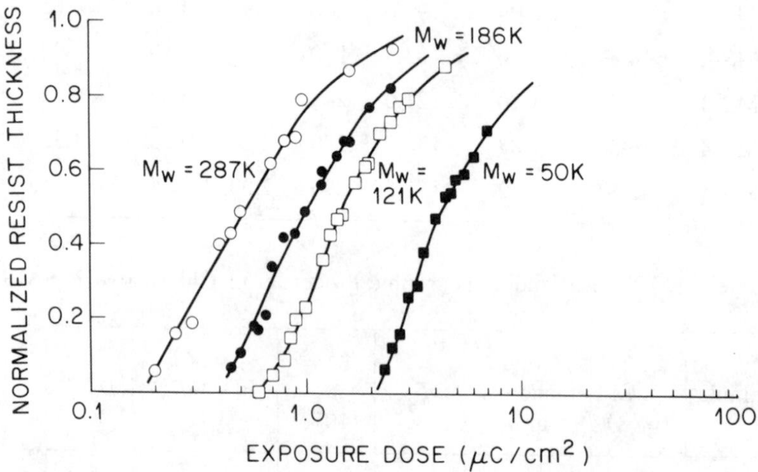

Figure 15. Exposure response curves of a series of PCMS with different molecular weights.

Table V. Effect of Molecular Weight of PCMS on Lithographic Parameters

Sample	M_w (10^{-3})	PD	$D_g^{0.5}$ $(\mu C/cm^2)$	γ^a
PCMS-1	50	1.86	4.3	1.72
PCMS-2	120	2.04	1.51	1.42
PCMS-3	185	2.39	1.0	1.29
PCMS-4	290	2.24	0.54	1.06

$$^a\ \gamma = \left[\log \frac{D_g^{1.0}}{D_g^i}\right]^{-1}$$

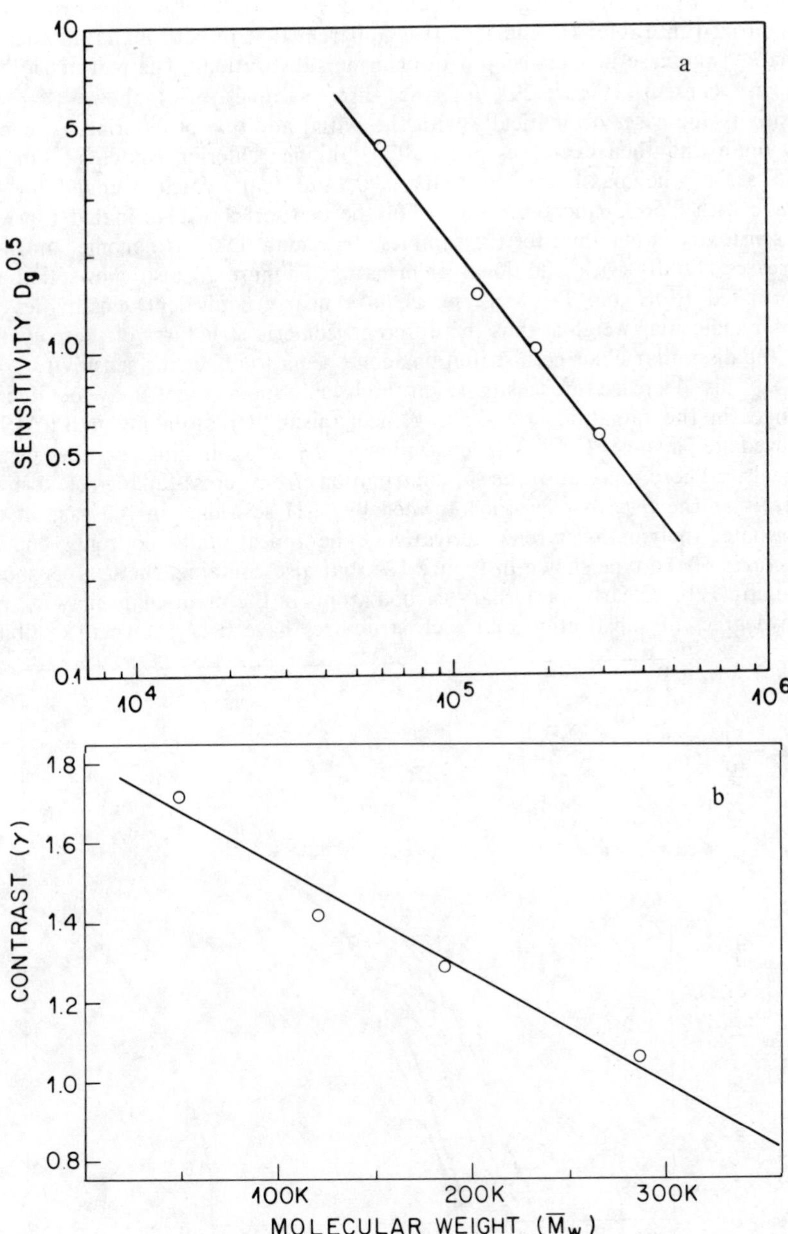

Figure 16. a) Plot of $D_g^{0.5}$ ($\mu C/cm^2$) vs. molecular weight of PCMS. b) Plot of contrast (γ) vs. molecular weight of PCMS.

The sensitivity curves for the series of chlorinated PMS, P-p-MS1, and P-p-MS2 are shown in Figures 17, 18 and 19 respectively. The results are summarized in Tables III and IV. It is apparent that in both materials there is a marked increase in sensitivity with chlorine substitution. The plot of $D_g^{0.5}$ vs. chlorine content (Figure 20) for the three sample series shows that the sensitivity increases dramatically with the initial addition of chlorine, reaches a maximum and then decreases gradually at higher chlorine contents. For the PMS series, the maximum sensitivity is 0.94 $\mu C/cm^2$ which is found for the sample with 0.56 Cl/monomer unit. For the two series of chlorinated P-p-MS, the sensitivity is maximal for the samples containing 0.90 Cl/monomer unit and decreases slightly with additional chlorination. Figure 20 also shows that the chlorinated PMS and P-p-MS2 series have nearly equivalent sensitivities for similar molecular weights, thus the different isomeric structures of the polymers and the dissimilar chlorine distributions don't seem to affect the sensitivity.

This decrease in sensitivity at high chlorine content may be due to changes in the radiation cross-linking mechanism. Cross-linking in PCMS is believed to involve C-Cl scission followed by cross-linking of the benzyl radicals. There may also be a contribution from cross-linking of tertiary radicals at the α-carbon position formed by C-H scission. In the case of the chlorinated polymethylstyrene derivatives, the repeat units not only contain structures of the type shown in Figure 12a, but also contain structures as shown in Figure 12b. Clearly quaternary carbon atoms in the main chain may exist at high degrees of substitution and such structures have been shown to exhibit a

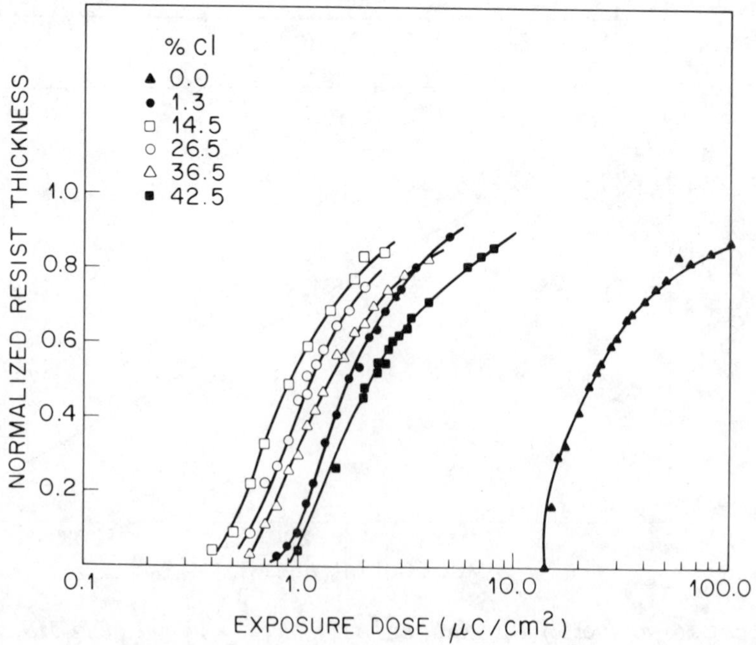

Figure 17. Exposure response curves of the chlorinated PMS series.

propensity for scission (*22*). The introduction of a scission mechanism at high chlorine contents would adversely affect cross-linking efficiency and may be responsible for the decrease in sensitivity at higher degrees of chlorination.

Contrast (γ) values are also summarized in Tables III and IV. A plot of γ vs. % Cl (Figure 21) shows that in all cases there is a slight reduction in contrast as the degree of chlorination increases. Factors which affect contrast

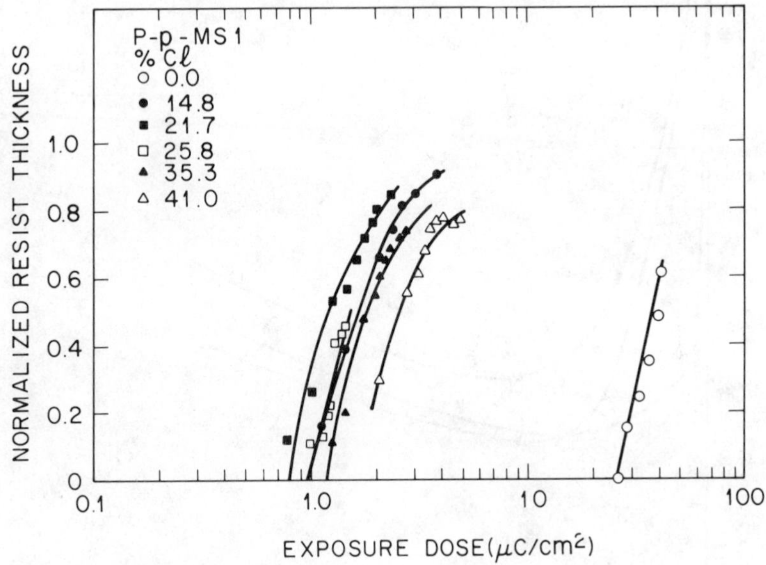

Figure 18. Exposure response curves of the chlorinated P-p-MS1 series.

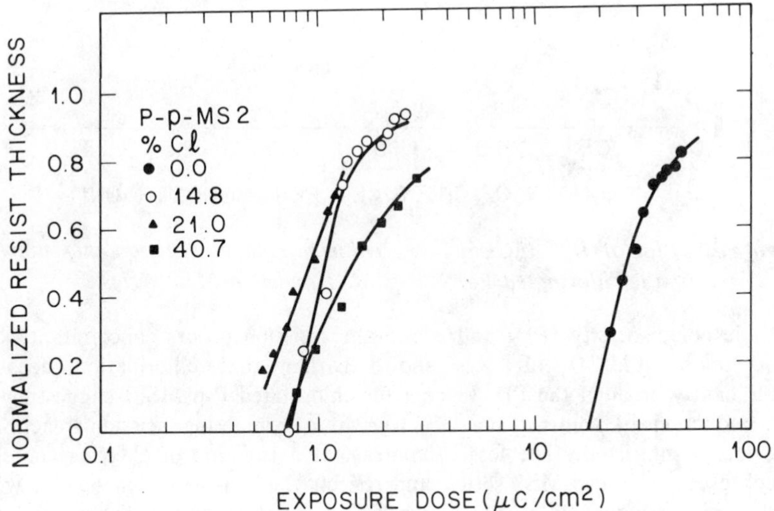

Figure 19. Exposure response curves of the chlorinated P-p-MS2 series.

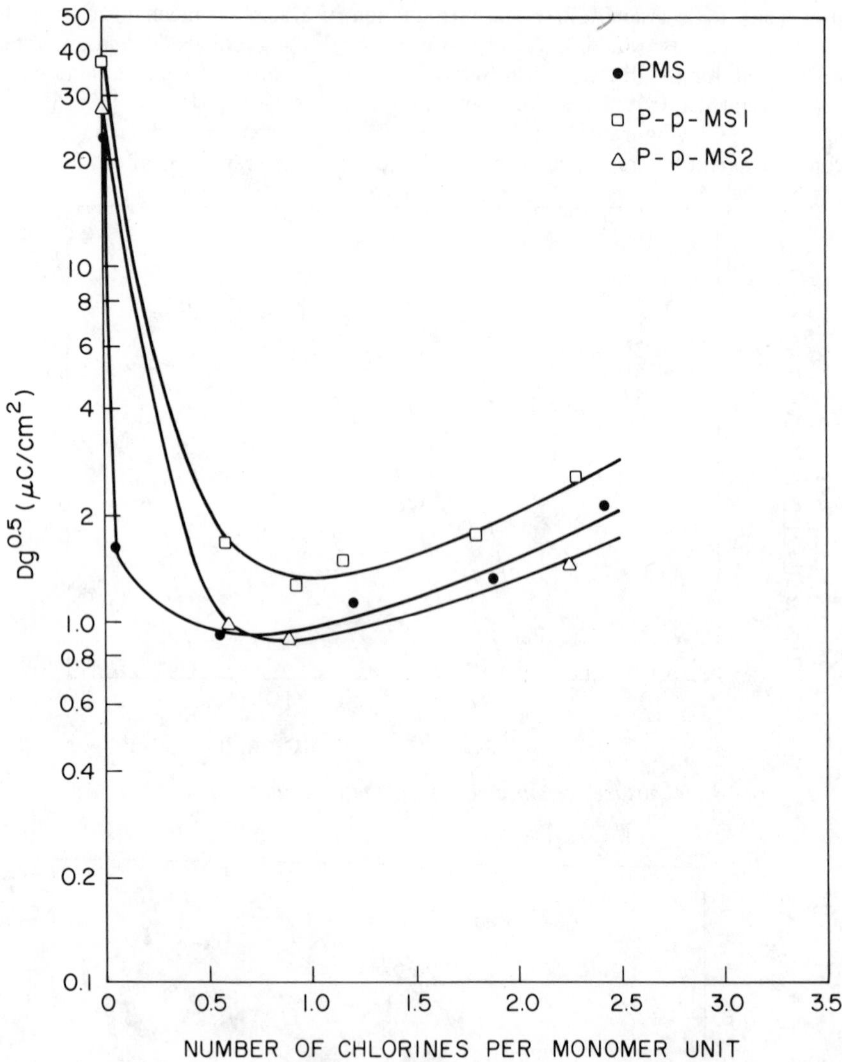

Figure 20. Plot of $D_g^{0.5}$ (μC/cm²) vs. chlorine content (Cl/monomer unit) for the chlorinated PMS, P-p-MS1 and P-p-MS2 series.

include polydispersity (PD) and/or whether scission occurs concomitantly with cross-linking (*23,24*). It was shown earlier that chlorination does not significantly broaden the PD, (except for chlorinated P-p-MS2) suggesting that the reduction in contrast may be due to a competing scission reaction as previously suggested. The loss in contrast as a function of chlorine content is much greater for P-p-MS2 series, and reflects the broadening molecular weight distribution. Figure 22 shows contrast as a function of polydispersity for the three sets of materials. The molecular weight distributions of the various

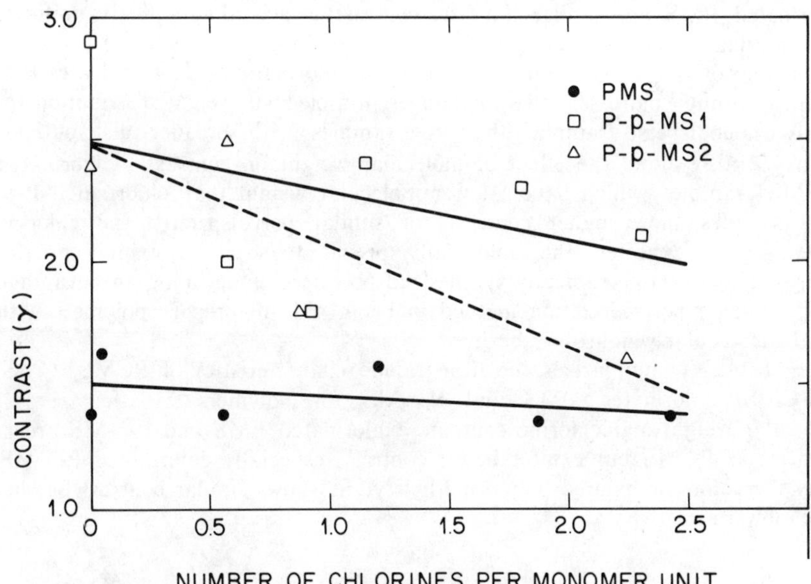

Figure 21. Plot of contrast (γ) vs. chlorine content (Cl/monomer unit) for the chlorinated PMS, P-p-MS1 and P-p-MS2 series.

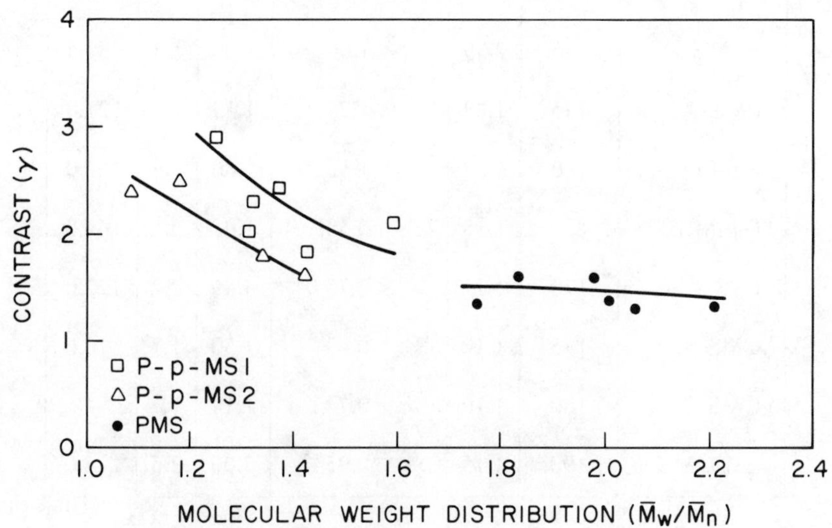

Figure 22. Plot of contrast (γ) vs. polydispersity for the chlorinated PMS, P-p-MS1 and P-p-MS2 series.

chlorinated PMS are similar and the contrast is accordingly fairly uniform. The chlorinated P-p-MS series both show a loss in contrast as a result of the broadening of the distribution with degree of chlorination. This indicates that contrast is much more sensitive to changes in molecular weight distribution for nearly monodisperse samples than for samples with broader distributions. Figure 22 also shows the effect of molecular weight on contrast. Chlorinated P-p-MS1 samples, which have a lower molecular weight than chlorinated P-p-MS2 samples show higher contrast for similar polydispersity and chlorine contents. As expected the anionically prepared materials exhibit greater contrast than the free radically synthesized polymers, underlining an advantage of the anionic polymerization method in being able to prepare polymers with narrow molecular weight distributions.

Table VI summarizes the lithographic characteristics of PCMS, CMS, αM-CMS, chlorinated PMS and P-p-MS for polymers with comparable molecular weight and chlorine content. Chlorinated PMS and P-p-MS are as sensitive as PCMS but exhibit better contrast, especially chlorinated P-p-MS. Only the chloromethylated poly-α-methylstyrene shows similar contrast but has much lower sensitivity.

Table VI. Comparison of the Lithographic Parameters of PCMS, Chlorinated PMS and P-p-MS, CMS and αM-CMS

Sample	M_w (10^{-3})	PD	Chlorine/ monomer unit	$D_g^{0.5}$	γ	T_g (°C)
PCMS	185	2.39	1.00	1.00	1.3	81.5
PMS-3	245	1.82	1.20	1.15	1.6	121.0
P-p-MS1-3	160	1.37	1.15	1.50	2.4	164.0
P-p-MS2-1	230	1.18	0.58	1.0	2.5	144.0
CPS	170	1.21	<0.75	4.45	2.2	122.0
CMS	145	1.10	0.43	1.30	1.7	110.0
CMS	180	1.10	0.74	1.14	1.7	–
αM-CMS	190	1.15	0.95	3.00	2.5	174.0

PCMS = polychloromethylstyrene
CPS = chlorinated polystyrene
CMS = chloromethylated polystyrene
αM-CMS = chloromethylated poly-α-methylstyrene

Resolution between 0.75 and 1.0 μm (equal lines and spaces), was obtained in preliminary lithographic evaluation of chlorinated PMS and P-p-MS. Typical patterns are shown in Figure 23. The small amount of residual material could be removed by descumming if desired.

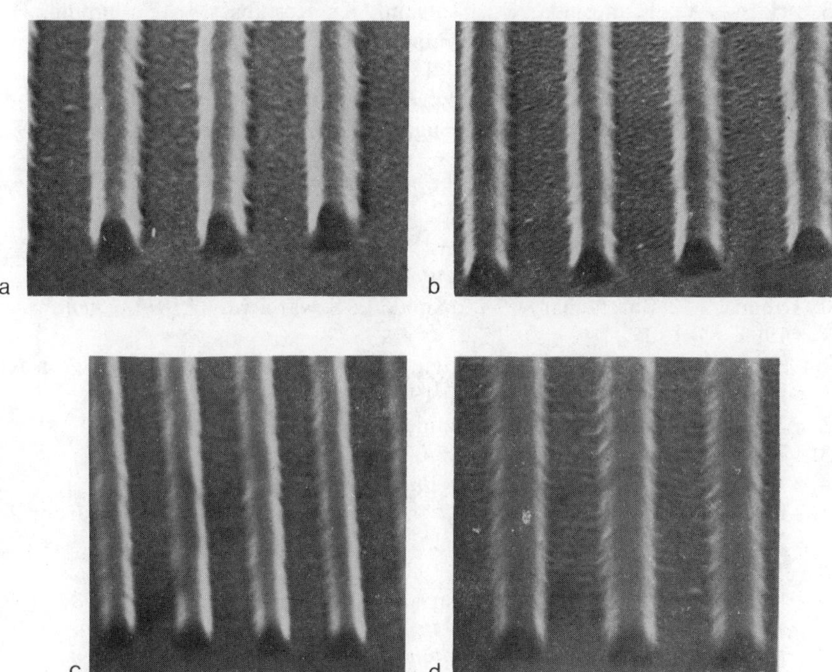

Conclusion

Two synthetic procedures have been developed which allow the reproducible chlorination of polymethylstyrene. Although chlorination with tertiary-butyl hypochlorite and sulfuryl chloride gives rise to some minor structural differences, the lithographic properties of both materials are essentially equivalent. Both chlorination techniques yield polymers with similar sensitivity but improved contrast (particularly for anionically prepared materials) compared with analogous compounds synthesized by other methods. Both materials exhibit superior lithographic properties compared with PCMS possibly reflecting the contribution of substitution on the main chain and/or higher degrees of chlorination on the pendant methyl group.

Literature Cited

1. Choong, H. S.; Kahn, F. J. *J. Vac. Sci. Technol.* 1981, *19(4)*, 1121.
2. Imamura, S.; Tamamura, T.; Harada, K.; Sugawara, S. *J. Appl. Polym. Sci.* 1982, *27*, 937.
3. Harita, Y.; Kamoshida, Y.; Tsutsumi, K.; Koshiba, M.; Yoshimoto, H.; Harada, K. SPSE 22nd Fall Symposium, Washington, D.C., 1982.

Literature Cited

1. Choong, H. S.; Kahn, F. J. *J. Vac. Sci. Technol.* 1981, *19(4)*, 1121.
2. Imamura, S.; Tamamura, T.; Harada, K.; Sugawara, S. *J. Appl. Polym. Sci.* 1982, *27*, 937.
3. Harita, Y.; Kamoshida, Y.; Tsutsumi, K.; Koshiba, M.; Yoshimoto, H.; Harada, K. SPSE 22nd Fall Symposium, Washington, D.C., 1982.
4. Imamura, S. *J. Electroch. Soc.* 1979, *126*, 1628.
5. Feit, E. D.; Stillwagon, L. E. *Polym. Eng. in Sci.* 1980, *20(16)*, 1058.
6. Tamamura, T.; Sukegawa, K.; Sugawara, S. *J. Electrochem. Soc.* 1982, *129*, 1831.
7. Sukegawa, K.; Sugawara, S. *Japan. J. of Appl. Phys.* 1981, *20(8)*, L583.
8. Poutsma, M. L. in "Free Radicals"; Kochi, J. K., Ed.; Vol. II, Chap. 15.
9. O'Driscoll, K. F.; Patsiga, R.; *J. Polym. Sci. A* 1969, *3*, 1037.
10. Shima, M.; Bhattacharyva, D.; Smid, J.; Szwarc, M. *J. Am. Chem. Soc.* 1963, *15*, 1306.
11. Szwarc, M. "Carbanions, Living Polymers and Electron Transfer Processes"; Wiley and Sons, Ed.; 1968.
12. Novembre, A. E., private communication.
13. Barrett, J. H. U.S. Patent 3 812 061, 1975.
14. Walling, C.; Jacknow, B. B. *J. Am. Chem. Soc.* 1960, *82*, 6108.
15. Lee, W. A.; Rutherford, R. A. in "Polymer Handbook"; Brandrup, J.; Immergut, E. H., Eds.; 1975; Vol. III, p. 139.
16. Tada, T. *J. Electrochem. Soc.* 1982, *129*, 1070.
17. Kawamura, T.; Uryu, T.; Matsuzaki, K. *Makromol. Chem.* 1982, *103*, 125.
18. Ebdon, J. R.; Huckerby, T. N. *Polymer* 1976, *17*, 170.
19. Colthup, N. B.; Daly, L. H.; Wiberley, S. In "Introduction to Infrared and Raman Spectroscopy"; Academic Press: 1975.
20. Hummel, D. O. In "Polymer Spectroscopy"; Hummel, D. O., Ed.; Verlag Chemie, 1974, Chap. 2, p. 112.
21. Nielsen, L. E. in "Mechanical Properties of Polymers and Composites"; Dekker, M., Ed.; 1974, Vol. 1, Chap. 1.
22. Charlesby in "Atomic Radiation and Polymers"; Vol. 1, Chap. 12.
23. Feit, E. D.; Thompson, L. F.; Wilkins, C. W., Jr.; Wurtz, M. E.; Doerries, E. M.; Stillwagon, L. E. *J. Vac. Sc. Technol.* 1979, *16(6)*, 1997.
24. Feit, E. D.; Heidenreich, R. D.; Thompson, L. F. *Appl. Polym. Symp.* 1974, *23*, 125.

RECEIVED October 8, 1984

Photochemistry of Ketone Polymers in the Solid Phase: Thin Film Studies of Vinyl Ketone Polymers

J. E. GUILLET and S.-K. L. LI[1]

Department of Chemistry, University of Toronto
Toronto, Canada M5S 1A1

S. A. MACDONALD and C. G. WILLSON

IBM, 5600 Cottle Road, San Jose, CA 95193

The study of the photochemistry of ketone polymers provides a useful means of elucidating the mechanisms of photophysical and photochemical processes in polymeric media. Early studies of small organic molecules containing the ketone carbonyl group have shown that a variety of chemical reactions can occur from both the excited singlet and triplet states induced by the absorption of a UV photon. It was also shown by Guillet and Norrish (*1*) that these same reactions occurred in copolymers of methyl vinyl ketone (MVK). In the solid phase, one of these reactions (the Norrish type II) was shown to be inefficient below the glass transition (T_g) of the polymer, because of the restrictions on conformational mobility which precluded the formation of the six-membered transition state required for the scission process to occur (*2*). At that time it was expected that this would be a general effect in polymer photochemistry, but it now appears to be the exception rather than the rule.

For example, in Part I of this paper it was shown that the Norrish type I reaction could be quite efficient in the solid phase, and that a reduction in polymer molecular weight could be achieved through β scission via the reaction sequence:

hν

C=O
R

C=O
R
+

(1)

β-SCISSION

This is Part II in a series.

[1]Current address: Facelle Co. Ltd., 1551 Weston Road, Toronto, Ontario M6M 4Y4.

0097-6156/84/0266-0389$06.00/0

The efficiency of the type I reaction usually depends on the relative stability of the radicals produced. For this reason, higher type I quantum efficiencies would be expected for methyl isopropenyl ketone (MIPK) styrene copolymers because the polymer chain radical formed is tertiary rather than secondary, i.e.,

(2)

Similar improvements would also be expected if the *R* group gives a secondary or tertiary radical. Thus, one would expect an enhancement of type I yields in copolymers of t-butyl vinyl ketone (tBVK), i.e.,

I

(3)

In this work, studies were made on the photochemistry of styrene copolymers containing minor amounts (2-7%) of MVK, MIPK and tBVK. The properties of these polymers are summarized in Table I. The low concentrations of carbonyl units present should minimize the effects of energy

Table I. Properties of Styrene-Vinyl Ketone Copolymers

Polymer	Ketone Content Mol-%	Av. Molecular Weight M_v	M_n	Polydispersity M_v/M_n	$\epsilon_{313\text{ nm}}$ M^{-1} cm^{-1}
PS-MVK	2.9	210,000	92,800	2.3	12.
PS-MIPK	6.1	174,000	71,400	2.9	19
PS-tBVK	6.7	341,000	145,000	2.9	20

Table II. Various Quantum Yields Upon Irradiation of Poly(styrene-co-vinyl aliphatic ketone)s at 313 nm at 23°C in N_2

Comonomer	Mol-%	Film			Solution
		ϕ_{-CO}	ϕ_{OH}	ϕ_S	ϕ_S'
MVK	2.9	0.093	0.018	0.030	0.20
MIPK	6.1	0.26	0.01	0.082	0.22
tBVK	6.7	0.45 (0.45)	0.015	0.088	–

transfer and migration in these polymers. The polymers were photolyzed in thin (~0.1 mm) solution-cast films and in benzene solution. In the latter case the rates could be followed by automatic viscometry using the procedure described by Kilp et al. (*3,4*). Chemical changes in the solid-state were followed by IR spectroscopy. The results are summarized in Table II. The major chemical changes which occur are the loss of ketone carbonyl function (ϕ_{-CO}), the formation of hydroxyl (ϕ_{OH}) and changes in molecular weight (ϕ_S). In solution the major change in molecular weight is due to chain scission by the Norrish type II process:

C=O R C=O R (4) +

There is some contribution due to β-scission of the alkyl radical formed by the type I process, particularly in the MIPK and tBVK polymers. Loss of carbonyl occurs from photoreduction or the formation of cyclobutanol rings, and also from vaporization of the aldehyde formed by hydrogen abstraction by acyl radicals formed in the Norrish type I process. As demonstrated previously (*2*) the quantum yields for chain scission are lower in the solid phase than in solution. Rates of carbonyl loss are substantially different for the copolymers, being fastest for tBVK, slower for MIPK, and least efficient for MVK copolymers (Table I and Figure 1).

These polymers are of potential interest as photoresists. Accordingly, their photochemistry was also studied in very thin (1-4 μm) films which were spin-coated onto polished salt plates. After irradiation in a standard xenon arc photoilluminator, the loss of carbonyl could be determined from FTIR measurements. Experiments were carried out both at 254 nm (deep-UV) and 313 nm (mid-UV). Typical rate curves are shown in Figures 2 and 3. The

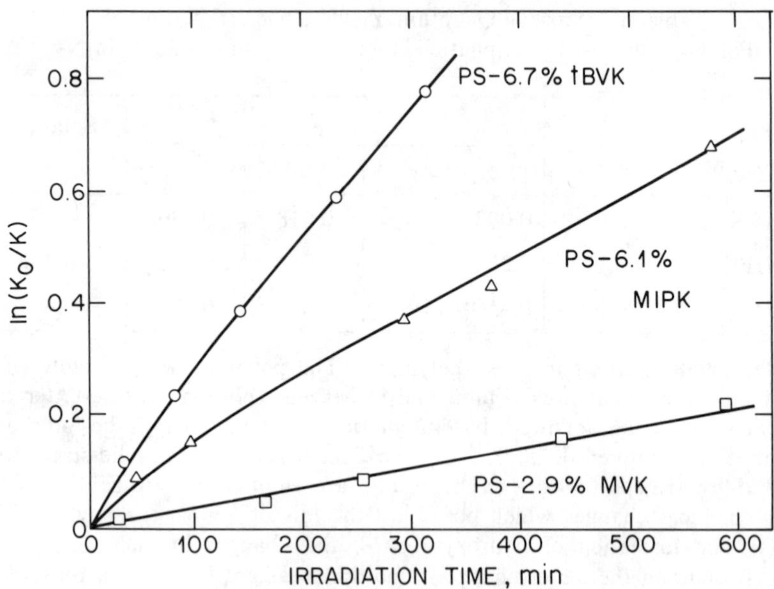

Figure 1. Irradiation of PS-vinyl aliphatic ketone copolymers in N_2 at 313 nm in the solid phase.

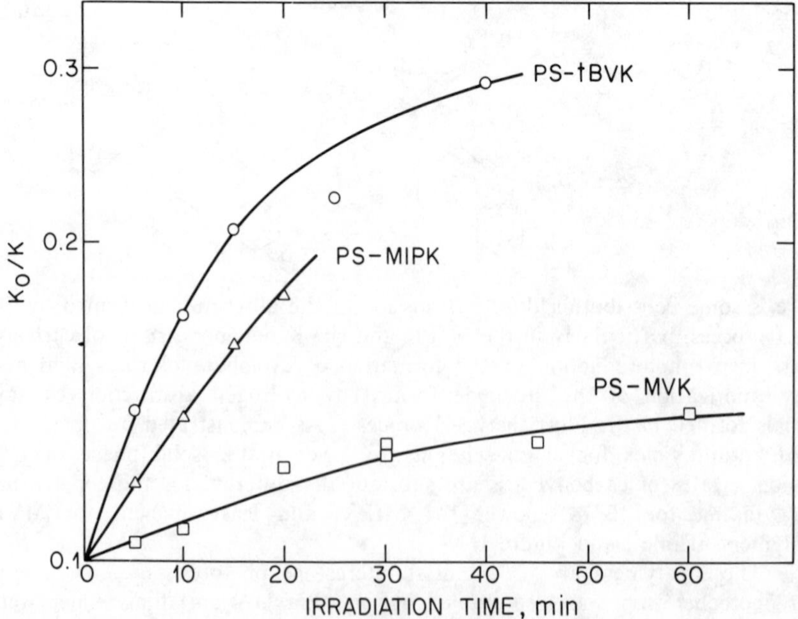

Figure 2. Photolysis of thin films of PS copolymers: 3-5 μ films on NaCl plates, λ = 313 nm.

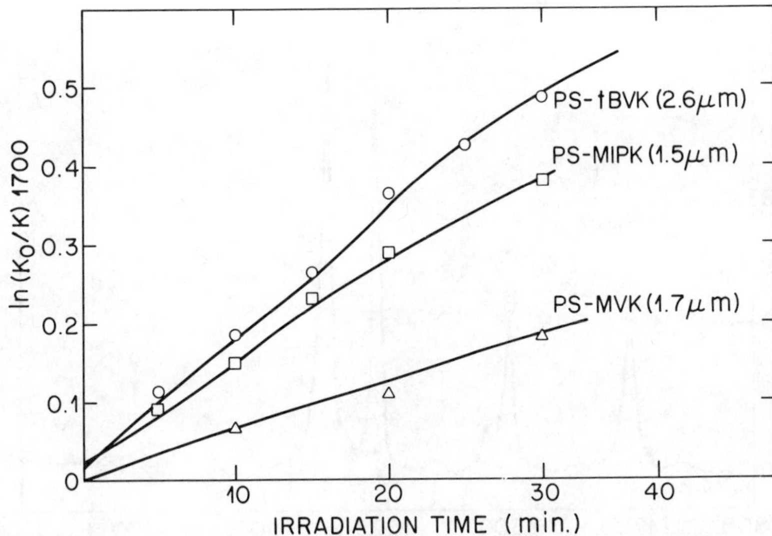

Figure 3. Irradiation of ketone copolymers at 254 nm: I_0 = 0.024 mwatt cm^{-2}.

results at 254 and 313 nm are qualitatively similar, in that the same order of sensitivity is observed. However, the structure-dependent difference in rate is greater at 313 nm than for irradiation at 254 nm. The relative slopes of the curves in Figures 2 and 3 (which are proportional to the sensitivity) are 1:4.3:7.1 at 313 vs. 1:2.7:3.4 at 254 nm. This is probably because much of the light at 254 nm is being absorbed by the aromatic groups in the styrene part of the copolymer.

The FTIR spectrum of PS-MIPK is shown in Figure 4. Band A is the ketone carbonyl absorption at 1700 cm^{-1} and is used to monitor changes in photochemistry. Bands B (1600 cm^{-1}), C (1495 cm^{-1}) and D (1455 cm^{-1}) are well resolved bands from the styrene portion of the copolymer. Provided that they are not involved in the photochemistry, which seems unlikely, they can be used as a practical measure of film thickness. Measurements were made on the actual thickness of a number of PS and PS copolymer films using a Talley-step apparatus, followed by FTIR measurements. Based on these results, the relative absorbance values were found to be: B, 2.56; C, 7.40; and D, 6.83 absorbance units per micron. The UV absorbance was also measured for films of various thickness at 254 nm, and the data are summarized in Table III. The constancy of these data suggests that this also could be used as a simple method of determination of film thickness.

Photolysis of PS-tBVK

FTIR measurements on the PS-tBVK copolymer showed that after irradiation, a new absorbance in the carbonyl stretching region at 1730 cm^{-1} appeared in the polymer. This absorbance increased with time of irradiation while that of the original ketone absorbance at 1700 cm^{-1} decreased, as shown in Table IV.

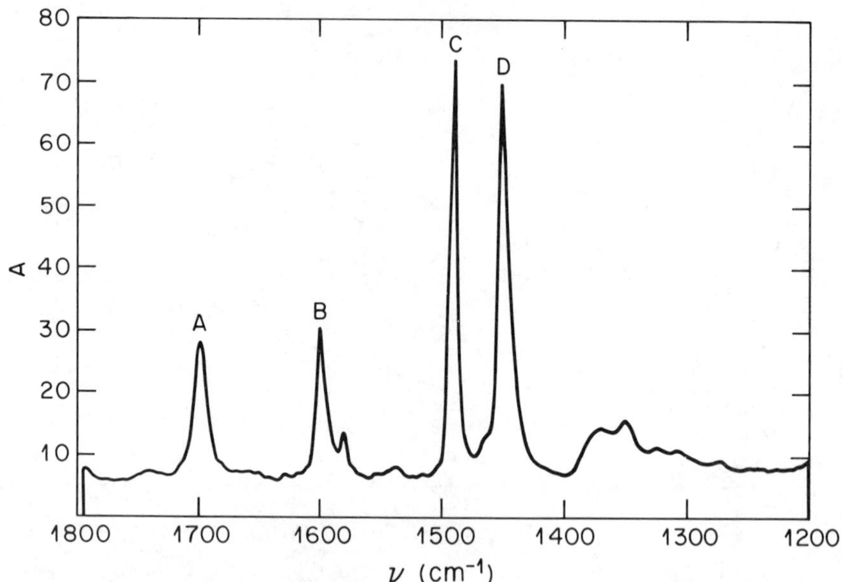

Figure 4. FTIR spectrum of 9.6 μ film of PS-MIPK.

Table III. UV Absorbance for Copolymer Films at 254 nm

Polymer	Absorbance per μ
polystyrene	0.21
PS-MVK	0.19
PS-MIPK	0.20
PS-tBVK	0.21

Table IV. Irradiation of PS-tBVK (2.56μ film) $I = 0.024$ mwatt cm^{-2}

Time (min)	$A_{1700\ cm^{-1}}$	$A_{1730\ cm^{-1}}$
0	4.90	0.25
5	4.38	0.40
10	4.08	0.65
15	3.75	0.83
20	3.40	0.91
25	3.20	0.98

We attribute this new absorbance to the formation of aldehyde groups on the polymer chain by hydrogen abstraction by the acyl radical formed in the primary photolysis of PS-tBVK (Equation 3).

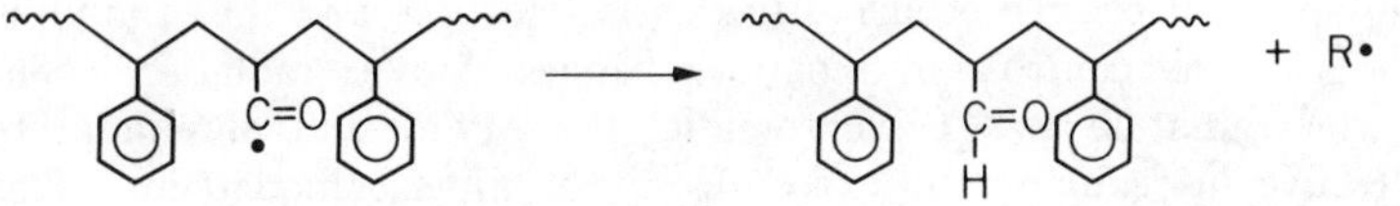

Assuming that the two carbonyl species have similar molar extinction coefficients, a simple calculation suggests that about half of the acyl radicals formed abstract hydrogen before decarbonylation. This process would thus be expected to reduce the efficiency of main-chain scission via the β-scission mechanism.

It is clear from these results that the relative sensitivity of the copolymers is unchanged in the very thin photoresist films. Quite high quantum yields can be obtained for photoprocesses which do not require large amounts of free volume.

Exposure to Synchrotron Radiation

The use of soft X-ray radiation from a synchrotron source has certain advantages for the production of microcircuitry. In particular, the short wavelength of the X-ray photons (1-50 Å) should provide higher pattern resolution in production devices, thereby increasing the density of circuit elements on the chips with a concomitant improvement in speed. It was therefore of interest to see if the same chemical selectivity observed in the photo response of these polymers in the near- and deep-UV extended to processes induced by the absorption of soft X-ray photons with energies two to three orders of magnitude greater than those in the UV region. In fact, earlier experiments by Slivinskas and Guillet (*5,6*) using γ-radiation suggested that this might indeed be the case.

Accordingly, by courtesy of the Stanford Synchrotron Radiation Laboratory, some time was made available for exposure of film samples to synchrotron radiation. The window size on the beam line was 2 mm × 12 mm. During exposure the machine was running at 10-13 ma and at an energy of 2.07 GeV. The small size of the window severely restricted the amount of material which could be exposed. However there was enough to measure changes in the IR spectrum by FTIR. After this measurement the molecular weight changes were estimated by high pressure liquid exclusion chromatography (GPC) using polystyrene standards for calibration.

In early exposures of 1 to 60 min no changes were observed in the spectra of the copolymer films, and it was subsequently determined that the beam intensity at short wavelength was much lower than originally predicted. Our measurements indicated an effective intensity of the order of 0.01 mwatt cm^{-2}, and to obtain measurable effects an exposure of ~20 h would be required. Accordingly, thin (~10 μm) films of various copolymers were prepared and exposed in a "film pack" ~0.5 × 1.0 inch in size. In the first two sections, each film was separated from the next by an 0.2 mil film of Mylar (poly(ethylene terephthalate)). Between the first and second sections was a 5 mil Mylar film, and between the second and third, 1 mil of aluminum foil (Reynolds 650). A fourth section was protected by a 3 mil layer of aluminum foil. No apparent change occurred in this last section, and thus the films there could be used as controls to monitor changes due to radiation. Since the synchrotron radiation was polychromatic, the Mylar and aluminum barriers were effective in reducing the exposures of the films behind them. Thus in a single experiment one could achieve three different exposure levels.

Measures of the sensitivity were made in two ways. (1) Loss of ketone carbonyl was determined by FTIR on the exposed samples by measuring the relative absorbance A at 1700 cm^{-1}. The ratio $(A_0/A)_{1700}$, was adjusted for film thickness using the styrene bands at 1600, 1495, and 1455 cm^{-1}. This value is proportional to the rates of the Norrish type I and photoreduction processes in the copolymer (*2*). Changes in molecular weight result from scission in the backbone of the polymer chain. A measure, Z, of the sensitivity to main-chain scission can be derived as follows.

The number of photons absorbed per unit time

$$n = IAla \tag{5}$$

where I is incident intensity per unit area, A is the absorption coefficient, l is the film thickness, and a is the area exposed. a is the same for all films irradiated and $la = w$, the weight of film exposed. Therefore

$$n = IAw \tag{6}$$

The quantum yield ϕ is the number of bond scissions, S, which occur per photon absorbed. From stoichiometric considerations (*7*),

$$S = \frac{w}{M_n^0}\left(\frac{M_n^0}{M_n} - 1\right) \tag{7}$$

where M_n^0 and M_n are the original and final number average molecular weight of the polymer. Therefore,

$$\phi = \frac{S}{IAw} = \left(\frac{1}{1A}\right)\left(\frac{1}{M_n^0}\right)\left(\frac{M_n^0}{M_n} - 1\right) \tag{8}$$

and we can define a response function as,

$$Z = \frac{1}{M_n^0}\left(\frac{M_n^0}{M_n} - 1\right) \tag{9}$$

which is related to the quantum yield ϕ by

$$\phi = Z/IA \tag{10}$$

Since the X-ray absorbance, A, is likely to be identical for the sytrene copolymer films used in this experiment, Z will be proportional to the quantum yield when the films are exposed to equal intensities, I, of radiation. In the film pack experiment, this will not always be the case, since some attenuation will occur in passing through preceding films. To monitor this, two identical PS-MIPK films were put at the beginning and end of each section of the pack. Variations in the response of these two should give an idea of the relative attenuation.

The results of these experiments are summarized in Table V, which shows values of A^0/A and Z for the ketone copolymer films. All of the polymers in the first section underwent significant damage and were very brittle. However, there was not much attenuation in the first section as indicated by the similar Z values of the PS-MIPK films placed at the beginning and the end of the section. It should be pointed out that the errors in determination of M_n by GPC are substantial – considerably greater than the FTIR measurements. Nevertheless, the Z values correlate quite well with A^0/A, indicating the importance of the Norrish type I process in causing chain scission.

Table V. Sensitivity of Various Polymers to Synchrotron Radiation

Thickness μ	Sample	Polymer	$(A_0/A)_{1700\ cm^{-1}}$	Z	Barrier*
					0.2 mil Mylar
9.6	A	PS-MIPK	1.24	6.1	
4.6	C	PS-MVK	1.11	0	
18	F	PS-tBVK	1.44	7.1	
9.6	G	PS-MIPK	—	5.6	
					5 mil Mylar
	H	PS-MIPK	1.15	4.8	
	I	PS-MVK	1.05	2.7	
	J	PS-tBVK	1.34	4.9	
	K	PS-MIPK	1.18	2.4	
					1 mil Al Foil
	L	PS-MVK	1.09	0	
	M	PS-tBVK	1.15	2.4	
	N	PS-MIPK	1.09	0	
					3 mil Al Foil
	O	PS-MIPK	No detectable change		
	P	PS-MVK	No detectable change		
	Q	PS-tBVK	No detectable change		
	R	PS-MIPK	No detectable change		

*Samples A through K separated by 0.2 mil Mylar film.

However, the most important conclusion of this work is that the polymers show the same relative sensitivity to synchrotron radiation as they do to mid and deep UV. This has important significance in the design of photoresists for use with soft X-ray sources.

Literature Cited

1. Guillet, J. E.; Norrish, R. G. W. *Proc. R. Soc., London, Ser. A*, 1955, *233*, 153.
2. Dan, E.; Guillet, J. E. *Macromolecules* 1973, *6*, 230.
3. Kilp, T.; Guillet, J. E. *Macromolecules* 1977, *10*, 90.
4. Kilp, T.; Houvenaghel-Defoort, B.; Panning, W.; Guillet, J. E. *Rev. Sci. Instr.* 1976, *47*, 1496.
5. Slivinskas, J. A.; Guillet, J. E. *J. Polym. Sci., Polym. Chem. Ed.* 1973, *11*, 3057.
6. Slivinskas, J. A.; Guillet, J. E. *J. Polym. Sci., Polym. Chem. Ed.* 1974, *12*, 1469.

RECEIVED October 8, 1984

20

Polymers of α-Substituted Benzyl Methacrylates and Aliphatic Aldehydes as New Types of Electron-Beam Resists

K. HATADA, T. KITAYAMA, Y. OKAMOTO, and H. YUKI

Department of Chemistry
Faculty of Engineering Science
Osaka University,
Toyonaka, Osaka 560, Japan

H. ARITOME and S. NAMBA

Department of Electrical Engineering
Faculty of Engineering Science
Osaka University
Toyonaka, Osaka 560, Japan

K. NATE, T. INOUE, and H. YOKONO

Production Engineering Research Laboratory
Hitachi, Ltd., 292 Yoshida-machi
Totsuka-ku, Yokohama 244, Japan

Many papers have been published on positive electron-beam resists. These resists are mostly polymers which are degraded upon electron-beam irradiation. The resulting lower molecular weight polymer in the exposed area can be selectively removed by a solvent under certain developing conditions. The development is accomplished by the difference in the rate of dissolution between the exposed and unexposed areas, which is a function of the molecular weight of the polymer. Recently, Willson and his co-workers reported the new type of positive resist, poly(phthalaldehyde), the exposure of which in the presence of certain cationic photoinitiators resulted in the spontaneous formation of a relief image without any development step (*1*).

In this article we will describe two different types of positive electron-beam resists, which were briefly reported in our previous communications (*2,3*). One is the homopolymer or copolymer with methyl methacrylate and α-substituted benzyl methacrylate, which forms methacrylic acid units in the polymer chain on exposure to an electron-beam and can be developed by using an alkaline solution developer. In this case, the structural change in the side group of the polymer effectively alters the solubility properties of the exposed polymer, and excellent contrast between the exposed and unexposed areas is obtained. The other is a self developing polyaldehyde resist, which is depolymerized into a volatile monomer upon electron-beam exposure. The sensitivity was extremely high without using any sensitizer.

0097–6156/84/0266–0399$07.00/0

Experimental

Materials. Benzyl methacrylate was obtained commercially. α-Methylbenzyl (*4*) and α,α-dimethylbenzyl (*5*) methacrylates were prepared in diethyl ether from methacryloyl chloride and the corresponding alcohols in the presence of triethylamine. α,α-Diphenylethyl methacrylate was synthesized from methacryloyl chloride and lithium α,α-diphenylethoxide in tetrahydrofuran (*6*). Triphenylmethyl methacrylate was prepared from silver methacrylate and triphenylmethyl chloride in diethyl ether (*7*). The first three methacrylates were purified by fractional distillation under reduced nitrogen pressure. The monomers thus purified were dried over calcium dihydride and then distilled under high vacuum just before use. α,α-Diphenylethyl methacrylate and triphenylmethyl methacrylate were purified by recrystallization from hexane.

Ethanal, propanal, butanal, heptanal and 3-phenylpropanal were obtained commercially. 3-Trimethylsilylpropanal was prepared from acrolein and trimethylsilyl chloride in tetrahydrofuran at about −50°C (*8*).

Polymerization. Poly(methyl methacrylate) was obtained commercially. The polymers of other methacrylates and their copolymers were prepared in toluene with 2,2'-azobisisobutyronitrile (AIBN) at 60°C. All the polymers prepared free radically were syndiotactic or atactic. Isotactic poly(α,α-dimethylbenzyl methacrylate) was obtained using C_6H_5MgBr as the initiator in toluene at 0°C. Poly(methacrylic acid) was prepared in water using potassium persulfate at as the initiator 60°C. The molecular weights, glass transition temperatures and tacticities of the polymethacrylates are summarized in Table I.

Table I. Polymethacrylates[a] Used for Electron-Beam Resists[b]

Ester group	M_n	M_w/M_n	T_g (°C)	Tacticity(%)		
				I	H	S
$-CH_3$[c]	125000	3.25	123	8	37	55
$-CH_2C_6H_5$	529000	2.34	63	5	34	61
$-CH(CH_3)C_6H_5$	460000	3.13	106	5	35	60
$-C(CH_3)_2C_6H_5$	847000	2.07	101	10	47	43
$-C(CH_3)_2C_6H_5$[d]	15000	3.47	85	86	9	5
$-C(CH_3)(C_6H_5)_2$	43000	3.33	98	31	42	27

a Polymerization conditions: monomer 10mmol, AIBN 0.1mmol, polymn temp. 60°C, polymn time 24 hr.

b The data are taken from the Table I of Ref. 2.

c Commercial product.

d Isotactic polymer. Polymerization conditions: monomer 10mmol, C_6H_5MgBr 0.25mmol, polymn temp. 0°C, polymn time 24 hr.

Copolymerizations of aldehydes were carried out in toluene at −78°C using diethylaluminum diphenylamide as an initiator (*9*). Some of the results are shown in Table II. Unlike the homopolymers of aliphatic aldehydres, the copolymers were soluble in organic solvents such as toluene, xylene or chloroform in a certain range of copolymer composition (*10-12*). The soluble copolymers containing approximately equal amounts of both monomer units were used for making the resist films. The weight average molecular weights of the poly(ethanal-co-butanal)s prepared in toluene at −78°C by diethylaluminum diphenylamide were reported to be more than 1,000,000 (*13*). It is reasonable to assume that the molecular weight of the aldehyde copolymers used in this work is at least the order of 10^6.

The compositions of copolymers of methacrylates or aldehydes were determined by ^{1}H NMR spectroscopy or elemental analysis.

Table II. Copolymerization of Aldehydes by Diethylaluminum Diphenylamide in Toluene at −78°C for 24 hr

Monomer[a]		$[M]_0$[b]	Initiator	Toluene	Polymer		
M_1–M_2	(mmol/ mmol)	(mol/1)	(mmol)	(ml)	Yield (%)	M_1[c] (mol%)	$[\eta]$[d] (dl/g)
AA-PA	(25/25)	2	0.25	21	52.4	51.2	–
AA-BA	(50/50)	4.5	1.00[e]	15	72.9	50.0	14.3
AA-HA	(25/25)	2	0.25[e]	19	58.3	54.4	18.2
PA-PhPA	(15/15)	1.3	0.30	20	43.1	52.0	10.3

a AA: ethanal, PA: propanal, BA: butanal, HA: heptanal, PhPA: 3-phenylpropanal.

b Initial concentration of total monomer.

c Determined from elemental analysis.

d Measured in toluene at 30.0°C.

e The solutions of initiator and monomer were mixed at 0°C and then the mixture was cooled to −78°C.

Lithographic Measurements. The polymers and copolymers of methacrylates were dissolved in toluene and the solution was spun onto a silicon substrate. Poly(methacrylic acid) was used in a pyridine solution. The thickness of the resist films was about 0.3 μm.

The resist films of polyaldehydes were made from xylene, toluene or chloroform solutions by spin-coating or dip-coating, and were 0.04 ~ 3.4 μm in thickness.

The coated silicon wafer was prebaked and then exposed to 20 KV electron-beam using JEOL JBX-5B or Elionix ERE-301 electron-beam exposure system. The exposed polymethacrylates were developed using a mixture of methyl isobutyl ketone (MIBK) and isopropyl alcohol (IPA) or a dilute solution of sodium methoxide in methanol, and subsequently rinsed with IPA or methanol, respectively. In the case of the polyaldehyde resists almost complete development was accomplished by exposure alone and the development process was not needed. The film thickness was measured on a Talystep instrument or by optical interference.

The molecular weights of the polymethacrylates were measured with a JASCO FLC-A10 GPC chromatograph with a Shodex GPC column A-80M (50 cm) with a maximum porosity of 5×10^7 using tetrahydrofuran as a solvent. The chromatogram was calibrated against standard polystyrene samples.

Glass transition temperatures of the polymers were measured with a Rigaku Denki Calorimeter, Model 8001 SL/C, at a heating rate of 10°C/min using an aluminum sample pan with lid.

Infrared spectra were recorded on the resist film spun onto a silicon wafer using a JASCO IR-810 spectrometer equipped with a JASCO BC-3 beam condenser or a JASCO A-3 spectrometer. In the measurements on the latter spectrometer an uncoated silicon wafer was placed in the reference beam in order to balance the silicon absorption band. The subtraction between the spectra was carried out on a built-in micro-processor attached to the IR-810 spectrometer, and the resulting difference spectrum was used to detect structural changes in the polymer molecule upon exposure. The subtraction technique was also used to balance the silicon absorption band.

The ^1H NMR spectra were taken on a JEOL JNM-MH-100 (CW) spectrometer using tetramethylsilane as an internal standard. ^{13}C spin-lattice relaxation time of the polymer was measured by the inversion-recovery Fourier transform method on a JNM-FX100 FT NMR spectrometer operating at 25 MHz.

Results and Discussion

Positive Electron-beam Resist of Poly(α-substituted Benzyl Methacrylate). The electron-beam resist behaviors of poly(α-substituted benzyl methacrylate)s are given in Table III. When the exposed resist films were developed with a mixture of MIBK and IPA, the sensitivities of these polymers were on the order of 10^{-4} C/cm^2. When a dilute solution of sodium methoxide in methanol was used as a developer, the sensitivity was enhanced as compared with the former case, and increased with an increase in the bulkiness of the ester group of the polymer except for poly(α,α-diphenylethyl methacrylate).

Figure 1 shows the exposure characteristics of atactic and isotactic poly(α,α-dimethylbenzyl methacrylate) resists with CH_3ONa development together with those of the poly(methyl methacrylate) resist with MIBK/IPA development. Poly(α,α-dimethylbenzyl methacrylate)s showed high sensitivity and very good contrast between exposed and unexposed areas. The atactic polymer with alkaline development was improved in the sensitivity and γ-value by a factor of more than three over poly(methyl methacrylate) with MIBK/IPA development.

Infrared spectra of unexposed and exposed poly(α,α-dimethylbenzyl methacrylate) are given in Figure 2 together with the spectrum of

Table III. Electron-beam Exposure Characteristics of Polymethacrylates[a,b]

Ester group	Prebake (°C)	MIBK/IPA (1/5 v/v) Sens. (C/cm²)	γ	CH_3ONa/CH_3OH (1/20 w/v) Sens. (C/cm²)	γ
$-CH_3$	170	167×10^{-6}	3.6	226×10^{-6}	2.1
$-CH_2C_6H_5$	170	172×10^{-6}	—	103×10^{-6}	—
$-CH(CH_3)C_6H_5$	170	250×10^{-6}	2.4	97×10^{-6}	—
$-C(CH_3)_2C_6H_5$	142	85×10^{-6}	4.4	36×10^{-6}	11.3
$-C(CH_3)_2C_6H_5$[c]	142	110×10^{-6}	4.8	28×10^{-6}	7.4
$-C(CH_3)(C_6H_5)_2$	120	150×10^{-6}[d]	—	150×10^{-6}	—

a Film thickness 0.3 μm, prebake 1 hr, development time 30 sec, rinse time 30 sec.

b The data are taken from the Table II of Ref. 2.

c Isotactic polymer

d Developer MIBK/IPA = 1/1 (v/v).

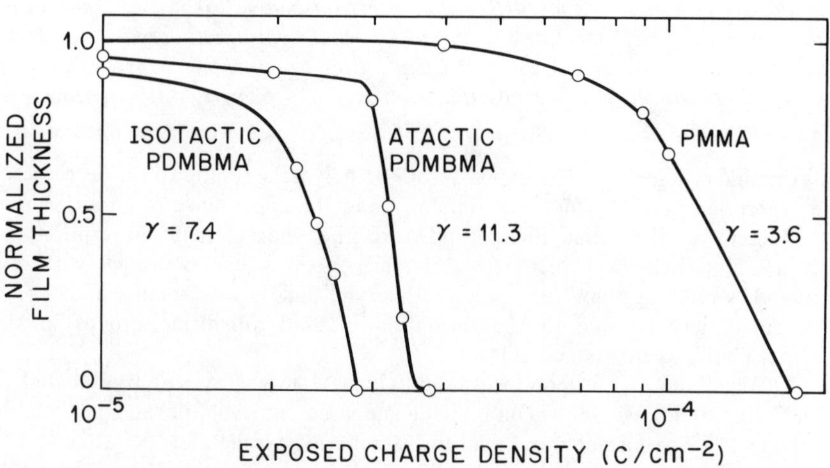

Figure 1. Exposure characteristics of poly(methyl methacrylate) (PMMA), and atactic and isotactic poly(α,α-dimethylbenzyl methacrylate)s (PDMBMA). Reproduced with permission from Ref. 2. Copyright 1983, "Springer Verlag".

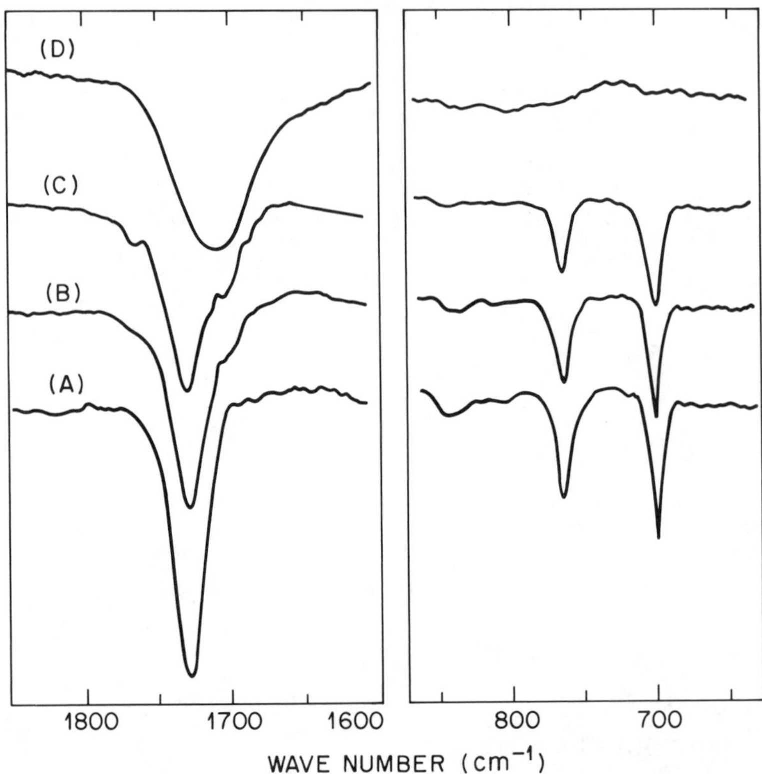

Figure 2. Infrared spectra of atactic poly(α,α-dimethylbenzyl methacrylate)s unexposed(A) and exposed(B) to electron-beam, isotactic poly(α,α-dimethylbenzyl methacrylate) exposed(C) and poly(methacrylic acid)(D). Exposure charge density 1.6×10^{-4} *C/cm², film thickness 0.5 μm, prebake at 142°C. Reproduced with permission from Ref. 2. Copyright 1983, "Springer Verlag".*

poly(methacrylic acid). The exposure of the polymers resulted in a decrease of the absorption at 1729, 764, and 700 cm^{-1} and the appearance of an absorption at 1700 cm^{-1}. The first three bands are the characteristic absorptions of poly(α,α-dimethylbenzyl methacrylate) and the last one coinsides with the carbonyl stretching band of poly(methacrylic acid). A small shoulder at 1760 cm^{-1} may be due to the formation of acid anhydride groups in the polymer chain, as mentioned below.

When the atactic poly(α,α-dimethylbenzyl methacrylate) was heated at 170°C for 30 min under vacuum, it decomposed into volatile and nonvolatile components. The former was found to be α-methylstyrene and the latter was to be very similar to poly(methacrylic acid) as determined by ^{1}H NMR spectroscopy. Figure 3 shows the infrared spectra of atactic and isotactic poly(α,α-dimethylbenzyl methacrylate)s heated at 174°C under vacuum for various times. In the spectra of the atactic polymer, the absorption of the ester carbonyl at 1729 cm^{-1} decreased and that of the acid carbonyl at 1700 cm^{-1} increased as the heating time increased. After heating for a period of 30 min

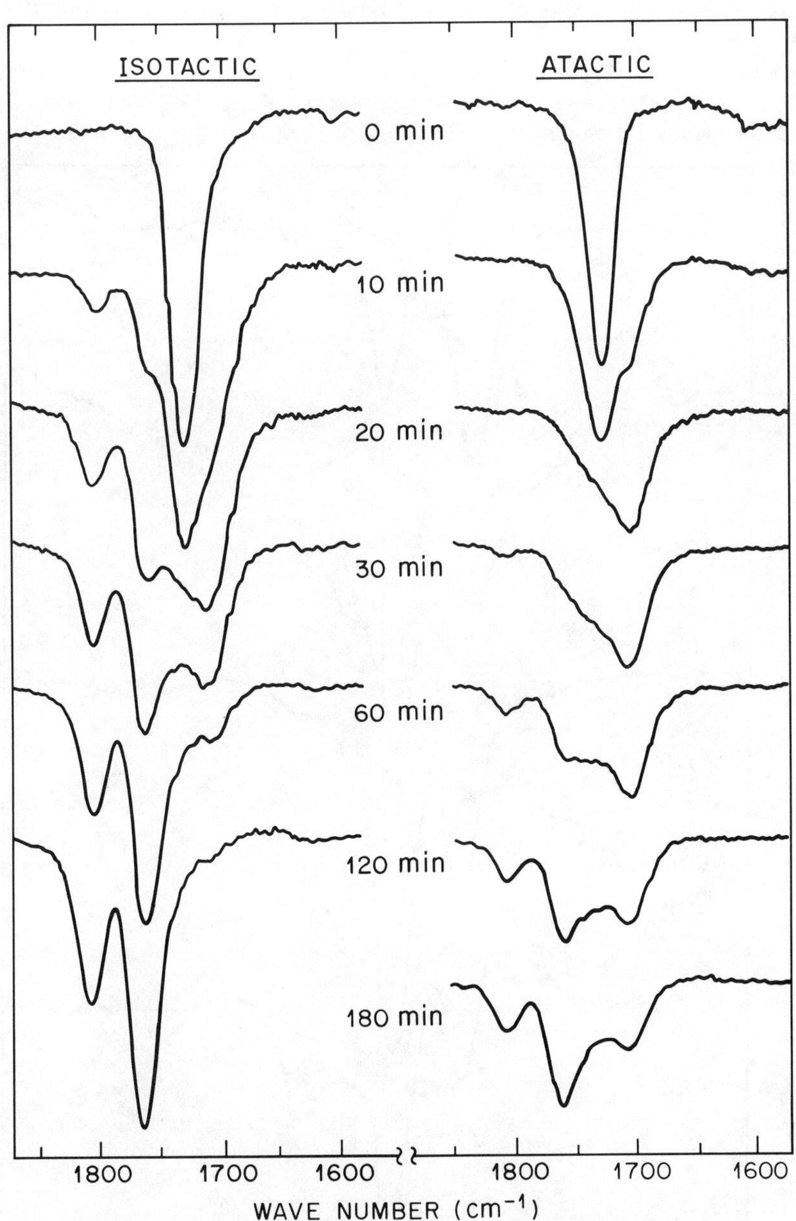

Figure 3. Infrared spectra of isotactic and atactic poly(α,α-dimethylbenzyl methacrylate)s heated at 174° C under vacuum for various times.

two new bands appeared at 1805 and 1760 cm^{-1}, and their intensities increased with an increase in time. A concomitant decrease in the absorption of the acid carbonyl group at 1700 cm^{-1} was observed. Glutaric anhydride shows two carbonyl stretching bands at nearly the same positions as 1805 and 1760 cm^{-1}, while the bands for isobutyric anhydride are slightly different (Figure 4). It

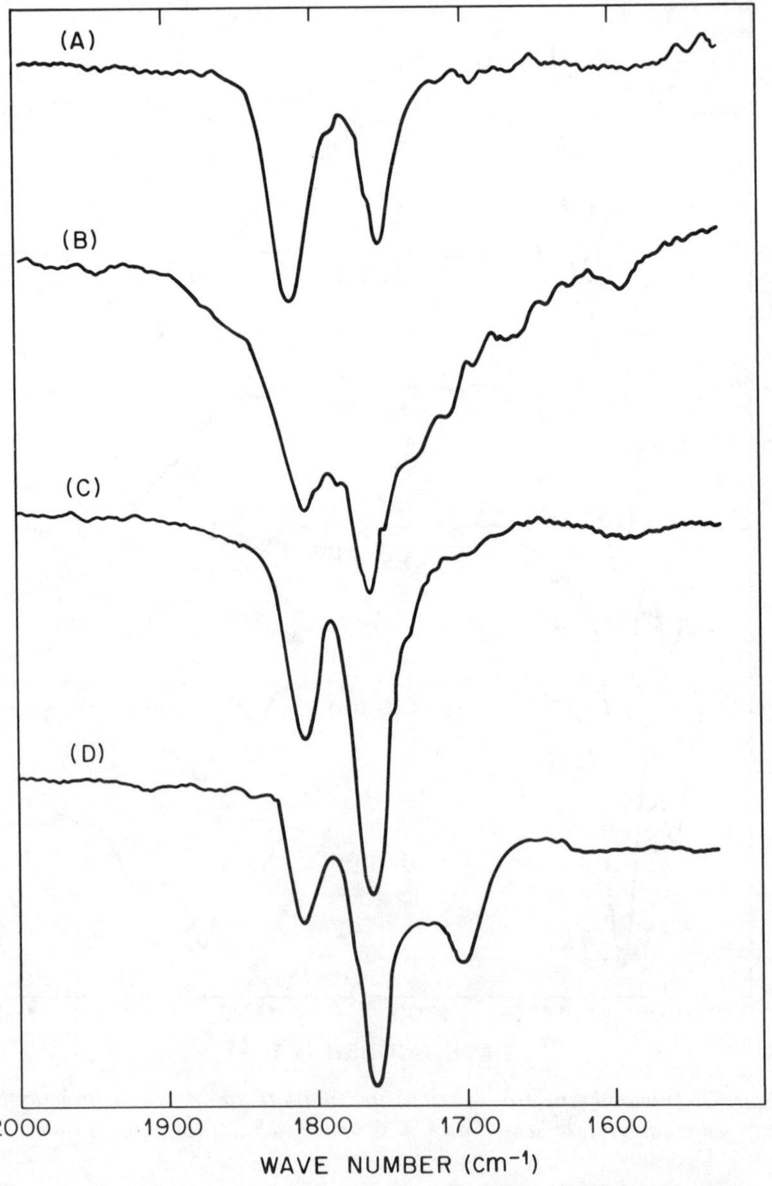

Figure 4. Infrared spectra of (A) isobutyric anhydride, (B) glutaric anyhdride, and (C) isotactic and (D) atactic poly(α,α-dimethylbenzyl methacrylate)s heated at 174° C under vacuum for 2 and 3 hr, respectively.

has been shown that the separation of two carbonyl frequencies in acid anhydrides varies from 40 to 80 cm^{-1} and those with six-membered rings usually have the least separation. The higher frequency peak is more intense for open chain anhydrides and this reverses for the five- or six-membered ring anhydrides (*14,15*). Therefore, the two peaks at 1805 and 1760 cm^{-1} should be assigned to the six-membered cyclic acid anhydride, which was formed intramolecularly from neighboring methacrylic acid units. Autocatalytic character was observed in the ester decomposition of poly(t-butyl methacrylate) into poly(methacrylic acid) and isobutene, for which a mechanism of olefin elimination involving an adjacent acid unit was proposed (*16*).

The results mentioned here clearly indicate that the enhancement in the sensitivity and γ-value of the poly(α,α-dimethylbenzyl methacrylate) resist over poly(methyl methacrylate) is mainly due to facilitated formation of methacrylic acid units on electron-beam exposure. The exposed area, which contains

CH3 CH3 CH3 CH3 CH3
~CH2–C–CH2———C~ → ~CH2–C–CH2–C~ + CH2=C
C=O C=O C=O C=O
O O O O
CH3–C–CH3 CH3–C–CH3 CH3–C–CH3 H

methacrylic acid units, easily dissolves in the alkaline developer but the unexposed area does not. In this case, the factor which determines the resist properties, is the change in solubility characteristics upon exposure, and not the rate of dissolution during the development process. This is the reason why high sensitivity and contrast are obtained with poly(α,α-dimethylbenzyl methacrylate). Some of the adjoining methacrylic acid units form acid anhydride groups, which exhibit a small shoulder at 1760 cm^{-1} in the spectrum of exposed resist film. The formation of acid anhydride may decrease the sensitivity and this will be discussed later.

Spectral subtraction usually provides a sensitive method for detecting small changes in the sample. Figure 5 shows the difference spectra between the atactic poly(α,α-dimethylbenzyl methacrylate)s unexposed and exposed to electron-beam at several doses. The positive absorption at 1729 cm^{-1} is due to the ester carbonyl group consumed on the exposure and the negative ones at 1700 and 1760 cm^{-1} to the acid and acid anhydride carbonyl groups formed, respectively. The formation of methacrylic acid units was more easily detected using the difference spectrum However, these difference spectra could not be used for the quantitative determination because the absorptions overlap somewhat.

In order to estimate the amount of methacrylic acid units formed on the exposure, the infrared spectra for the mixtures of known amounts of poly(α,α-dimethylbenzyl methacrylate) and poly(α,α-dimethylbenzyl methacrylate-co-

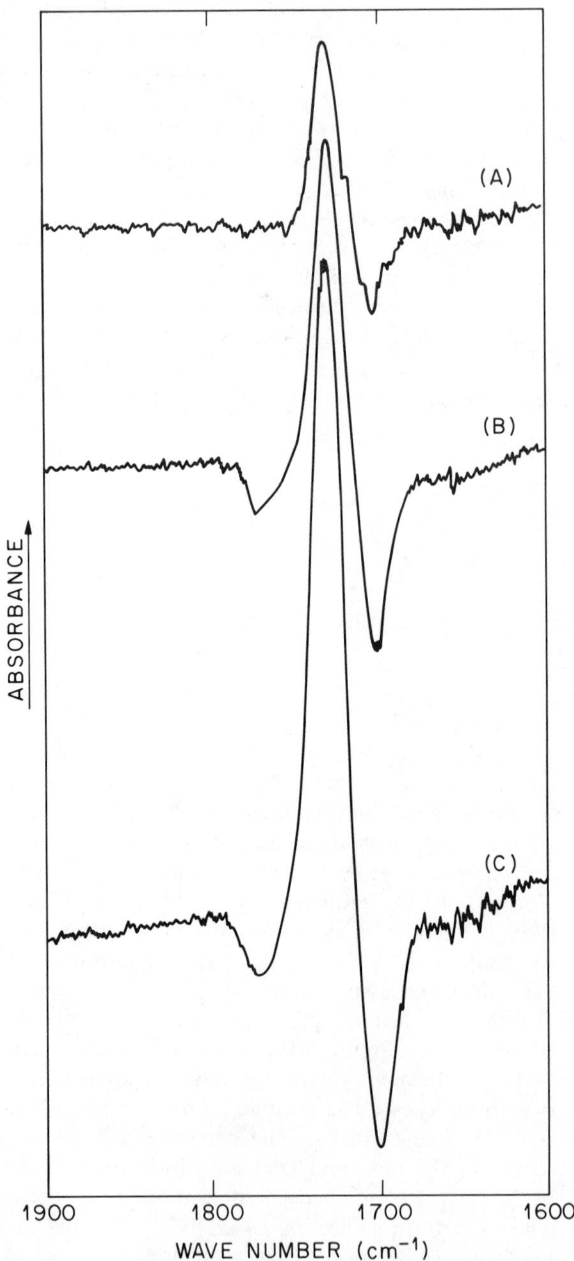

Figure 5. Difference infrared spectra between the atactic poly(α,α-dimethylbenzyl methacrylate)s unexposed and exposed to electron-beam of several doses. (A) 4×10^{-5}, *(B)* 1.6×10^{-4}, *(C)* 7×10^{-4} *C/cm*^{2}.

methacrylic acid) were measured on films cast from chloroform solution. The relative intensities of $A_{1700cm^{-1}}/A_{1729cm^{-1}}$ were plotted against the contents of the acid units. The copolymer was prepared in dimethylformamide with AIBN at 60°C and contained 15.0 mol% of methacrylic acid units. The results of this plot are shown in Figure 6. Using this calibration curve and the spectra of the exposed polymer, the methacrylic acid content in the atactic polymers exposed to electron-beam doses of 4×10^{-5} and 1.6×10^{-4} C/cm^2 were estimated to be 5.5 and 15.2 mol%, respectively.

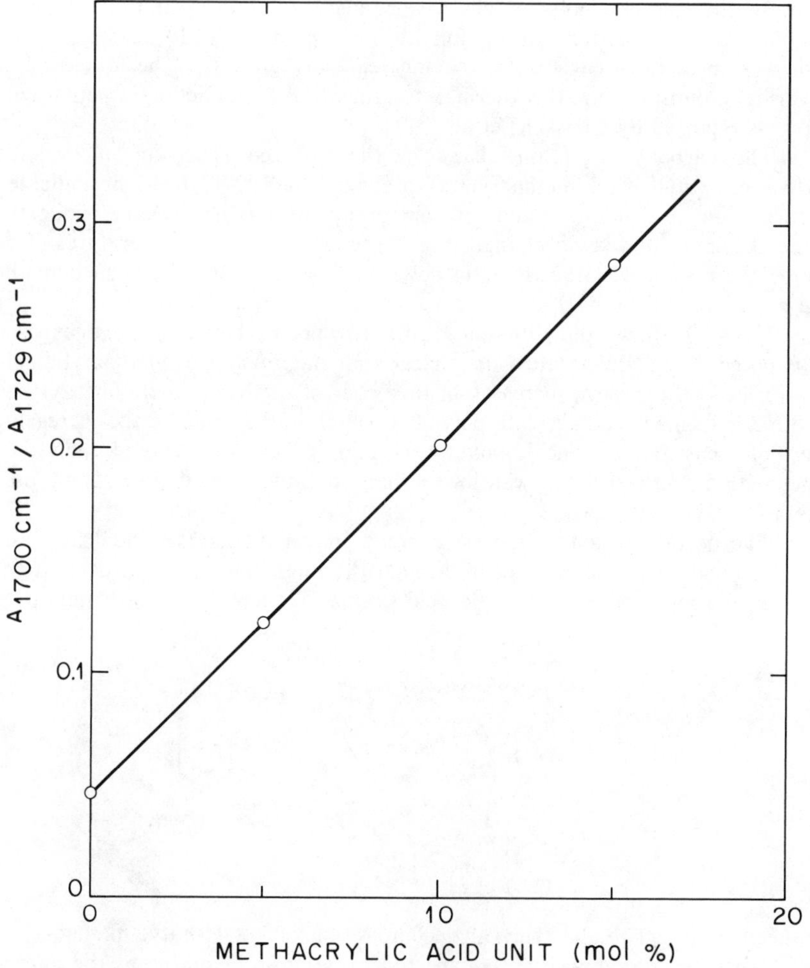

Figure 6. Correlation between the relative intensity of $A_{1700cm^{-1}}/A_{1729cm^{-1}}$ and the fractions of methacrylic acid units.

From the comparison of the spectra of electron-beam exposed atactic and isotactic poly(α,α-dimethylbenzyl methacrylate)s (Figure 2), it is clear that the isotactic polymer forms acid anhydride groups as well as acid groups more easily than the atactic polymer upon exposure. This should enhance the sensitivity of the polymer. However, the conversion of acid groups into acid anhydrides will depress the dissolution rate of the exposed polymer in the alkaline developer, which results in a decrease in sensitivity. Consequently, the sensitivity of the isotactic polymer is slightly higher than that of the atactic polymer as shown in Table III. Enhanced formation of acid and acid anhydride groups in the isotactic polymer chain were also observed upon heating under vacuum at 174°C. After heating for 120 min, predominantly acid anhydride bands were present in carbonyl stretching region (Figure 3). The formation of anhydride groups during the thermal treatment of isotactic poly(methacrylic acid) was reported by Geuskens et al. (*17*).

The carbonyl stretching band in the infrared spectrum of isotactic poly(α,α-dimethylbenzyl methacrylate) prebaked at 142°C for 1 hr indicated the formation of a small amount of acid group during the prebake, while the atactic polymer showed no change in the spectrum at this temperature. This may be the reason why the isotactic polymer showed a lower γ-value than the atactic polymer (Table III).

Figure 7 shows the difference infrared spectra between the unexposed and exposed polymethacrylates. It is clear that the amount of methacrylic acid units formed on exposure increases in the order of methyl, α-methylbenzyl and α,α-dimethylbenzyl methacrylate polymers, which is the same as the increasing order of sensitivity. The amount of acid group in exposed poly(α,α-diphenylethyl methacrylate) was lower than that in exposed polymer of α,α-dimethylbenzyl methacrylate.

The decomposition of the ester group into a methacrylic acid unit and the corresponding olefin may occur through the migration of β-hydrogen atoms in the ester group to the carboxylic acid group. Similar β-hydrogen migration

$$\sim CH_2-C(CH_3)(C(=O)O-C(CH_3)(C_6H_5)-CH_2-H)\sim \;\rightarrow\; \sim CH_2-C(CH_3)(C(=O)OH)\sim + CH_2=C(CH_3)(C_6H_5)$$

has been proposed to the thermal decomposition of poly(t-butyl methacrylate) (*18*). Then the decomposition is expected to be more favorable as the number of β-hydrogen atoms is larger. This is the case for the poly(α-substituted benzyl methacrylate)s as shown in Figure 7. However, when poly(t-butyl methacrylate) containing nine β-hydrogen atoms was exposed to an electron-beam, the amount of acid group formed was smaller than that for poly(α,α-

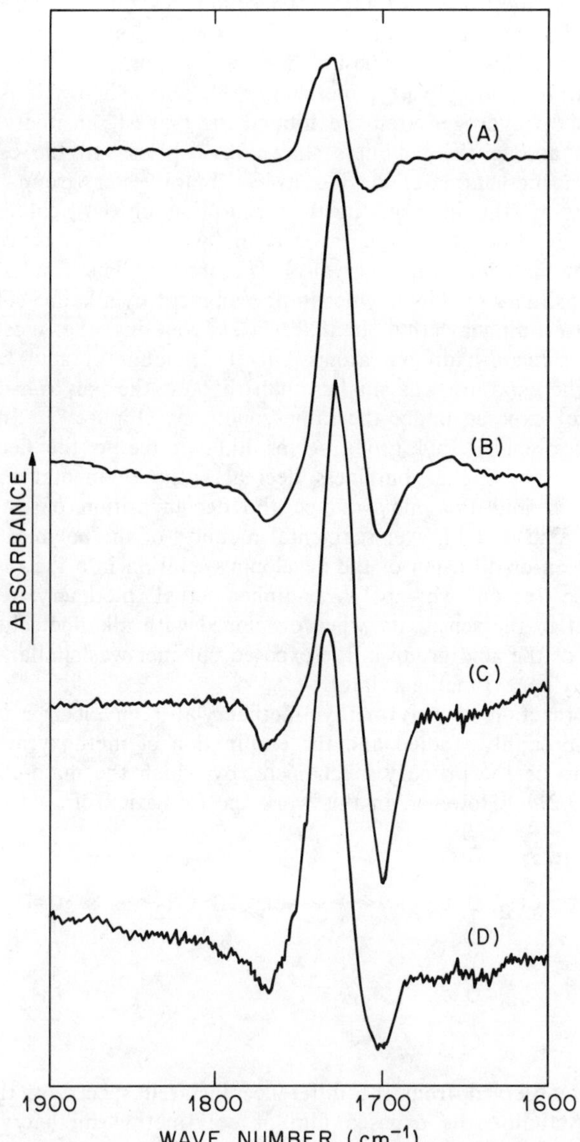

Figure 7. Difference infrared spectra between the polymethacrylates unexposed and exposed to the electron-beam of 1.6×10^{-4} *C/cm*^{2}*. (A) poly(methyl methacrylate) (B) poly(α-methylbenzyl methacrylate) (C) atactic poly(α,α-dimethylbenzyl methacrylate) (D) poly(α,α-diphenylethyl methacrylate).*

dimethylbenzyl methacrylate) (*19*). This indicates that the bulkiness of ester group as well as the number of β-hydrogen atoms is an important factor for the formation of acid groups on exposure to electron-beams.

Poly(α,α-dimethylbenzyl methacrylate), which had a bulky ester group containing six β-hydrogen atoms, exhibited the highest sensitivity upon alkaline development among the polymers shown in Table I. In the case of poly(α-methylbenzyl methacrylate) having a less bulky ester group with three β-hydrogen atoms, the intensity of the absorption of the acid carbonyl group formed on exposure to an electron-beam was smaller as compared with poly(α,α-dimethylbenzyl methacryalte) (Figure 7). The results parallel with the lower sensitivity of this polymer in development by alkaline solution.

Poly(α,α-diphenylethyl methacrylate) contains the most bulky ester group with three β-hydrogen atoms but the amount of acid carbonyl group formed on the exposure was smaller than that for the poly(α,α-dimethylbenzyl methacrylate) exposed under the same conditions (Figure 7). In this case the bulkiness of diphenylethyl group seems unfavorable to the decomposition of ester group. The higher bulkiness decreases the segmental mobility of the polymer chain and this may reduce the decomposition of ester group into methacrylic acid unit. Lower segmental mobility of the polymer also decreases the penetration or diffusion of the developing solution into the resist film. This may be the reason why poly(α,α-diphenylethyl methacrylate) showed no improvement in the sensitivity when developed with alkaline solution, although the amount of the acid group in the exposed polymer was similar to that for the poly(α-methylbenzyl methacrylate).

The reaction of poly(methyl methacrylate) on electron-beam exposure has been thoroughly studied and the elimination of methoxycarbonyl group is considered to be the primary mechanism, by which the main-chain scission is initiated (*20,21*). However, in this work the formation of acid carbonyl group

$$\sim CH_2-\underset{\underset{\underset{\underset{CH_3}{|}}{O}}{\underset{|}{C=O}}}{\overset{\overset{CH_3}{|}}{C}}-CH_2\sim \longrightarrow \sim CH_2-\overset{\overset{CH_3}{|}}{\dot{C}}-CH_2\sim + \cdot\underset{\underset{\underset{CH_3}{|}}{O}}{\underset{|}{C}}=O$$

was clearly observed from the difference infrared spectrum, though in small amount. Actually, the exposed film of poly(methyl methacrylate) could be developed by alkaline solution. This indicates that either of the following reactions may occur to some extent on exposure. Formation of methacrolein units on exposure by the following reaction was also proposed by Geuskens et al. (*21*). Similar reactions may occur in the case of poly(benzyl methacrylate), which showed higher sensitivity than poly(methyl methacrylate) with alkaline development.

$$\sim CH_2-\underset{\underset{\underset{\underset{CH_3}{|}}{O}}{|}}{\overset{\overset{CH_3}{|}}{\underset{\underset{C=O}{|}}{C}}}-CH_2\sim \xrightarrow{-(\cdot OCH_3)} \sim CH_2-\overset{\overset{CH_3}{|}}{\underset{\underset{\dot{C}=O}{|}}{C}}-CH_2\sim \xrightarrow{+(H\cdot)}$$

$$\sim CH_2-\overset{\overset{CH_3}{|}}{\underset{\underset{\underset{H}{|}}{C=O}}{C}}-CH_2\sim \xrightarrow{O_2} \sim CH_2-\overset{\overset{CH_3}{|}}{\underset{\underset{\underset{\underset{H}{|}}{O}}{|}}{\underset{C=O}{|}}}{C}}-CH_2\sim$$

$$\sim CH_2-\overset{\overset{CH_3}{|}}{\underset{\underset{\underset{\underset{CH_3}{|}}{O}}{|}}{\underset{C=O}{|}}}{C}}-CH_2\sim \xrightarrow{-(\cdot CH_3)} \sim CH_2-\overset{\overset{CH_3}{|}}{\underset{\underset{\underset{\dot{O}}{|}}{C=O}}{|}}{C}}-CH_2\sim \xrightarrow{+(\cdot H)} \sim CH_2-\overset{\overset{CH_3}{|}}{\underset{\underset{\underset{\underset{H}{|}}{O}}{|}}{\underset{C=O}{|}}}{C}}-CH_2\sim$$

Positive Electron-beam Resists of Poly(Methyl Methacrylate-co-α-substituted Benzyl Methacrylate). In the previous section the low sensitivity of poly(α,α-diphenylethyl methacrylate) on alkaline development was attributed to the low segmental mobility of polymer chain containing a bulky ester group. Then, it was expected that the introduction of a methacrylate having a smaller ester group as a comonomer into the poly(α,α-diphenylethyl methacrylate) chain provides a means to increase in the segmental mobility, which may result in an improvement in sensitivity. Copolymers of α,α-diphenylethyl methacrylate and methyl methacrylate of various compositions were prepared with radical initiation and electron-beam sensitivities were examined. The results are given in Table IV. The sensitivity increased with increasing content of methyl methacrylate in the copolymer and up to a maximum obtained with a copolymer containing 90.3 mol% methyl methacrylate. The sensitivity then decreased probably due to the decrease in the α,α-diphenylethyl group.

In order to study the segmental mobility of the copolymer, the ^{13}C spin-lattice relaxation times (T_1) were measured on the copolymer containing 48.5 mol% of methyl methacrylate and the results were compared with those for poly(α,α-diphenylethyl methacrylate) (Table V). All the carbons of α,α-

Table IV. Electron-beam Exposure Characteristics of Copolymers of α,α-Diphenylethyl Methacrylate(M_1) with Methyl Methacrylate(M_2)[a]

Copolymer			Sensitivity
M_2(mol%)	M_n	M_w/M_n	(C/cm^2)
0.0	43000	3.33	150×10^{-6}[b]
48.5	22800	3.62	125×10^{-6}
74.7	34500	2.73	75×10^{-6}
90.3	42100	2.13	50×10^{-6}
95.0	37500	2.34	125×10^{-6}
98.0	38200	2.41	125×10^{-6}
100.0	125000	3.25	226×10^{-6}[c]

a Film thickness 0.4 μm, prebake 130°C × 1 hr, development CH_3ONa/CH_3OH (1/20 w/v) 30 sec.

b Prebake 120°C × 1 hr.

c Prebake 170°C × 1 hr.

Table V. $^{13}C-T_1$'s (msec) for Poly(α,α-diphenylethyl methacrylate) and for α,α-Diphenylethyl Methacrylate Units of Poly(α,α-diphenylethylmethacrylate-co-methyl methacrylate)[a] in Toulene-d_8 at 100°C and 25 MHz[b]

Main Chain			Ester Group		
Group	Polymer	Copolymer	Group	Polymer	Copolymer
$-CH_2-$	37	45[c]	-O-C-	2100	2580
α-CH_3	100	200[c]	$-CH_3$	1460	2020
-C-	677	698	C-X	164	272
C=0	1902	2380	o,m	122	171
			p	176	212

a Content of methyl methacrylate units 48.5 mol%.

b Concentration of sample solution 15 w/v%.

c The values are for the overlapped signals of both monomeric units.

diphenylethyl methacrylate units in the copolymer showed longer T_1's than those of the corresponding carbons in the homopolymer of α,α-diphenylethyl methacrylate. This is a clear indication that the segmental mobility of α,α-diphenylethyl methacrylate units is enhanced by the introduction of methyl methacrylate units in the polymer chain. In the copolymer of higher sensitivity, most of the α,α-diphenylethyl methacrylate units should be flanked by methyl methacrylate units on one or both sides and their segmental mobility should be enhanced. This results in easier formation of methacrylic acid units in the copolymer chain as well as the easier penetration of the developing solution into the exposed resist film.

The formation of methacrylic acid units was clearly observed in the difference spectrum for the copolymer containing 74.7 mol% of methyl methacrylate units as shown in Figure 8. By comparing this spectrum with the

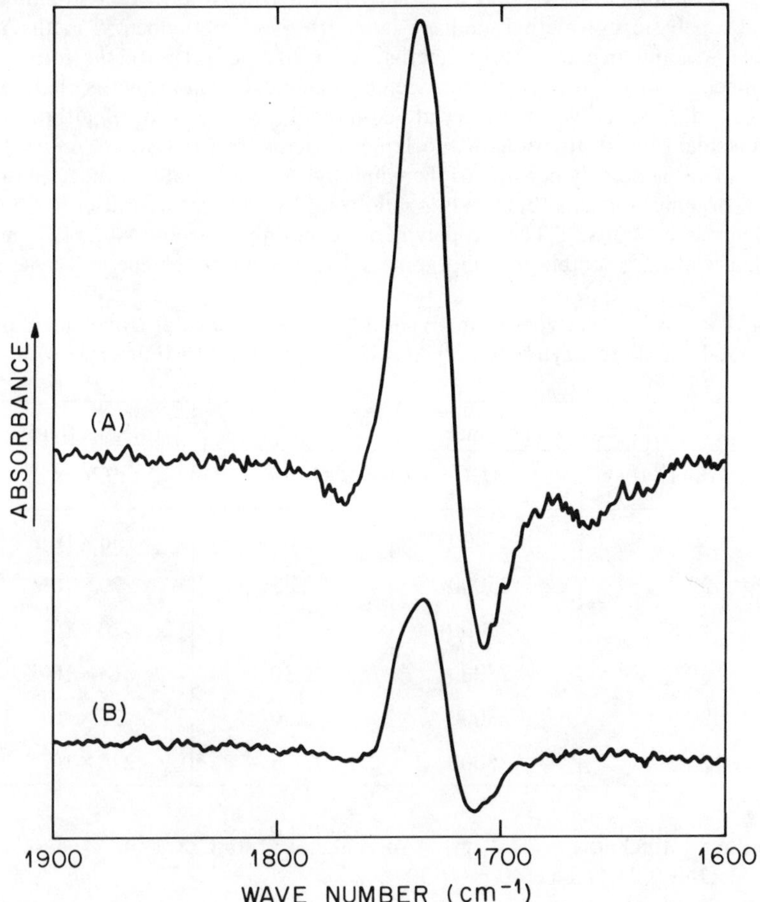

Figure 8. Difference infrared spectra between the polymers unexposed and exposed to electron-beam of 1×10^{-4} *C/cm*^{2}. *(A) Poly(α,α-diphenylethyl methacrylate-co-methyl methacrylate) containing 74.7 mol% of methyl methacrylate units (B) Poly(methyl methacrylate).*

difference spectra for the poly(α,α-dimethylbenzyl methacrylate)s exposed to several doses and also by referring to the acid contents of the latter polymers, the amount of methacrylic acid units in the exposed copolymer was estimated as approximately 3 mol%. This value should include the amounts of acid groups from both monomer units. The contribution from the methyl methacrylate units was estimated as 0.8 mol% on the basis of the content of acid group in poly(methyl methacrylate) exposed under the same conditions. Then, the yield of methacrylic acid units from the α,α-diphenylethyl methacrylate units in the copolymer was calculated to be 9%, while that for the homopolymer was 5%. The results clearly indicate that the incorporation of methyl methacrylate units into the poly(α,α-diphenylethyl methacrylate) chain enhances the decomposition of diphenylethyl ester group owing to the increase in the segmental mobility.

A similar improvement in sensitivity via copolymerization was observed in the copolymers of methyl methacrylate with α,α-dimethylbenzyl methacrylate (Table VI) and triphenylmethyl methacrylate (Table VII). In the former case the mechanism of sensitivity enhancement should be the same as that for the poly(α,α-diphenylethyl methacrylate-co-methyl methacrylate), although the enhancement of sensitivity with copolymerization is rather low.

The homopolymer of triphenylmethyl methacrylate is not soluble in usual organic solvents but it is solubilized by the incorporation of methyl methacrylate units. The copolymers containing over 90 mol% methyl methacrylate are soluble in the organic solvents such as toluene and xylene. In

Table VI. Electron-beam Exposure Characteristics of Copolymers of α,α-Dimethylbenzyl Methacrylate(M_1) with Methyl Methacrylate(M_2)[a]

Copolymer			Sensitivity
M_2(mol%)	M_n	M_w/M_n	(C/cm^2)
0.0	847000	2.07	36×10^{-6}[b]
22.7	25000	2.73	20×10^{-6}
45.8	21000	2.75	20×10^{-6}
80.6	23000	2.30	18×10^{-6}
90.5	23000	2.26	35×10^{-6}
100.0	125000	3.25	226×10^{-6}[c]

a Film thickness 0.4 μm, prebake 140°C × 1 hr, development CH_3ONa/CH_3OH (1/40 w/v) 30 sec.

b Prebake 142°C × 1 hr.

c Prebake 170°C × 1 hr.

Table VII. Electron-beam Exposure Characteristics of Copolymers of Triphenylmethyl Methacrylate(M_1) with Methyl Methacrylate(M_2)[a]

Copolymer			Sensitivity	
M_2(mol%)	M_n	M_w/M_n	(C/cm^2)	γ
90.6	58000	1.88	37×10^{-6}	–
93.7	88000	1.72	3×10^{-6}	6.3
96.1	101000	2.01	20×10^{-6}[b]	–
100.0	125000	3.25	226×10^{-6}[c]	2.1[c]

a Film thickness 0.3 μm, prebake 100°C × 1 hr, development CH_3ONa/CH_3OH=1/40 (w/v) 30 sec.

b Without prebake.

c Prebake 170°C × 1 hr, developer CH_3ONa/CH_3OH = 1/20 (w/v).

Table VII the electron-beam exposure characteristics are given for the soluble poly(triphenylmethyl methacrylate-co-methyl methacrylate)s. The sensitivity on alkaline development was strongly influenced by the copolymer composition. The highest sensitivity was obtained on the copolymer containing 93.7 mol% methyl methacrylate. The copolymer of highest sensitivity showed the γ-value of 6.3, which was nearly twice as large as that for poly(methyl methacrylate). Formation of methacrylic acid units on exposure is obvious from the infrared spectrum. However, the mechanism of the occurrence should be different from the case of the α,α-dimethylbenzyl methacrylate polymer since there are no β-hydrogen atoms in the triphenylmethyl group, and may be similar to the case of poly(methyl methacrylate). This will be explored in the near future.

Aldehyde Copolymer Self Developing Electron-beam Resists. The ceiling temperature for the copolymerization of aliphatic aldehydes is usually below 0°C and the copolymers are easily depolymerized into monomeric aldehydes above 150°C under vacuum. This depolymerization into monomers also occurs on electron-beam or X-ray exposure as evidenced by combined gas-liquid partition chromatography-mass spectrometry. As a result, the copolymers of aldehydes behaved as self-developing positive resists and almost complete development was accomplished without any solvent treatment. Electron-beam exposure characteristics of the aliphatic aldehyde copolymers studied here are shown in Table VIII. The sensitivities were in the order of $10^{-7} \sim 10^{-6}$ C/cm^2 for the resist films of 0.04 ~ 0.7 μm in thickness, and increased with decreasing film thickness. Poly(ethanal-co-propanal) of 0.04 *μm* in thickness showed a sensitivity of 1×10^{-7} C/cm^2.

As mentioned before, it was reported that the electron-beam exposure of poly(phthalaldehyde) resulted in spontaneous formation of a relief image with

Table VIII. Electron-beam and X-ray Exposure Characteristics of Aldehyde Copolymers[a,b]

Copolymer[c]	(mol%)[d]	Thickness (μm)	Sensitivity (C/cm^2)	γ
AA-PA	51.3	0.04	1×10^{-7}	0.74
AA-PA	51.3	0.4	8×10^{-7}	0.70
AA-BA	50.0	0.65	7×10^{-7}	1.3
AA-BA	50.0	1.8	1.7×10^{-6}	0.61
AA-BA[e]	50.0	2.4	30 mJ/cm^2	1.0
AA-HA	54.4	0.38	3.3×10^{-7}	1.0
AA-HA	54.4	0.70	1.7×10^{-6}	0.75
PA-PhPA	52.0[f]	0.65	6.5×10^{-6}	0.86
PA-PhPA	52.0[f]	1.2	2.1×10^{-5}	0.72
PA-PhPA	52.0[f]	3.4	5.0×10^{-5}	0.57

a Prebaked at 80°C for 0.5 hr and without any development step.

b The data are taken from the Table I of Ref. 3.

c AA: ethanal, PA: propanal, BA: butanal, HA: heptanal, PhPA: 3-phenylpropanal.

d The content of AA unit in the copolymer.

e Exposure to X-ray (Al-K_α).

f The content of PA units in the copolymer.

10 ~ 60% loss in the film thickness in the exposed region. The addition of certain cationic photoinitiators to this polymer allowed imaging of a 1 μm thick film of the polymer at 1.0×10^{-6} C/cm^2 without any development process (*1*). Self development also occurred in the polyolefinsulfones but complete development to the substrate can be achieved only in thin films using high temperatures and low accelerating voltages (*22*). Our aldehyde copolymers act as self developing resists even at a film thickness of more than 1 μm and almost complete development is accomplished without any sensitizer.

Recently it was disclosed in a Japanese patent that the copolymers of hexanal with propanal, butanal and isobutanal could be used as self-developing X-ray resists of 200 ~ 400 mJ/cm^2 sensitivity (*32*). Our poly(ethanal-co-butanal) showed the sensitivity of 30 mJ/cm^2 on the exposure to X-ray radiation without requiring a wet development process (Table VIII). Other copolymers also functioned as a positive self developing X-ray resist.

The solutions of aldehyde copolymers prepared by diethylaluminum diphenylamide were usually highly viscous owing to their high molecular weights and where rather difficult to spin-coat onto a silicon wafer. The molecular weight could be controlled by changing the amount of initiator used in the polymerization. Table IX shows the electron-beam exposure

Table IX. Electron-beam Exposure Characteristics[a] of Poly(ethanal-co-butanal)s of Various Molecular Weights Prepared in Toluene at −78 °C with Et_2AlNPh_2

Copolymer				
AA[b] (mol%)	$[\eta]^c$ (dl/g)	Thickness (μm)	Sensitivity (C/cm^2)	γ
58.9	42.3	1.3	5.7×10^{-6}	0.70
63.8	26.6	2.1	7.0×10^{-6}	0.52
67.5	14.5	2.2	5.0×10^{-6}	0.70
68.5	4.5	2.6	1.2×10^{-5}	0.54

a Prebake 50 °C × 1 hr (1 mmHg).

b The content of ethanal units in the copolymer.

c Measured in toluene at 30.0 °C.

characteristics of poly(ethanal-co-butanal)s of various molecular weights. The copolymers, whose intrinsic viscosities in toluene were 42.3 to 14.5 dl/g, showed a similar order of sensitivities (~10^{-6} C/cm^2. Unfortunately, the sensitivity for the copolymer of low molecular weight ($[\eta]$ = 4.5 dl/g) decreased to the order of 10^{-5} C/cm^2.

Poly(propanal-co-3-phenylpropanal) (Table VIII) and the copolymers of 3-trimethylsilylpropanal with other aliphatic aldehyde (Table X) did not exhibit the problem of high solution viscosity. In the former case slight decrease in sensitivity occurred probably owing to the incorporation of the phenyl group. However, most of the copolymers of 3-trimethylsilylpropanal exhibited high sensitivity, $10^{-6} \sim 10^{-7}$ C/cm^2. The addition of Ph_2IPF_6 to poly(3-phenylpropanal-co-3-trimethylsilylpropanal) improved the sensitivity of this polymer but the effect was not remarkable.

Another inconvenience of polyaldehyde resists was the softness of the films. The copolymer of 4,4,4-triphenylbutanal and butanal prepared using a C_2H_5MgBr-(-)-sparteine complex (*24*) dissolves easily in certain organic solvents and forms a fairly hard resist film. The copolymer containing 8 mol% of the former monomer units showed a sensitivity of 1.7×10^{-6} C/cm^2 (*25*).

Table X. Electron-beam Exposure Characteristics for the Copolymers of 3-Trimethylsilylpropanal(M_1) with Aliphatic Aldehydes(M_2)

Copolymer[b]		Thickness[a] (μm)	Sensitivity (C/cm^2)	γ
M_2	M_1(mol%)			
Ethanal	50.2	2.8	8.2×10^{-7}	0.58
		3.8	2.3×10^{-6}	0.49
Propanal	47.7	0.4	8.0×10^{-7}	0.49
		1.9	5.0×10^{-6}	0.44
Butanal	48.3	1.9	3.3×10^{-6}	0.60
3-Phenylpropanal	42.6	0.4	3.2×10^{-6}	0.70
		0.8	1.7×10^{-5}	0.79
		1.1	8.0×10^{-6}[c]	1.09[c]

a Prebake 80°C × 0.5 hr.

b Initiator: Et_2AlNPh_2, solvent: toluene, temp. -78°C, time 24 hr, $[M_1]_0/[M_2]_0 = 1$ mol/mol, $([M_1]_0+[M_2]_0)$ $[Et_2AlNph_2]_0$ = 50 mol/mol.

c Ten wt% of Ph_2IPF_6 was added to the resist film.

Acknowledgments

The authors wish to thank Messrs. S. Danjo, M. Nishikawa and Y. Kobayashi for their experimental contributions and also Mrs. F. Yano for her clerical assistance in preparing this manuscript. They are also indebted to Miss C. Jin of Japan Spectroscopic Co., Ltd. for her help in the measurements of difference infrared spectra.

Literature Cited

1. Willson, C. G.; Ito, H.; Frechet, J. M. J.; Houlihan, F. *Proceeding of 28th IUPAC Macromolecular Symposium*, July 12-16, 1982, Amherst, p. 448.
2. Hatada, K.; Kitayama, T.; Danjo, S.; Tsubokura, Y.; Yuki, H.; Moriwaki, K.; Aritome, H.; Namba, S. *Polym. Bull.* 1983, *10*, 45.
3. Hatada, K.; Kitayama, T.; Danjo, S.; Yuki, H.; Aritome, H.; Namba, S.; Nate, K.; Yokono, H. *Polym. Bull.* 1982, *8*, 469.
4. Yuki, H.; Ohta, K.; Ono, K.; Murahashi, S. *J. Polym. Sci., A-1* 1968, *6*, 829.
5. Yuki, H.; Ohta, K.; Hatada, K.; Okamoto, Y. *Polymer J.* 1977, *9*, 511.

6. Okamoto, Y.; Yashima, E.; Hatada, K., unpublished data.
7. Adrova, N. A.; Prokhorova, L. K. *Vysokomol. Soedin* 1961, *3*, 1509.
8. Picard, J. P.; Ekouya, A.; Dunogues, J.; Duffaut, N.; Calas, R. *J. Organometallic Chem.* 1972, *93*, 51.
9. Hatada, K. *Macromolecular Syntheses* 1982, *8*, 65.
10. Tanaka, A.; Hozumi, Y.; Hatada, K.; Endo, S.; Fujishige, R. *Polymer Letter* 1964, *2*, 181.
11. Tanaka, A.; Hozumi, Y.; Hatada, K.; Fujishige, R. *Kobunshi Kagaku* 1963, *21*, 694.
12. Tanaka, A.; Hozumi, Y.; Hatada, K. *Kobunshi Kagaku* 1965, *22*, 216.
13. Tanaka, A.; Hozumi, Y.; Hatada, K. *Kobunshi Kagaku* 1965, *22*, 317.
14. Cooke, R. G. *Chem. & Ind. (Rev.)* 1955, 142.
15. Grant, D. H.; Grassie, N. *Polymer* 1960, *1*, 125.
16. Grant, D. H.; Grassie, N. *Polymer* 1960, *1*, 445.
17. Geuskens, G.; Hellinckx, E.; David, C. *Europ. Polym J.* 1971, *7*, 561.
18. Grassie, N.; Johnston, A.; Scotney, A. *Europ. Polym. J.* 1981, *17*, 589.
19. Hatada, K.; Kitayama, T.; Kobayashi, Y., unpublished results.
20. Dole, M. "The Radiation Chemistry of Macromolecules"; Dole, M., Ed.; Academic Press: New York, 1973; Vol. II, p. 97.
21. Geuskens, G.; David, C. *Makromol Chem.* 1973, *165*, 273.
22. Bowden, M.; Thompson, L. F. *Polymer Eng. & Sci.* 1974, *14*, 525.
23. Sukegawa, K.; Sugawara, S.; Murase, K. Japanese Patent 82-24801, 1982.
24. Okamoto, Y.; Shohi, H.; Yuki, H. *Polymer Preprints, Japan* 1981, *30*, 102.
25. Hatada, K.; Okamoto, Y.; Kitayama, T.; Toyota, H.; Nishikawa, M., unpublished results.

RECEIVED September 28, 1984

21

Radiation Chemistry of Phenolic Resin Containing Epoxy and Azide Compounds

H. SHIRAISHI, T. UENO,
O. SUGA, and S. NONOGAKI

Central Research Laboratory, Hitachi Ltd.
Kokubunji, Tokyo 185, Japan

Various kinds of polymers have been used as resist materials for electron beam lithography. Poly(methyl methacrylate) (PMMA) (*1*) was the first polymer used as a positive electron beam resist. Although it exhibits high resolution, it has low sensitivity. On the other hand, polymers containing epoxy groups were found to be highly sensitive to electron beams (*2*) and on the basis of this finding, epoxidized *cis* 1,4-polybutadiene (EPB) (*3*) was developed as a highly sensitive negative electron beam resist. Since then, many kinds of polymers containing electron sensitive functional groups have been developed as negative electron beam resists. Poly(glycidyl methacrylate) (PGMA) (*4*) and its copolymer with ethyl acrylate (COP) (*5*) are now widely used as negative electron beam resists in the making of photomasks. These are acrylic polymers containing an electron beam sensitive epoxy group. Chloromethylated polystyrene (CMS) (*6*), iodinated polystyrene (IPS) (*7*), and chlorinated polystyrene (CPS) (*8*) have been developed as dry-etching-durable negative electron beam resists for direct device fabrication. All of these are polystyrene derivatives sensitized by the introduction of radiation sensitive substituents into the benzene ring. These conventional negative electron beam resists utilize radiation-induced gel formation by cross-linking and fractional dissolution in the developing solvent. Because the molecular weights of these polymers are relatively high, only a slight change in chemical structure is necessary for gel formation. In such a case, swelling of the resist gel in the developer is inevitable and this swelling limits the resolution of these negative resists.

Recently, some novel negative resist systems have been proposed. Hofer and co-workers (*9*) have reported on charge transfer X-ray and electron beam negative resists consisting of polystyrene to which tetrathiafulvalene structures are attached and carbon tetrabromide used to form insoluble charge transfer complexes. In this resist system, the charge transfer complex composed of an ionic pair is formed during irradiation. The irradiated areas are rendered insoluble in non-polar solvents because of the formation of the polar ionic species. Swelling does not take place during development in non-polar solvents. Tomkiewics and co-workers (*10*) have reported on organic conductors as electron beam resists that also utilize the charge transfer complex. Mochiji and

0097–6156/84/0266–0423$06.00/0

co-workers (*11*) have obtained high resolution negative patterns by using conventional positive photoresist, AZ-1350J, as an electron beam resist. In their process, AZ-1350J resist film is desensitized by electron beam exposure and then flood-exposed by UV light and developed in aqueous alkaline developer. The negative resist patterns thus obtained show no swelling-induced deformation. Oldham and Hieke (*12*) have reported on the same process. These resist systems require a large chemical structure change, i.e., high exposure-induced product density, to cause a large change in solubility to the developer. Therefore, these non-swelling resist systems exhibit low sensitivity to electron beam exposure. Iwayanagi and co-workers (*13*) have developed a novel deep-UV negative resist (MRS) consisting of an azide compound and a phenolic resin matrix. MRS shows no swelling induced pattern deformation during development in an aqueous alkaline solution. They have reported that the development proceeds in the manner of an etching process.

From this result on MRS, we expected that a combination of phenolic-resin-based resist and aqueous alkaline developer would lead to etching-type dissolution and non-swelling resist patterns. In this paper, we report on a new non-swelling negative electron beam resist consisting of an epoxy novolac, azide compound and phenolic resin matrix (EAP) and discuss the radiation chemistry of this resist.

Experimental

Materials. Epoxy novolac, DEN-431, obtained from Dow Chemical Co. was selected as the epoxy component. A 3,3′-diazidodiphenyl sulfone synthesized in our laboratory (*5*) was used as the azide compound. Poly(*p*-vinyl phenol) obtained from Maruzen Oil Co. was used as the phenolic resin matrix. The coating solvent was cyclohexanone. The developer used in this study was 0.1 N tetramethylammonium hydroxide aqueous solution.

Characteristics. The infrared spectra of films were measured with a Hitachi 260-10 infrared spectrophotometer using an IRR-3 reflection instrument. To obtain the exposed film, each sample was spin-coated onto an aluminum coated silicon wafer to a thickness of 2.0 to 3.0 μm, and exposed to an undeflected, nearly-collimated electron beam at an acceleration voltage of 25 kV, in a modified Hitachi electron microscope. The molecular weight distributions were measured by gel permeation chromatography (GPC) with a Hitachi 635 liquid chromatography system equipped with Shodex A-804, A-802, and A-801 GPC columns. The GPC solvent was tetrahydrofuran. Finely delineated patterns and electron beam exposure characteristics were obtained by using a vector scanning type shaped electron beam writing machine designed by Hitachi Central Research Laboratory. The acceleration voltage was 30 kV.

Results and Discussion

Vacuum Curing Effect. In the early stage of this work, we investigated a mixture of epoxy novolac and poly(*p*-vinyl phenol) (EP) to obtain an electron beam sensitive non-swelling resist. Epoxy novolac was chosen as the sensitizing component, because epoxy groups are known to be electron-beam-sensitive substituents (*2*). However, it is also known that electron beam resists containing an epoxy group show vacuum curing effect, as exemplified in the case of PGMA (*4*). This effect was also observed in the case of EP resist. The exposure characteristics of EP consisting of phenolic resin and epoxy novolac (100 wt% based on the resin) are shown in Figure 1. Sensitivity is enhanced by

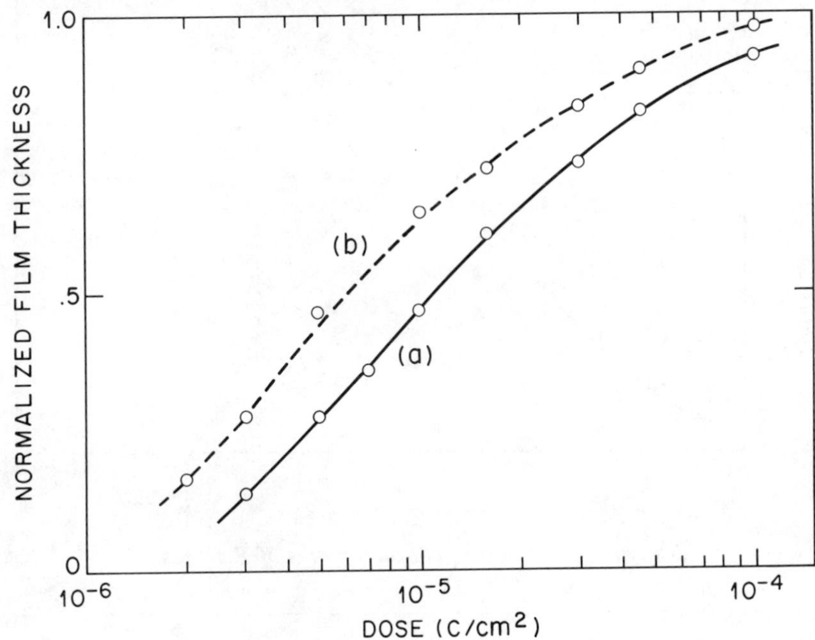

Figure 1. Exposure characteristics of EP resist. (a) no curing, (b) cured in vacuum for 150 min.

vacuum curing, as shown in the figure, but curing does not saturate even after 17 hours. In order to reduce this effect, we examined several kinds of additives. It was found that the addition of an azide compound is remarkably effective in reducing curing effect. Figure 2 shows exposure characteristics of EAP consisting of phenolic resin, epoxy novolac (33 wt% based on the resin), and 3,3′-diazidodiphenyl sulfone (20 wt% based on the resin). No vacuum curing effect is observed in this EAP resist.

Radiation-induced Reactions in EAP. To elucidate the reason no vacuum curing effect is observed in EAP, we measured the infrared spectra of EAP films under various conditions. The results are summerized in Figures 3 to 5. The infrared spectrum of a standard EAP film is shown in Figure 3. Absorbance at 910 cm^{-1} (Aep) caused by the epoxy group and that at 2110 cm^{-1} (Aaz) caused by the azide group were chosen as measures for concentrations of these groups in the films.

The possible reactions related to vacuum curing termination in the EAP resist are 1: exposure-induced ring-opened epoxy radical with an azide, 2: exposure-induced ring-opened epoxy ion with an azide, 3: exposure-induced

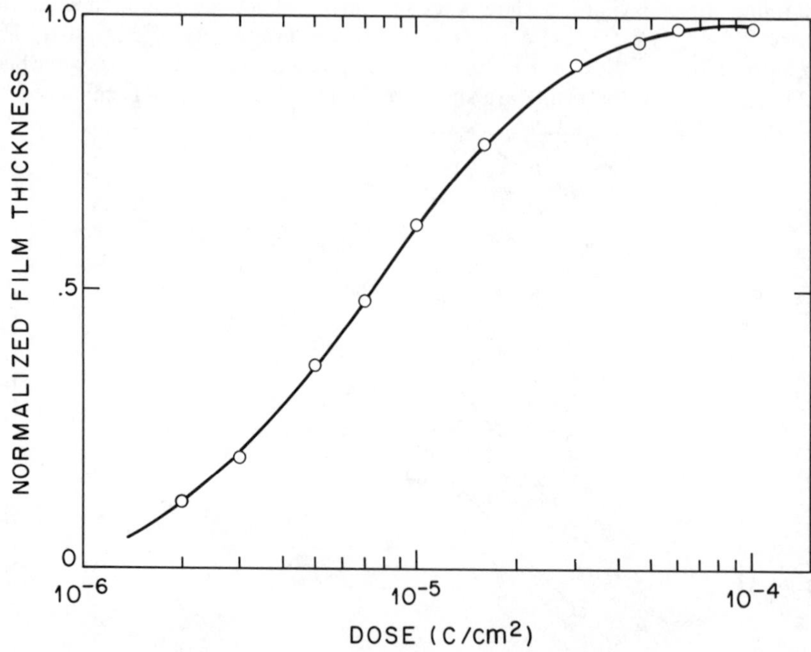

Figure 2. Exposure characteristics of EAP resist.

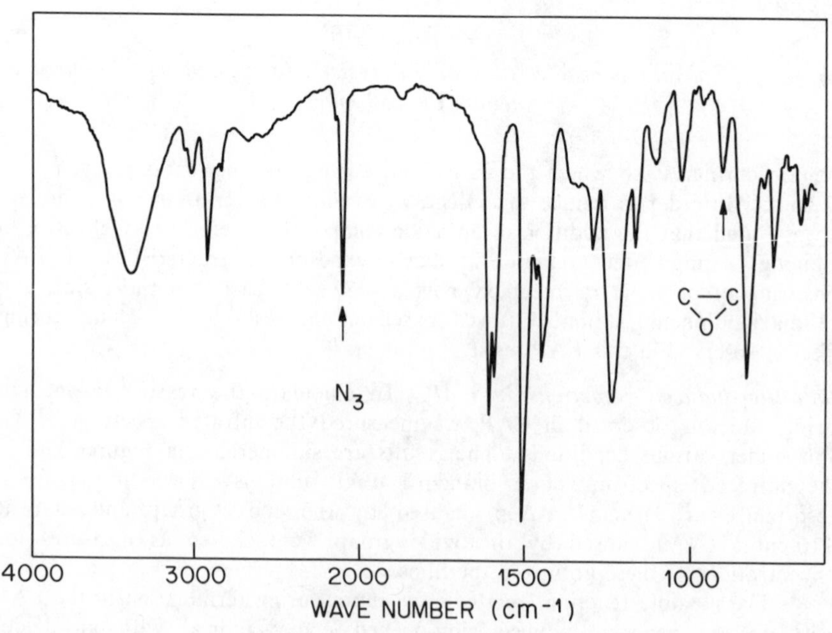

Figure 3. Infrared spectrum of standard EAP film sample.

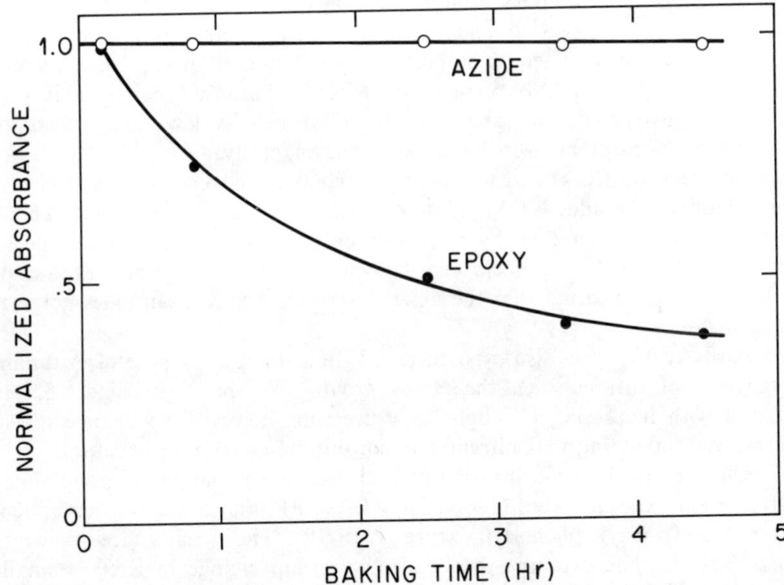

Figure 4. Infrared absorbance of epoxy and azide groups in EAP film containing triethlylamine, as a function of baking time (baking temperature: 80° C).

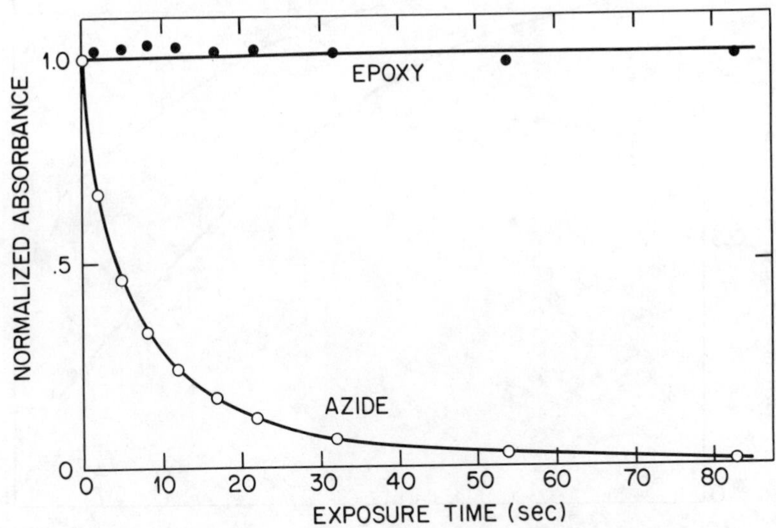

Figure 5. Infrared absorbance of epoxy and azide groups in EAP film, as a function of UV light exposure time.

nitrene reaction with an epoxy group, and 4: exposure-induced nitrene with an exposure-induced ring-opened epoxy group. Each reaction was examined by using model reactions. There is no proper model reaction for the No. 1 reaction and the possibility of No. 4 reaction is low, because the possibility of encountering two kinds of exposure-induced species is low. Therefore, the No. 2 and No. 3 reactions were examined experimentally.

In order to see the reaction of the ring-opened epoxy group with an azide, triethylamine-added EAP film was baked at 80°C. As is well known, amines initiate successive epoxy ring opening reactions through ionic species. As shown in Figure 4, Aep decreased with increasing baking time, but Aaz did not change thus indicating that the ionically-opened epoxy group does not react with the azide.

An EAP film was also exposed to UV light for the purpose of examining the reaction of nitrene with the epoxy group. As shown in Figure 5, Aaz decreased with increasing UV light exposure time however, no change in Aep was observed indicating that nitrenes cannot initiate epoxy ring opening.

Changes in the concentration of both the epoxy and azide groups under electron beam exposure were observed for EAP, EP, and a 3,3′-diazidodiphenyl sulfone- poly(*p*-vinyl phenol) mixture (MRS). The results are shown in Figures 6 to 7. The exposure-induced epoxy group change in EAP is smaller than that in EP, as shown in Figure 6. Additionally, the exposure-induced azide group concentration change in EAP is larger than that in MRS, as shown in Figure 7.

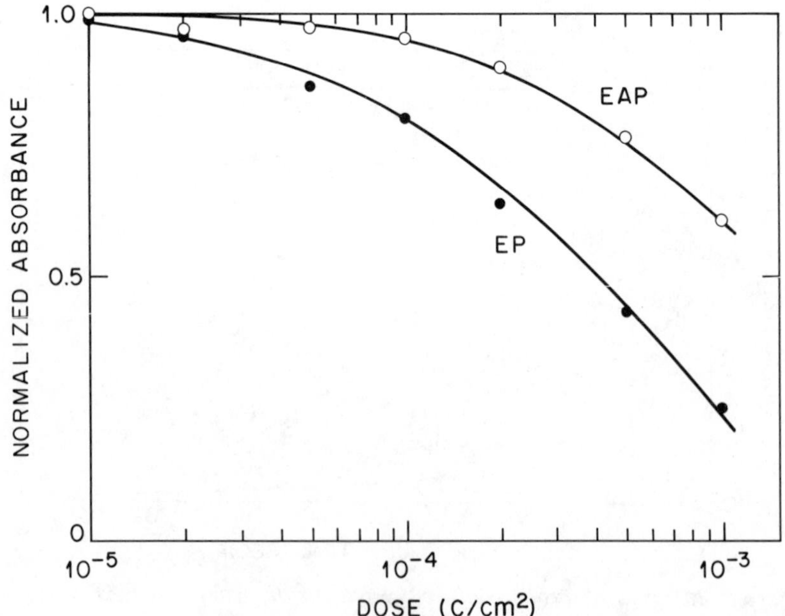

Figure 6. Infrared absorbance of epoxy group in phenolic resin matrix, as a function of electron beam exposure dose.

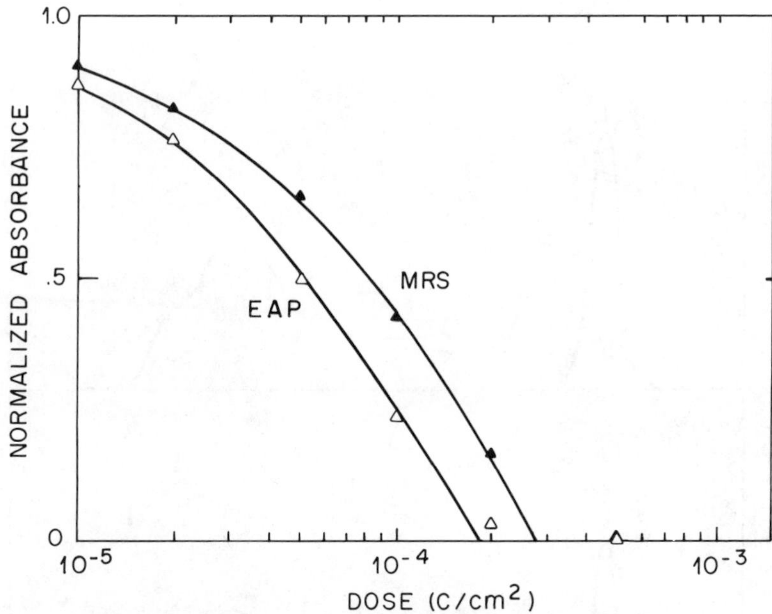

Figure 7. Infrared absorbance of azide group in phenolic resin matrix, as a function of electron beam exposure dose.

These results indicate that vacuum curing occurs through a radical reaction mechanism and is terminated by reaction of the ring-opened epoxy group with the azide group (not nitrene) under exposure. There is a possibility that polymerization initiated by an exposure-induced radical cation may occur. Furthermore, it is thought that reaction products from both the azide and epoxide serve as dissolution inhibitors, because the sensitivity of EAP is almost the same as that of EP, as shown in Figures 1 and 2.

Exposure-induced Reaction Products. Gel permeation chromatograms of EP, MRS, and EAP were measured before and after exposure at 20 μC/cm^2. The results are summerized in Figures 8 to 10. In the case of EP resist, shown in Figure 8, peaks 1, 2, and 3 represent epoxy novolac dimer, trimer, and tetramer, respectively. Peak 4 represents the main component of the poly(*p*-vinyl phenol) resin and peak 5 indicates the presence of exposure-induced high molecular weight components in the resin.

The gel permeation chromatograms of MRS are shown in Figure 9, where peaks 1 and 2 represent the azide compound and the main component of poly(*p*-vinyl phenol), respectively. Peak 3 corresponds to a primary amine, which is one of the exposure-induced reaction products from the azide compound (*14*). Peak 4 indicates the same exposure-induced high molecular weight components as for EP.

The result obtained from EAP is shown in Figure 10, where peak 1 represents the azide compound and epoxy novolac dimer, and peaks 2, 3, and 4 represent epoxy novolac trimer, tetramer, and the main component of poly(*p*-

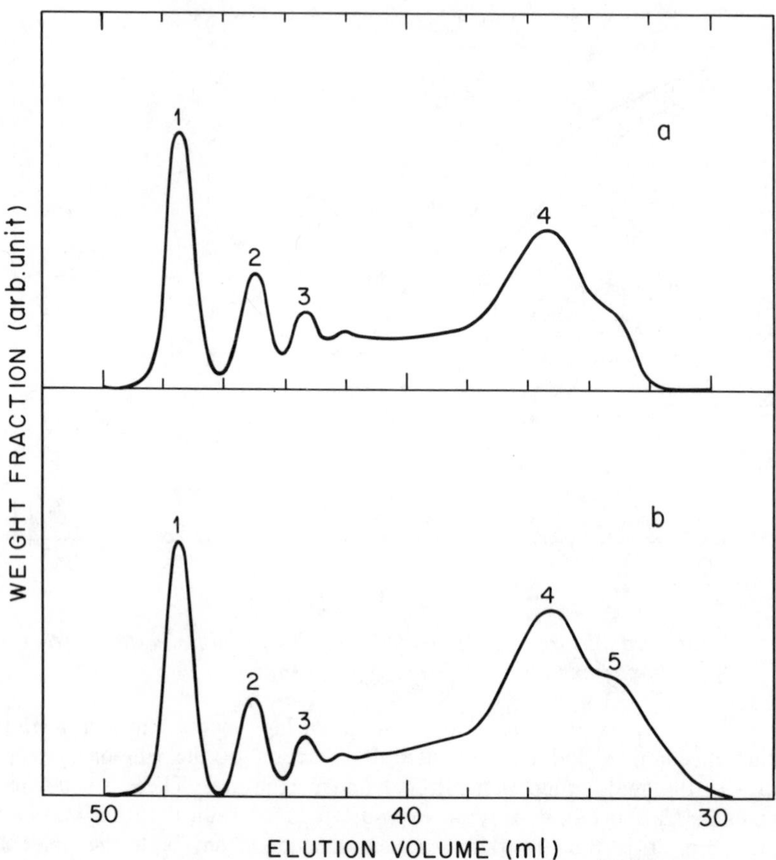

Figure 8. Gel permeation chromatogram of EP resist, before and after exposure. (a) before exposure, (b) after exposure, dose: 20 μC/cm².

vinyl phenol), respectively. Peak 5 indicates exposure-induced high molecular weight components of poly(*p*-vinyl phenol).

In general, a high-molecular-weight phenolic resin is less alkaline-soluble than a low-molecular-weight one. Therefore, these high molecular weight poly(*p*-vinyl phenol) components may contribute to the decrease in solubility. Furthermore, we assume that the etching type dissolution of these resists is non-fractional, as was previously shown in the case of poly(methyl methacrylate-co-acrylonitrile) (*15*).

Lithographic Performance. Figure 11 shows an SEM photomicrograph of EAP resist patterns. The dose was 15 $\mu C/cm^2$ at 30 kV. The sensitivity is not particularly good but there is no evidence of swelling. Since EAP is a non-swelling type resist, it can form fine patterns even if the relative remaining thickness is small. EAP resist was used as the top imaging resist in a tri-level resist process (*16*) for direct device fabrication. An SEM microphotograph of

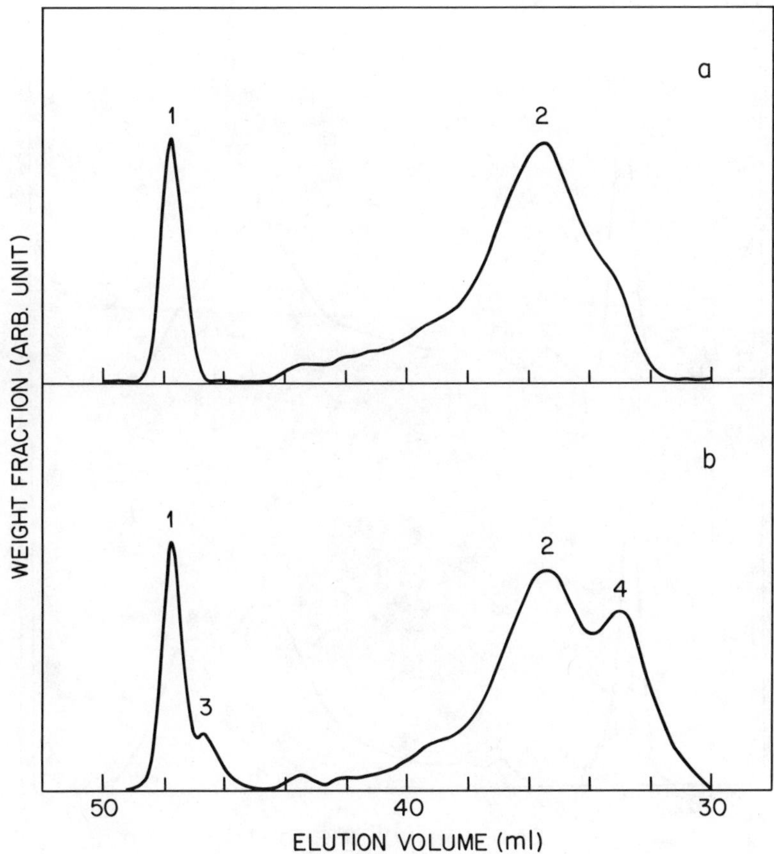

Figure 9. Gel permeation chromatogram of MRS resist, before and after exposure. (a) before exposure, (b) after exposure, dose: 20 μC/cm².

the tri-level resist patterns are shown in Figure 12. The coated EAP film thickness was 0.5 μm. The design dimension of the smallest space was 0.3 μm, and the dose was 2.2 $\mu C/cm^2$ at 30 kV which indicates practical sensitivity for direct wafer fabrication.

Summary

A mixture of epoxy novolac and poly(*p*-vinyl phenol) shows vacuum curing effect when it is used as an electron beam resist. The addition of an azide compound to the mixture remarkably reduces this effect. A negative electron beam resist consisting of an epoxy novolac, azide compound, and poly(*p*-vinyl phenol) matrix (EAP) has been prepared. It was clarified that vacuum curing in this resist is terminated by reaction of the ring-opened epoxy group with the azide group. Exposure-induced high molecular weight poly(*p*-vinyl phenol) has

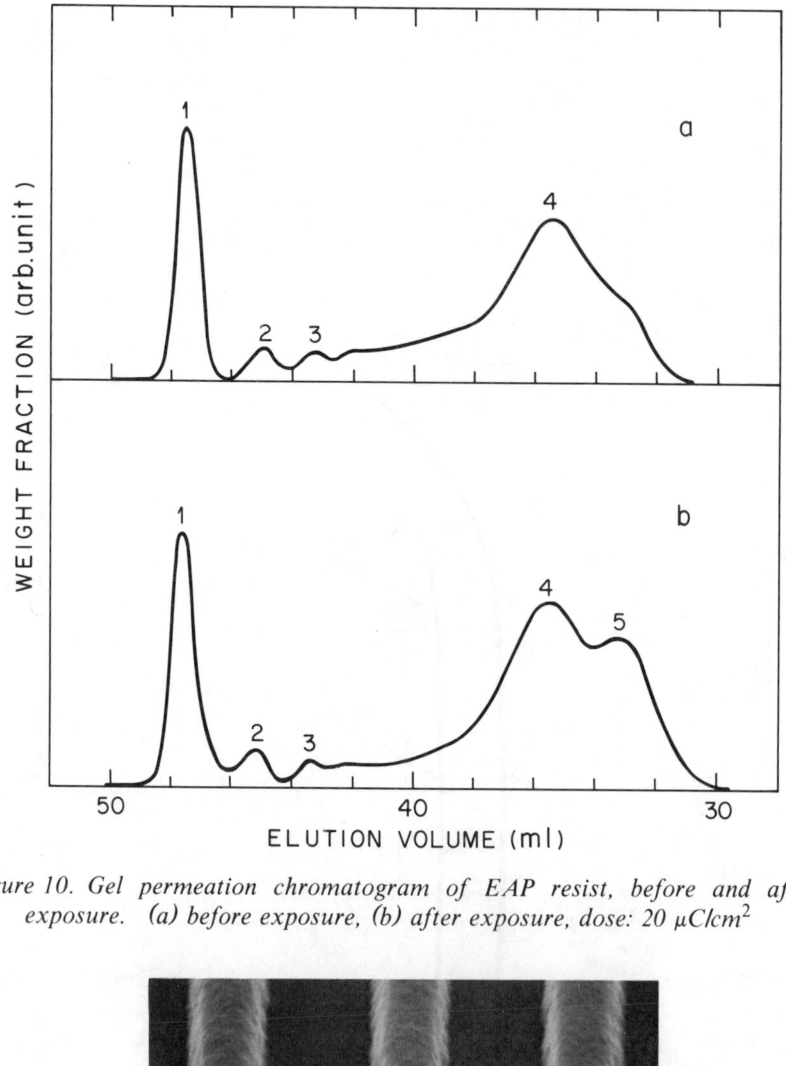

Figure 10. Gel permeation chromatogram of EAP resist, before and after exposure. (a) before exposure, (b) after exposure, dose: 20 μC/cm^2

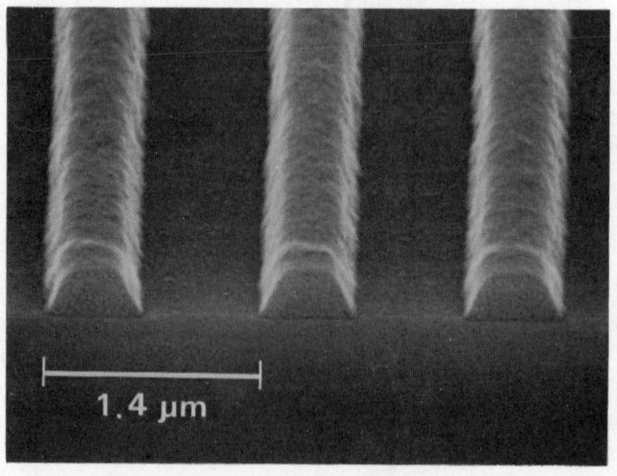

Figure 11. SEM microphotograph of EAP resist patterns.

Figure 12. SEM microphotograph of tri-level resist patterns using EAP as a top imaging resist.

been found by gel permeation chromatography. The EAP resist does not swell in aqueous alkaline developer and exhibits high resolution capability. The practical sensitivity of this resist as a top imaging layer in a tri-level structure is 2.2 $\mu C/cm^2$ at 30 kV.

Acknowledgments

The authors would like to thank S. Okazaki and Y. Takeda for their assistance in the electron beam fabrication experiment. They also thank Dr. T. Iwayanagi for his helpful discussion on the dissolution mechanism.

Literature Cited

1. Haller, I.; Hatzakis, M.; Srinivasan, R. *IBM J. Res. Dev.* 1968, *12*, 251-6.
2. Hirai, T.; Hatano, Y.; Nonogaki, S. *J. Electrochem. Soc.* 1971, *118*, 669-72.
3. Nonogaki, S.; Morishita, H.; Saitou, N. *Appl. Polymer Symp.* 1974, *23*, 117-23.
4. Taniguchi, Y.; Hatano, Y.; Shiraishi, H.; Horigome, S.; Nonogaki, S.; Naraoka, K. *Japanese J. Appl. Phys.* 1979, *18*, 1143-8.
5. Thompson, L. F.; Ballantyne, J. P.; Feit, E. D. *J. Vac. Sci. Technol.* 1975, *12*, 1280-3.
6. Imamura, S. *J. Electrochem. Soc.* 1979, *126*, 1628-30.
7. Shiraishi, H.; Taniguchi, Y.; Horigome, S.; Nonogaki, S. *Polymer Eng. Sci.* 1980, *20*, 1054-7.
8. Feit, E. D.; Stillwagon, L. E. *Polymer Eng. Sci.* 1980, *20*, 1058-63.
9. Hofer, D. C.; Kaufman, F. B.; Kramer, S. R.; Aviram, A. *Appl. Phys. Lett.* 1980, *37*, 314-6.
10. Tomkiewics, Y.; Engler, E. M.; Kuptsis, I. D.; Schad, R. G.; Patel, V. V.; Hatzakis, M. *Appl. Phys. Lett.* 1982, *40*, 90-2.
11. Mochiji, K.; Maruyama, Y.; Murai, F.; Okazaki, S.; Takeda, Y.; Asai, S. *Proc. 12th Conf. Solid-State Devices, Tokyo*, 1980, p. 63-7.

12. Oldham, W. G.; Hieke, E. *IEEE Electron Devices Lett.* 1980, *EDL-1*, 217-9.
13. Iwayanagi, T.; Kohashi, T.; Nonogaki, S.; Matsuzawa, T.; Douta, K.; Yanazawa, H. *IEEE Trans. Electron Devices* 1981, *ED-28*, 1306-10.
14. Iwayanagi, T.; Hashimoto, M.; Nonogaki, S.; Koibuchi, S.; Makino, D. *Polymer Eng. Sci.* 1983, *23*, 935-40.
15. Hatano, Y.; Shiraishi, H.; Taniguchi, Y.; Horigome, S.; Nonogaki, S.; Naraoka, K. *Proc. Symp. Electron and Ion Beam Sci. Technol. 8th Int. Conf.*, 1978, p. 332-40.
16. Moran, J. M.; Maydan, D. *Polymer Eng. Sci.* 1980, *20*, 1097-101.

RECEIVED August 6, 1984

22

Organic Direct Optical Recording Media

L. ALEXANDRU, M. A. HOPPER, R. O. LOUTFY, J. H. SHARP, and P. S. VINCETT

Xerox Research Centre of Canada
2600 Speakman Drive
Mississauga, Ontario

G. E. JOHNSON and K. Y. LAW

Xerox Corporation
Webster Research Center
Webster, NY 14580

The interaction of laser light with a wide variety of imaging materials has been actively explored over the past few years because of its potential in high density information storage and retrieval systems. These systems are applicable to the storage of x-ray and video pictures as well as to the storage of financial, satellite, computer and government statistical data. An optical recording material, for example, can store up to 10^{11} bits of information on a 12" disk. This represents an increase in packing density of two orders of magnitude over a magnetic disk.

The key component of the optical recording system is the recording medium, which is *marked* in one step and read in another step with a laser beam. Many materials have been considered for the purpose, including metal-films, organic dyes, dye-loaded polymers, metal-loaded polymers, discontinuous metal films, thermal coloration systems and bilayers (*1-19*). Tellurium- and gold-based optical disks have been the most widely studied materials to date.

A major requirement of a suitable optical recording material is resolution since a spot size of 1 μm or less is required. This corresponds to a resolution of 500 lp/mm (lp = line pairs). A second requirement is marking speed. Information must be recorded at ~10 MHz, so that the recording process must be essentially instantaneous. Based on current laser sources a power density of 1 mW/cm^2 can be delivered to the surface of the recording material. Hence for the generation of a 1 μm spot, recorded at 10 MHz, the imaging material must have an exposure sensitivity of at least 10^{-2} J/cm^2. With the advent of the development of high power solid-state diode lasers, which offer low power consumption, compactness, increased reliability and direct modulation capability and economy over gas lasers, a further requirement of the optical recording material will be infrared sensitivity. A summary of typical requirements of the optical recording material is given in Table I. The

0097-6156/84/0266-0435$06.00/0

TABLE I. Typical Requirements of the Optical Recording Material

REQUIREMENT	VALUE
Resolution	1000 LP/MM
Writing Speed	10 MHz
Exposure Sensitivity	10^{-2} J/cm^2
Storage time	1 - 10 Years
Spectral Sensitivity	Visible, I.R.
Defect Density	1 : 10^6
Readout Contrast	> 0.5

most intensively investigated materials to date (*4-7*) are those that involve a photoinduced thermal process. In this case the recording laser beam is focussed onto a thin film of the recording material that has been deposited onto a substrate. The absorbed radiation raises the temperature above a threshold value for pit formation. The marking process usually involves melting or vaporization. The recorded marks have a different reflectivity or transmissivity than the unmarked surface.

The simplest recording medium is a bilayer structure. It is constructed by first evaporating a highly reflective aluminum layer onto a suitable disk substrate. Next, a thin film (15-50 nm thick) of a metal, such as tellurium, is vacuum deposited on top of the aluminum layer. The laser power required to form the mark is dependent on the thermal characteristics of the metal film. Tellurium, for example, has a low thermal diffusivity and a melting point of 452°C which make it an attractive recording material. The thermal diffusivity of the substrate material should also be as low as possible, since a significant fraction of the heat generated in the metal layer can be conducted to the substrate. For this reason, low cost polymer substrates such as poly(methylmethacrylate) or poly(vinyl chloride) are ideal.

Relatively less work has been reported on ablative organic materials. These materials, which generally combine low thermal diffusivity with low melting points suggest that they may be more sensitive for ablative optical recording applications. Bell and Spong described bilayer structures consisting of dye films evaporated onto a reflective substrate (*5*). Subsequent activities based on Bell and Spong's bilayer structure using various dyes such as 6,6'-diethoxythioindigo (*7*), di-indeno[1,2,3-cd:1',2',3'-lm]perylene (*20*), metallophthalocyanines (*21-23*), and platinum complexes of bis-(dithio-α-diketones) (*24*) have appeared in the patent literature. Trilayer structures using squarylium dyes have also been reported very recently (*25*). Apparently Eastman Kodak Company has been or is working on a dye/polymer optical disk approach (*26*).

During the past few years, two novel concepts, both involving the use of organic dye/polymer composites as the marking media, have been investigated at the Xerox Research Centers. The first, described as Laser Induced Dye Amplification (LIDA), is based on the principle of dye diffusion. The second, described as Dye in Polymer (DIP), involves ablative optical marking using dissolved dyes in polymer films. Each of these concepts will be discussed in detail in the following sections.

Laser Induced Dye Amplification (LIDA)

Several investigators have explored the use of organic dyes for use in high sensitivity light marking processes. For example, Braudy (*4*) and Bruce and Jacobs (*27*) have studied the light induced transfer of volatile organic dyes from one polymer substrate to another.

The concept of Light Induced Dye Amplification was first proposed and reduced to practice in 1976 by Alexandru and Novotny (*9*). The imaging process involves the simultaneous exposure and fixing of an image by photoinducing dye diffusion from a solid film of dye into a compatible polymer substrate. The simple structure of the recording film is shown in Figure 1. The laser radiation is absorbed by the solid dye and converted via non-radiative processes into thermal energy. The dye subsequently melts, flows in a lateral direction and also diffuses into the polymer substrate. A depression or pit is created in the solid dye film due to the diffusion of the dye into the polymer as well as the lateral flow. The imaging contrast results from the difference in the spectral characteristics of the diffused dye, which is similar to that for the dye in solution, and that of the solid dye film.

In order to demonstrate the feasibility of the LIDA concept in high density optical recording systems, a number of dyes have been investigated and the results reported (*9,10*). In this paper, the results for two of these dyes, Disperse Red II, and Disperse Blue 60 are described in detail. Disperse Red II is manufactured by ICI and Disperse Blue 60 is manufactured by Dupont. Both are derivatives of 1,4-diaminoanthraquinone. Their molecular structures are shown in Figure 2.

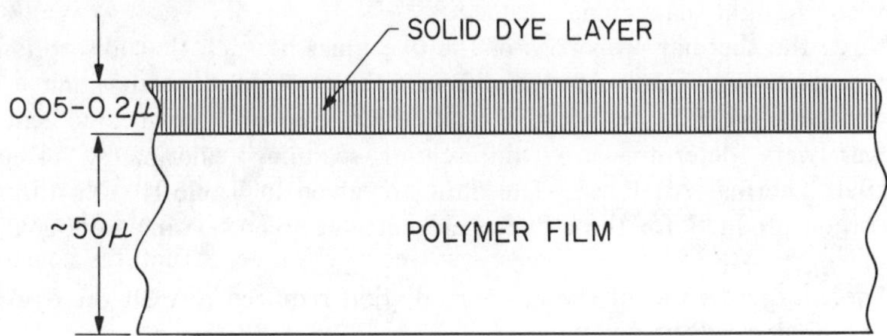

Figure 1. LIDA optical disk structure.

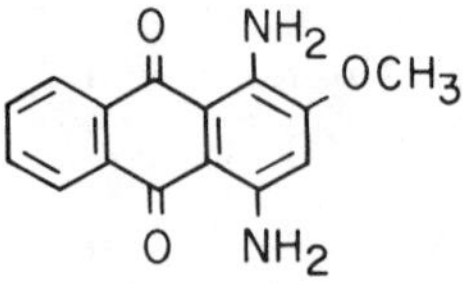

Figure 2. Molecular structure of Disperse Red and Disperse Blue dyes.

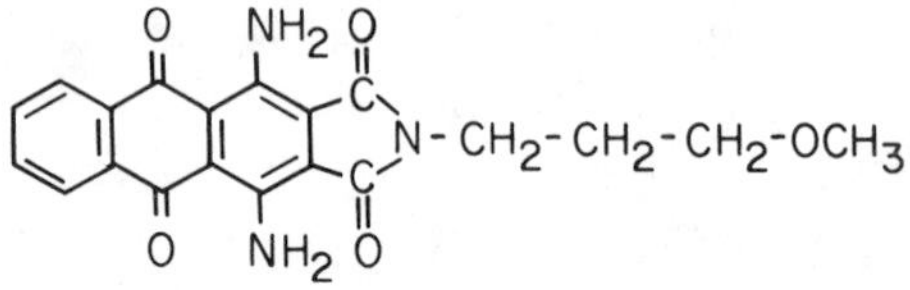

LIDA Dye Characteristics. There are several important characteristics which a dye must posses in order to be useful in the LIDA concept. First, it must match the lasing frequency of the recording laser. The laser utilized in this work was an argon ion laser which could deliver up to 32 mW of power to the film surface at a wavelength of 5145A. The absorption spectrum of a thin evaporated solid film of Disperse Red 11 on a substrate of polyethyleneterephthalate is shown in Figure 3a and indicates that it is an excellent match to the 5145A emission of the argon ion laser. As stated earlier, the imaging contrast in the LIDA concept results from the difference in the spectral characteristics of the diffused dye and those of the solid film. Figure 3 also shows the absorption spectra of the polyethyleneterephthalate after dye diffusion has occurred. Figure 3b is the result of thermally induced diffusion while Figure 3c is the result of laser-induced dye diffusion. A blue shift is observed in each case. Both are similar to the spectrum of Disperse Red 11 in solution and indicate that the dye has diffused into the polymer substrate on the molecular level and has formed a solid solution with the polymer. Figure 4 shows the same spectra for the Disperse Blue 60 dye. The difference between the spectral characteristics of the diffused dye and that of the solid film is obvious in this case as well. The solid film exhibits a diffused absorption probably due to light scattering.

Next, the thermal properties of the dye must be such that absorption of the laser energy will result in dye diffusion but not in decomposition. The melting temperature T_m, the latent heat of fusion, ΔH, and the specific heat for these dyes were determined by differential scanning calorimetry using a DuPont 990 Thermal Analyzer. The data are given in Table II. No thermal decomposition products for these dyes were detected upon heating to 600°C for 20 msec.

The energy density of the laser irradiation required to melt the dye and to induce diffusion will be given by:

$$E_D = mC_p\Delta T + m\Delta H \qquad (1)$$

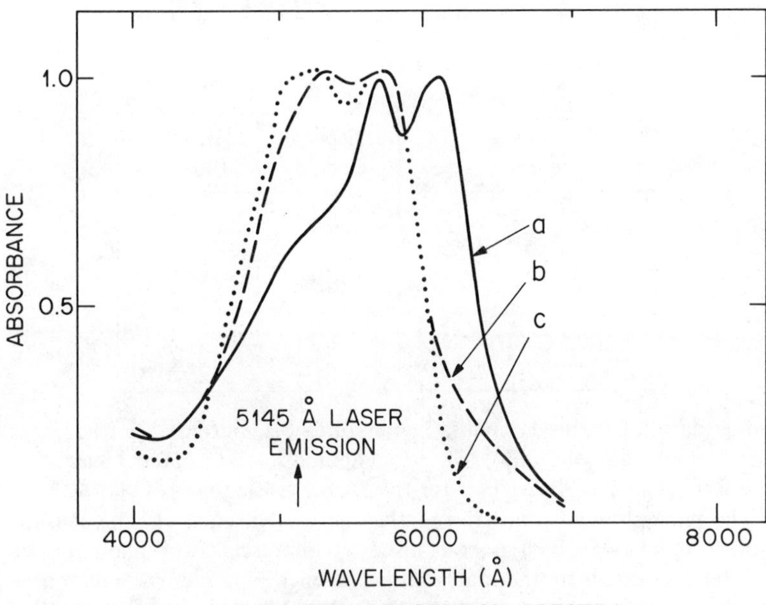

Figure 3. Absorption Spectra of the Disperse red/polyethyleneterephthalate system. (a) Evaporated film, (b) Thermally diffused, (c) Laser diffused.

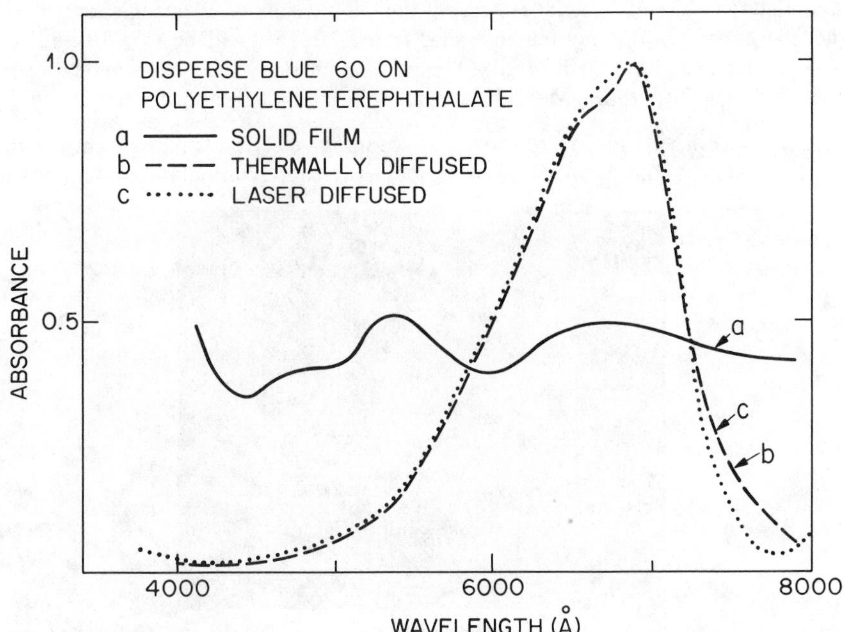

Figure 4. Absorption Spectra of the Disperse Blue 60/polyethyleneterephthalate system. (a) Evaporated film, (b) Thermally diffused, (c) Laer diffused.

Table II. Thermal Properties of the Disperse Dyes

Thermal Property	DYE TYPE	
	Disperse Red 11	Disperse Blue 60
T_m (°C)	250	200
ΔH (J/g)	120	36
C_p @ 160°C (J/g-°C)	1.74	1.80

Using the data in Table II and a value of ~0.1 mg/cm^2 for the mass of dye/cm^2, m, in the solid film, E_D is computed to be ~50 mJ/cm^2 for the Disperse Red 11 and ~30 mJ/cm^2 for the Disperse Blue 60.

The photophysical properties of the chosen dye must also be considered. Any absorbed energy which is re-emitted as fluorescence or phosphorescence can not be converted to thermal energy. Hence ϕ_F and ϕ_P must be low. Therefore, the amount of absorbed energy that can be converted to thermal energy will be governed by the quantum yields of internal conversion, ϕ_{ic}^I and intersystem crossing, ϕ_{isc}^T. Since the optical recording medium must have the capability of read after write verification, the imaging process must be completed within 10^{-7} sec. However, the conversion of electronic energy to thermal energy by intersystem crossing from T to So will be slow due to the long triplet lifetime (~10^{-3} sec). Hence, the ideal dye for our application is one which has a quantum yield of internal conversion, ϕ_{ic}^i, approaching unity.

Marking Results. Figure 5 shows a scanning electron micrograph of the marked bits on the Disperse Red 11 polyethyleneterephthalate system. The

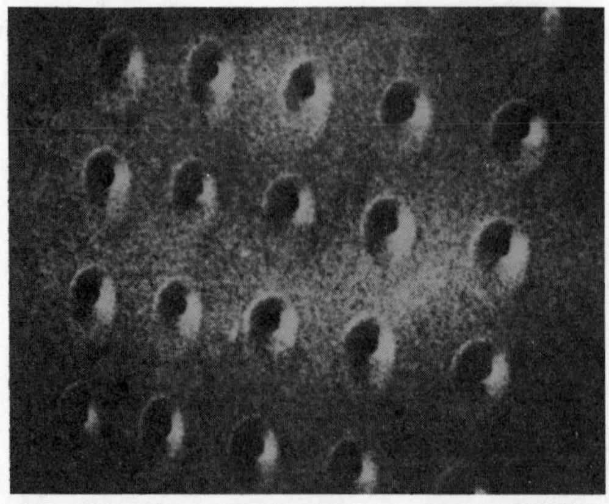

Figure 5. SEM of Marked Bits (×4800) in the LIDA disk. Laser Power = 32 mW; Exposure time = 1 μs.

spots were marked using a power of 32 mW and an exposure time of 1 μs.

For a high density optical recording medium, two important parameters which characterize its capability are the size of mark and the threshold energy density required for mark creation. The size of the mark was found to be dependent on both the incident power and the exposure time. The size of the marked areas is shown in Figure 6 for different levels of power and exposure

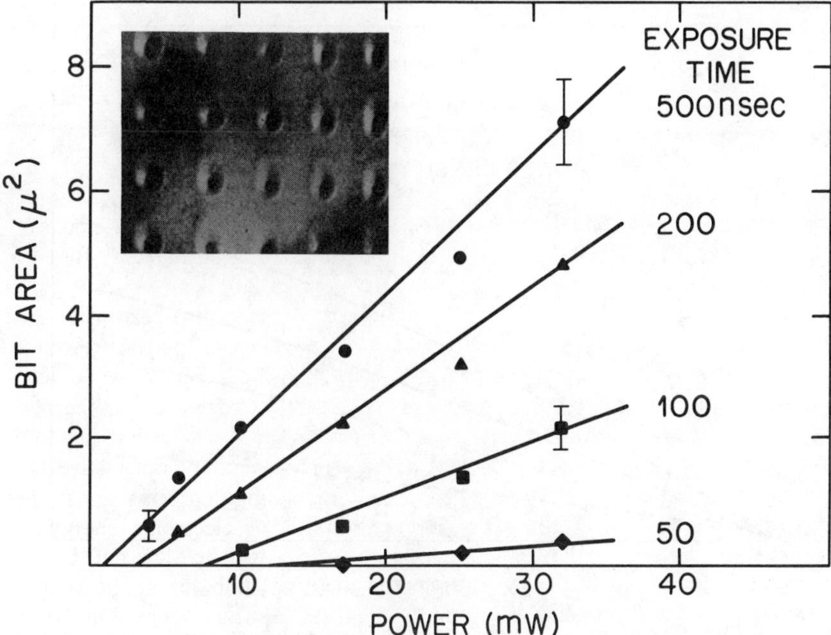

Figure 6. Bit sizes as a function of laser power and exposure time in a 0.2 μm LIDA disk.

times. Although the diameter of the laser beam spot size was kept constant at 1 μm, the marks vary in size from less than a micron to several microns in diameter. As might be expected, increasing the laser beam power or the exposure time, increased the mark size.

The power threshold, P_t, defined as the minimum power to induce dye diffusion, is obtained by extrapolating the curves in Figure 5 to zero bit area. When P_t is plotted as a function of exposure times less than 1 μs, reciprocity is maintained, demonstrating that the marking process is independent of exposure time and depends only on the marking energy. The threshold energy density required for marking, E_D, is defined by

$$E_D = \frac{P_t t}{A} \tag{2}$$

where P_t is the laser power, t is the exposure time and A is the area of the laser beam. The experimentally determined value for E_D is ~100 mJ/cm^2. This is in good agreement with the value of 50 mJ/cm^2 calculated from the thermal properties of Disperse Red 11.

It is worthwhile to discuss another interesting aspect of dye diffusion into polymer which is governed by polymer morphology. When the bit sizes marked with Disperse Blue 60 film are compared, relatively large systematic differences are observed between bits marked on Mylar (from Dupont) and on Kodar (from Kodak) substrates. The sizes of the bits obtained on Kodar substrate are substantially smaller (by a factor of up to 2) than the ones obtained on Mylar under the same conditions. Figure 7 shows the dependence of bit size marked

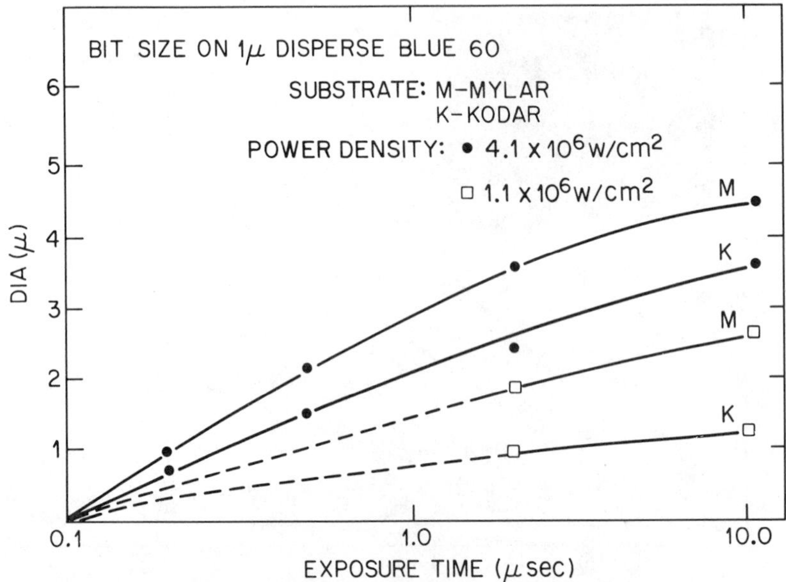

Figure 7. The dependence of Disperse Blue dye, mark-size on laser exposure time on Mylar and Kodar substrates.

on 1 micron film of Disperse Blue 60 on these substrates. It is clear that the polymer substrate has an influence on the diffusion process and the geometry of the recorded bit. The Mylar and Kodar used in this study have non-identical terephthalate polyester structures. However, their thermal properties are practically identical. What seems important for this application is that their morphological properties are very different, as is shown in Table III. It was found that the thermal diffusion rate of the dye into the amorphous Kodar film is much higher than the crystalline Mylar film. As a result during the marking process the perpendicular dye diffusion rate into Kodar is faster than the lateral diffusion giving a smaller bit compared with those in Mylar.

Summary of LIDA. An optical recording medium, based on the concept of laser induced dye diffusion, has been studied. One micron diameter marks can be generated in less than 10^{-7} sec with an exposure sensitivity of 100 mJ/cm^2. The marks are permanent and a readout contrast density of unity ($OD_i - OD_F/OD_F$, where OD_i and OD_F are the optical density of the image and film, respectively) is obtained in the transmission mode. Due to the lateral

Table III. Comparison of Mylar and Kodar Properties

PROPERTY	FILM TYPE	
	MYLAR	KODAR
Structure	Homopolymer	Copolymer
T_g (°C)	100	90
T_m (°C)	260	260
Density	1.37 - 1.38	1.20 - 1.22
Film Isotropy	Anisotropic (oriented)	Isotropic (unoriented)
Order	Crystalline	Amorphous
Dye Thermal Diffusion	175°C, 2-4 mins	120°C, 10-20 secs

flow of the melted dye, a surface relief structure is also obtained which may allow a readout contrast in the reflection mode.

Dye in Polymer (DIP)

The LIDA concept, discussed above, has demonstrated the potential of organic dyes for use in high density optical recording systems. However, it is apparent that the contrast density is marginal, particularly in the reflection mode. Furthermore, vacuum deposited dyes are also subject to crystallization effects which lead to focussing and archival problems.

The Dye-in-Polymer concept was first proposed and reduced to practice in our laboratory (*12*). It involves the ablative marking with an appropriate laser, of a polymer film containing a dissolved dye. The dye-in-polymer film can be easily prepared by spin-coating a dye/polymer solution onto a reflective substrate. The structure of the DIP recording medium is shown in Figure 8.

The Blue DIP Disc. K. Y. Law, et al. (*12*) have recently reported the use of a yellow dye in the DIP structure as an optical recording medium. The yellow dye, Polyester Yellow, was obtained from Eastman Chemicals. Its molecular structure and the absorption spectrum of a 20% by wt solid solution of the dye in a thin film of polyvinylacetate are shown in Figure 9. The solid solution has an absorption maximum at 446 mm which matches the 457 nm output of the dye laser used for the marking experiments. Two concentrations of Polyester Yellow in polyvinylacetate (PVA) were used in the preparation of the marking media. High absorption efficiency of the marking laser beam can be achieved by using a thick film or by using a thinner film having a high loading of the dye. Excessive loading however, may lead to dye aggregation and microcrystallization which results in non-uniform reflectivity, poor film quality and archival problems. Another consideration which will affect the reflection

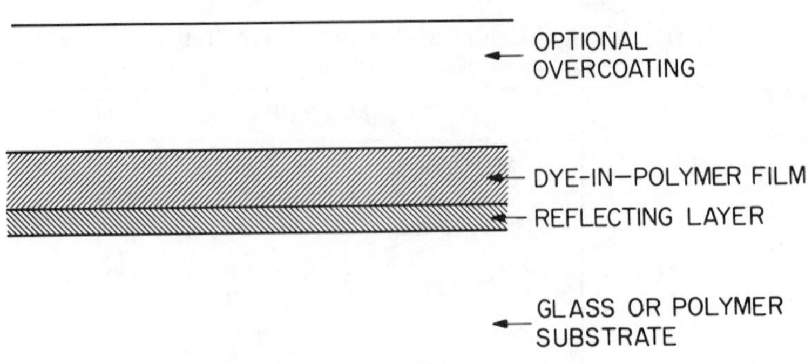

Figure 8. DIP optical disk structure.

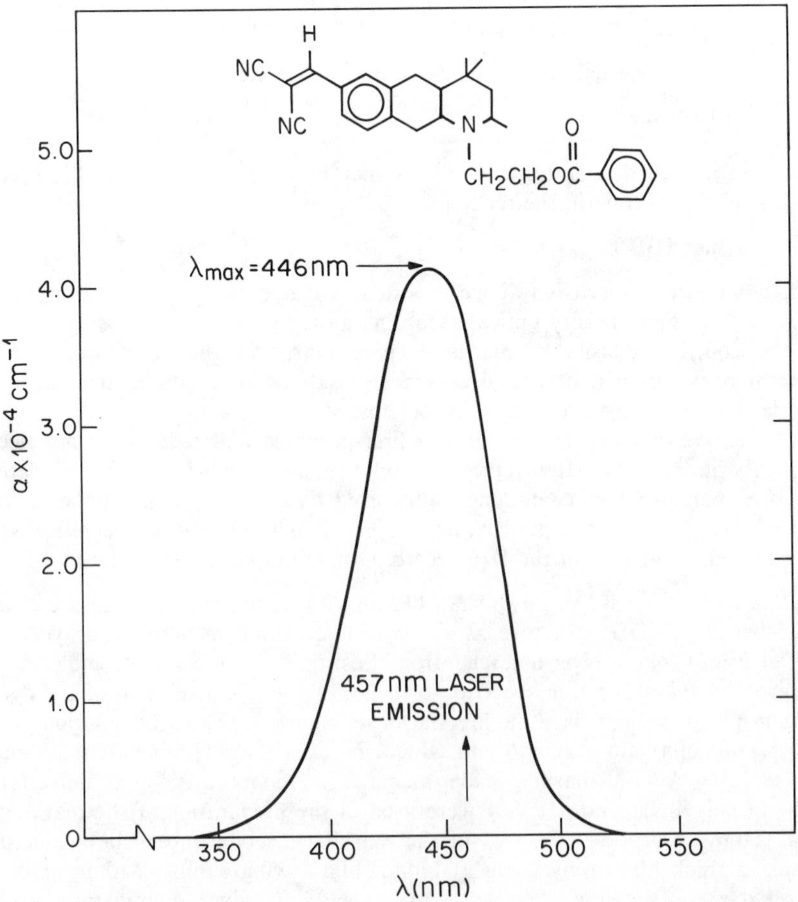

Figure 9. Molecular structure and DIP absorption spectrum of the Polyester Yellow/Polyvinylacetate system. Dye loading in PVA is 20% by weight.

characteristics of thin films is the interference criterion. The reflectance of the laser radiation (457 nm) at the marking wavelength for the two concentrations of dye, 5% and 20%, as a function of film thickness is shown in Figure 10. For both concentrations, reflection minima due to interference are observed. An enhancement in absorption will occur at these minima.

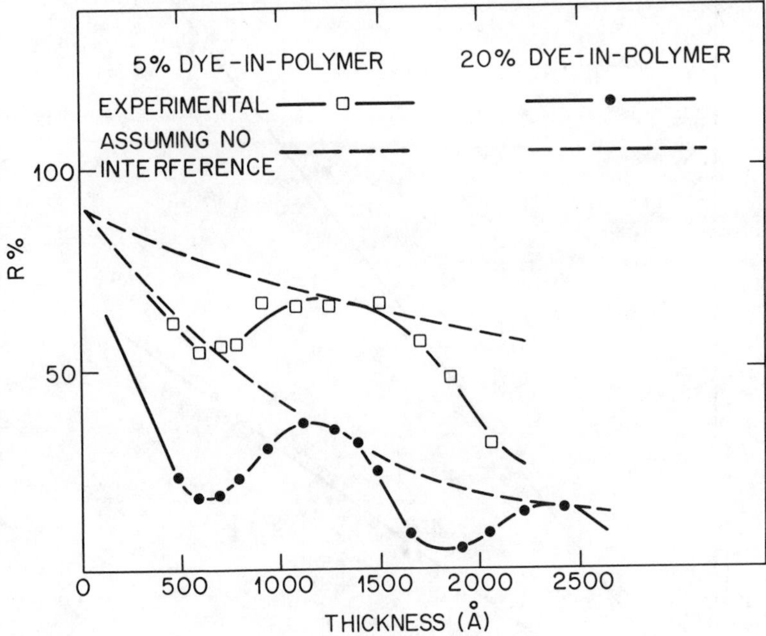

Figure 10. Reflectance of the Blue DIP disk at 457 nm as a function of thickness.

In addition to having high absorption of the marking radiation, the recording film must also have sufficient reflectivity (> 10% is desirable) for focussing and sufficient signal contrast, which is defined as:

$$C = \frac{(R_f - R_i)}{(R_f + R_i)} \tag{3}$$

where R_i is the reflectance of the film and R_f is the reflectance of the imaged bit. In the case of the 20% film at a thickness of 600 A, the initial reflectivity is ~17% and C is ~0.7, assuming that the reflectivity of the imaged area is equal to that of the aluminum substrate (90%). Hence, the 20% film meets all of these requirements and was chosen for the marking experiments.

Marking Results — Blue Disk. The 457 nm marking radiation was obtained using a nitrogen laser. The power of the laser pulse, which had a width of 8 nsec, was varied by rotating a prism polarizer through which the beam

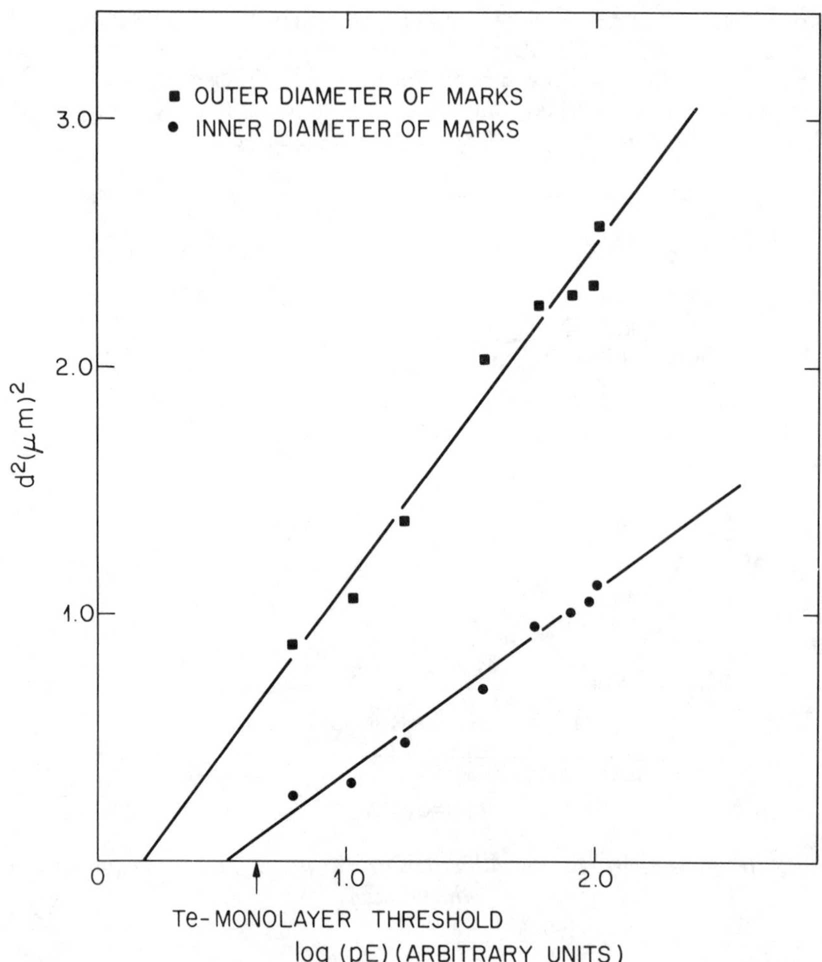

Figure 11. Bit size as a function of laser pulse energy for the blue DIP disk.

passed. Although the absolute energy of the pulse was not measured, it was calibrated with respect to the threshold exposure sensitivity of a 150 Å tellurium film on polymethylmethacrylate under identical marking conditions.

Figure 11 shows a plot of the mark diameter as a function of pulse energy. As expected, the bit size is a function of the marking energy. These results also show that the Polyester Yellow/PVA marking medium is more sensitive than a 150 Å tellurium film at threshold and throughout the bit size range.

The reflectivity of the marks, $(R_f - R_i)$, was measured as a function of pulse energy. At high pulse energies, the reflectivity saturates indicating that all the dye in polymer material is removed. Furthermore, the reflectivity of very large marks was approximately 90% (identical to that of the aluminum reflecting layer). The signal contrast, C, was therefore, as expected, equal to 0.7.

In summary, a blue DIP optical disk has been fabricated and evaluated. The system has high resolution, it has greater imaging sensitivity than a 150 Å tellurium disk, and good readout contrast in the reflection mode.

The Infra-Red DIP Disk. As indicated previously, solid-state diode lasers, such as the GaAs laser, are potentially much more attractive for optical recording applications than gas lasers. A major problem in the development of an optical recording system utilizing a diode laser, however, is the spectral response of the marking medium. Recently Asbeck et al. (*28*) have reported ablative optical recording on thin tellurium films using a diode laser, but the decrease in the tellurium absorption in the near infrared resulted in a decrease in marking sensitivity. Further work on the use of anti-reflecting tellurium tri-layer films to increase the near infrared absorption has also been reported (*29*).

The DIP concept is readily applicable to any part of the UV/visible/near infrared spectrum by the appropriate choice of the dye. Law, Vincett and Johnson (*19*), from these laboratories have reported the development of a infrared sensitive DIP disk.

The dye used in the fabrication of the IR DIP disk was a carbocyanine (3,3'-diethyl-12-acetyl-thiatetracyanine perchlorate); it is designated NK1748 and was obtained from the Japanese Research Institute for Photosensitizing Dyes Co. Ltd. Its molecular structure is shown in Figure 12 along with its absorption spectra in methanol and in a thin solid film of polyvinylacetate

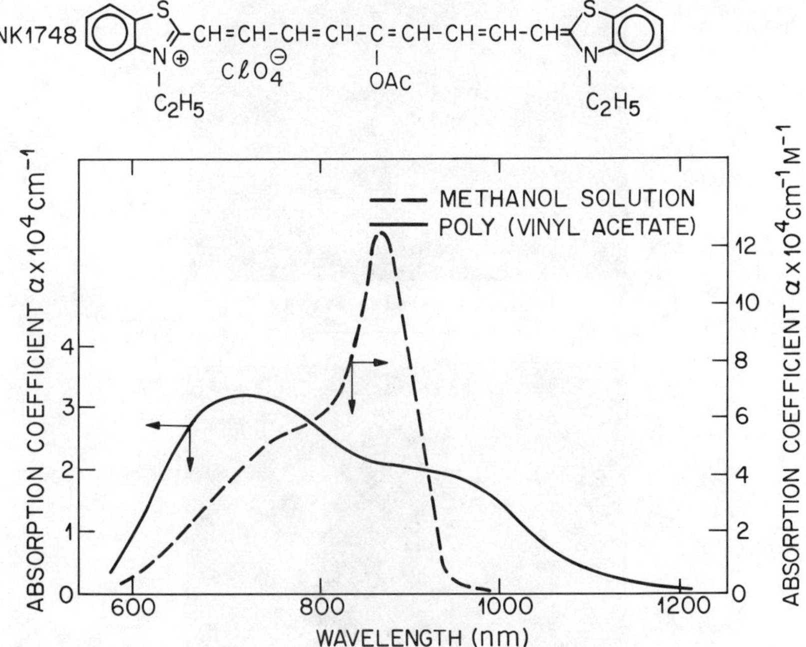

Figure 12. Molecular structure and absorption spectra of NK1748 in methanol solution and in polyvinylacetate.

(PVA). In methanol, an intense monomer absorption maximum is located at 870 nm. In the polymer film, at a dye polymer loading ratio of 1 to 5, the 870 nm solution maximum is broadened and split into two bands due to aggregation, centered at ~700 and 950 nm respectively. However, no evidence of microcrystal formation in the film was detected upon examination with an optical microscope and thus, if present, microcrystals are less than 1000 Å in diameter.

The thickness of films of the dye in PVA, cast on aluminum coated glass substrates; was optimized to obtain a partially anti-reflecting condition, and hence maximized the absorption of infrared light. A reflection minimum of 30% was observed for 835 nm light (the marking laser wavelength) using a 1050 Å thick film. Using a film of this thickness enhances the absorption of the film by a factor of ~2, which enhances writing sensitivity while ensuring adequate focussing and readout contrast.

Marking Results – Infrared Disk. The 835 nm laser beam was obtained from a dye laser which was pumped by a pulsed nitrogen laser. The energy of the laser pulse was varied as before. The marking threshold was determined (*19*) with respect to the threshold energy required for marking of a 150 Å tellurium film on a polymethylmethacrylate substrate (a commonly-used benchmark system) under identical conditions.

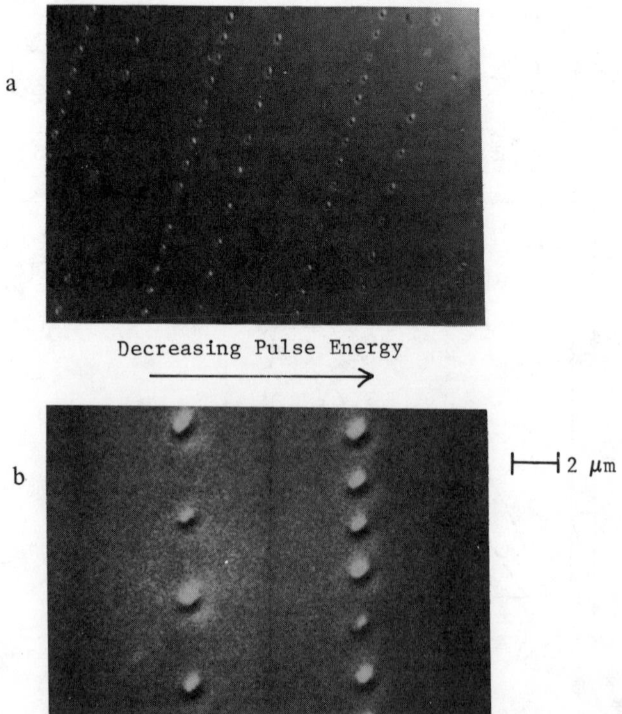

Figure 13. a. Interference-contrast optical micrograph of marked bits on the infrared DIP disk as a function of laser pulse energy. b. Scanning electron micrograph of marked bits on the infrared DIP disk.

Figures 13a and 13b show the optical interference-contrast and SEM micrographs, respectively, of the marked bits at various energy levels. The high spatial resolution of the Infrared disk Disc is apparent from the uniform, submicron dimensions of the marks.

Figure 14 shows a plot of the square of the mark diameter as a function of the pulse energy. Extrapolation of the plot to zero gives a threshold marking energy which is ~3.5 times lower than that for the 150 Å tellurium film used as the reference.

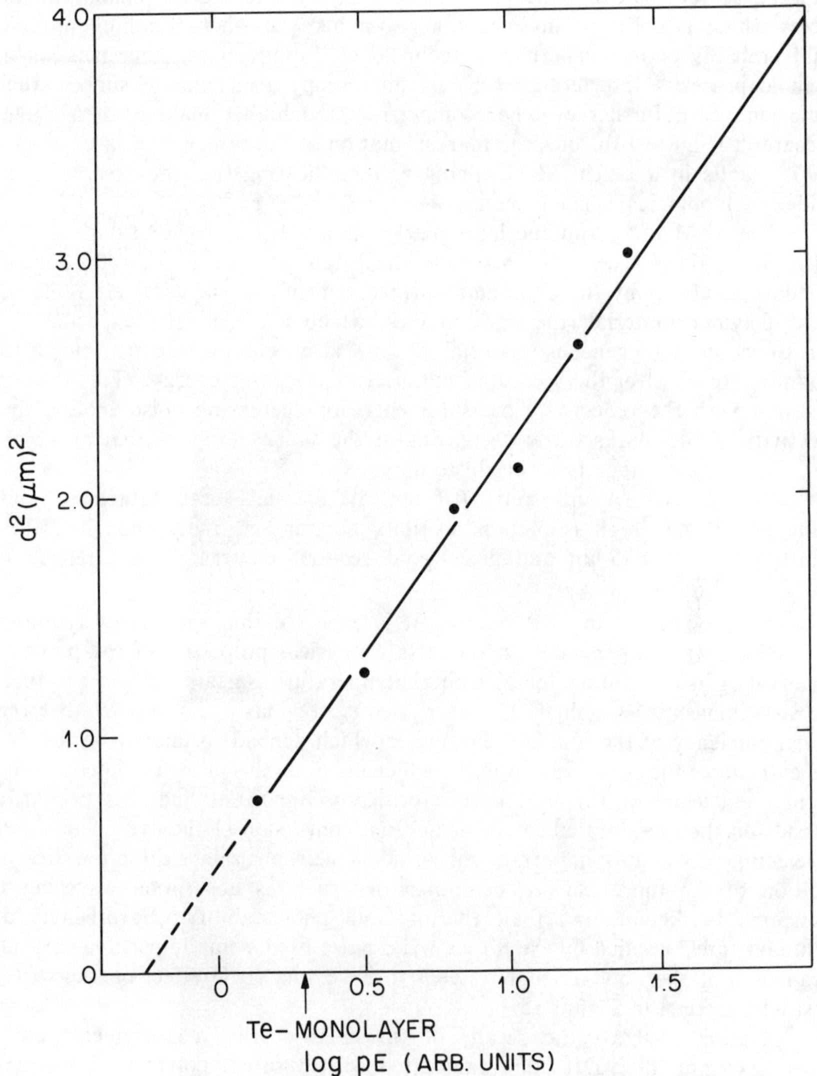

Figure 14. Bit size as a function of laser pulse energy for the infrared DIP disk.

The reflectivity change $(R_f - R_i)$, upon marking with a 1 μm beam, was also measured as a function of the pulse energy. It is significant to note that the reflectivity change saturates at high pulse energies indicating that all the DIP material is removed. Since the reflectivities of the infrared DIP disk and the aluminized glass substrates are ~30% and 90% at 835 nm, the signal contrast, defined as before by Equation 3, is 0.5. The morphology of the marks was studied with a SEM (*19*) using an uncoated sample observed at a tilt angle of 45°. The visibility of the marks in the SEM confirms that the mark formation is topological rather than simply photobleaching, a photochemical process which is observed in some other systems. (A photobleaching process would probably be unacceptable for technological applications, since it is not a threshold process.) Interference contrast microscopy also tends to support this conclusion. Still further evidence comes from the higher magnification SEM micrograph (Figure 13b) of large marks; rotation of the same set of laser marks by 90° results in identical SEM micrographs, indicating that the observed rim is indeed a topological phenomenon.

The SEM images of the laser marks change from bright to dark as the writing energy decreases. We attribute the bright images to the reflection of secondary electrons by the aluminum surface, a result of the complete removal of dye-polymer material; the dark images, which are generally expected for depressions in homogeneous material at low laser energy, are attributed to incomplete removal of the recording material at low laser energy. This is very consistent with the reflectivity measurements, for decreasing pulse energy, the reflectivity of the marks starts decreasing at the same energy as that at which the SEM images change from bright to dark.

In summary, a infra-red DIP optical disk has been fabricated and evaluated. It has high resolution, exhibits greater sensitivity than a 150 Å tellurium disk at 835 nm and gives good readout contrast in the reflection mode.

Mechanistic Studies. In DIP systems, it is expected that the energy required for marking will be dependent upon certain physical properties of the polymer such as its glass transition (or softening) temperature, surface tension and melt viscosity (molecular weight). However, it was previously shown (*30*) that the optical efficiency of the film is a parameter which depends primarily on the dye concentration, the dye absorption coefficient and the film thickness. The thermal efficiency of the film is also critically important and this primarily depends on the quantum efficiency of internal conversion of the dye. Therefore, in selecting dyes, it is important to consider their photophysical properties in addition to the more obvious characteristics such as absorption wavelength, absorption coefficient, and their thermal and photostability. Surprisingly at high data rates, implied by the 8 nsec write pulse used, some important physical parameters of the polymer matrix seem to have very little effect on the write-sensitivity as seen in Figure 15.

Figure 15 shows the results of threshold energy measurements on a series of 60 nm thick DIP films fabricated with different polymeric materials, each containing 20 weight percent Polyester Yellow. The solid line, about which the data points are scattered, corresponds to the variation of the

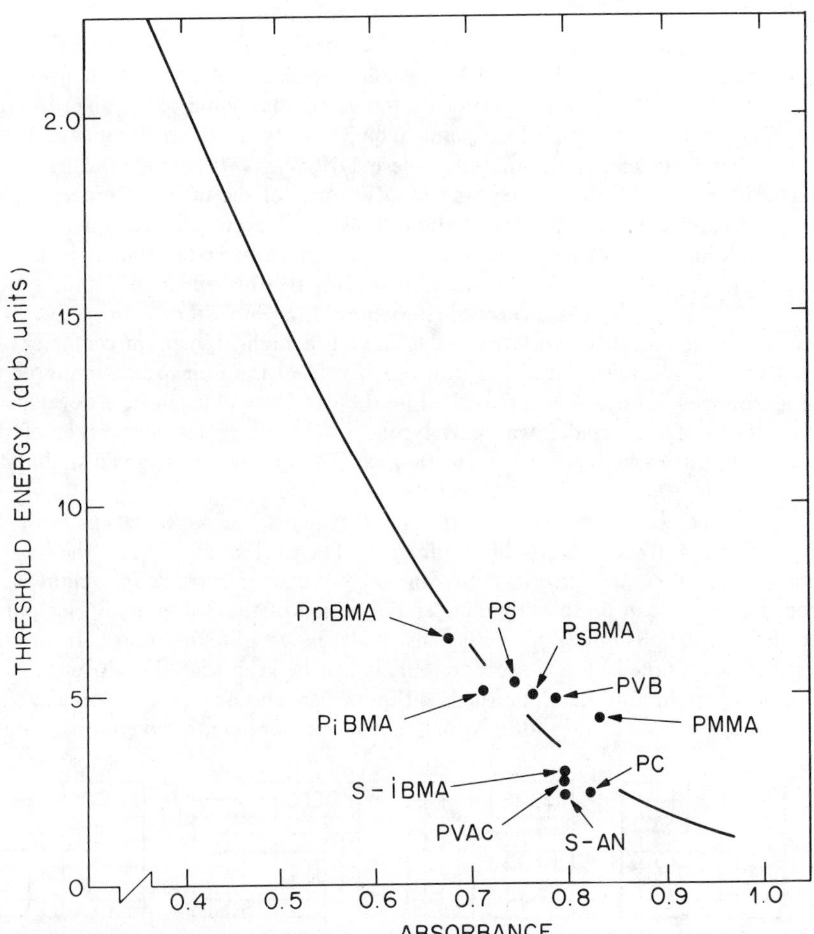

Figure 15. Influence of the Polyester Yellow dye film absorbance and polymer binder material on the marking threshold energy. PnBMA = poly(n-butyl methacrylate); PiBMA = poly(isobutyl methacrylate); PS = polystyrene; PsBMA = poly(sec-butyl methacrylate); PVB = polyvinylbutyl; PMMA = polymethyl methacrylate; PVAC = polyvinylacetate, S-iBMA = poly(styrene-co-isobutyl methacrylate), PC = polycarbonate; S-AN = poly(styrene-co-acrylontrite).

threshold energy with the film absorbance. The dependence of the threshold energy on absorbance was determined from measurements on Polyester Yellow/PVAc films of varying dye loading and from marking experiments at different wavelengths. (Note that although each of these films was of the same thickness and contained the same dye concentration, there are some differences in the absorbance which is due to differences in the refractive index of the polymers and solvatochromatic shifts of the dye.) The glass transition temperatures (T_g) of the polymers tested ranges from 18°C for PnBMA to 109°C for PS. In spite of these differences in T_g values, it appears that the dominant factor in determining the threshold energy is the film absorbance.

Although the above experiments indicated that some physical properties of the polymer matrix such as T_g and melt viscosity exert no significant effect on the threshold energy for marking these DIP films. It as noted, during the course of an investigation concerning the dynamics of the marking process, (*31*) that in certain cases complex relaxation phenomena occurred subsequent to the initial hole opening. These effects were found to be most prominent in higher molecular weight polymers. In order to investigate this effect in detail, a new series of DIP films was prepared in which the Polyester Yellow dye was replaced by a specially synthesized dye having a high absorption coefficient at 632.8 nm. This allows the increase in reflectance of the film to be monitored as the aluminized substrate is revealed while the dye laser induced hole develops in time. The polymer binder was polystyrene where the molecular weight of the polystyrene varies between 2.1 K to 2000 K. The molecular weight distribution, M_w/M_n, was < 1.1 for all polystyrene samples used.

The results of the hole-opening measurements are presented in the series of oscilloscope traces shown in Figure 16. Traces 16a and b are the results obtained on a sample fabricated with polystyrene of molecular weight 4 K. From trace a it can be seen that under the action of the 8 nsec dye laser pulse, the holes or pits develop fully in ~30 nsec. Furthermore, from the measurements taken at 100 nsec/div (b), it can be seen that the pit, once fully developed, maintains its dimensions. Similar conclusions can be drawn from oscilloscope traces c, d and e which show measurements taken on samples

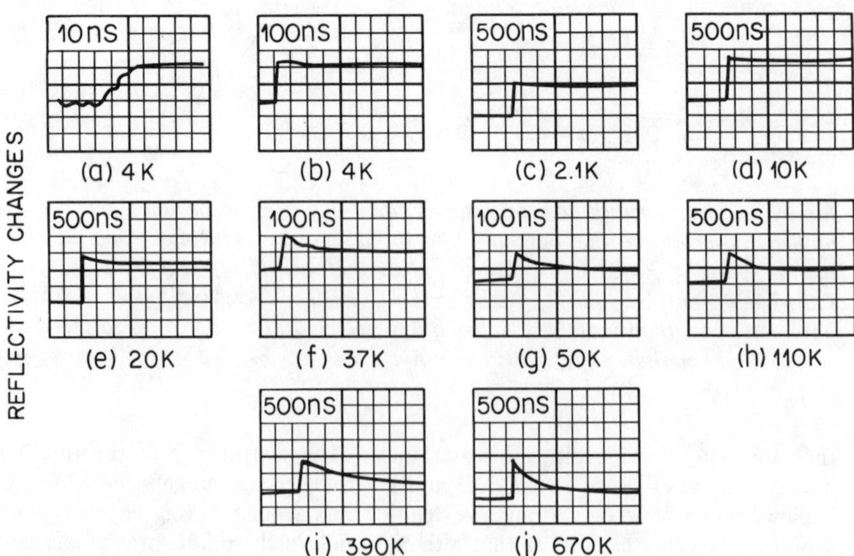

Figure 16. Reflectivity change in the dye-in-polymer films as a function of time for a series of polystyrene samples of different molecular weights.

fabricated with 2.1 K, 10 K and 20.4 K polystyrene respectively. These measurements were each taken at a sweep rate of 100 or 500 nsec/div to emphasize the point that over this range of molecular weights, the laser induced pits open cleanly, and experience no further change subsequent to the initial opening process. It should be noted, however, that at a molecular weight of 20.4 K, there is evidence for a small relaxation effect which can be interpreted as a flow back of molten polystyrene into the initially created mark.

The oscilloscope traces shown in Figures 16f-16j are the results of measurements made on DIP films in which the binder is polystyrene with molecular weights of 37 K, 50 K, 110 K, 390 K and 670 K. The behavior of these samples is clearly significantly different from those fabricated with lower molecular weight polystyrene. These traces are interpreted as indicating that subsequent to the initial rapid hole opening, a much slower process occurs in which some initially displaced polymeric material flows back into the pit yielding a reduction in the read-out signal. The rate of the relaxation process appears to be dependent on the polystyrene molecular weight. For example, at 37 K, the relaxation from the initial to final hole configuration occurs in approximately 250 nsec whereas at the much higher molecular weights of 390 K and 670 K, this process occurs on the order of 2 μsec. Polymer chain entanglement can be invoked to explain these results. Thus selection of suitable polymer matrix material should be directed primarily towards those materials with the best film forming properties, compatible with the dye and with controlled molecular weight.

Conclusion

In this paper we have described the potential of optical recording systems in high density information storage and retrieval systems. It is clear that optical recording results in an increase in storage density of two orders of magnitude over conventional magnetic disk technology. Although many imaging materials have been investigated as the marking media for optical disks, the most successful systems have used thin metallic layers, such as tellurium, which rely on an ablative process for bit creation with a marking laser.

We have reviewed the material requirements for a viable optical disk and have described two novel materials concepts, each based on the use of an organic dye-polymer interaction. The dyes used in these concepts must be thermally and photochemically stable. In each case the imaging mechanism involves an efficient and rapid conversion of the absorbed electronic energy to heat via internal conversion.

The LIDA concept, which is based on laser induced dye diffusion, can produce 1 μm diameter marks in less than 10^{-7} seconds with an exposure sensitivity of 100 mJ/cm^2. The marks are permanent and a readout contrast density of unity can be obtained in the transmission mode.

The DIP concept is based on an ablative marking process and blue and an infra-red DIP disk have been evaluated. Both DIP disks have high resolution and both are more sensitive than a 150 Å tellurium disk to which they were compared. Acceptable readout contrast was obtained in the reflective mode for each.

The infra-red DIP disk is considered to be the most attractive in optical recording applications because of its match to inexpensive, reliable solid-state diode lasers. Although its ease and economy of fabrication demonstrate the potential of organic materials in this emerging new technology, its long-term stability in applications requiring archival storage, has to be demonstrated.

Literature Cited

1. Maydan, D. *Bell Syst. Tech. J.* 1971, *50*, 1761.
2. Gufield, R. J.; Chaudhari, P. *J. Appl. Phys.* 1972, *43*, 4688.
3. Deryugin, L. N.; Komotskii, V. A.; Fridman, G. Kh. *Sov. J. Quantum Electron.* 1973, *2*, 555.
4. Braudy, R. S. *Proc. IEEE*, 1969, *57*, p. 1771; Braudy, R. S. *J. Appl. Phys.* 1974, *45*, 3512.
5. Tolle, H. J.; Memming, R. *Appl. Phys. Lett.* 1975, *26*, 349.
6. Bartolini, R. A.; Weakliem, H. A.; Williams, B. F. "Optical Engineering"; 1976, Vol. 15, No. 2, p. 99.
7. Congleton, D. B.; Smith, M. R.; Diamong, A. S. *J. Appl. Photographic Eng.* 1977, *3*, 97.
8. Bell, A. E.; Spong, F. W. "IEEE J. of Quantum Electronics"; 1978, Vol. QE-14, No. 7, p. 487.
9. Novotny, V.; Alexandru, L. *J. Appl. Polymer Sci.* 1979, *24*, 1321.
10. Novotny, V.; Alexandru, L. *J. Appl. Phys.* 1979, *50*, 1215.
11. Bell, A. E. *Nature* 1980, *287*, 583.
12. Law, K. Y.; Vincett, P. S.; Loutfy, R. O.; Alexandru, L.; Hooper, M. A.; Sharp, J. H.; Johnson, G. E. *Appl. Phys. Lett.* 1980, *36*, 884.
13. Jipson, V. B. and Jones, C. R. *SPIE Proc.*, 1981, *263*, 105.
14. Howe, D. G.; Wrobel, J. J. *SPIE Proc.*, 1981, *263*, p. 92.
15. Drexler, J. *SPIE Proc.*, 1981, *263*, p. 87.
16. Smith, T. W.; Ward, A. T. *Material Research Society Meeting*, Boston, Massachusetts, November 1980.
17. Jipson, V. B.; Lynt, H. N.; Graczyk, J. F. *Appl. Phys. Lett.* 1983, *43*, 27.
18. Morinaka, A.; Oikawa, S.; Yamazaki, H. *Appl. Phys. Lett.* 1983, *43*, 524.
19. Law, K. Y.; Vincett, P. S.; Johnson, G. E. *Appl. Phys. Lett.* 1981, *39*, 718.
20. U.S. Patent 4 242 689.
21. U.S. Patent 4 244 355.
22. Kivits, P.; de Bont, P.; Vander Veen, J. *Appl. Phys.* 1981, *A-2B*, 101.
23. Goto, Y.; Kohino, N.; Goto, H.; Ogawa, K. "European Patent Application"; 1983, No. 0 083, 991.
24. U.S. Patent 4 219 826.
25. Jipson, V. B.; Sons, C. R. *J. Vac. Sci. Techn.* 1981, *18*, 105; *IBM Techn. Disc. Bull.* 1981, *24*, 298.
26. Kurtz, C. N.; Tuite, R. J. "J. Inform. and Image Management"; October 15, 1983.
27. Bruce, C. A.; Jacobs, J. T. *J. Appln. Photogr. Eng.* 1977, *3*, 40.

28. Asbeck, P. M.; Commack, J. J.; Daniele, D.; Low, D.; Heemskerek, J. P. J.; Kleuters, W. J.; Aphey, W. H. *Appl. Phys. Lett.* 1979, *34*, 275.
29. Bell, A. E.; Bartolini, R. A. *Appl. Phys. Lett.* 1979, *34*, 275.
30. Law, K. Y.; Johnson, G. E. *J. Appl. Phys.* 1983, *54*, 4799.
31. Johnson, G. E.; Law, K. Y. *SPIE Proc.*, 1984, *420*.

RECEIVED August 6, 1984

23

Primary and Secondary Reactions in Photoinitiated Free-Radical Polymerization of Organic Coatings

A. HULT and B. RANBY

Department of Polymer Technology
The Royal Institute of Technology, S-10044
Stockholm Sweden

During the past decade there has been, and still is, a growing interest in applications of polymer photochemistry. Many of these applications depend on photo-cross-linking of polymers. Industrial applications include photoimaging (*1*), photocuring of coatings (*2*) and printing inks (*3*). Photocuring is essentially the use of ultraviolet radiation to convert monomeric or oligomeric substances into polymers. These processes are based on the photogeneration of radical or cationic species which initiate polymerization and cross-linking. Photogeneration of free radicals can be achieved by at least two different mechanisms (*4,5*):

1. Homolytic fragmentation of a photo-excited molecule, giving a radical pair (intramolecular process).
2. Intermolecular hydrogen abstraction by a triplet excited photosensitizer from a hydrogen donor.

Initiators that operate on the basis of the first mechanisms are commonly different acetophenone derivatives (*6*) and are mainly used in clear lacquers. The second mechanism employs derivatives of aromatic ketones and quinones (*7*) as photosensitizers, usually in combination with tertiary amines as hydrogen donors. These systems are often used in pigmented coatings.

Photoinitiated free radical polymerization is a typical chain reaction. Oster and Nang (*8*) and Ledwith (*9*) have described the kinetics and the mechanisms for such photopolymerization reactions. The rate of polymerization depends on the intensity of incident light (I_o), the quantum yield for production of radicals (ϕ), the molar extinction coefficient of the initiator at the wavelength employed (ϵ), the initiator concentration $[S]$, and the path length (l) of the light through the sample. Assuming the usual radical termination processes at steady state, the rate of photopolymerization is often approximated by

$$R_p = k_p/k_t^{1/2}(I_o \phi \epsilon [S] l)^{1/2}[M] \qquad (1)$$

0097-6156/84/0266-0457$06.00/0

Joshi (*10*) has shown that there are deviations from the postulated square root dependence at higher initiator concentrations. If the concentration is higher than 1 mM, the rate of initiation is given by

$$R_i = \phi I_o (1 - e^{-\epsilon [S] l}) \tag{2}$$

McGinnis, Porvder, Kuo and Gallopo (*11*) have also reported on this deviation. They investigated photopolymerization initiated by 4.4'-bis(N,N-diethylamino) benzophenone. Gothe and Ranby (*12*) made the same observation for polymerizations initiated by 2-isopropylthioxanthone/tertiary amine systems.

Photocured organic coatings always contain a small amount of unreacted photoinitiator. This residue serves as a free radical source when the coating is exposed to sunlight. It is well known that triplet excited aromatic ketones abstract hydrogen atoms from polymers (*13,14*) and thereby, initiate photodegradation of the polymer. This paper will discuss both the primary reactions leading to photo-cross-linking and the secondary reactions that cause photodegradation. It will also show how the amount of initiator required for curing can be decreased by the use of a surface active photoinitiator thereby producing more stable films with acceptable cure rates.

Primary Reactions of Type 1 Photoinitiators

Among the most widely used photofragmenting initiators are alkoxyacetophenones and hydroxy-alkylacetophenones (Figure 1). The primary reaction of these initiators is a Norrish Type I cleavage leading to the formation of a benzoyl radical and a fragment radial moiety both of which may initiate polymerization. Sander and Osborn (*15*) have shown that 2,2-dimethyoxy-2-

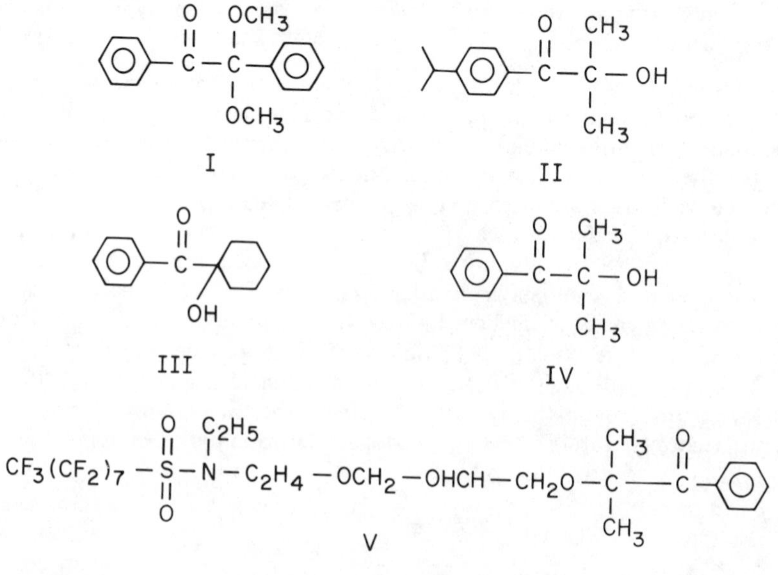

I

II

III

IV

$CF_3(CF_2)_7-SO_2-N(C_2H_5)-C_2H_4-OCH_2-OHCH-CH_2O-C(CH_3)_2-CO-C_6H_5$

V

Figure 1. Type 1 photoinitiators.

phenylacetophenone (initiator I) is decomposed into a benzoyl radical and a dimethoxybenzyl radical. They also showed that this dimethoxy radical is rearranged into a methyl radical and methylbenzoate [Scheme 1]. Further they

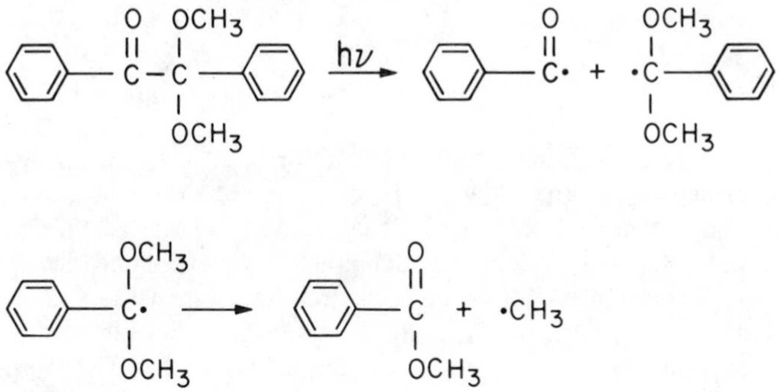

$$\mathrm{Ph{-}C(=O){-}C(OCH_3)_2{-}Ph \xrightarrow{h\nu} Ph{-}\dot{C}(=O) + \cdot C(OCH_3)_2{-}Ph}$$

$$\mathrm{Ph{-}\dot{C}(OCH_3)_2 \longrightarrow Ph{-}C(=O){-}OCH_3 + \cdot CH_3}$$

Scheme 1.

showed that acetophenone was formed by an in-cage or near-cage combination of methyl and benzoyl radicals. These observations were also made by Borer, Kirchmayr and Rist (*16*) who used the CIDNP technique in their work and determined the structures of many secondary recombination products resulting from cage reactions.

Eichler, Herz, Naito and Schnabel (*17*) have investigated the photochemistry of some hydroxy-alkylacetophenones (initiators II-IV). They showed that these initiators are decomposed from their triplet state with a quantum yield is between 0.2-0.3. The main difference between the hydroxy-alkylacetophenones and initiator I is that they only form one initiating radical. The other radical only acts as a chain terminator at relatively high radical concentrations but does not initiate polymerization. In the case of initiator I, both the benzoyl and the methyl radical can initiate polymerization of vinyl monomers. The useful absorption spectra of these initiators range from ~300-380 nm, below 300 nm most of the light is absorbed by the coating itself. In this region initiator I has a higher ϵ than II-IV. Hence, I is significantly more efficient than the hydroxy-alkylacetophenones.

Primary Reactions of Type 2 Photoinitiators

Type 2 initiators are bimolecular systems and consist of a photosensitizer and a hydrogen donor. The most well known system is benzophenone/tertiary amine. Because of the relatively weak absorption of benzophenone at 360 nm the efficiency of this system is rather low. A more efficient aromatic ketone is thioxanthone (TX) and its derivatives simply because of the increased ϵ at the exposure wave length.

The primary reaction of Type 2 photoinitiators is a hydrogen abstraction from the tertiary amine by a triplet excited ketone. The amino radical thus formed is sufficiently active to initiate the polymerization of vinyl monomers Scheme 2.

$$>C=O \xrightarrow{h\nu} [>\dot{C}-O\cdot]^{S} \xrightarrow{ISC} [>\dot{C}-O\cdot]^{T}$$

$$[>\dot{C}-O\cdot]^{T} + :N-(C_2H_5)_3 \longrightarrow [\text{exciplex}] \longrightarrow >\dot{C}-OH + :N(\cdot CHCH_3)-(C_2H_5)_2$$

Scheme 2.

Capitano (*18*) has shown that the absorption band for TX around 380 nm corresponds to the lowest lying $\pi \rightarrow \pi^*$ electronic transition. Herkstroeter, Lamola and Hammond (*19*) calculated the triplet energy for TX to be 65.5 kcal/mol. Amirzadeh and Schnabel (*20*) estimated the quantum yield for triplet formation of TX to be 0.85 and determined the rate constants for a number of the photochemical reactions that take place during the photocuring reaction. We have outlined a reaction scheme for TX based on our studies (Figure 2) which includes several pathways that occur during the curing reaction.

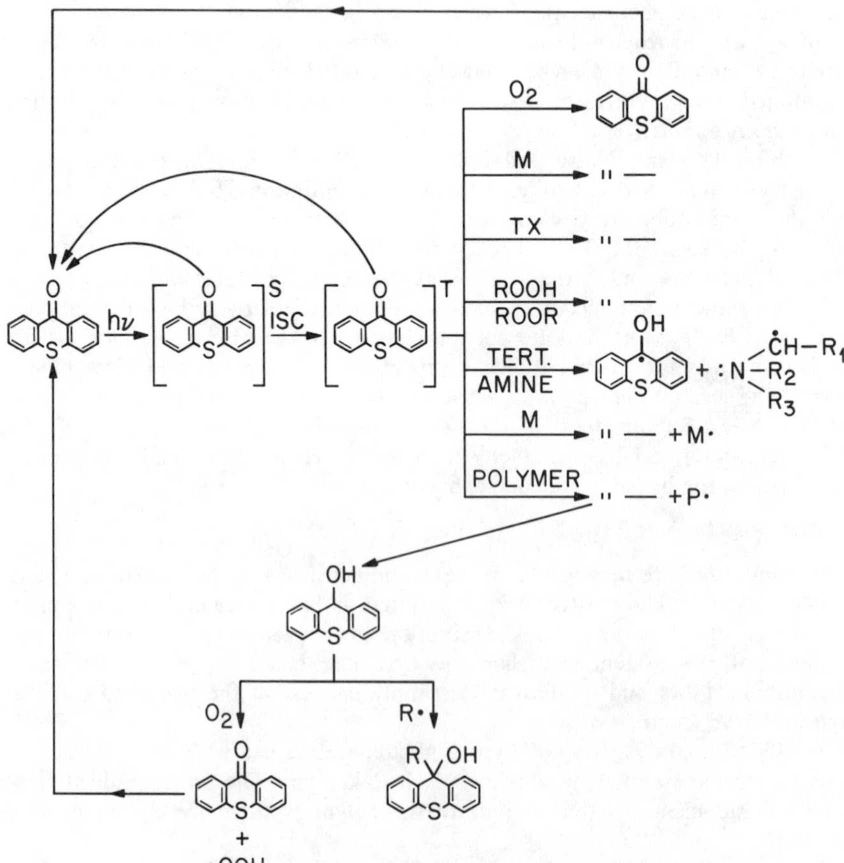

Figure 2. Thioxanthone reaction scheme.

Secondary Reactions of Type 1 Photoinitiators: Model Studies with Polystyrene

The amount of unreacted initiator that remains in a coating after photocuring is dependent on the initial concentration, the quantum yield, the irradiation time and the light source. To investigate how unreacted initiator affects the photostability of coatings, model studies with polystyrene were conducted (*21*).

Rabek and Ranby (*22*) have shown that a free radical induced degradation of polystyrene occurs in the presence of oxygen, that leads to rapid chain scission. Benzoyl radicals derived from Type 1 initiators abstract hydrogen atoms from the polymer and thereby start a chain degradation process. Berner, Kirchmayr and Rist (*6*) and others have shown that when initiator I is irradiated in solution, the benzoyl radical abstracts a hydrogen from the surrounding solvent to form benzalkdehyde and a free radical.

Figure 3 shows the effect on the viscosity of polystyrene samples irradiated ($\lambda \geqslant 305$ nm) in benzene solution in the presence of initiator I-III. The degradation of the polymer is caused by free radicals generated from the initiators. There is also a significant difference between the three initiators. Initiator I has a stronger absorption band between 300 and 400 nm than the

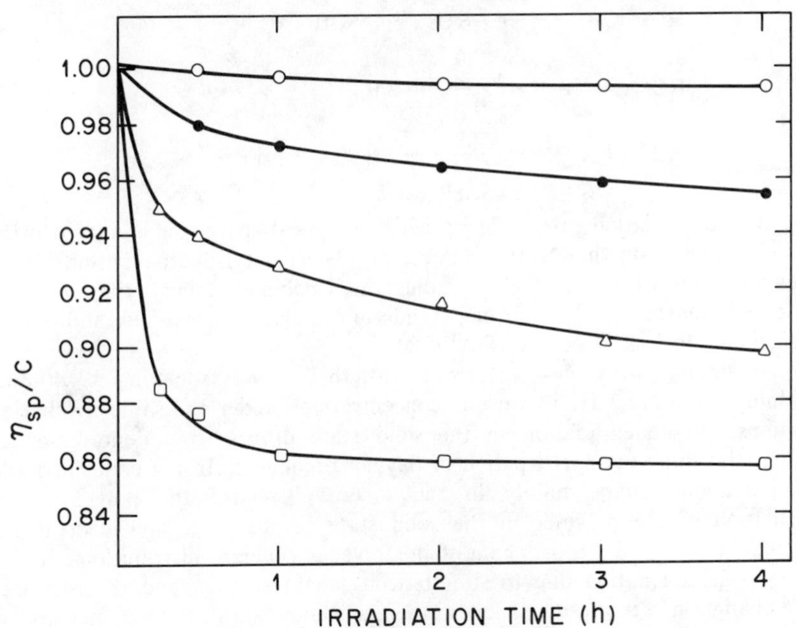

Figure 3. Intrinsic viscosity of polystyrene samples, irradiated in benzene solution in the presence of initiator I-III. Polystyrene concentration 7.69×10^{-2} *M, photoinitiator concentration* 2.31×10^{-3} *M/ p Pure polystyrene;* □ *initiator I* ● *initiator II* Δ *initiator III. (Reproduced with permission from Polym. Deg. Stability Ref. 21).*

others and may also produce acetophenone as mentioned above. Triplet excited acetophenone can abstract a hydrogen from the polymer and thereby initiate a chain degradation process.

The benzoyl radical from initiator III is more reactive than the parasubstituted benzoyl radical from initiator II. The isopropyl substituent stabilizes the benzoyl radical and make it less reactive.

The degradation rate of polystyrene in the presence of photoinitiators is strongly enhanced by the presence of oxygen. Oxygen adds to radicals to form peroxy radicals which then abstract labile hydrogens from the polymer. The resulting hydroperioxides decompose to form reactive alkoxy and hydroxyl radicals; autoxidation starts and the concentration of free radicals in the system increases [Scheme 3]. The result is an increase in the degradation rate as is shown in Figure 4.

$$R\cdot + O_2 \longrightarrow R-OO\cdot$$

$$R-OO\cdot + P-H \longrightarrow R-OOH + P\cdot$$

$$R-OOH \longrightarrow RO\cdot + \cdot OH$$

$$RO\cdot + P-H \longrightarrow R-OH + P\cdot$$

$$HO\cdot + P-H \longrightarrow H_2O + P\cdot$$

Scheme 3.

The degradation rate is decreased if a radical scavenger is added to the solution. Figure 4 shows the effect of 3-tert-butyl-hydroxyanisol. This scavenger reacts with both benzoyl radicals and polymer radicals thus reducing the degradation rate. The rate of degradation is also decreased by addition of 1.4-diazobicyclo (2.2.2)-octane (DABCO).

A similar study was performed with thin polystyrene films (~40 μm) containing initiators I-III in various concentrations under the same irradiation conditions. Photodegradation in the solid state differs from degradation in solution. In films, the participation of oxygen becomes diffusion controlled and photo-oxidation occurs mainly in the surface layer of the polymer film. Degradation of the polymer in the solid state results in a slow decrease in molecular weight and a broadening of the molecular weight distribution. It also results in gel formation due to simultaneous chain scission and cross-linking. These changes are the result of low oxygen concentration in the polymer matrix and the low mobility of the polymer chains in the solid state. The results from this study are listed in Table I.

From the studies described above, it was concluded that the presence of unreacted photoinitiator in a cured polymer initiates a free radical degradation of the polymer upon irradiation with UV light. It was further concluded that the rate of degradation is dependent on the structure of the initiator and the initiator concentration.

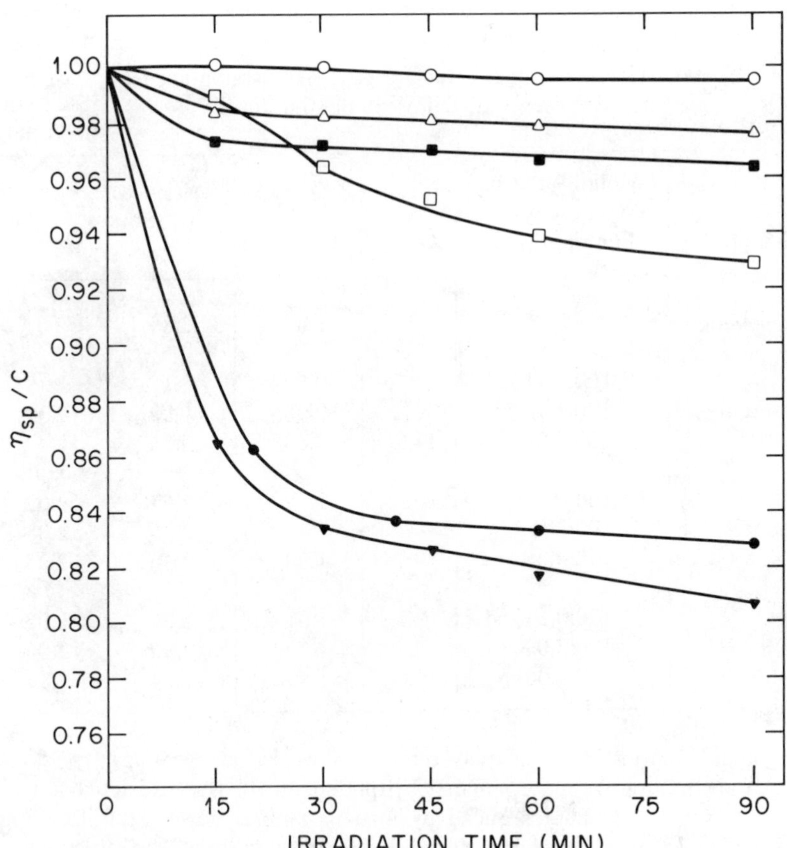

Figure 4. Intrinsic viscosity of polystyrene samples irradiated in benzene solution under various conditions. Polystyrene concentration 7.69×10^{-2} *M and initiator I concentration* 3.12×10^{-3} *M. o vacuum;* □ *nitrogen saturated;* ● *air saturated;* ▽ *oxygen saturated;* Δ 3.12×10^{-3} *M 3-tert-butyl-4-hydroxyanisole;* ■ 3.12×10^{-3} *M 1.4-diazobicyclo(2.2.2)-octane (DABCO). (Reproduced with permission from Polym. Stability Ref. 21).*

Secondary Reactions of Type 2 Initiators: Model Studies with PMMA

To study the secondary reactions in ketone/amine photocured organic coatings 2-isopropylthioxanthone (ITX) and poly(methyl methacrylate) (PMMA) were used as a model system (*23*).

There are many reports in the literature describing photodegradation sensitized by aromatic ketones and quinones. Beachell and Chang (*24*) showed that triplet excited benzophenone accelerates degradation of polyurethanes by hydrogen abstraction. Harper and McKellar (*25*) showed the same effect with benzophenone on opolypropylene, and Amin and Scott (*26*) on polyethylene. Rabek and Ranby (*27*) have demonstrated that triplet excited quinones enhance

Table I. Amount of Gel, Molecular Weight Averages (M_n,M_w) and Molecular Weight Distributions (M_w,M_n) of Polystyrene Samples after a 2 hour UV-Irradiation in the Presence of Different Photoinitiators

Sample	Molar Ratio of Styrene Units to Photoinitiator	Gel%	$M_n \times 10^5$	$M_w \times 10^5$	M_w/M_n
Pure polymer			1.43	2.89	2.01
	100:0.813		0.994	2.18	2.19
Initiator I	100:2.03		0.785	1.82	2.32
	100:4.06	11.2	0.643	1.56	2.42
	100:0.813		1.32	2.67	2.02
Initiator II	100:203		1.14	2.36	2.08
	100:4.06	9.74	1.04	2.30	2.25
	100:0.813		1.31	2.66	2.02
Initiator III	100:2.03		1.29	2.66	2.06
	100:4.06	11.6	1.17	2.54	2.17

the photodegradation rate of polystyrene. This is, at least partly, do to hydrogen abstraction from the polymer by quinone in the triplet excited state.

TX is not only photoreduced by tertiary amines however, Wilson (*28*) showed that TX can be photoreduced by irradiation in a dimethoxyethane solution and Marteel et al (*29*) have demonstrated that TX can be reduced even by irradiation in cyclohexane.

Irradiation of a PMMA film containing ITX results in rapid photoreduction of ITX as is shown in Figure 5. Similar experiments in a deaerated benzene solution containing PMMA and ITX also result in photoreduction of ITX, albeit at much lower rate. The most likely interpretation of these observations is that triplet excited ITX abstracts hydrogen from the polymer to form a ketyl radical and a macro alkyl radical. The higher photoreduction rate in the film may result from deactivation reactions of triplet excited ITX which are faster in solution than in a glassy film. The triplet lifetime for ITX in a PMMA film is 10^{-3} sec compared with 10^{-4} sec in a benzene solution (*20*). In solution triplet-triplet annihilation ($k_{TT} = 7 \pm 3 \times 10^8$ M^{-1} sec^{-1}) and selfquenching ($k_{sq} = 5 \times 10^7$ M^{-1} sec^{-1}) compete with hydrogen abstraction from the polymer. In a PMMA film the probability of these deactivation reactions decreases since the diffusion rate is low compared to the diffusion in a liquid solution.

The average distance of displacement of an excited molecular is related to its diffusion coefficient and the lifetime of the excited state. As the viscosity

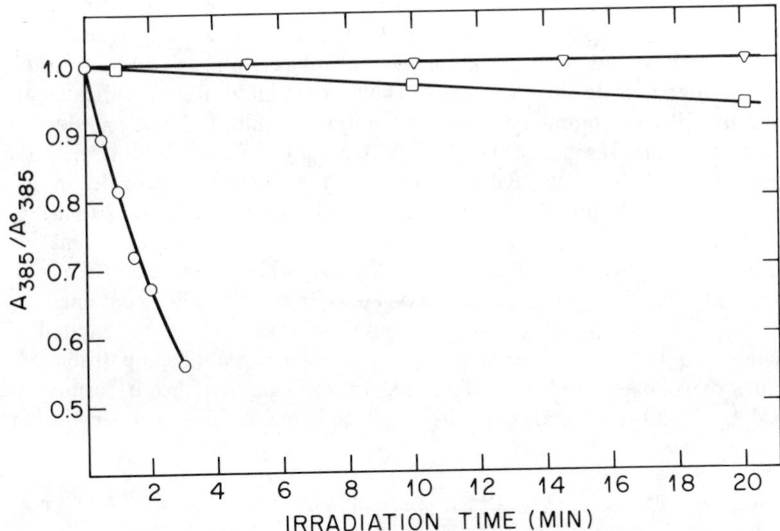

Figure 5. Photoreduction of ITX measured as change in absorbance at 385 nm. ▽ 10^{-4} *M ITX in benzene;* □ 10^{-4} *M ITX + 0.1 g PMMA/100 ml benezene; o* 4×10^{-3} *M ITX in a PMMA film. (Reproduced with permission from Polym. Deg. Stability Ref. 23).*

of the solution increases the diffusion coefficient and the diffusion length decreases. In a polymer matrix ($D \sim 10^{-1}$ cm^2/sec) a typical molecular can be transferred approximately 45 A in 10^{-3} sec (*30*). A quenching reaction in a polymer matrix can take place only if th donor and acceptor molecules are sufficiently close to each other. The distance between the molecules is, of course, determined by their concentrations. If the donor-acceptor solution is less concentrated than 10^{-2} M, the donor and the acceptor molecules are too far apart to react in a polymer matrix unless they can diffuse toward one another with a time constant less than their deactivation rate.

In an ITX/amine photocured organic coating the amine concentration is normally about 0.1 M. Here it will be a competition between hydrogen abstraction from the amine and from the polymer.

Free radical polymerizations performed in the presence of oxygen result in incorporation of peroxides and hydroperoxides in the polymer structure. Perioxides play an important role in the photodegradation of most polymers. The weak peroxy bond PO-OH (42 kcal/mol) can be broken by thermal dissociation, by direct absorption of light (*31*) or by energy transfer from some excited sensitizer molecule. Alkyl peroxides have only a weak absorption above 300 nm (*32*) but Ng and Guillet have reported that the primary quantum yield for photolysis of polymer hydroperoxides is high (0.8). Walling and Gibian (*33*) have shown that alkylhydroperoxides like tert-butylhydroperoxide may be decomposed by energy transfer from triplet excited benzophenone. Ng and Guillet (*34*) showed that also alkylperoxides like di-tert-butylperoxide were decomposed by energy transfer from triplet excited ketones however, a a rate lower than for hydroperoxides.

The triplet energy of ITX (62.2 kcal/mol) is high enough for decomposition of peroxides and hydroperoxides through energy transfer. The reaction result would be formation of an alkoxy and a hydroxy radical. Mackor, Wajer and de Boer (*35*) have shown that alkyl alkoxy radicals can be trapped by nitroso compounds like tert-butylnitroxide (BNO) and identified with electron spin resonance (ESR) spectroscopy. To study this reaction a mixture of ITX, tert-butylhydroperoxide and di-tert-butylperoxide in a de-aerated benzene solution was irradiated with filtered light (λ ⩾ 320 nm) (*37*). The solution contained BNO to trap the formed radicals. This same experiment was also carried out with PMMA which had been free radical polymerized in the presence of oxygen. This polymer contained both hydroperoxides and peroxides which could be decomposed by excited ITX molecules. In both experiments a rapid photosensitized decomposition of the peroxides took place and the alkyl alkoxy radicals that were formed were trapped by BNO and analyzed with ESR Scheme 4. The splitting constant

```
  CH3                          CH3
  |                 hν         |
~~C—OOH  +  ITX  ———————>   ~~C—O•   +   •OH   +   ITX
  |                            |
  C=O                          C=O
  |                            |
  OCH3                         OCH3

  CH3       CH3                        CH3
  |         |                hν        |
~~C—O—O—C~~        +  ITX  ———————>  2~~C—O•   +   ITX
  |         |                          |=O
  C=O       C=O                        C
  |         |                          |
  OCH3      OCH3                       OCH3

  CH3         O   CH3              CH3   Ō   CH3
  |           ||  |                |     |   |
 —C—O•   +    N—C—CH3  ———>      —C—O—N—C—CH3
  |               |                |    +•   |
  C=O             CH3              C=O       CH3
  |                                |
  OCH3                             OCH3
```

Scheme 4.

found, a_n = 26.6 G, is in agreement with the literature (26.6 G) (*36*). Irradiation in the absence of ITX did not result in formation of alkoxy radicals that could be trapped by BNO.

From these studies it was concluded that the main secondary reactions for TX/amine systems are hydrogen abstraction from the surrounding polymer and decomposition of peroxides through energy transfer.

Photocuring in Air Using a Surface Active Photoinitiator

One of the main advantages provided by photocuring is a high curing rate. The rate may be increased by an increased initiator concentration. There exists, however, an upper limit for the initiator concentration. This limit depends on the film thickness, the coating formulation, the initiator absorption and the spectral distribution of the light source. The curing rate also depends on the presence of oxygen. Oxygen is both an efficient free radical scavenger and an excited state quencher. Consequently photocuring reactions performed in the presence of oxygen require a higher initiator concentration than coatings cured under inert atmosphere in order to achieve the same cure rate.

Photo-oxidation studies followed by IR and ESCA analysis show that secondary reactions involving oxygen are mainly a surface effect (*38*). Figures 6 and 7 show IR transmission and reflection (MIR) spectra for an acrylic coating photocured in air and aged in an Atlas UVCON for 96 h. The penetration depth for the reflection measurement was less than 1 μm clearly demonstrating that oxygen reacts mainly in the outer surface layer of the coating. Therefore, it was proposed that the problem of undesirable, degrading, secondary reactions that occur when using a high initiator concentration in the presence of oxygen might be overcome by using a very low concentration of surface active photoinitiator. To test this hypothesis a surface active acetophenone derivative containing a fluorinated carbon chain was synthesized (initiator V).

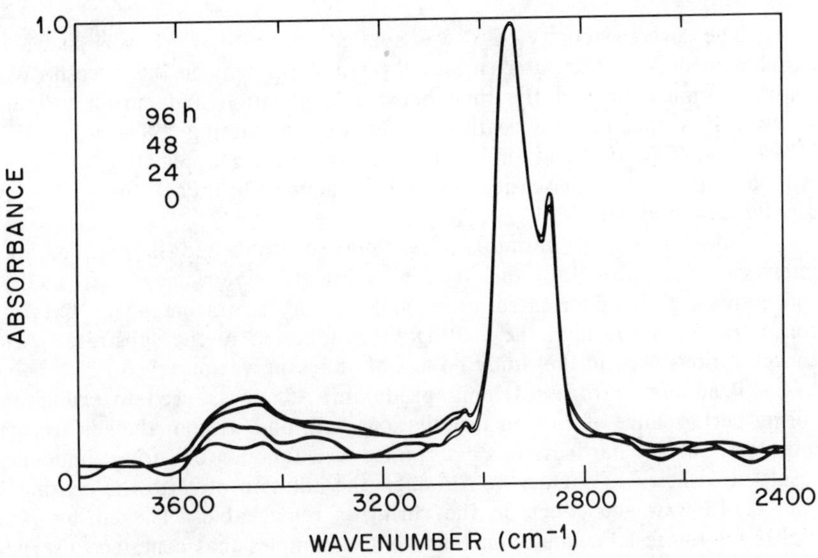

Figure 6. Infrared transmission spectra of a photocured epoxyacrylate film (30 μm) before and after aging in an Atlas UVCON. (Reproduced with permission from Polym. Deg. Stability Ref. 37).

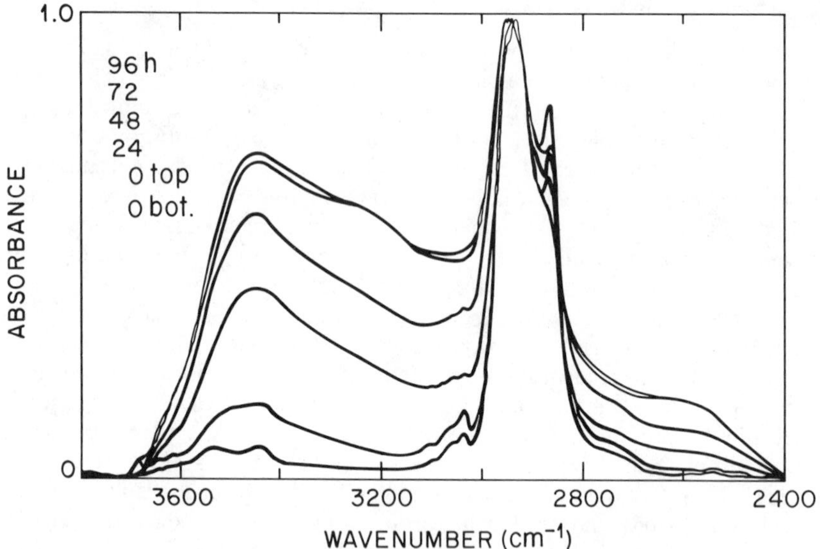

Figure 7. Infrared reflection (MIR) spectra of a photocured epoxyacrylate (same as Figure 6) before and after aging in an Atlas UVCON. (Reproduced with permission from Polym. Deg. Stability Ref. 37).

The surface activity of V was studied with ESCA. It was shown that enrichment of V in the outer surface layer is dependent on the viscosity of the coating formulation and the time between application and curing. Figure 8 shows ESCA spectra of a coating formulation containing 2% w/w of IV and 0.14% w/w of V taken at different intervals between application and curing. Although the coating viscosity was rather high (530 mPas) the surface was rapidly saturated with V.

The fluorinated initiator is not only concentrated in the coating-air interface but also in the coating-substrate interface. At moderate concentrations the fluorinated compound acts as a wetting agent but if the concentration is too high the coating loses adhesion to the substrate. At the concentrations used in this study no loss of adhesion was observed.

Pendulum hardness (Konig pendulum) (*38*) was used to evaluate the curing performance of coating formulations with and without the surface active initiator. Surface hardness was determined as a function of curing time and the results are shown in Figure 9. Although the addition of V to the coating was only 0.14% w/w the effect on the curing is remarkable. The curing rate is higher for the test forumulation than control samples that contained nearly 6% more initiator. This shows that to a large extent the problem with using a high initiator concentration in the presence of oxygen can be overcome by the use of a surface active photoinitiator. The initiator concentration in the coating-air interface is high enough to provide fast curing while the concentration in the coating matrix can be held low enough for reasonable photostability.

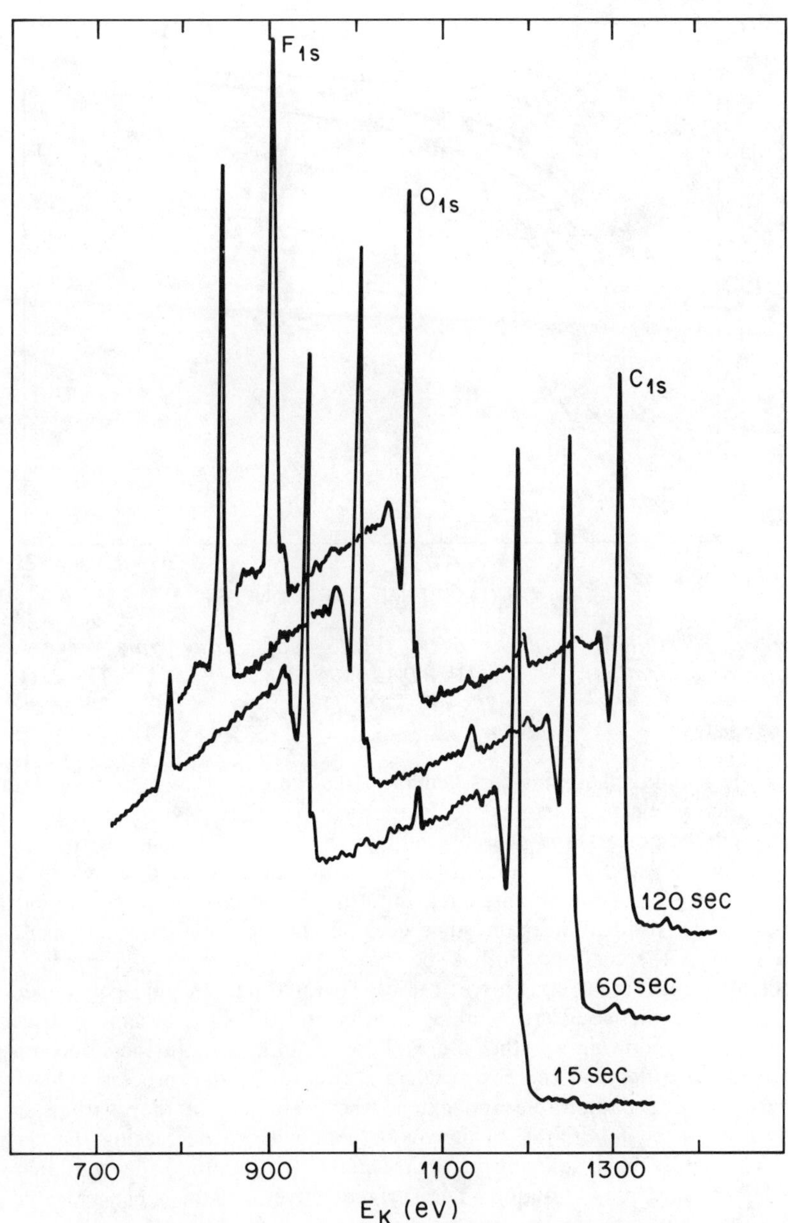

Figure 8. ESCA spectra of an acrylic resin containing 2% w/w of IV and 0.14% w/w of V taken at 15,60 and 120 sec intervals between application and curing.

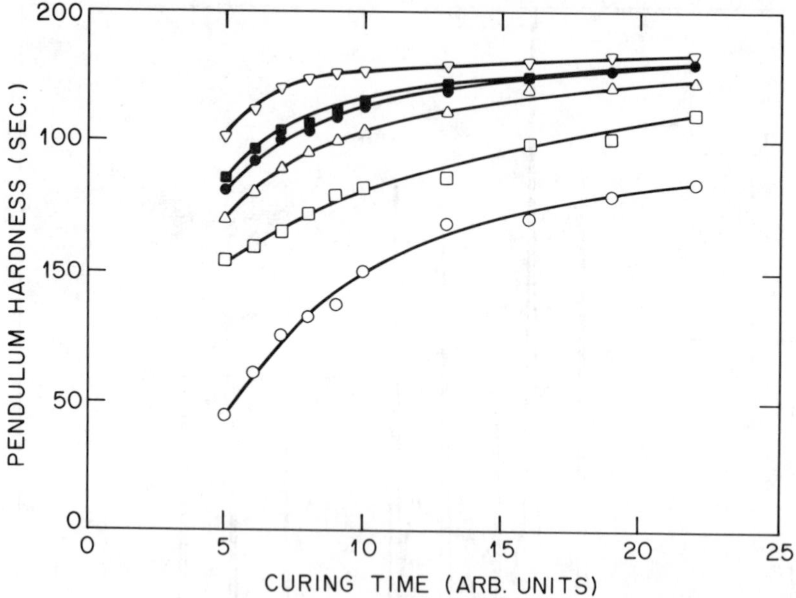

Figure 9. Pendulum hardness as a function of curing time for several initiator concentrations. ▽ *2% IV + 0.14% V;* ■ *8% IV;* ● *6% IV;* Δ *IV;* □ *2% IV; o 1% IV.*

Conclusions

Both Type 1 and 2 photoinitiators generate free radicals from their triplet state and are therefore sensitive to excited state quenchers like oxygen. Oxygen also scavenges free radicals and thereby inhibits the polymerization. To overcome this problem a fairly high amount of photoinitiator is needed. This in turn leads to a large amount of unreacted initiator in the coating after the curing reaction. This residual initiator undergoes secondary reactions that result in degradation of the coating. Both Type 1 and 2 abstract hydrogen atoms from the polymer and thereby start a free radical degradation. In solid polymer films this leads to an increased cross-linking density and the films become brittle. It also initiates yellowing of the film. Type 2 initiators further decompose performed peroxides and thereby accelerate an autoxidation process. One way to decrease these undesirable and extend the usefulness of the coatings is to simply use less photoinitiator. This can be done without decreasing the curing rate most simply by employing a surface active photoinitiator in combination with the standard photoinitiator. The surface active initiator is concentrated in the coating-air interface where the interference from oxygen is largest.

Literature Cited

1. "Introduction to Microlithography"; Thompson, L. F.; Willson, C. G.; Bowden, M. J., Eds.; ACS SYMPOSIUM SERIES No. 219, American Chemical Society: Washington, D.C., 1983.

2. "UV Curing: Science and Technology"; Pappas, S.P., Eds.; Technology Marketing Corp., Norwalk, Conn., 1978.
3. "UV Curing in Screen Printing for Printed Circuits and the Graphic Arts"; Wentik, S. G.; Kock, S. D., Eds.; Technology Marketing Corp., Norwalk, Conn., 1981.
4. Pappas, S. P. *Prog. Org. Coatings* 1973/74, *2*, 333.
5. McGinnis, V. D. "Ultraviolet Light Induced Reactions in Polymers"; Labana, S. S., Eds.; ACS SYMPOSIUM SERIES No. 25, American Chemical Society: Washington, D.C., 1976.
6. Berner, G.; Kirchmayr, R.; Rist, G. *Oil Col. Chem. Assoc.* 1976, *61*, 105.
7. McGinnis, V. D. *Photogr. Sci. Eng.* 1979, *23*, (3) 124.
8. Oster, G.; Yan, N. *Chem. Rev.* 1968, *68*, (2) 125.
9. Ledwith, A. *Pure Appl. Chem.* 1977, *49*, 431.
10. Joshi, M. G. *J. Appl. Polym. Sci.* 1981, *26*, 3946.
11. McGinnis, V. D.; Porvder, T.; Kuo, C.; Gallopo, A. *Macromolecules* 1978, *11*, (2) 405.
12. Goethe, S.; Ranby, B. *J. Polym. Photochem.*, in press.
13. Ranby, B.; Rabek, J. "Photodegradation, Photo-Oxidation and Photostabilization of Polymers"; Wiley: New York, 1975.
14. McKellar, J. F.; Allen, N. S. "Photochemistry of Man Made Polymers"; Applied Science Publ.: London, 1979.
15. Sander, M. R.; Osborn, C. L. *Tetrahedron Letters* 1974, 415.
16. Borer, A.; Kirchmayr, R.; Rist, G. *Helv. Chim. Acta* 1978, *61*, (24) 305.
17. Eichler, J.; Herz, C. P.; Naito, I.; Schnabel, W. *J. Photochem.* 1980, *12*, 225.
18. Capitanio, D. A. *Diss. Abstr. Int. B* 1976, *37*, (6) 2874.
19. Herkstroeter, W. G.; Lamola, A. A.; Hammond, G. S. *J. Am. Chem. Soc.* 1964, *86*, 4537.
20. Amirazdeh, G.; Schnabel, W. *Makromol. Chem.* 1981, *182*, 2821.
21. Hult, A.; Ranby, B. *J. Polym. Deg. Stability*, Part 3, in press.
22. Rabek, J. F.; Ranby, B. *J. Polym. Sci. A-1* 1974, *12*, 273.
23. Hult, A.; Ranby, B. *J. Polym. Deg. Stability*, Part 1, in press.
24. Beachell, H. C.; Chang, I. L. *J. Polym. Sci. A-1* 1972, *10*, 503.
25. Harper, D. J.; McKellar, J. F. *J. Appl. Polym. Sci.* 1973, *17*, 3503.
26. Amin, M. U.; Scott, G. *Europ. Polym. J.* 1974, *10*, 1019.
27. Rabek, J. F.; Ranby, B. *J. Polym. Sci., Chem.* 1974, *12*, 295.
28. Wilson, R. *J. Chem. Soc. (B)*, 1968, *12*, 1531.
29. Marteel, J. P.; Decock, P.; Groudmand, P.; Derolder, P. *Bull. Soc. Chim. Fr.* 1975, 1767.
30. Turro, N. J. "Modern Molecular Photochemistry"; The Benjamin Cummings Publishing Company: Menlo Park, CA, 1978.
31. Ng, H.C.; Guillet, J. E. *Macromolecules* 1978, *11*, 929.
32. Lala, D.; Rabek, J. *Europ. Polym. J.* 1981, *17*, 7.
33. Walling, C.; Gibian, M. J. *J. Am. Chem Soc.* 1965, *87*, 3413.
34. Ng, H. C.; Guillet, J. E. *Macromolecules* 1978, *11*, 937.

35. Mackor, A.; Wajer, T. A. J. W.; deBoer, T. J. *Tetrahedron Letters* 1967, 385.
36. Perkins, M. J. *Chem. Soc., Spec. Publ.* 1970, *24*, 97.
37. Hult, A.; Ranby, B. *J. Polym. Deg. Stability*, Part 2, in press.
38. International Standardization Organization, #ISO/R1522-1971.

RECEIVED August 6, 1984

24

Radiation Stability of Silicon Elastomers

G. C. CORFIELD

Humberside College of Higher Education
Hull, HU6 7RT, England

D. T. ASTILL and D. W. CLEGG
Sheffield City Polytechnic
Sheffield, S1 1WB, England

Silicones form a class of chemical materials intermediate between organic polymers and inorganic glasses. The silicon-oxygen backbone provides good thermal stability, allowing unusually high operating temperatures (up to 270°C) as well as flexibility at low temperatures (down to −70°C). However, in general polysiloxanes are inferior to other elastomers in radiation resistance (*1*,*2*). On irradiation polydimethylsiloxane (PMDS) rapidly hardens, due to cross-linking, with evolution of ethane, methane and hydrogen. The introduction of aromatic substituents into the siloxane structure is found to increase resistance to cross-linking (*3-6*). The relative ease with which cross-links form in a series of polysiloxanes being:

$- Si(Me)_2O -$ 20

$- Si(Ph)(Me)O -$ 2

$- Si(Ph)_2O -$ 1

In polymethylphenylsiloxane, cross-linking does occur through the aromatic moiety as well as through the methyl group, but to a much lesser extent. When aromatic groups having greater conjugation are present, such as biphenyl or naphthyl, even greater stability is observed.

In copolymers having both a readily cross-linked structure and an aromatic component, such as styrene-butadiene copolymers, it is found that radiation protection is greatest when the aromatic units are randomly dispersed in the copolymer rather than in segregated units, as with a block copolymer (*7*,*8*). In polysiloxanes, the methyl group receives the greatest protection when an aromatic component is attached to the same silicon atom (*9*).

This study is investigating the possibility of obtaining a silicone polymer having good radiation resistance, with retention of elastomeric properties. The main area of interest is the resistance to radiation of blends and block copolymers in which an aromatic component can form a separate microphase

0097-6156/84/0266-0473$06.00/0

from the elastomeric silicone component. The influence of this type of morphology upon radiation resistance in silicones has not been studied to date, although the effect of related morphological features and crystallinity have been investigated in some other cases (*10,11*). In this paper we report preliminary results on the irradiation behavior of blends of polystyrene (PS) and PDMS.

Experimental

Blends of various compositions of PDMS (M_w, 650,000; M_n, 253,000) and PS (M_w, 100,000) were prepared by dissolution of the homopolymers in tetrahydrofuran, mixing the solutions and then rapidly removing the solvent using a freeze-drier. Blends were dried in a vacuum oven at 55°C, sealed in glass tubes under a nitrogen atmosphere and irradiated, using a ^{60}Co source, with an absorbed dose rate of 0.2 Mrad h^{-1} (0.56 Gy s^{-1}) at 30°C.

Soxhlet extraction, with toluene as solvent, of the soluble fraction from the cross-linked gels produced by irradiation, was used as a method to determine the radiation cross-linking efficiency, $G(X)$, for the homopolymers and blends under investigation. Samples were extracted to constant weight and two or three replicate extractions were carried out.

Results and Discussion

Homopolymer PDMS. Results obtained from the extractions of irradiated samples of the homopolymer PDMS are given in Table I and Figure 1.

Extrapolation of this data yields a value for the incipient gelation (r_{gel}) of 0.3 Mrad. Substitution of this value into Equation 1 gives $G(X) = 2.48$.

$$G(X) = \frac{4.83\times10^5}{M_w \times r_{gel}} \quad (1)$$

Use of the Charlesby-Pinner relationship (Equation 2) with this data (Table II, Figure 2) gives $G(X) = 2.86$

$$S + \sqrt{S} = \frac{2r_{gel}}{r} \quad (2)$$

Table I. Percent Gel after Soxhlet Extraction of Irradiated PDMS

Dose/Mrad	% Gel
0.55	23.3 ± 0.6
0.99	63.6 ± 2.1
1.67	74.6 ± 0.8
3.41	81.1 ± 1.5
5.03	81.9 ± 1.5
10.1	83.5 ± 0.1
20.2	84.0 ± 1.0

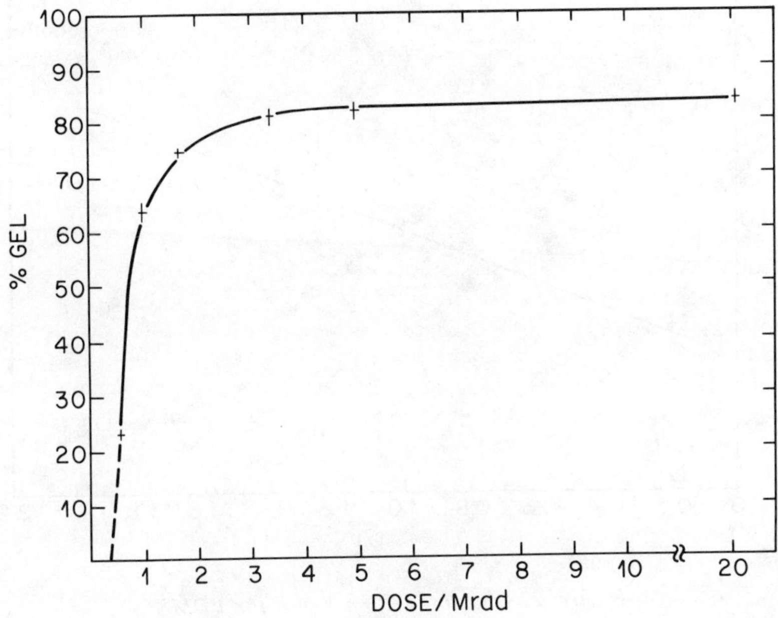

Figure 1. Irradiation of PDMS

Table II. Charlesby-Pinner Data for PDMS

Dose/Mrad	$\frac{1}{r}$	Gel Fraction	Sol Fraction (S)	$\sqrt{S}$	$S + \sqrt{S}$
0.55	1.818	0.233	0.767	0.876	1.643
0.99	1.010	0.636	0.364	0.603	0.967
1.67	0.600	0.746	0.254	0.504	0.758
3.41	0.293	0.811	0.189	0.435	0.624
5.03	0.199	0.819	0.181	0.425	0.606
10.0	0.100	0.835	0.165	0.406	0.571
20.1	0.049	0.840	0.160	0.400	0.560

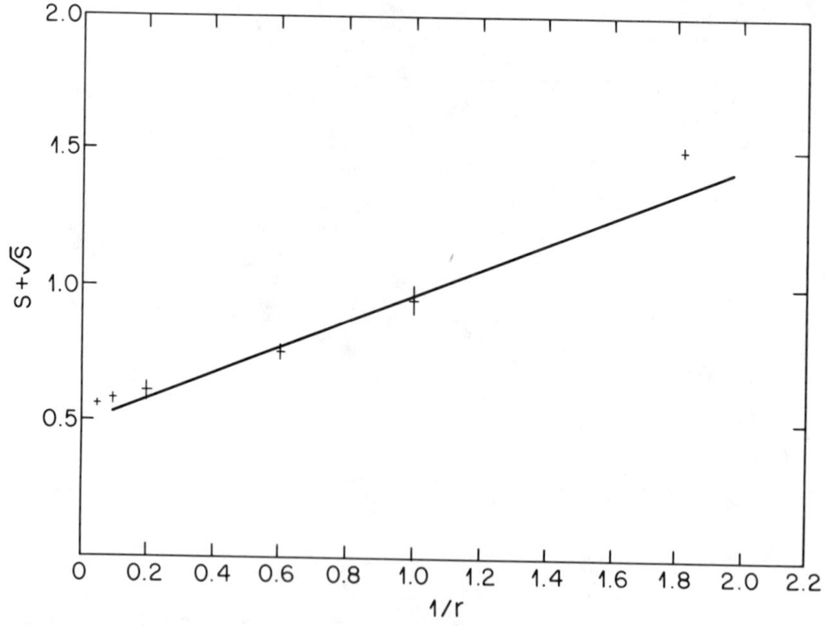

Figure 2. Charlesby-Pinner plot for PDMS

These results for $G(X)$ in PDMS are in good agreement with other reported values which range from 2.5-2.8 (*12*).

The Charlesby-Pinner relationship is regarded as empirical, applicable in polymers that have a random molecular weight distribution, and where the cross-link density is proportional to dose. When the distribution is broader than the random value, the plot deviates from linearity (*13,14*). For the PDMS under investigation $M_w/M_n = 2.57$, and a deviation from linearity is evident. If irradiation is continued to high doses, a linear relationship is observed. The gradient in the high dose region has been used to calculate r_{gel}. Extrapolation of the data to $\frac{1}{r} = 0$ have an intercept at $S + \sqrt{S} > 0$. This deviation of the intercept from zero is generally taken as an indication of simultaneous main chain scission in addition to cross-linking.

Homopolymer PS. In contrast to PDMS, for doses up to 200 Mrad the PS sample under investigation remained completely soluble (Table III), emphasizing the resistance to irradiation of PS compared to PDMS. Calculation of $G(X)$ from samples of a PS of higher molecular weight (M_w, 150,000) gave a value of 0.028 which is in agreement with other reported results (*12*). In the blends produced from these homopolymers it would be expected that the component to undergo cross-linking would be the polysiloxane.

Table III. Solubility Data For Polystyrene

Dose/Mrad	% Gel After Extraction	
	M_w = 100,000	150,000
10	0	0
20	0	0
30	0	0
50	0	0
75	0	51.0 ± 0.5
100	0	73.0 ± 1.0
200	0	80.2 ± 0.6

Blends of PS and PDMS. Irradiation of blends of PS and PDMS revealed interesting results. As expected r_{gel} increased, hence $G(X)$ decreased as increasing amounts of PS were incorporated into the blends. This is attributed to protection of the PDMS by PS. However, at low % PS an enhanced protecting effect was observed (Table IV).

Further, when blends of various compositions were irradiated at a constant dose, the results indicated that the components do not act independently, i.e. they depart from an "additive" behavior.

For example, irradiation of blends at 1.55 Mrad shows that at compositions of >7% PS, gel formation decreases with increasing PS content, but not as predicted by an additive rule (Table V, Figure 3). At compositions of <7% PS a significant decrease in gel formation is observed.

Table IV. Variation of r_{gel} and $G(X)$ With Polymer Composition

Polymer	r_{gel}/Mrad	$G(X)$
PDMS	0.3	2.48
3% PS blend	1.4	0.53
5% PS blend	1.1	0.67
7% PS blend	0.9	0.82
20% PS blend	0.6	1.24

Table V. Gel Formation In Blends Irradiated at a Constant Dose (1.55 Mrad)

Polymer	% Gel	% Gel if "Additive"
PDMS	74.5	74.5
3% PS blend	49.5	72.5
5% PS blend	60.0	71.0
7% PS blend	83.5	69.5
15% PS blend	58.5	63.5
25% PS blend	30.5	56.0
31% PS blend	21.0	52.0
50% PS blend	0	37.5
PS	0	0

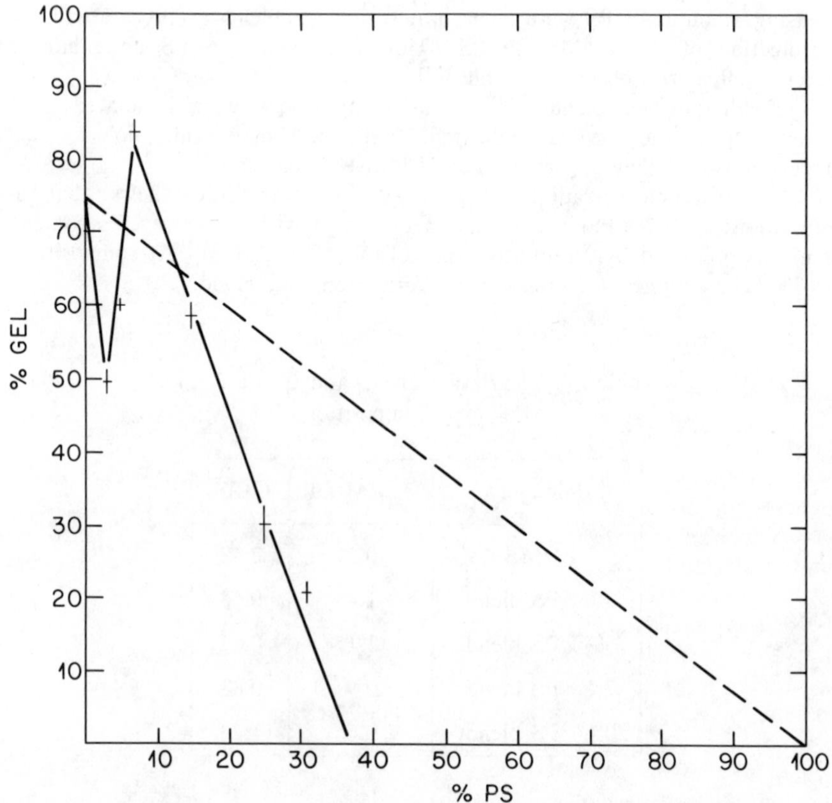

Figure 3. Irradiation of PS/PDMS blends at 1.55 Mrad. Experimental data (______); "additive" line (- - - -)

This effect is observed at other low absorbed doses (Figure 4) and in each case a minimum of gel formation occurs at around 3% PS. However, it is noticeable that as the absorbed dose is increased the extent of this protective effect decreases.

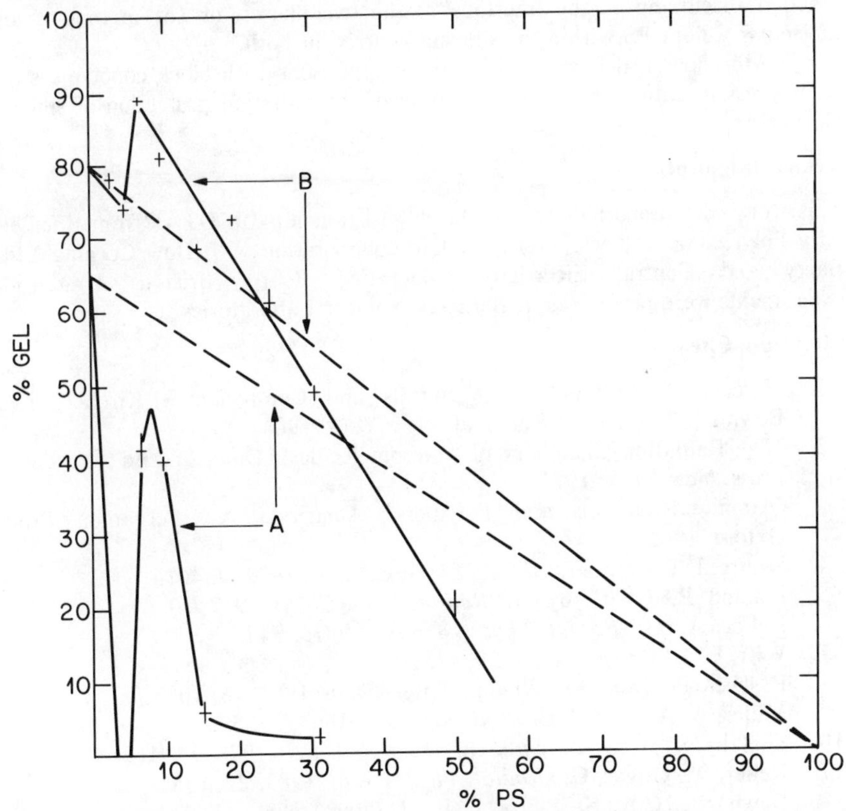

Figure 4. Irradiation of PS/PDMS blends at: A, 1.0 Mrad; B, 3.28 Mrad. Experimental data (______); "additive" line (- - - -)

Also, it is apparent that the gel content of irradiated blends can be higher than that expected if the gelation was entirely due to the reaction of homopolymer PDMS. We can eliminate the possibility that PS is responsible for this deviation since previous results indicate that gel formation in the PS used does not take place below 200 Mrad. There are two possible explanations; the first is that some PS is "trapped" (not bound) within the PDMS gel, and cannot be extracted from the matrix. The other is that the PDMS is undergoing a higher degree of cross-linking, as a result of a reduction in chain scission.

The considerable stabilizing effect observed at low % PS, possibly is related to the compatibility of the components in the blends. From scanning electron microscopic studies, it is clear that all compositions, a two phase system exists with PS dispersed in the PDMS matrix as very small regular spheres. Initial observations suggest that as the PS content increases, the PS spheres initially increase in number, then increase in size. It is possible that the enhanced protective effect at low % PS is due to either (a) stabilizing chemical reactions occurring at the interface of the two phases, or (b) mixing of low molecular weight PS within the siloxane matrix, or both.

Morphological investigations and comparisons with block copolymers are in progress to gain further understanding of the radiation protection of silicone elastomers.

Acknowledgments

This work was supported by a research assistantship (to DTA) from Sheffield City Polytechnic and was carried out in collaboration with Dow Corning Ltd., Barry and Central Electricity Generating Board, Bristol. Irradiation experiments were performed at Berkeley Nuclear Laboratories.

Literature Cited

1. "Effects of Radiation of Materials and Components"; Kircher, J.F.; Bowman, R. E., Eds.; Reinhold, New York, 1964.
2. "The Radiation Chemistry of Macromolecules"; Dole, M., Ed.; Academic Press, New York, 1973.
3. "Atomic Radiation and Polymers"; Charlesby, A. Pergamon Press, Oxford, 1960.
4. Squire, D.R.; Turner, D. T. *Macromolecules* 1972, *4*, 401.
5. Folland, R.; Charlesby, A. *Radiat. Phys. Chem.* 1977, *10*, 61.
6. Delides, C. G. *Radiat. Phys. Chem.* 1980, *16*, 345.
7. Witt, E. *J. Polym. Sci.* 1959, *41*, 507.
8. Basheer, R.; Dole, M. *Radiat. Phys. Chem.* 1981, *18*, 1053.
9. Miller, A. A. *I. & E.C. Prod. Res. Dev.* 1964, *3*, 252.
10. Shalaby, S. W. *J. Polym. Sci., Macromol. Rev.* 1978, *14*, 419.
11. Keller, A.; Ungar, G. *Radiat. Phys. Chem.* 1983, *22*, 155.
12. Dewhurst, H. A.; St. Pierre, L. E. *J. Phys. Chem.* 1960, *1063*.
13. Mitsui, H.; Hosoi, F.; Ushirokawa, M. *J. Appl. Polym. Sci.* 1975, *19*, 361.
14. Shimizu, Y.; Mitsui, H. *J. Polym. Sci.* 1979, *17*, 2307.

RECEIVED August 6, 1984

AUTHOR INDEX

SUBJECT INDEX

A

D

E

Q

R

S

Z

RECENT ACS BOOKS

"Computers in the Laboratory"
Edited by Joseph Liscouski
ACS SYMPOSIUM SERIES 265; 136 pp.; ISBN 0-8412-0867-0

"The Chemistry of Low-Rank Coals"
Edited by Harold H. Schobert
ACS SYMPOSIUM SERIES 264; 328 pp.; ISBN 0-8412-0866-2

"Resonances in Electron-Molecule Scattering, van der Waals Complexes, and Reactive Chemical Dynamics"
Edited by Donald G. Truhlar
ACS SYMPOSIUM SERIES 263; 536 pp.; ISBN 0-8412-0865-4

"Seafood Toxins"
Edited by Edward P. Ragelis
ACS SYMPOSIUM SERIES 262; 473 pp.; ISBN 0-8412-0863-8

"Computers in Flavor and Fragrance Research"
Edited by Craig B. Warren and John P. Walradt
ACS SYMPOSIUM SERIES 261; 157 pp.; ISBN 0-8412-0861-1

"Polymers for Fibers and Elastomers"
Edited by Jett C. Arthur, Jr.
ACS SYMPOSIUM SERIES 260; 448 pp.; ISBN 0-8412-0859-X

"Treatment and Disposal of Pesticide Wastes"
Edited by James N. Seiber and Raymond F. Krueger
ACS SYMPOSIUM SERIES 259; 384 pp.; ISBN 0-8412-0858-1

"Stable Isotopes in Nutrition"
Edited by Judith R. Turnlund and Phyllis E. Johnson
ACS SYMPOSIUM SERIES 258; 240 pp.; ISBN 0-8412-0855-7

"Bioregulators: Chemistry and Uses"
Edited by Robert L. Ory and Falk R. Rittig
ACS SYMPOSIUM SERIES 257; 286 pp.; ISBN 0-8412-0853-0

"Polymeric Materials and Artificial Organs"
Edited by Charles G. Gebelein
ACS SYMPOSIUM SERIES 256; 208 pp.; ISBN 0-8412-0854-9

"Pesticide Synthesis Through Rational Approaches"
Edited by Philip S. Magee, Gustave K. Kohn, and Julius J. Menn
ACS SYMPOSIUM SERIES 255; 351 pp.; ISBN 0-8412-0852-2

"The Chemistry of Solid Wood"
Edited by Roger M. Rowell
ADVANCES IN CHEMISTRY SERIES 207; 588 pp.; ISBN 0-8412-0796-8

"Polymer Blends and Composites in Multiphase Systems"
Edited by C. D. Han
ADVANCES IN CHEMISTRY SERIES 206; 400 pp.; ISBN 0-8412-0783-6

Production by Deborah Corson and Meg Marshall
Indexing by Deborah Corson
Jacket design by Pamela Lewis

Elements typeset by Hot Type Ltd., Washington, D.C.
Printed and bound by Maple Press Co., York, Pa.